README!

README!

FILTERED BY

NETTIME

ASCII CULTURE AND THE REVENGE OF KNOWLEDGE

EDITED BY

JOSEPHINE BOSMA, PAULINE VAN MOURIK BROEKMAN, TED BYFIELD

MATTHEW FULLER, GEERT LOVINK, DIANA McCARTY, PIT SCHULTZ

FELIX STALDER, McKENZIE WARK, AND FAITH WILDING

AUTONOMEDIA

FOR MORE INFORMATION ABOUT NETTIME:
HTTP://WWW.NETTIME.ORG (MAIN WEBSITE)
HTTP://WWW.DESK.NL/~NETTIME (OLD HOMEPAGE & ZKP PUBLICATIONS)
HTTP://WWW.FACTORY.ORG/NETTIME (ARCHIVE)
HTTP://THING.NET/LOGIN.THING (ARCHIVE IN NEW YORK)
HTTP://TAO.CA/FIRE/NETTIME/INDEX.HTML (ARCHIVE IN TORONTO)
NEWS://NEWS.THING.AT/ALT.NETTIME (NEWSFEED)
HTTP://WWW.FACTORY.ORG/NETTIME-NL (DUTCH LIST ARCHIVE)

WEBSITE WITH EXTENDED VERSION OF THIS BOOK
HTTP://WWW.NETTIME.ORG
COMMENTS, DISCUSSION AND ADDITIONS WELCOME!

E-MAIL ADDRESSES:
NETTIME@DESK.NL (INFO)
NETTIME-L@DESK.NL (FOR POSTINGS)

IF YOU WANT TO SUBSCRIBE, SEND AN EMAIL TO:
MAJORDOMO@DESK.NL
WITH NO SUBJECTLINE AND THE SINGLE LINE:
SUBSCRIBE NETTIME-L YOU@EMAIL.ADDRESS
IN THE BODY OF THE MAIL.

IF YOU WANT TO UNSUBSCRIBE, SEND AN EMAIL TO:
MAJORDOMO@DESK.NL
WITH NO SUBJECTLINE AND THE SINGLE LINE:
UNSUBSCRIBE NETTIME-L YOU@EMAIL.ADDRESS
IN THE BODY OF THE MAIL.

AUTONOMEDIA
POB 568 WILLIAMSBURGH STATION
BROOKLYN, NY 11211-0568 USA

FAX: 718-963-2603
E-MAIL: AUTONOBOOK@AOL.COM
WEBSITE: WWW.AUTONOMEDIA.ORG

PUBLISHED WITH THE SUPPORT OF:
THE ARS ELECTRONICA FESTIVAL (A)
THE ARTS COUNCIL OF ENGLAND (UK)
THE ART AND TECHNOLOGY FOUNDATION (ES)
DEAF/THE V2_ORGANISATION (NL)
THE LJUBLJANA DIGITAL MEDIA LAB (SLO)
THE SOCIETY FOR OLD AND NEW MEDIA (NL)

PRODUCED AT THE SOCIETY FOR OLD AND NEW MEDIA
COVER AND BOOK DESIGN BY MIEKE GERRITZEN
LOGO BY HEATH BUNTING

PRINTED IN THE NETHERLANDS
DRUKKERIJ HOOIBERG, EPE

"HOW IS A BOOK MADE? IT IS NO LESS INTERESTING THAN KNOWING HOW A PIN AND A BUTTON ARE MADE.": AN UNIMAGINABLE OPENING LINE FOR ECONOMISTS OF HIS DAY -- AND, PERHAPS, OF OUR OWN -- BUT FAR LESS SO FOR US, SINCE THE PRODUCTION OF A BOOK MAY BE THOUGHT OF AS A PARADIGM FOR POST-FORDIST PRODUCTION. (MAURIZIO LAZZARATO)

WE MUST BE ABLE TO PASS ON TO THE COMING GENERATIONS - IF NOT AS THE LEGACY OF THESE TIMES THEN AS A KIND OF MESSAGE IN A BOTTLE WHAT COMPUTER-TECHNOLOGY MEANT TO THE FIRST GENERATION IT EFFECTED. (FRIEDRICH KITTLER)

IT JUST CAN'T WORK THAT WAY -- IF I PAID EVERYONE WHO'S EVER CONTRIBUTED SOMETHING TO MY WORK, THEY'D ALL END UP GETTING ABOUT HALF A CENT EACH AND I'D END UP WITH NOTHING TO PAY MY RENT WITH AND THE ADDED BURDEN OF KNOWING THAT EVERY WORD I WRITE MIGHT END UP COSTING ME MORE MONEY THAN IT'S EVER LIKELY TO MAKE FOR ME. (BETH SPENCER)

WE DO HAVE THE ADVANTAGES OF BEING NAIVE VISIONARIES WITH COLLECTIVE POLITICAL EXPERIENCE, THE DESIRE TO SHARE SKILLS AND RESOURCES, AND THE COLLECTIVE ABILITY TO OPEN ANY DESIRED FIELD OF KNOWLEDGE. PRAISE BE TO THE TINKERERS, TO THE TOY MAKERS, AND TO THE AMATEURS. NEW VERSIONS OF EXPERTISE MUST BE CONSTRUCTED. (CRITICAL ART ENSEMBLE)

THE RAPIDLY APPROACHING MILLENNIUM OFFERS A UNIQUE CULTURAL OPPORTUNITY. AFTER MANY YEARS OF CUT-AND-PASTE, APPROPRIATION, DETOURNEMENT AND NEO-RETRO AHISTORICALITY, POSTMODERNITY IS ABOUT TO END. IMMEDIATELY AFTER THE END OF THE FIN DE SIECLE, THERE WILL BE A SUDDEN AND INTENSE DEMAND FOR GENUINE NOVELTY. (BRUCE STERLING)

FABRICATED SUBJECTS ARE FRACTURED SUBJECTS, AND NO INJECTION OF STRAIGHT SCIENCE WILL FIX THEM WHERE THEY ARE BROKEN. IT IS TIME TO MOVE BEYOND SCIENTIFICALLY ENGINEERING AN ABSTRACT SUBJECTIVITY, TO HOOK AUTONOMOUS AGENTS BACK INTO THE ENVIRONMENTS THAT CREATED THEM AND WISH TO INTERACT WITH THEM. (PHOEBE SENGERS)

KNOWLEDGE IS THE REALM OF NON-SCARCITY, AS OPPOSED TO THE ECONOMY. TRUE, KNOWLEDGE HAS ALWAYS BEEN CLASSIFIED AS A RARE GOOD. BUT WHO SAYS THAT THE KNOWLEDGE NECESSARY TO FIX A SCOOTER IS LESS IMPORTANT THAN KNOWLEDGE ABOUT QUANTUM PHYSICS? IN A SOCIETY WHERE GARBAGE-MEN ARE MORE IN DEMAND THAN NATURAL SCIENTISTS, KNOWLEDGE IS ON AN EQUALIZATION TRAJECTORY. (MICHEL SERRES)

THE METAPHOR OF THE 'INFORMATION HIGHWAY' RELATES ALSO TO THE TRADITION OF DRINKING, PROSTITUTION AND GAMBLING AT HALTING PLACES AND PORTS, FROM ROMAN TIMES TILL OUR CENTURY, AND THE CONSTANT FIGHT OF AUTHORITIES TRYING TO BAN SUCH DEBAUCHERY. (TJEBBE VAN TIJEN)

ITS HARD, FROM THE PERIPHERY, TO SHARE THE ENTHUSIASM FOR ANY OF THE REIGNING DISCOURSES OF CYBERSPACE, AS THEY ALL SEEM TO ME IMPLICATED IN THE UNEVEN SPATIAL DISTRIBUTION OF WHAT I WOULD CALL VECTORAL POWER. BUT FROM THE TELEGRAPH TO TELECOMMUNICATIONS, IT HAS ALWAYS BEEN EXPERIENCED IN THE PERIPHERY AS AN UNEQUAL FLOW. HOW CAN YOU GET ENTHUSIASTIC IN THE PERIPHERY ABOUT NEW IMPERIAL VECTORS? HOW CAN YOU GET ENTHUSIASTIC IN THE PERIPHERY ABOUT NEW RHETORICS ABOUT THE POWER OF NEW MODES OF COMMUNICATION? (MCKENZIE WARK)

WHAT HOLDS FOR CHICKEN IN A PRIMITIVE BARTER ECONOMY HOLDS ALSO FOR INTANGIBLES SUCH AS IDEAS AND REPUTATION IN THE PART OF THE ECONOMY THAT OPERATES ON THE INTERNET. AND SOME OF THESE INTANGIBLES, IN THE RIGHT CIRCUMSTANCES, CAN CERTAINLY BE CONVERTED INTO THE SORT OF MONEY THAT BUYS CARS, LEAVE ALONE PIZZAS TO KEEP HUNGER AWAY. (RISHAB AYER GHOSH)

WITHIN THE DISCOURSE NETWORKS OF ART, INCLUDING CRITICAL TECHNIQUE; LICENSE TO IRRESPONSIBILITY; COMPOSITIONS-IN-PROGRESS OF TASTE STRATIFICATION AND BREAKS; INSTITUTIONS; FINANCE; INDIVIDUAL SURVIVAL STRATEGIES; MEDIA; SOCIAL NETWORKS; LEGITIMATION DEVICES; AT LEAST POTENTIAL OPENNESS TO NEW FORMS; AND AVOWED ATTENTIVENESS TO MANIFESTATIONS OF BEAUTY, THERE WERE DYNAMICS THAT WERE USEFUL TO MOBILISE IN ORDER TO OPEN UP POSSIBILITIES OF CIRCULATION AND EFFECT. (MATTHEW FULLER)

THE NEW WORLD IS ALWAYS ALREADY THE OLD ONE IN A REVERSED FORM. THE OTHER YOU DISCOVER IS ALWAYS ALREADY THE SAME IN A REVERSED AND THEREBY SLIGHTLY REARRANGED FORM. THERE IS NO WAY OF GRASPING THE RADICAL OTHER, BECAUSE AS SOON AS YOU MANAGE TO GRASP IT, IT IMMEDIATELY BECOMES PART OF YOUR OWN. THAT'S WHY CYBERSPACE IS DISCURSIVELY CONSTRUCTED AS A NEW YET UNAPPROACHABLE CONTINENT. (OLIVER MARCHART)

PARADOXICALLY, IT IS THE GLOBAL FINANCIAL MELTDOWN THAT MAY OFFER THE FIRST REAL CHANCE FOR TRANSNATIONAL CIVIL SOCIETY TO HAVE A SIGNIFICANT IMPACT ON WORLD POLITICS. IN THE CONTEXT OF A WORLD-WIDE ECONOMIC CRISIS, NETWORKERS MAY BE ABLE TO USE AN UNDERSTANDING ACQUIRED BY DIRECT PARTICIPATION IN GLOBAL INFORMATION FLOWS TO EFFECTIVELY CRITICIZE THE INSTITUTIONS, IDEOLOGIES, AND ECONOMIC POLICIES OF THEIR OWN COUNTRIES. (BRIAN HOLMES)

WE HAVE THE OPPORTUNITY TO EXPLAIN PRODUCTION AND THUS ORGANIZE HUMAN LIFE WITHIN THIS WEALTH OF POWERS THAT CONSTITUTE THE TOOL: LANGUAGES AND AFFECTS. (ANTONIO NEGRI)

DESPITE ORIGINALLY BEING INVENTED FOR THE U.S. MILITARY, THE NET WAS CONSTRUCTED AROUND THE GIFT ECONOMY. WITHIN SMALL TRIBAL SOCIETIES, THE CIRCULATION OF GIFTS ESTABLISHED CLOSE PERSONAL BONDS BETWEEN PEOPLE. IN CONTRAST, THE ACADEMIC GIFT ECONOMY IS USED BY INTELLECTUALS WHO ARE SPREAD ACROSS THE WORLD." (RICHARD BARBROOK)

NO TO MONOPOLIZATION OF TECHNOLOGY BY NARCISSISTIC SUBJECTS--FOR A DRAMATICS OF THE DIFFERENCE (SIEGFRIED ZIELINSKI)

STOP READING

START BROWSING

ART

LOCAL

NEIGHBORS

SOUND

SUBJECTS

YOU ARE ENTERING
<NETTIME>
A HIGH DENSITY CONTENT ZONE

INTRO

NOTHING IS SPECTACULAR IF YOU AREN'T PART OF IT

Welcome. Bienvenue. Guten Tag. This is an anthology of Nettime, an internet mailing list—an attempt to transform thousands of emails, articles, and comments into book form. But what is "Nettime"? Once upon a time, an unlikely group of people gathered around a table in a house somewhere in a German forest. Around the table sat a group of men, all eating, talking, drinking, sampling each other's ideas. The language was German. The hours passed, and the table burgeoned under a mass of papers, notes, books. At the end, they cleared the table, taking various notes with them as they returned to their own desks, scattered across Europe, from Amsterdam to Budapest. The months passed; email was exchanged. Another meeting was planned for late spring 1995—this time in Venice, the floating city, during the Biennale in the Teatro Malibran. By night it housed an imported Berlin club scene; by day, the men—and now a few women—gather. The languages are English, fast and slow, sometimes broken, and also some Italian. The days pass, and once again the table disappears under the papers, notes, books, scribbles. It was at this second meeting of the Medien Zentralkomittee (ZK) that the Nettime mailing list is conceived. The ZK itself was a parasite attached to the main body of the Biennale; it had a small budget to invite a eclectic group of international activists, artists, organizers, theoreticians, and writers, all involved with the net, for an intense three-day, closed meeting. The name: Nettime. The topics: the city metaphor versus the life metaphor, the labyrinths of real and virtual worlds, wandering websites, the city-state, a critique of the political agenda that would come to be called the "Californian Ideology," and the perennial question of art. Nettime became a reality at this meeting. Or so one version of the story goes. Since this is the story of a network, there is a network of stories about the its multiple beginnings. Some day someone will think of a way to write a history of such a network. For the time being, this fable will have to do. The Venice group cleared the table and departed for the desks and screens back home. The passing days turn into weeks, then a month—traffic began to rise on the Nettime list. Over a series of meetings, festivals and events—in Budapest, Amsterdam, Madrid, New York, Ljubljana, and countless railway stations in between—the social networks began to self-organize to launch a new type of discourse for probing the space of the media networks, carving out niches for mixed modes of autonomous living and working. The list grew from 20 to 30 and to 100, 300, on to 850 subscribers as of November 1998. Not a whole lot, now that the internet hits the final curve on the way to mass-medium status, but Nettime never really cared about numbers. Nettime isn't much concerned with the mass distribution of a product. It's more about the self-organization of a process. We tentatively call the process "collaborative text filtering." Who are we? Who is Nettime? A saloon? Journal? Bulletin board? Billboard? Web archive? Community? System? Soapbox? Warehouse? Parasite? Real-time oral history? Spittoon? Bitbucket? Open-mike night? A small world after all? A splintery glory hole? A modest means of self-promotion? A dead weight oppressing fresh blood? Net.crit chicken hawks? An invisible dictatorship? A typing pool? All of those and more. It's a collective subjectivity with no fixed identity, made up of the people who come and go from the Nettime list, who contribute more or less to its characteristic ideas and expressions. Nettime is always different from what it was a moment ago; it's always discovering something new about itself. As such, it is a working implementation of what subjectivity might become in an online environment. Then again, some or many of the participants whose ideas form parts of Nettime will almost certainly dispute this. Nettime is made up of the differences between the ideas as to what it is or might become. Send a message to the majordomo software that runs the Nettime list and it will promptly respond with this very out-of-date message in reply: "Nettime is not only a mailing list, but an attempt to formulate an international, networked discourse, that is neither promoting the dominant euphoria (in order to sell some product), nor to continue with the cynical pessimism, spread by journalists and intellectuals working in the 'old' media, who can still make general statements without any deeper knowledge on the specific communication aspects of the so-called 'new' media. We intend to bring out books, readers and floppies and web sites in various languages, so that the 'immanent' net critique will not only circulate within the internet, but can also be read by people who are not on-line" Geert Lovink, Pit Schultz, 27th February, 1996 Another version of this trajectory might go like this: Once upon a time there was a rather tired and ailing political agenda called leftism. It had some fixed ideas in its collective head about the media, about the arts, about theory and practice. It got itself stuck in academic ways of thinking sometimes, and other times it snorted too much art. The mash of papers on the tables, the lives of the people around them and the emails going between them pointed toward something else. The purpose of the undertaking, was "net critique," a species of radical pragmatism (or perhaps of pragmatic radicalism) for working late and deep in the "information age." This type of critique would seek—in a way that is by no means necessarily an innovation—involvement at the root level rather than getting stuck in endless repetitions of formal introductions and quack diagnoses. The theories of the media the leftism relied upon were the product of a certain kind of history, with political, cultural, intellectual, and technological dimensions. Net critique aimed to rethink the legacy of leftist media theory and practice. Nettime was a vector for experimenting with net critique that would confront it with the possibility of inventing new forms of discourse and dialogue in a new medium. Consensus is not the goal. There's no governing fantasy according to which the differences within this "group" will on some ever-deferred day be resolved. The differences are Nettime; they might be dialectical, implying each other, or they might be differential, making absolutely no reference whatsoever to each others' terms. Net critique, if understood as a shared practice in and against a never pre-defined techno-local environment, contains many modes of possible participation. Conventional cultural criticism, as an academic discipline, contains no imperative to actually do anything beyond the continuation of polite footnoted complaint. Nevertheless, libraries contain sources of knowledge that can be newly selected and contextualized to gain momentum. Nettime will always contain the writings of genuine insects trapped in the amber of their own writing-habit, but it is also very much about the examination and development of other bugs in the system. One discovery is that the relatively closed system of a moderated mailing list can be a good environment for developing a rich set of ideas. It is a certain kind of milieu, a plane upon which certain kinds of work flourishes. The best moments on Nettime are perhaps those when contributors cultivate and differentiate their language and internal reference system without becoming completely obscure. The discursive interactions on Nettime appear as a fluid process that can't be simulated or staged. The list is a milieu that encourages a certain radicalism of approach: miscellaneous ex-East going on ex-West ancien-regime misfits turned NGO-perfect-fits, fun-guerrilla playgirls, connected autonomists, entrepreneurial molto-hippies, squatters turned digital imperialists, postcynical berks, slacktivists and wackademics, minimalist elitist subtechnodrifters, name-your-cause party people, name-your-price statists, can-do cyberindividualists, can't–won't workers, accredited weird-scientists, and assorted other theoretical and practical avant-gardeners, senders, receivers, and orphans. Over the years, Nettime has mutated, survived, and escaped its Oedipal relations to leftism by oozing along new vectors. Nettime always distanced itself, sometimes dialectically, sometimes absolutely, from the "cyberhype" propagated by *Wired* magazine, which in any case exhausted itself and declined intellectually. Neither the emergency rhetorics of the old militants about the threat of the internet, not the technoboosting of the military-entertainment complex appeared, in the end, to be all that intellectually interesting. As Nettime continually suggests, the ac-

`tion is elsewhere.` Instead, Nettime has created a milieu in which a collective process of thinking, or sometimes just a collective migraine, could pose again some questions of itself and to itself. What is actual? What is possible? What can we hope for? What seemed important was to maintain of a milieu that enabled a certain continuity and reliable instability. Full-time, or even part-time, Nettime requires a certain intellectual modesty. It avoided the sillier behavior of the net's "teen years"—flamewars, axe-grinding, and the spiraling noise of chat—through light moderation semidemocratized (or at least randomized) by a rotating group. It's hardly the first list to work through issues of openness and closure, democracy and justice, free speech and fair speech; but it doesn't seem as though most participants have fetishized these issues. Since its early days as a parasite event on various art festivals, Nettime has thrived as a mixed economy. It isn't a commercial project, although its participants certainly have mixed motives for contributing, and those motives don't at all exclude gain. Various kinds of economy sustain it, and this hybridity may be a contributing factor to its sustainable autonomy. The way to avoid capture by the state or the market is to be neither one thing nor the other. Not every kind of difference can be accommodated directly within Nettime. Projects dip in and split off. Cyberfeminism logs in and logs out, a sometimes parallel, sometimes intersecting project. Ideas, concepts, experiences are given away in large quantities and uncertain results. Rarely new, sometimes stolen, and often borrowed, ideas, concepts, and experiences are given away in large quantities, with uncertain results. Some fall on deaf ears and spark no reaction whatsoever; others drift off into other channels, and disappear from the radar for a while, to return morphed as something else; still others provoke heated debates, some of which have been quickly quoted in the mass media as "the voice of the net." But the voice of the net is a silly idea: it has much more to do with broadcasters' need to represent than with what is represented. The Nettime project moves in the opposite direction: not a voice, but voicings, less a melody, than a sound. Net Critique isn't dogmatic—it can't be, because it isn't even a synthetic set of ideas, let alone a twelve-step program for instant cyberculture. Rather, it's a series of interventions, some theoretical, some aesthetic, some technical, even some with a soldering iron—a network of ideas-in-process. As a topology, the Nettime network is a mix of a ring and a star—it's hybrid in many ways. Open and closed, academic and nonacademic, bits and atoms, theory and practice. Most Nettime subscribers are in Europe. In the U.S., Nettime is stronger in New York than on the West Coast. There are also many active subscribers in Australia. Asia is coming on line, and subscribers from Japan, Taiwan, India, even China are dropping in. There is a different style in using language online, which has mostly to do with the fact that English isn't the native language of many subscribers. English becomes Englishes, and different norms for writing it rub against each other. A plural standard, emerges where nonnative Englishes are recognized as valid and coherent standards of English, rather than a hierarchical one, where native English is assumed superior to other variants. One hope early on was that Nettime could help to shift media theory and practice into a new communication vectors, to see how they might perform itself differently in a different spaces. Part of the purpose of this book is to shift some of the results of that experiment back into the vector of print media, to see how these efforts looks when re-imagined at a different speed. The practices of collaborative filtering developed on Nettime became the basis for a practice of editing and publishing. This book was produced as a collaborative process, by people working on different continents, in different time zones, at different intensities. It documents the process not just of Nettime but of net critique applied to itself. It follows the twists and folds in the information landscape as it is being created, discovering that things which were remote have suddenly become strange neighbors. This is what a bottom-up, international, networked discourse might look like. A book of Nettime might seem retrograde. Between old and new media, it cultivates a zone of fertile textuality which can take the form of a book, a xerox publication, a private collection of printouts, or an electronic archive of Nettime emails. Vectors of different texts intersect at surprising places. Different aggregates of etexts, interviews, announcements, essays, replies, commentaries, reports, calls, letter, letters, lists, poems, ascii art, articles, reviews, manifestoes, sermons, have been cut and remixed. The joy of text finally results in an eclectic blend of the elements of discourse and dialogue. Social intensities find a common platform, to differentiate, articulate into an alchemy of desires. Giving away time spent on the net and into text, it becomes a collective source of social, immaterial labor, a "text mine," as well and a source of elements for new ideas. This book is the transformation of Nettime as a time–space into a different level, where the relative fixity of print allows one (or many) to measure time in months and years, rather than the minutes and hours of the net. What this book is *not* is an adequate representation of Nettime. Some of the authors included have never participated Nettime. Some are dead. They belong to Nettime because they provide important reference points, historical depth, and continuity. Nettime still has centers and peripheries. It has not solved the structural inequality of global information flows, nor could it. But it is at least a space that tries to learn through experiment how to overcome the imperial past of the architecture of global media vectors. Part of the impetus for Nettime was the desire, after 1989, to create a milieu for that could pass between Eastern and Western Europe, and to some extent, as this book shows, that process has produced results. Nettime is part of the practice of realizing the potential of the net as a means of communicating otherwise. Nettime has often been accused of being a white Eurocentric boys' club. And so it is, to a certain degree. But this perception is superficial. It is certainly beyond even Nettime's pragmatic utopian capability to solve all problems of difference and representation. Nettime's open structure encourages participation and a variety of voices, expressions, lines of flight. Whoever wants to do the work and share in the joys of text can simply join in. The male culture of scientific-, business-, and military-based structures and biases built into communications technology is daunting and alien to many people from different cultural, racial, and class sectors. The kind of intellectual and critical text-based virtual communication represented by Nettime may be wholly unsatisfying and irrelevant to many whose voices we need to hear. Even women with full online access, good educations, and excellent English writing skills, can find Nettime a difficult forum to crack. Yet Nettime has made a strong effort to include and address cyberfeminist issues and texts. The Nettime editorial group has strong feminist representation and this is reflected in the quality and variety of texts by women included in the book, as well as in texts from other cultural constituencies which deal with issues of difference, work, net politics, access, and the struggle against discrimination of all kinds. Nettime will never be politically correct; to practice its process it will travel along vectors, desires, political liquidities, inventive interventions—rich texts of all kinds. READ ME! is structured into several sections which represent some of the major whirls in the text flows of Nettime. *Software* examines the tools with which we build our media environments, not all of them are computer-based. *Markets* is a collection of theory and experiences of living in and out of the grip of this ambiguous and poorly understood beast. *Work* presents new theoretical approaches to knowledge production and some tales from the shady underbelly of the brave new world of the knowledge workers. *Art* presents reflections on art and what it licenses going on and through the net. *Local* samples the diversity of living realities, of struggles that are carried out in specific places along trajectories that are influenced as much by local history as they are by global media. *Neighbors* presents other lists, some of which overlap, some of which are friendly. *Sound* examines the acoustic properties and potentials of the net. *Subjects* ranges across the translucent landscapes of overlaid histories. *Maze* is a collection of third-person eat-em-ups for first-person thinkers. *Virus* is where critique finally gives up, kicking off its boots into pure invention.

The Editors.

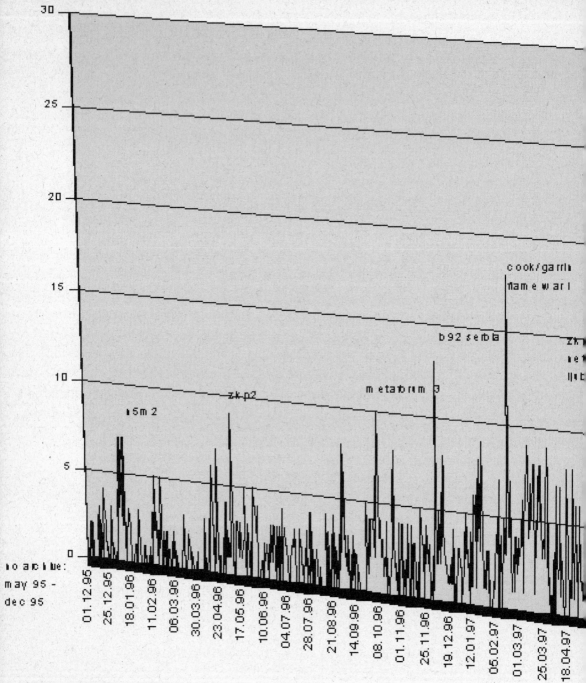

cook/garrin
flame war !

b92 serbia

zk
ne
ljub

metaform 3

zk-p2

s5m 2

30
25
20
15
10
5
0

no archive:
may 95 -
dec 95

01.12.95
25.12.95
18.01.96
11.02.96
06.03.96
30.03.96
23.04.96
17.05.96
10.06.96
04.07.96
28.07.96
21.08.96
14.09.96
08.10.96
01.11.96
25.11.96
19.12.96
12.01.97
05.02.97
01.03.97
25.03.97
18.04.97

nettime mailinglist stats
www.nettime.org
01.12.1995 - 18.10.1998
max. postings: 26
average per day: 2.8

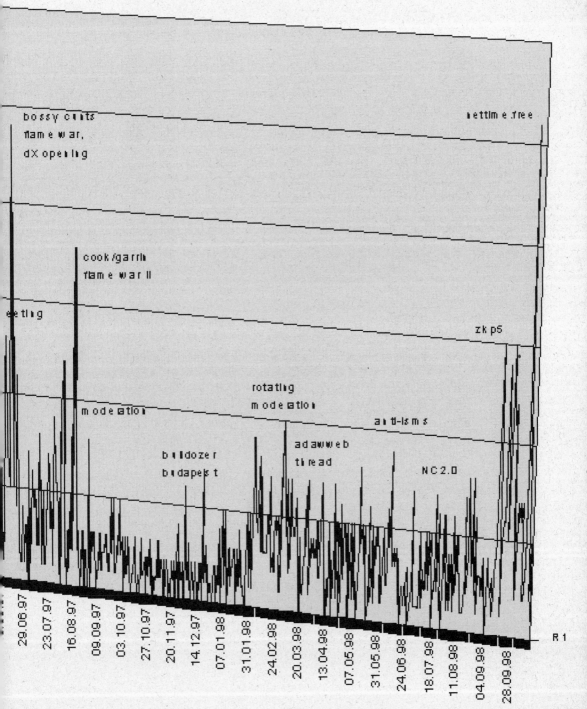

bossy cunts
flame war,
dX opening

nettime.free

cook/garrin
flame war II

eeting

zk pS

rotating
moderation

moderation

anti-isms

buldozer
budapest

adawweb
thread

NC 2.0

29.06.97
23.07.97
16.08.97
09.09.97
03.10.97
27.10.97
20.11.97
14.12.97
07.01.98
31.01.98
24.02.98
20.03.98
13.04.98
07.05.98
31.05.98
24.06.98
18.07.98
11.08.98
04.09.98
28.09.98

R 1

pixxelbulz, mcg van cdca l 998

SOFTWARE

SUBJECT: THE BAG

FROM: VILÉM FLUSSER (BY WAY OF MICHAEL STAPLEY
<MSTAPLEY@MIDAT.DE>)
DATE: MON, 12 OCT 1998 10:23:03 +0100

Omnia me mecum porto. In plain English: Everything that I've written and published in the last eighteen months is kept in a bag. The bag was stolen recently from a car parked outside a Paris hotel. It was found again in a nearby
street with the contents intact. The thief found no value in them. A discarded literary judgment.

The bag can be seen as a part of my *memory*. Whoever reads the papers it
contains and the way they are ordered will recognize me, in a limited though
intense way. I intend here to examine and analyze the bag. Not as if I myself
were interesting but because the thief, if he had inspected the contents more
carefully, would have found himself in the company of historians, archeologists, paleontologists, psychoanalysts, and similar researchers.

What is at issue here is a yellow leather bag equipped with a zipper. It contains different colored folders. One contains my correspondence from June
1972 until now, including copies of my letters and letters addressed to me.
Some of my letters have remained unanswered, and some of those that I
have received I have never replied to. The letters are ordered chronological-
ly. Another folder is titled: "unpublished papers." It contains about thirty
essays in Portuguese, English, or German concerning art criticism and phenomenology, the originals of which were sent to newspapers. These papers
are unordered. Another folder is titled "published papers." It contains about
ten essays published during my stay in Europe. They are arranged according
to their date published. A further folder is titled "*La Force du Quotidian*" and
contains a book manuscript—fifteen essays about things in our environment—it will be released in December in Paris. Another is titled "*Ça existe, la
Nature?*" and contains eight essays. Both folders are arranged according to
their content. A further is titled "New York" and contains outlines for a lecture about the future of television that I plan to hold next year at the
Museum of Modern Art. Another is titled "Rio" and contains essays that my
publisher in Rio de Janeiro will bring out soon. Another is titled "Talks" and
contains outlines for lectures that I have held and will hold in Europe. They
are not ordered. Another is titled "*Bodenlosigkeit*" and contains a hundred
pages of an autobiography that I began and never completed. Another is
titled "*Biennal*" and contains references to the "*XII Bienal des Arts*" in Sao
Paulo. The last has the title "*Documentatos*" and contains "self-referential" cer-
tificates from government offices, universities, and other institutions. This is
then the semantic and syntactical dimension of the bag.

The folders are firstly arranged *syntactically*. They are arranged in three classes:

(A) Dialogues (the correspondence folder)
(B) Discourses to others (lectures and manuscripts)
(C) Discourses about myself (documents)

The first class would have given the thief a view into the structure of my relationships with others, what connects me to them, who rejects me, and who I reject. The second class would have allowed the thief to see me from "within," and how I try to make myself public. The third class would have allowed him to see me in the way the establishment does, my mask, via which I play my public role.

The knowledge that the thief thus gains would be problematic for the following reasons: (1) The authenticity of the papers would need to be checked (2) The authenticity of the documents contained therein would have to be checked. The thief would be required to make a close reading of the texts and of their contexts. The folders are also arranged semantically.

Again they are arranged into three classes:

(A) Factual information (documents, sections of letters, lectures, and manuscripts).
(B) Interpretations of facts (lectures and manuscripts)
(C) Expressions of emotion and value (letters, and beneath the surface in most manuscripts).

The first class would have offered the thief a view into my "objective-being-in-the-world." The second the way in which I maintain a distance therefrom. The third a view of my "subjective and intersubjective-being-in-the-world." From this he might have held the keys to the subjective and objective position we find ourselves in. All this, of course, cautiously. The facts could be misunderstood or misinterpreted, and the emotions and values expressed dishonestly, as much by me as by others. The thief would have to "decode" and "de-ideologize" the messages contained in the bag.

The folders are also arranged structurally. Again there are three classes:

(A) Chronological arrangement
(B) Logical arrangement
(C) Disorder.

The first structure puts us in mind of geological and botanical formations. The second of encyclopedia and computers. The third of genetic information. Together they reveal a picture of the structure of the human memory. What is missing however is a "formal structure" of the kind found in "alphabetical arrangement." Without this the thief might have concluded a defect in my way of thinking. The interaction of the ordered and disordered structures in the bag would have given the thief the opportunity to contribute to Jaques Monods problem "coincidence and necessity." The bag is a fertile hunting ground for "structural analysis."

Finally the folders are arranged according to their relationship to the bag itself. Two classes result:

(A) Folders that are in the bag so that they can be kept in mind.
(B) Folders that are there to keep things that are not there in mind.

The letters, manuscripts, and essays belong to the first class, the unfinished autobiography to the second. This reveals two functions of the bag(and of memory): to keep things in the present and to bring things into the present. The real situation is nevertheless much more complex. Some papers in the bag point to the future (the "New York" folder and the unpublished manuscript); thus proving the function of memory, namely to construct designs for the future. The thief could have recognized all of this. Not, however, this: This article itself which the reader has before him is found in the bag in the folder titled "published papers." The article is not only concerned with the bag, it is not just a "metabag" but a part that the thief could not have studied. The thief could never have recognized this aspect of the bag.
I always carry the bag with me. We all do this only my bag is more readily available. The question is: can our bags be stolen from us? Or would they always be found again a few blocks away, intact? Put differently; firstly: are we lighter and therefore progress more quickly into the future when our bags are lifted from us? And secondly; are these living or dead weights in our bags? The bag is too complicated to give a satisfactory answer to these questions. In any case it's good that from now on the questions themselves are kept safely in the bag.

[This text dates from 1972; it first in appeared in *Nachgeschichten* (Düsseldorf: Bollmann, 1990). Translated from German (1998) by Michael Stapley.]

SUBJECT: SEARCH ENGINES: METAMEDIA ON THE INTERNET?

FROM: HARTMUT WINKLER
<WINKLER@TFM.UNI-FRANKFURT.DE>
DATE: TUE, 29 SEP 1998 16:15:07 +0200

We use them daily, and don't know what we're doing. We don't know who operates them or why, don't know how they're structured, and little about the way they function. It's a classic case of the black box—and all the same, we're abjectly grateful for their existence.
Where, after all, would we be without them? Now that the expanse of web offerings has proliferated into the immeasurable, isn't anything that facilitates access useful? After all, instantly available information is one of the fundamental Utopias of the data universe.
Nevertheless, I think the engines are worth some consideration, and propose research should concentrate on the following points. First, the specific impetus of blindness that determines our handling of these engines. Second, the

conspicuously central, even "powerful" position the engines meanwhile occupy on the net—and this question is relevant if one wants to forecast the medium's development trends. Third, I am interested in the structural assumptions on which the various search engines are based. Fourth, and finally, a reference to language and linguistic theory that shifts the engines into a new perspective and a different line of tradition.

1

The main reason search engines occupy a central position on the net is that they are started infinitely often; in the case of Altavista, accessed 32 million times per workday, if the published statistics can be trusted. Individual users see the entry of a search command as nothing more than a launching pad to get something else, but to have attracted so many users to a single address signifies a great success. The direct economic consequence is that these contacts can be sold, making the search engines eminently suitable for the placement of advertising and therefore among the few net businesses that are in fact profitable. With remarkable openness, Yahoo writes: "Yahoo! also announced that its registered user base grew to more than 18 million members...reflecting the number of people who have submitted personal data for Yahoo!'s universal registration process.... 'We continued to build on the strong distribution platform we deliver to advertisers, merchants, and content providers.'"

Second, and even more important, the frequency of access means the overall net architecture has undergone considerable rearrangement. Thirty-two million users per day signify a thrust in the direction of centralization. This should put on the alert all those who recently emphasized the decentral, anti-hierarchic character of the net, and link its universal accessibility with far-reaching hopes for basis democracy.

All the same—and that brings me to my second point—this centralization is not experienced as such. The search engines can occupy such a central position only because they are assumed to be neutral in a certain way. Offering a service as opposed to content, they appear as neutral mediators. Is the mediator in fact neutral?

2

The question must be addressed first of all to the design of the search engines. Steve Steinberg, my main source for the factual information in the following text, described the things normal users don't know about the search engines and, even more important, what they think they don't need to know in order to use them expediently ("Seek and Ye Shall Find (Maybe)," *Wired* 4.05 [May 1996], 108ff.). Steinberg's first finding is that providers keep secret the exact algorithm on which their functioning is based (ibid., 175). Since the companies in question are private enterprises and the algorithms are part of their productive assets, the competition has, above all, to be kept at a distance; only very general information is disclosed to the public, the details remain in the dark of the black box. So if we operate the search engines with relative blindness, there are good economic reasons for this.

Three basic types of search engine can be distinguished. The first type is based on a system of predefined and hierarchically ordered keywords. Yahoo, for instance, employs human coders to assign new websites to the categories; the network addresses are delivered by email messages or hunted down by a search program known as a spider. In 1996, the company registered 200,000 web documents in this way.

The above figure alone indicates that coding through human experts is quick to meet its quantitative limitations. Of the estimated total volume of 30–50 million documents available on the net in 1996 (ibid., 113), Yahoo was offering some 0.4 percent; current estimates suggest that the total volume has meanwhile grown to 320 million websites.

However, the problems of the classification system itself are even more serious. The twenty thousand keywords chosen by Yahoo are known in-house (with restrained self-irony?) as "the ontology." But what or who would be in a position to guarantee the uniformity and inner coherence of such a hierarchy of terms. If pollution, for example, is listed under "Society and Culture"/"Environment and Nature"/ "Pollution", then the logic can be accepted to some degree, but every complicated case will lead to classificatory conflicts that can no longer be solved even by supplementary cross-references.

The construction of the hierarchy appears as a rather hybrid project, but its aim is to harness to a uniform system of categories millions of completely heterogenous contributions from virtually every area of human knowledge. Without regard to their perspectivity, their contradictions and rivalries.

Yahoo's "ontology" is thus the encumbered heir of those real ontologies whose recurrent failure can be traced throughout the history of philosophy. And the utilitarian context alone explains why the philosophical problem in new guise failed to be identified, and has been re-installed yet again with supreme naiveté. If the worst comes to the worst, you don't find what you're looking for—that the damage is limited is what separates Yahoo from problems of philosophy.

The second type of search engine manages without a predefined classification system and, even more important, without human coders. Systems like AltaVista, Inktomi, or Lycos generate an "inverted index" by analyzing the texts located. The search method employed is the full-text variant, word for word, meaning that in the end every single term used in the original text is contained in the index and available as a search word. This is less technically demanding than it might appear. For every text analyzed, a row is created in a huge cross-connected table, while the columns represent the general vocabulary; if a word is used in the text, a bit is set to "yes," or the number of usages is noted. An abstract copy of the text is made in this way, condensed to roughly 4 percent of its original size. The search inquiries now only make use of the table.

Since the system is fully automatic, the AltaVista spider can evaluate 6 million net documents every day. At present, some 125 million texts are represented in the system.

The results of a search are, in fact, impressive. AltaVista delivers extremely useful hit lists, ordered according to an internal priority system. And those who found what they were looking for are unlikely to be offended by the fact that AltaVista too keeps its algorithm under wraps.

There are some problems nevertheless. It is conspicuous that even slight variations in the query produce wholly different feedback; if you try out various queries for a document you already know, you will notice that one and the same document is sometimes displayed with high priority, sometimes with lower priority, and sometimes not at all. This is irritating, to say the least.

The consequence, in general terms, is that often one does not know how to judge the result of a search objectively—it remains unclear which documents the system does not supply because either the spider has failed to locate them or because the evaluation algorithm does indeed work otherwise than presumed. Even if the program boastfully claims to be "searching *the web*," the singular form of the noun is illusory, of course, if you consider the fact that even 125 million texts are only a specific section of the overall expanse. Furthermore, users for their part can register only the first 10, 50 or, at most, 100 entries. They too scarcely have the possibility of estimating how this section relates to the rest of the expanse in terms of content.

The second and main problem is however present already in the basic assumption. A mechanical keyword search presupposes that only such questions will be posed as are able to be clearly formulated in words, and differentiated and substantiated through further keywords. Similarly, nobody will expect that the system is able to include concepts of similar meaning alongside the query, or can exclude homonyms. Search engines of this type are wholly insensible to questions of semantics or, to make it more clear: their very point is to exclude semantic problems of the type evident with Yahoo. Yet that is not to say that the problems themselves are eradicated. They are imposed on the users through the burden of having to reduce their questions to unambiguous strings of significants, of having to be satisfied with the mechanically selected result. All questions unable to be reduced to keywords fall through the screen of the feasible. Technical and scientific termini are relatively suitable for such a search, humanistic subjects are less suitable, and once again this emerges as that "soft"—all too soft—sphere that should be circumvented from the outset, if one is unwilling to fall into the abyss.

But the problem of semantics has not been ignored, and efforts in this direction have led to the third type of search engine. Systems like Excite by Architext, or "Smart," claim to search no longer mechanically with strings of significants, but on the basis of a factual semantic model. In order to be able to discriminate between articles on oil films and ones on cinema films, such programs examine the context in which the respective concepts figure.

"The idea is to take the inverted index of the Web, with its rows of documents and columns of keywords, and compress it so that documents with roughly similar profiles are clustered together—even if one uses the word 'movie' and one uses 'film'—because they have many other words in common" (Steinberg, 175). The result is a matrix where the columns now represent concepts instead of mechanical keywords. The exciting thing about this type of engine is that it progresses from mechanical keywords to content-related concepts; and also that it obtains its categories solely on the basis of the entered texts, of a statistical evaluation of the documents.

[The engine] learns about subject categories from the bottom up, instead of imposing an order from the top down. It is a self-organizing system.... To come up with subject categories, Architext makes only one assumption: words that frequently occur together are somehow related. As the corpus changes—as new connections emerge between, say O. J. Simpson and murder—the classification scheme automatically adjusts. The subject categories reflect the text itself"; "this eliminates two of the biggest criticisms of library classification: that every scheme has a point of view, and that every scheme will be constantly struggling against obsolescence. (Ibid.)

Other designs, such as the Context system by Oracle, attempt to incorporate analyzes of the syntax, and by doing so find themselves in the minefield of how to model natural language—a problem that has been worked upon in the field of AI since the sixties, without convincing results having been produced so far. The evaluation of such systems is more than difficult; and it is even more difficult to make forecasts about the possible chances of developments.

For that reason, I would like to shift the focus of the question from the presented systems' mode of function and their implications and limitations to the sociocultural question of what their meaning is, what their actual project is in the concurrence of discourses and media.

3

The path from the hierarchic ontologies over the keyword search and on to the semantic systems shows, in fact, that it is a matter of a very fundamental question beyond the pragmatic usage processes. The search engines are not a random "tool" that supplements the presented texts and facilitates their handling. On the contrary, they appear as a systematic counterpart on which the texts are reliant in the sense of a reciprocal and systematic interrelation. My assertion is that the search engines occupy exactly that position which—in the case of non-machine-mediated communication—can be claimed by the system of *language*. (And that is the main reason why search engines interest me.)

Language, as Saussure clearly showed, breaks down into two modes of being, two aggregate states. Opposite the linear, materialized texts in the external world—utterances, speech events, written matter—exists the semantic system that, as a knowledge, as a language competence, has its spatially distributed seat in the minds of the language users. Minds and texts are therefore always opposite each other.

If access to the data network is now organized over systems based on vocabulary, and if these systems are being advanced in the direction of semantically qualifying machines, then this means that language itself, the semantic system, the lexicon, is to be liberated from the minds and technically implemented in the external world. In other words: not just the texts are to be filed in the computerized networks, but the entire linguistic system. The search engines, with all their flaws and contradictions, are a kind of advance payment on this project.

Search engines, then, represent *language* in the network. And this has completely changed the emphasis. The engines face the texts not as additional tools but as the "actual" structure that the texts merely serve; a machine for opening up, but at the same time a condensation that represents the body of texts as a whole.

4

The conjecture that it is a matter of the language admits a new perspective on the internal organization of search engines. And it becomes clear that engines have prominent predecessors in the history of knowledge and historical notions of language.

It is difficult not to see in the hierarchically composed structure of the Yahoo pyramid of concepts those medieval models of the world described for us by writers such as Bolzoni in her history of mnemonics (L. Bolzoni, "The Play of Images," in P. Corsi, ed., *The Enchanted Loom*, NY: Oxford, 1991, 16–65). A large fourteenth-century panel shows the figure of Jesus in the center of the tree of life, whose branches and leaves all contain stations in his earthly existence, his path to the Cross and his transfiguration. A second picture, this time from the thirteenth century, shows a horse-mounted knight who is riding, sword drawn, toward the Seven Deadly Sins, which are divided up into a scheme of fields branching of step-by-step into the infinite diversity of the individual sins (ibid., 27–29). Bolzoni explains that such schemes initially served didactic mnemonic purposes; order and visualization made it easier to note the complex connections. But their actual meaning goes further. The implicit ambition of these systems was to bring the things of the world into a consistent scheme, namely into a necessarily hierarchic scheme that no less necessarily culminated in the concept of God. Only the concept of God was capable of including all other concepts and furnishing a stable center for the pyramidal order. The linguistic structure (the cathedral of concepts) and the architecture of knowledge were superimposed over each other in this "order of things." This metaphysical notion of language has become largely alien to us today. But is it really alien?

As far as Yahoo's surface is concerned, if you will permit the abrupt return to my subject, it manages without an organizing center. The user faces fourteen, not one, central categories from which the subcategories branch off. Thus, the pyramid has lost its tip. Or would it be more appropriate to ask what has taken God's place?

In a model of the world created by Robert Fludd, an English encyclopedist of the Renaissance, God had already abandoned the center position ("*Integrae Naturae speculum artisque imago*" [1617], British Library). Retained has been a system of strictly concentric rings that contains the things of the world, encompassing a range from minerals to the plants and animals of nature up to the human arts and finally the planetary spheres. The center is occupied by a schematic diagram of the earth, a forerunner of that blue ball the astronauts radio-relayed to earth. The representation looks like a mandala in which viewers can absorb themselves in order to take up contact with a cosmic whole. The new, secularized solution becomes even more distinct in the memory theater of the Italian Camillo, which, frequently discussed in the meantime, itself belongs to the history of technical media. At the beginning of the sixteenth century, Camillo built a wooden construction resembling a small, round theater (see, for example, F. A. Yates, *Gedächtnis und Erinnern*, Weinheim 1991, 123ff.). Those who ventured inside were confronted by a panel of 7 x 7 pictures Camillo had commissioned from highly respected painters of the period. The horizontal division corresponded to the seven planetary spheres, the vertical division to seven stages of development from the first principles up to the elements, to the natural world, to the human being, to the arts and, finally, the sciences. In this way, every field in the matrix represented a certain aspect of the cosmos. The images were merely there to convey the general picture, whereas behind them were compartments with the texts written by the great writers and philosophers. It was in these compartments, then, that the user looked for sources, concepts and further information. To this extent, the whole thing was a system of access, and the analogy with search engines becomes evident in the clear separation between the access to the texts and the texts themselves.

Camillo's theater has finally brought the human being, the viewer, into the center of the construction. The surface of the images is oriented to his view, and solely the beholder's perspective joins up the forty-nine fields in the matrix. Exactly that appears to me to be the logic on which Yahoo is based. The very lack of the pinnacle in the pyramid of concepts defines the position taken by the user. Like in the optical system of the central perspective, the "royal overlooking position" is reserved for the user/beholder.

Yahoo is indeed an "ontology"; but not because Yahoo and likewise ontologies are arbitrary. It is more because they keep things in their place, and define for the user a position relative to this place. Its ontology offers an ordered world. And anything threatening to be lost in the chaotic variety of available texts can take one final respite in the order of the search engine.

The solution, however, is historically outdated, and has been abandoned in the history of philosophy. Because any positively defined hierarchy of concepts is perspectival and arbitrary, it soon reveals those points of friction that represent the beginning of its end. Does this make the solution of the keyword- or semantics-based engines more modern?

It must indeed appear to be so at first glance. The strategy of making the search words dependent on the empirically collected content of the network documents—the texts—imitates the mechanism of language itself. Or the mechanism, to be more precise, by which language arrives at its concepts.

Linguistic theory tells us that the synchronous system of language is created through the accumulation and condensation of an infinite multitude of concrete utterances. The place where condensation takes place is the language user's memory, where the concrete utterances are submerged; linear texts are obliviated into the structure of our language capability; on the basis of concrete texts, this structure is subject to constant modification and differentiation. Our faculty of language is an abstract copy of speaking—speech and language (discourse and system) are systematically cross-linked. (For a more detailed analysis, see my book *Docuverse*, Munich: Boer, 1997.) What this means for the isolated concept is that it accumulates whatever the tangible contexts provide as meaning. It isn't a one-time act of definition that assigns it a place in the semantic system, but the disorderly chain of its usages; concepts stand for and typify contexts, concepts encapsulate past contexts.

The semantic search engines imitate this accumulation by typifying contexts in order to arrive at concepts—in this case the search concepts. As outlined above, the table of search words is created as a condensed, cumulated copy of the texts. A statistical algorithm draws together comparable contexts, typifies them, and assigns them to the search concepts as the equivalent of their meaning.

A system imbued with such dynamism is superior to the rigidly predefined systems, even if the statistical algorithm only imperfectly models the mechanisms of natural language. More complex, closer to intuition, it is bound to offer less centers of friction. So, once again, what's the objection?

5

It's important to remember that, despite all the advances made, the actual fundamental order has remained constant. Just as in Camillo's wooden theater, we are dealing not with only two instances—a set of reading/writing/searching subjects approaching a second set of written texts—but also with a third instance, namely a system of access that has placed itself between the first two like a grid, or raster.

And if the access system in Camillo's media machine served to break down the infinite expanse of texts into a manageable number of categories from which the position—from a strictly central perspective—was defined for the observing subject, then this fundamental order remains intact also.

This image makes it clear that it is not necessarily better if the raster cannot be felt. It's almost the other way round: the less resistance offered by the access system, the more neutral, transparent, and weightless it seems, and the more plausible appears the suspicion that it cannot be a question of the nature of thing, but of a naturalization strategy.

The raster of categories *must* purport to be transparent if it does not want to rouse the problems that Yahoo rouses. To avoid the reproach of being arbitrary and exercising a structuring influence on the contents accessed, the raster must instill in the users the impression of being purely a "tool" subject only to utility—the key in the customers' hand that opens any Sesame, a compliant genie with no ambitions of its own.

This puts the veil of secrecy cast over the algorithms in a somewhat different light. Far more important than the rivalry between different product suppliers is the wish to actually dispose over a neutral, transparent access machine—and this wish is something the makers share with their customers, and probably with us all. At the basis of the constellation emerges an illusion that organizes the discourse.

Since there is no such thing as algorithms without their own weight, the metadiscourse has to help them out and salvage transparency by means of mere assertions. In the usage of the salutary singular ("searching the web"), in the way the algorithms are kept under wraps, in the emphasis on the performance as opposed to the limitations that might be more defining, and in the routine promises that, thanks to Artificial Intelligence, new and even more powerful systems are in the pipeline (see, for example, PointCast

<http:// www.pointcast.com>). In the unawareness and unwillingness to know on the part of the customers, and in the primacy of a practice that mostly, in any case, doesn't know what it's doing.

Data processing—and one feels almost cynical in bringing up this point—was propagated with the ideal of creating a very different type of transparency. The promise was to create only structures that were in principle able to be understood—the opposite, in fact, of natural language; to confine itself to the structural side of things, but to escribe this in a way that would not only admit analysis, but apparently include the latter from the outset. If programs have now, as Kittler correctly notes, begun to proliferate like natural-language texts, then this is not because the programs (and already even the search engines) have been infected by the natural-language texts. It is because of our need for both: for unlimited complexity and the narcissistic pleasure of having an overview, the variety of speaking and the transparency with regard to the objects, a language without metaphysical hierarchic centering that still maintains its unquestionable coherence.

That our wish is once again doomed to failure is clear from the fact that any number of search engines of different design are competing with each other in the meantime, and that metasearch engines are now said to be able to search through search engines. So there we sit on God's deserted throne, opposite us the infinite universes of texts, in our hands a few glittering, but deficient, machines. And we feel uneasy.

[Translated from German by Tom Morrison.]

SUBJECT: A MEANS OF MUTATION: NOTES ON I/O/D 4 THE WEB STALKER

FROM: MATTHEW FULLER <MATT@AXIA.DEMON.CO.UK>
DATE: MON, 12 OCT 1998 01:16:59 +0100

During 1997 and 1998 a series of legal and media confrontations were made in the United States and elsewhere. Amongst those involved were Microsoft, Netscape, and the U.S. Department of Justice. The key focus of contention was whether Microsoft, a company that has a near complete monopoly on the sale of operating systems for personal computers, had—by bundling its own web browser, Internet Explorer with every copy of its Windows '95/98 OS—effectively blocked Netscape, an ostensible competitor in browser software, from competing in a "free" market ("ostensible" because the nearly identical browsers toegther form if not an economic then a technical-aesthetic monopoly). This confrontation ran concurrently with one between Microsoft and Sun Microsystems, developers of the language Java.

The "browser wars" involved more than these three relatively tightly constructed and similar actors, though. Millions of internet users were implicated in this conflict. The nature of the proprietary software economy meant that for any side, winning the browser wars would be a chance to construct the ways in which the most popular section of the internet—the world wide web—would be used, and to reap the rewards. The conflict took place in a U.S. court and was marked by the deadeningly tedious superformalized rituals that mark the abstraction of important decisions away from those in whose name they are made. Though the staging of the conflict was located within the legal and juridical framework of the U.S. it had ramifications wherever software is used.

On connecting to a URL, HTML appears to the user's computer as a stream of data. This data could be formatted for use in any of a wide variety of configurations. As a current, given mediation by some interpretative device, it could even be used as a flowing pattern to determine the behaviour of a device completely unrelated to its purpose. (Work it with tags? Every <HREF> could switch something on, every <P> could switch something off—administration of greater or lesser electric shocks for instance). Most commonly it is fed straight into a browser.

What are the conditions that produce this particular sort of reception facility? Three fields that are key amongst those currently conjoining to form what is actualized as the browser: economics, design, and the material. By material is meant the propensities of the various languages, protocols, and data types of the web.

If we ask, "What produces and reinforces browsing?" There is no suprise in finding the same word being used to describe recreational shopping, ruminant digestion and the use of the web. The browser wars form one level of consistency in the assembly of various forms of economy on the web.

Websites are increasingly written for specific softwares, and some elements of them are unreadable by other packages (for example, the I/O/D "shout" HTML tag). You get Netscape sites, Explorer sites, sites that avoid making that split and stay at a level that both could use—and therefore consign the "innovations" of these programs to irrelevance. This situation looks like being considerably compounded with the introduction of customizable (and hence unusable by web-use software not correctly configured) Extensible Markup Language tags.

What determines the development of this software? Demand? There is no means for it to be mobilized. Rather more likely, an arms race between on the one hand the software companies and the development of passivity, gullibility, and curiosity as a culture of use of software.

One form of operation on the net that does have a very tight influence—an ability to make a classical "demand"—on the development of proprietary software for the web is the growth of online shopping and commercial information delivery. For companies on the web this is not just a question of the production and presentation of "content," but a very concrete part of their material infrastructure. For commerce on the web to operate effectively, the spatium of potential operations on the web—that is everything that is described or made potential by the software and the network—needs to be increasingly configured toward this end.

Bloatware—49% of software features are never used, 19% rarely used, 16% sometimes, 13% often, 7% always. Size of software is growing, Windows went from 3M lines of code (Windows 3.1) to 14M lines (Windows 95) to 18M (Windows 98). [Ninfomania NewsFeed]

That there are potentially novel forms of economic entity to be invented on the web is indisputable. As ever, crime is providing one of the most exploratory developers. How far these potential economic forms, guided by notions of privacy; pay per use; trans- and supra-nationality; and so on. will develop in an economic context in which other actors than technical possibility, such as the state, monopolies, and so on is open to question. However, one effect of net-commerce is indisputable. Despite the role of web designers in translating the imperative to buy into a post-rave cultural experience, transactions demand contracts, and contracts demand fixed, determinable relationships. The efforts of companies on the web are focused on tying down meaning into message delivery. While some form of communication may occur within this mucal shroud of use-value-put-to-good-use the focal point of the communication will always stay intact. Just click here.

Immaterial labor produces "first and foremost a social relation [that] produces not only commodities, but also the capital relation" (see M. Lazzarato, "Immaterial Labor," in this volume, and P. Virno, *Radical Thought in Italy*, Minneapolis: Minnesota University, 1996, 142). If this mercantile relationship is also imperative on the immaterial labor being a social and communicative one, the position of web designers is perhaps an archetype, not just for the misjudged and cannibalistic drive for a "creative economy" currently underway in Britain, but also within a situation where a (formal) language—HTML—explicitly rather than implicitly becomes a means of production: at one point vaingloriously touted as "How to Make Loot."

Web design, considered in its wide definition: by hobbyists, artists, general purpose temps, by specialists, and also in terms of the creation of websites using software such as Pagemill or Dreamweaver, is precisely a social and communicative practice, as Lazzarato says, "whose 'raw material' is subjectivity." This subjectivity is an ensemble of preformatted, automated, contingent and "live" actions, schemas, and decisions performed by both softwares, languages, and designers. This subjectivity is also productive of further sequences of seeing, knowing, and doing.

A key device in the production of websites is the page metaphor. This of course has its historical roots in the imaginal descriptions of the Memex and Xanadu systems—but it has its specific history in that Esperanto for computer-based documents, Structured Generalized Markup Language (SGML) and in the need for storage, distribution, and retrieval of scientific papers at CERN. Use of metaphor within computer interface design is intended to enable easy operation of a new system by overlaying it or even confining it within the characteristics of a homely futuristic device found outside of the computer. A metaphor can take several forms. They include emulators where say, the entire workings of a specific synthesizer are mapped over into a computer where it can be used in its "virtual" form. The computer captures the set of operations of the synthesizer and now the term *emulation* becomes metaphorical. Allowing other modalities of use and imaginal refrain to operate through the machine, the computer now *is* that synthesizer—while also doubled into always being more. Metaphors also include items such as the familiar "desktop" and "wastebasket." This is a notorious

In the one-to-one future, companies will do their best to get their hands on as much of your original writing as they can. They'll subscribe to discussion lists, sort through usenet, hang out in chat rooms, and, depending on the scruples of your internet service provider, scour your outgoing email. Survey data is one thing. Your own language is another. At first, they'll be able learn something about you by virtue of where you choose to express yourself. Then, after they've compiled enough of your ASCII, they'll use natural-language-processing technology to add tidbits of psychographic data on you to their databases. And finally, when the technology matures, they'll be able to start using your language and vocabulary patterns to sell products back to you in highly personalized email messages. Individual-level statistics could be sold to direct marketers, couponers, publishers...or even health-insurance companies. In the one-to-one future, your insurance premiums could be adjusted on a near-real-time basis based on your recent food-purchase patterns. Buy a steak and sour cream, your premium goes up. Buy bran cereal and nonfat milk, and your premium goes down. And, given the appropriate networked calendar software, they could even schedule you for an appointment with your managed-care specialist should your purchases of foods with high levels of saturated fats reach a critical level. In the one-to-one future, however, consumers will not only have aggregated and personalized content. They'll have aggregated and personalized commerce. But in the one-to-one future, personalization won't be limited to just one product category. Instead, consumers will be able to find an online seller who sells a particular lifestyle, defined as a mix of products and services. The seller, in effect, becomes a "commerce editor," presenting the books, clothes, records, movies, shoes, cars, computers, electronics, home furnishings and personal-care products that define a particular lifestyle. The seller will be able to deploy a wide variety of technologies in order to reach the target customer (that is, text, graphics, audio, video, push, chat, discussion, and so on) and can create an online shopping experience that correlates with the customer's personal aesthetics, sense of taste and desired level of interactivity. In the one-to-one future, these thousands of online sellers will be able to focus on the act of selling—creating and maintaining relationships with customers. Meanwhile, the "traditional" e-commerce retailers will be able to focus on retailing—exploiting their economies of scale in sourcing, storing, transacting and fulfilling product. The one-to one-future will not only displace the creators of mass culture but also the creators of micro culture—fine artists. When every piece of information we consume becomes customized for our unique wants and needs, we will

case of a completely misapplied metaphor. A wastebasket is simply an instruction for the deletion of data. Data does not for instance just sit and rot as things do in an actual wastebasket. That's your backup disk. Actual operations of the computer are radically obscured by this vision of it as some cosy information appliance always seen through the rearview mirror of some imagined universal.

The page metaphor in web design might as well be that of a wastebasket. While things have gone beyond maintaining and re-articulating the mode of address of arcane journals on particle physics the techniques of page layout were ported over directly from graphic design for paper. This meant that HTML had to be contained as a conduit for channeling direct physical representation—integrity to fonts, spacing, inflections, and so on. The actuality of the networks were thus subordinated to the disciplines of graphic design and of graphical user interface simply because of their ability to deal with flatness, the screen. (Though there are conflicts between them based around their respective idealizations of functionality). Currently of course this is a situation that is already edging toward collapse as other data types make incursions onto, through, and beyond the page—but it is a situation that needs to be totaled, and done so consciously and speculatively.

Another metaphor is that of geographical references. Where do you want to go today? This echo of location is presumably designed to suggest to the user that they are not in fact sitting in front of a computer calling up files, but hurtling round an earth embedded into a gigantic trademark "N" or "e" with the power of some voracious cosmological force. The web is a global medium in the approximately the same way that the World Series is a global event. With book design papering over the monitor the real processes of networks can be left to the experts in computer science...

It is the technical opportunity of finding other ways of developing and using this stream of data that provides a starting point for I/O/D 4: The Web Stalker. I/O/D is a three-person collective based in London, whose members are Simon Pope, Colin Green, and myself. As an acronym, the name stands for everything it is possible for it to stand for. There are a number of threads that continue through the group's output. A concern in practice with an expanded definition of the techniques/aesthetics of computer interface. Speculative approaches to hooking these up to other formations that can be characterized as political, literary, musical, etc. The production of stand-alone publications/applications that can fit on one high-density disk and are distributed without charge over various networks.

The material context of the web for this group is viewed mainly as an opportunity rather than as a history. As all HTML is received by the computer as a stream of data, there is nothing to force adherence to the design instructions written into it. These instructions are only followed by a device obedient to them.

Once you become unfaithful to page-description, HTML is taken as a semantic mark up rather than physical markup language. Its appearance on your screen is as dependent upon the interpreting device you use to receive it as much as its "original" state. The actual "commands" in HTML become loci for the negotiation of other potential behaviours or processes.

Several possibilities become apparent. This data stream becomes a phase space, a realm of possibility outside of the browser. It combines with another: there are thousands of other software devices for using the world wide web, waiting in the phase space of code. Since the languages are pre-existing, everything that can possibly be said in them, every program that could possibly be constructed in them is already inherently pre-existent within them. Programming is a question of teasing out the permutations within the dimensions of specific languages or their combinations. That it is never only this opens up programming to its true power—that of synthesis.

One thing we are proposing in this context is that one of the most pressing political, technical, and aesthetic urgencies of the moment is something that subsumes both the modern struggle for the control of production (that is of energies), and the putative postmodern struggle for the means of promotion (that is of circulation) within the dynamics of something that also goes beyond them and that encompasses the political continuum developing between the gene and the electron that most radically marks our age: the struggle for the means of mutation.

A file is dropped into the unstuffer. The projector is opened. The hard drive grinds. The screen goes black. The blacked out screen is a reverse nihilist moment. Suddenly everything is there.

A brief description of the functions of the Web Stalker is necessary as a form of punctuation in this context, but it can of course only really be fully sensed by actual use (see <http://www.backspace.org/iod>). Starting from an empty plane of color, (black is just the default mode—others are chosen using a pop-up menu) the user begins by marqueeing a rectangle. Using a contextual menu, a function is applied to the box. The box, a generic object, is specialized into one of the following functions. For each function put into play, one or more box is created and specialized

Crawler: The Crawler is the part of the Web Stalker that actually links to the web. It is used to start up and to show the current status of the session. It appears as a window containing a bar split into three. A dot moving across the bar shows what stage the Crawler is at. The first section of the bar shows the progress of the net connection. Once connection is made and a URL is found, the dot jumps to the next section of the bar. The second section displays the progress of the Web Stalker as it reads through the found HTML document looking for links to other URLs. The third section of the bar monitors the Web Stalker as it logs all the links that it has found so far. Thus, instead of the user being informed that connection to the net is vaguely "there" by movement on the geographic TV-style icon in the top right hand corner the user has access to specific information about processes and speeds.

Map: Displays references to individual HTML documents as circles and the links from one to another as lines. The URL of each document can be read by clicking on the circle it is represented by. Once a web session has been started at the first URL opened by the Crawler, Map moves through all the links from that site, then through the links from those sites, and so on. The

lose the ability to enjoy, or even tolerate, the singular statement of an individual painter, sculptor, photographer, or printmaker. "If they're not going to paint what I like," the consumer of the future will ask, "why should I buy it?" This means that artists who intend to support themselves with their work will have to adopt market research techniques to make sure they're creating works that targeted segments of collectors will actually enjoy. Alternatively, the art market could become solely commission-based, where buyers work with artists to custom-create a piece that fits with their tastes, their politics, their personalized color scheme. [M. Sippey <msippey@≠viant.com>, One-to-one Future, Mon, 21 Sep 1998 08:02:29 -0700]

mapping is dynamic—"Map" is a verb rather than a noun.

Dismantle: The Dismantle window is used to work on specific URLs within HTML documents. URLs at this level will be specific resources such as images, email addresses, sound files, downloadable documents, and so on. Clicking and dragging a circle into the Dismantle window will display all URLs referenced within the HTML document you have chosen, again in the form of circles and lines.

Stash: The Stash provides a document format that can be used to make records of web use. Saved as an HTML file it can also be read by "browsers" and circulated as a separate document. Sites or files are included by dragging and dropping URL circles into a Stash.

HTML Stream: Shows all of the HTML as it is read by the Web Stalker in a separate window. Because as each link is followed by the crawler the HTML appears precisely as a stream, the feed from separate sites is effectively mixed.

Extract: Dragging a URL circle into an extract window strips all the text from a URL. It can be read on screen in this way or saved as a text file.

The Web Stalker performs an inextricably technical, aesthetic, and ethical operation on the HTML stream that at once refines it, produces new methods of use, ignores much of the data linked to or embedded within it, and provides a mechanism through which the deeper structure of the web can be explored and used.

This is not to say much. It is immediately obvious that the Stalker is incapable of using images and some of the more complex functions available on the web. These include for instance: gifs, forms, Java, VRML, frames, etc. Some of these are deliberately ignored as a way of trashing the dependence on the page and producing a device that is more suited to the propensities of the network. Some are left out simply because of the conditions of the production of the software—we had to decide what was most important for us to achieve with available resources and time. This is not to say that if methods of accessing this data were to be incorporated into the Stalker that they would have been done so "on their own terms." It is likely that at the very least they would have been dismantled, dissected, opened up for use in some way.

Another key factor in the shape of the program and the project as a whole is the language it was written in: Lingo, the language within Macromedia Director—a program normally used for building multimedia products and presentations. This is to say the least a gawky angle to approach writing any application. But it was used for two reasons—it gave us very good control over interface design and because NetLingo was just being introduced, but more importantly because within the skill base of I/O/D, that was what we had. That it was done anyway is, we hope, an encouragement to those who have the "wrong" skills and few resources but a hunger to get things done, and a provocation to those who are highly skilled and equipped but never do anything.

Previous work by artists on the web was channeled into providing content for websites. These sites are bound by the conventions enforced by browser-type software. They therefore remain the most determining aesthetic of this work.

"A Kansas City company, Applied Micro Technology Inc., is about to begin selling a device for censoring language in TV broadcasts (intended for the protection of children). It works only on closed-captioned broadcasts. If a banned word is found in the closed caption, the sound is muted and the closed caption displayed with a milder word substituted. The original design just matched on words, causing DICK VAN DYKE to turn into JERK VAN GAY. This was obviously inadquate, so it was extended to recognize context. The designer, Rick Bray, says that it now catches 65 out of 66 "offensive words" in the movie Men in Black (for example), and so he now allows his children to see it, and so they're pleased with the device." [T. Byfield <tbyfield@panix.com>, Four Allegories, Sat 14 Mar 1998, 10:51:41 -0500]

The majority of web-based art, *if it deals with its media context at all* can be understood by four brief typologies:

incoherence (user abuse, ironic dysfunctionality, randomness to mask pointlessness)

archaeology (media archaeology, emulators of old machines and software, and structuralist materialist approach)

retrotooling (integrity to old materials in "new" media, integrity as kitsch derived from punk/jazz/hip hop, old-style computer graphics, and "filmic references"—the Futile Style Of London; see the "FSOL" section of the IOD website)

deconstruction (conservative approach to analyzing-in-practice the development of multimedia and networks, consistently re-articulating contradiction rather than using it as a launching pad for new techniques of composition).

Within the discourse networks of art, including critical technique; license to irresponsibility; compositions-in-progress of taste stratification and breaks; institutions; finance; individual survival strategies; media; social networks; legitimation devices; at least potential openness to new forms; and avowed attentiveness to manifestations of beauty, there were dynamics that were useful to mobilize in order to open up possibilities of circulation and effect for the Web Stalker. However, at the same time as the project was situated within contemporary art, it is also widely operative outside of it. Most obviously it is at the very least, a piece of software.

Just as the Stalker is not-just-art, it can only come into occurrence by being not just itself. It has to be used. Assimilation into possible circuits of distribution and effect in this case means something approaching a media strategy.

Operating at another level to the Web Stalker's engagement within art were two other forms of media that were integral to the project: stickers (bearing a slogan and the I/O/D URL) and freeware. Both are good contenders for being the lowest, most despised grade of media. That the Web Stalker is Freeware has been essential in developing its engagement with various cultures of computing.

The Web Stalker has gone into circulation in the low hundreds of thousands. Responses have ranged from intensely detailed mathematical denunciations of the Map and a total affront that anyone should try anything different; to evil glee, and a superb and generous understanding of the project's techniques and ramifications.

While for many, the internet simply *is* what is visible with a browser, at the same time it is apparent that there is a widespread desire for new nonformulaic software. One of the questions that the Stalker poses is how program design is taken forward. Within the limitations of the programming language and those of time, the project achieved what it set out to do. As a model of software development outside of the superinvested proprietary one this speculative and interventional mode of production stands alongside two other notable radical models: that of free software and that derived from the science shops, (wherein software is developed by designers and programmers in collaboration with clients for specifically social uses). Unlike these others it is not so likely to find itself becoming a model that is widely adoptable *and* sustainable.

In a sense then, the web stalker works as a kind of "tactical software" but it is also deeply implicated within another kind of tacticity—the developing street knowledge of the nets (see G. Lovink and D. Garcia, "The ABC of Tactical Media" <http://www.waag.org/tmn/>). This is a sense of the flows, consistencies and dynamics of the nets that is most closely associated with hackers, but that is perhaps immanent in different ways in every user.

Bringing out and developing this culture however demands attention. In some respects this induction of idiosyncratic knowledges of minute effects ensures only that while the browser wars will never be won, they are never over. So long as there's the software out there working its temporal distortion effects on "progress." So long as there's always some nutter out there in the jungle tooled up with some VT100 web viewer, copies of Mosaic, Macweb, whatever.

At the same time we need to nurture our sources of this *ars metropolitani* of the nets. During recent times and most strongly because of the wider effects of specific acts of repression, hacking itself has often become less able to get things going because it has (a) been driven more underground, (b) been offered more jobs, and (c) been less imaginatively willing or able to ally itself with other social currents.

Software forges modalities of experience—sensoriums through which the world is made and known. As a product of "immaterial labor" software is a social, technical, and aesthetic relation that is embodied—and that is at once productive of more relations. That the production of value has moved so firmly into the terrain of immaterial labor, machine embodied intelligence, style as factory, the production of subjectivity, makes the evolution of what was previously sectioned as "culture" so much more valuable to play for—potentially always as sabotage—but, as a development of the means of mutation, most compellingly as synthesis.

Synthesis is explicitly not constitutive of a universe of synchronization and equivalence where everything connects to everything. Promising nothing but reconstitutive obliteration to "worlds" where everything means only one thing: virtual office, virtual pub, virtual gallery, virtual nightclub, however many more sonic gulags passing as virtual mixing decks. What is so repulsive about this nailed-down faithfulness is not so much that its darkside is about as disturbing as a blacklight lightbulb, or that it presents a social terrain that has been bounced clean by the most voracious of doormen—the miserable consciousnesses of its producers—but that it is continually dragging this space of composition, network, computer, user, software, socius, program production, back into the realm of representation, the dogged circular churning of avatars through the palace of mundane signs, stiffs reduced if at all possible to univocal sprites, rather than putting things into play, rather than making something happen.

Synthesis incorporates representation as a modality. Representation is not replaced but subsumed by the actualization of ideas and the dynamism of material through which, literally "in the realm of possibility," it becomes contingent. But this is not to trap synthesis within the "inherent" qualities of materials. "Truth to materials" functions at once as both a form of transcendence through which by the purest of imputations interpretative schema can pluck out essences and as a form of repressively arch earnestness. This is a process of overflowing all ideal categories.

The Map makes the links between HTML documents. Each URL is a circle, every link is a line. Sites with more lines feeding into them have brighter circles. Filched data coruscating with the simple fact of how many and which sites connect to boredom.com, extreme.net or wherever. (Unless it's been listed on the ignore.txt file customizable and tucked into the back of the Stalker). Every articulation of the figure composing itself on screen is simply each link being followed through. The map spreads out flat in every direction, forging connections rather than faking locations. It is a figuration that is *immutably* live. A processual opening up of the web that whilst it deals at every link with a determinate arrangement has no cutoff point other than infinity. Whilst the Browser just gives you history under the Go menu, the Map swerves past whichever bit of paper is being pressed up to the inside of the screen to govern the next hours of clickthrough time by developing into the future—picking locks as it goes.

From there, in unison with whichever of the other functions are applied, a predatory approach to data is developed. Sites are dismantled, stored, scanned to build up other cultures of use of the nets. That the software is cranky, that things become alien, that it is not the result of years of flow-charted teams, that it forces (horrific act) PC users to use alt, ctrl, delete to quit the program is not in question.

All the while, synthesis keeps running, keeps mixing. Producing sensoriums, modes of operation, worldviews that are downloadable (that is both traceable and open), mixable, measurable, assimilable (but not without risk of contamination), discardable, perhaps even immersive. This is a poetics of potential that is stringent—not just providing another vector for perpetually reactive opportunism—yet revelling in the possibility always also operating within the most intensified sounds: a hardcore methodology.

Aggregates are formed from the realm induced by the coherence of every possibility. Syntactics tweaks, examines, and customs them according to context. This context is not preformatted. It is up for grabs, for remaking. Synthesis determines a context within which it is constitutive and comes into composition within ranges of forces. Everything—every bit, every on or off fact—is understood in terms of its radical coefficiency, against the range of mutation from which it emerged and amongst the potential syntheses with which it remains fecund. It is the production of sensoria that are productive not just of "worlds" but of the world.

SUBJECT: THE LANGUAGE OF NEW MEDIA

FROM: LEV MANOVICH <HTTP://JUPITER.UCSD.EDU/~MANOVICH>
DATE: MON, 26 OCT 1998 20:10:54 -0800 (PST)

1. CATEGORIES

New media requires a new critical language—to describe it, to analyze it, and to teach it. Where shall this language come from? We can't go on simply using technical terms such as "a website" to refer to works radically different from each other in intention and form. At the same time, traditional cultural concepts and forms prove to be inadequate as well. Image and viewer, narrative and montage, illusion and representation, space, and time—everything needs to be redefined again.

To articulate the critical language of new media we need to correlate older cultural/theoretical concepts and the concepts that describe the organization/operation of a digital computer. As an example of this approach, consider the following four categories: *interface, database, navigation, and spatialization.* Each of these categories provides a different lens through which to inquire about the emerging logic, grammar, and poetics of new media; each brings with it a set of different questions.

Database: After the novel and later cinema privileged narrative as the key form of cultural expression of the modern age, the computer age brings with it a new form—database. What are the origins, ideology and possible aesthetics of a database? How can we negotiate between a narrative and a database? Why is database imagination taking over at the end of the twentieth century?

Interface: In contrast to a film, which is projected upon a blank screen and a painting which begins with a white surface, new media objects always exist within a larger context of a human–computer interface. How does a user's familiarity with the computer's interface structure the reception of new media art? Where does interface end and the "content" begin?

Spatialization: The overall trend of computer culture is to spatialize all representations and experiences. The library is replaced by cyberspace; narrative is equated with traveling through space (*Myst*); all kinds of data are rendered in three dimensions through computer visualization. Why is space being privileged? Shall we try to oppose this spatialization (that is, what about time in new media)? What are the different kinds of spaces possible in new media?

Navigation: We no longer only look at images or read texts; instead, we navigate through new media spaces. How can we relate the concept of navigation to more traditional categories such as viewing, reading, and identifying? In what ways do current popular navigation strategies reflect military origins of computer imaging technology? How do we demilitarize our interaction with a computer? How can we describe the person doing the navigation beyond the familiar metaphors of "user" and "flâneur"?

2. GENRES

The next step in articulating the critical language of new media involves defining genres, forms, and figures that persist in spite of constantly changing hardware and software, using the categories as building blocks. For example, consider two key genres of computer culture: a database and navigable space. (That is, creating works in new media can be understood as either constructing the right interface to a multimedia database or as defining a navigation method through spatialized representations.)

Why does computer culture privilege these genres over other possibilities? We may associate the first genre with work (postindustrial labor of information processing) and the second with leisure and fun (computer games), yet this very distinction is no longer valid in computer culture. Increasingly, the same metaphors and interfaces are used at work and at home, for business and for entertainment. For instance, the user navigates through a virtual space both to work and to play, whether analyzing financial data or killing enemies in Doom.

3. APPLICATION

New media theory also should trace the historical formation of these categories and genres. Here are examples of such an analysis.

Exhibit 1: Dziga Vertov, *Man with a Movie Camera*, USSR, 1928
Vertov's avant-garde masterpiece anticipates every trend of new media of the 1990s. Of particular relevance are its *database* structure and its focus on the camera's *navigation* through space.

Computer culture appears to favor a database ("collection," "catalog," and "library" are also appropriate here) over a narrative form. Most websites and CD-ROMs, from individual artistic works to multimedia encyclopedias, are collections of individual items, grouped together using some organising principle. Websites, which continuously grow with new links being added to already existent material, are particularly good examples of this logic. In the case of many artists' CD-ROMs, the tendency is to fill all the available storage space with different material: documentation, related texts, previous works, and so on. In this case, the identity of a CD-ROM (or of a DVD-ROM) as a storage media is projected onto a higher plane, becoming a cultural form of its own.

Vertov's film reconciles narrative and a database by creating narrative out of a database. Records drawn from a database and arranged in a particular order become a picture of modern life—and simultaneously an interpretation of this life. *A Man with a Movie Camera* is a machine for visual epistemology. The film also fetishizes the camera's mobility, its abilities to investigate the world beyond the limits of human vision. In structuring the film around the camera's active exp

Exhibit 2: Evans and Sutherland, *Real-time Computer Graphics for Military Simulators*, USA, early 1990s.
Military and flight simulators have been one of the main applications of real-time 3-D photorealistic computer graphics technology in the seventies

and the eigties, thus determining to a significant degree the way this technology developed. One of the most common forms of *navigation* used today in computer culture—flying through spatialized data—can be traced back to simulators representing the world through the viewpoint of a military pilot. Thus, from Vertov's mobile camera we move to the virtual camera of a simulator, which, with the end of the Cold War, became an accepted way to interact with any and all data, the default way of encountering the world in computer culture.

Exhibit 3: Peter Greenaway, *Prospero's Books*, 1991.
One of the few directors of his generation and stature to enthusiastically embrace new media, Greenaway tries to re-invent cinema's visual language by adopting computer's *interface* conventions. In *Prospero's Books*, cinematic screen frequently emulates a computer screen, with two or more images appearing in separate windows. Greenaway also anticipates the aesthetics of later computer multimedia by treating images and text as equals.
Like Vertov, Greenaway can be also thought of a *database* filmmaker, working on a problem of how to reconcile database and narrative forms. Many of his films progress forward by recounting a list of items, a catalog that does not have any inherent order (for example, different books in *Prospero's Books*).

Exhibit 4: Tamás Waliczky, "The Garden" (1992), "The Forest" (1993), "The Way" (1994), Hungary/Germany. Joachim Sauter and Dirk Lüsenbrink (Art+Com), *The Invisible Shape of Things Past*, Berlin, 1997.
Tamás Waliczky openly refuses the default mode of *spatialization* imposed by computer software, that of the one-point linear perspective. Each of his computer animated films "The Garden," The Forest," and "The Way" utilizes a particular perspective system: a water-drop perspective in "The Garden," a cylindrical perspective in "The Forest", and a reverse perspective in "The Way." Working with computer programmers, the artist created custom-made 3-D software to implement these perspective systems.
In "The Invisible Shape of Things Past" Joachim Sauter and Dirk Lüsenbrink created an original *interface* for accessing historical data about Berlin. The interface devirtualizes cinema, so to speak, by placing the records of cinematic vision back into their historical and material context. As the user navigates through a 3-D model of Berlin, he or she comes across elongated shapes lying on city streets. These shapes, which the authors call "filmobjects", correspond to documentary footage recorded at the corresponding points in the city. To create each shape the original footage is digitized and the frames are stacked one after another in depth, with the original camera parameters determining the exact shape.

Exhibit 5: Computer Games, 1990s.
Today computer games represent the most advanced area of new media, combining the latest in real-time photorealistic 3-D graphics, virtual actors, artificial intelligence, artificial life and simulation. They also illustrate the general trend of computer culture toward the *spatialization* of every cultural experience. In many games, narrative and time itself are equated with the

movement through space (that is, going to new rooms, levels, or words.) In contrast to modern literature, theater, and cinema that are built around the psychological tensions between characters, these computer games return us to the ancient forms of narrative where the plot is driven by the spatial movement of the main hero, traveling through distant lands to save the princess, to find the treasure, or to defeat the Dragon.

[This text is based on the program of the symposium "Computing Culture: Defining New Media Genres," which I and my collegues organized in the spring of 1988 at Center for Research in Computing and the Arts, University of California, San Diego. See <http://jupiter.ucsd.edu/~cul≠ ture/symposium.html>. Edited by Mathew Fuller.]

SUBJECT: PHENOMENOLOGY OF LINUX

FROM: ALAN JULU SONDHEIM <SONDHEIM@PANIX.COM>
DATE: TUE, 17 JUN 1997 20:29:54 -0400 (EDT)

Beyond the traditional division of graphic user interface (GUI) and text-based interface, the unix and linux system/s create a unique environment problematizing machine, boundary, surface, and structure.
The environment has implications far beyond a kinesic study of a particular technology; these tend toward an (un)accountancy of splintering or sputtering, stuttered linkages, microsutures, scanning intention across or among traditionally "isolated" platforms. Begin with the apparent file structure:

1. Working within the files, there are several domains: the formal tree-organization of the operating system (beginning with the root and ascending/descending); the *accumulation* or *heap* of files within the local directory (these files may or may not be related beyond their common path); and the *imminent domain*, the file or files currently open or in the process of being modified (these are nonexclusionary).

2. The graphic interface opens to shells as well, and since the interface devolves from a *blank screen*, there is simultaneously *potential* (click anywhere on it) and absence (nothing visible), reflecting upon the human operator / monitor interface as well.

3. *Errors* may or may not be characterized by *error messages*, which are inscribed by a process evolving from the root cause; there is then both the symptom (program *x* misbehaving) and the message (Error: <etc.>) that intersect: the message may be the (only visible) symptom, and the symptom itself may carry the message.

4. It is easy to assume that source code is equivalent to bones and operable

binaries to flesh; or the kernel as fundament, and file structure as slough. I would rather argue for a system of cubist plateaus of intersecting information regimes, with vectors/commands operating among them. In this sense it is information that is immanent within the operating system, not any particular plateau-architecture.

5. Language moves among performative, declarative, and neutral /dev/nul regimes; again, the boundaries are blurred, even on a technical level. Programs, more properly *scripts* (an apt word, since code is inscribed) call up different languages, shells, other programs, internal or external conduits (see below); internal and external interpenetrate here.

6. The division between GUI and text-based net access is blurred; shell accounts use IP and can open X Window and browsers, just as browser GUIs can share window space with shells.

7. The *space* of the operating system is problematized since machines carve out what I call *fractal channeling*, ports and commands rapidly shuttling back and forth between traditionally external netspace and internal vehicle space. Channels may open to other shells which may open to other channels; loopback channels operate within the local vehicle (internally), for example, and may be used to communicate with incoming on a local talk application. In shell-to-shell, both are equivalent on the screen: think of this as *screen-resonance* or system of strange attractors.

8. Furthermore, within the screen-resonance there are the spaces of the user/s on the system, partly application-dependent, shuttling among persons, tenses, and semantico-grammatical categories (Whorfian, in other words). Two linked talkers may be opened in relation to a net browser on an X Window while top (a program monitoring machine processes) is also running, and files are being transferred from a cdrom to hard drive. Attention moves among these spaces/applications, blurring distinctions; the talkers, for example, may demand considerable psychological investment, while anomalies in one or more of the other applications also call for immediate examination and response. If errors etc. appear, the anomalies (in relation to the normative ongoing chat) may best be described phenomenologically by Schutz's relevance theory, consider lifeworld strata, projects, and presentifications—in spite of the fact that all of this is primarily *read* and *written to*, inscribed and counterinscribed.

9. One might argue that the fractured domain in its entirety is never grasped—nor is there a "domain" and "entirety" at all. If we extend inscription and counterinscription, taking into account fuzzy and fractal channeling (deconstruction of category object/arrow theory), we can work toward a loosely defined sememe undergoing continuous and fairly rapid transformations, which are not necessarily charted from either interior or exterior (meta-) positions. The traditional metapsychology of the user splits, just as it splits beneath the sign of morphing gender in MOOs and IRC; it

is always already possible for theory to take morphing into account (as if morphing is *being-accounted-for* and therefore *accountable*), but this is a posteriori; in fact the splitting problematizes *any* metapsychology insofar as the mind is considered a somewhat closed (hydraulic model) frame, as opposed to a fuzzy communicative systemics paralleling the description herein of the operating system itself.

10. It is not difficult to see, not the operating system as mind, but both mind and operating system as challenging dyadic conventions of interior/exterior, grammatical tense and person, and so on. As I have mentioned before, Merlin Donald takes steps in this direction; one can also consider an accumulation or sememe of *flows* moving among bodies, organs, and so on, along the lines of Deleuze and Guattari.

11. Within and without all of this, the *cyborg* model, based on the suturing of disparate epistemes, becomes oddly antiquated; it accounts well for prosthetics, robotics, and machine/organism navigation, but remains based on traditionally separate ontological domains. Instead, think of spread epistemes and ontologies—for example, the distinction between declarative and performative becomes oddly confused in the case of basic HTML coding (that is. without "refresh" or JavaScript), which flows texts around screens.

12. Finally one might bring up postmodernisms, with their flows, part-objects, relativities, multiculturalisms, incommensurability of commensurable languages (and commensurability of incommensurable language)—as well the postmodern architectures, with their deconstructions, skewlines, and exposures/doublings, baring the systems, decomposing them. And it is true that such architectures have their equivalent among the operating system architectures; the operating system kernel for example may be equivalent to the control center of a building, and the communicative flow through a building has its equivalence with the fractal channeling described above. Nevertheless, I would not want to push this analogy, to the extent that the postmodern is representative of a *stage* (that is, post-Fordism among other things), and not necessarily the (de)construct of a broken episteme more or less permanently on the (broken) horizon and always-having-been-present. For the operating systems under consideration may be likened to the production of a *scanning electron microscope*, a case in which scanning is related to phenomenological intentionality instead of the discrete world of envisioned objects and flows described in, say, Gibson's work. The difference, yet to be accounted for, never to be accounted for, lies between the optical circularity of the phenomenology of the image produced by the light microscope, and the exaggerated dimensionality and exploratory scanning of any electron microscope, such as the tunneling or even the recent development of the scanning probe, which promises to "image single electrons," one might almost say, bits and their own architectures down to that very *level* (*Scientific American* July 1997.)

[See <http://jefferson.village.virginia.edu/~spoons/internet_txt.html>.]

Schizophrenia is the ego crisis of the cyborg; it is inevitable. Cyborgs are the fabrications of a science invested in the reproduction of subjects it takes to be real, a science whose first mistake was the belief that cyborg subjects were autonomous agents, that they existed outside any web of pre-existing significations. Prestructured by all comers, but taken to be pristine, the artificial agent is caught in the quintessential double bind. Fabricated by the techniques of mass production, the autonomous agent shares in the modern malady of schizophrenia. This piece tells the story of that cyborg, of the ways it has come into being, how it has been circumscribed and defined, how this circumscription has led to its schizophrenia, and the ways in which it might one day be cured.

THE BIRTH OF THE CYBORG: CLASSICAL AI

The cyborg was born in the fifties, the alter ego of the computer. It was launched into a world that had already defined it, a world whose notions of subjectivity and mechanicity not only structured it but provided the very grounds for its existence. It was born from the union of technical possibility with the attitudes, dreams, symbols, concepts, prejudices of the men who had created it. Viewed by its creator as pure potentiality, it was, from the start, hamstrung by the expectations and understandings that defined its existence.

Those expectations were, and are, almost unachievable. The artificial subject is an end point of science, the point at which knowledge of the subject will be so complete that its reproduction is possible. The twin births of Cognitive Science and Artificial Intelligence (AI) represent two sides of the epistemological coin: the reduction of human existence to a set of algorithms and heuristics and the re-integration of those algorithms into a complete agent. This resulting agent carries the burden of proof on its back; its "correctness" provides the objective foundation for a complex system of knowledge whose centerpiece is rationality.

Make no mistake, rationality is the central organizing principle of classical AI. The artificial agent is fabricated in a world where "intelligence," not "existence," is paramount, an "intelligence" identified with the problem-solving behavior of the scientist. For classical AI, the goal is to reduce intelligent behavior to a set of more or less well-defined puzzles, to solve each puzzle in a rational, ideally provably correct, manner, and, one day, to integrate all those puzzle-solvers into an agent indistinguishable (within a sufficiently limited framework) from a human.

That limited framework had better not exceed reason. Despite early dreams of agents as emotionally volatile as humans, the baggage of an engineering background quickly reduced agenthood to rationality. Allen Newell, one of the founders of AI, stated that the decision procedure of an agent must follow the "principle of rationality:" any agent worthy of its name must always pursue a set of goals, and may only take actions it believes help achieve one of its goals ("The Knowledge Level," *CMU CS Technical Report* CMU-CS-81-131, July 1981). In this system's narrow constraints, any agent that defies pure rationality is declared incomprehensible, and hence scientifically invalid.

Given these expectations, it was ironic when the artificial agent began to show signs of schizophrenia. Designing a rational decision procedure to solve a clearly defined puzzle was straightforward; combining these procedures to function holistically in novel situations proved to be nearly impossible. Bound in the straitjacket of pure rationality, the cyborg began to show signs of disintegration: uttering words it did not understand upon hearing, reasoning about events that did not affect its actions, suffering complete breakdown in situations that did not fit into its limited system of preprogrammed concepts. It could play chess like a master, re-arrange blocks on command in its dream world, configure computer boards; but it could not see, find its way around a room, or maintain routine behavior in a changing world. It was defined and fabricated in an ideal, Platonic world, and could not function outside the boundaries of neat definitions. Faced with an uncertain, incompletely knowable world, it ground to a halt.

THE PROMISE OF ALTERNATIVE AI

Understanding that the cyborg was caught in a rational, disembodied double bind, some AI researchers abandoned the terrain of classical AI. Alternative AI—a/k/a Artificial Life, behavior-based AI, situated action— sought to treat agents by redefining the grounds of their existence. No longer limited to the Cartesian subject, the principle of situated action shattered notions of atomic individualism by redefining an agent in terms of its environment. An agent should be understood in terms of interactions with its environment. "Intelligence" is not located in an agent but is the sum total of a pattern of events occurring in the agent and in the world. The agent no longer "solves problems," but "behaves"; the goal is not "intelligence" but "life."

Redefining the agent's conditions of existence breathed new life into the field, if not into the agent itself. Where once there had been puzzle-solvers and theorem-provers as far as the eye could see, there were now herds of walking robots, self-navigating cans-on-wheels and insect pets. Alternative AI gave the cyborg its body and lifted some of the constraints on its behavior. No longer required to be rational, the artificial agent found new vistas open to itself.

It did not, however, escape schizophrenia. Liberated from the constraints of pure reason, practitioners of alternative AI, unwittingly following the latest trends in postmodernism, embrace schizophrenia as a factor of life. Rather than creating schizophrenia as a side effect, they explicitly engineer it in: the more autonomous an agent's behaviors are, the fewer traces of Cartesian ego left, the better. May the most fractured win!

Dear "Nettimers": Allow me to introduce myself to you all. My name is Nathan Myhrvold, and I'm a Group Vice President at Microsoft. One of my responsibilities is identifying new opportunities for us to explore. Over the past several years, I have taken a special interest in the intersection of multimedia and networking. It is my impression that many of you understand just how complex this intersection will be. At Microsoft we take great pride in the quality of our work, but we occasionally find the answers to our quests off our Campus, in vibrant and creative communities of every kind. Nettime is one such community, I believe. You were brought to my attention by Mark Stahlman. In a recent meeting he recommended that I take a look at your mailing list. So I set aside some time to look through your archives; and overall I was extremely impressed with what I found. I was particularly struck by the "rhythm" of your conversations: so many people from so many countries able to identify and work toward a common set of goals, while accepting the necessary tensions and conflicts that advance entails. Believe me when I say that part of my job is to research these kinds of things; so I hope you will accept my compliment when I say it is very rare to find this dymanic on a mailing list. When I investigated the individual efforts of the Nettime list's contributors, I was even more impressed by this commonality, because you are individually pursuing very disparate efforts. At Microsoft, we believe that a very important transformation is taking place. Things that many people take for granted—I am thinking of the mass media at one extreme, and individual creators at the other—will give way to a more complex and multivalent publishing environment. In this developing world, distributed efforts such as Nettime will be truly decisive. That is why I am writing to you now. On behalf of the Microsoft Corporation, I would like to extend an invitation to you Nettime to hold your next meeting on our Redmond Campus. Needless to say, we will make arrangements for travel and lodging costs for up to a total of 300 attendees. As a token of our gratitude, Microsoft will underwrite reasonable expenses incurred in the process of publishing your proceedings. (Mark mentioned to me that the list's maintainers would like to publish a hard-copy anthology of your prior publications.) Of course, I recognize that some members of your group may be hesitant about this invitation for various reasons. I trust that your experiences here will convince you that these reservations can be set aside for the time being in favor of the common pursuit of a better world for all. That is what I am working toward, and I believe the same is true of all "Nettimers." Thank you for your consideration, and I look forward to hearing from you soon. I hope you will accept my apologies in advance when I request that you refrain from publicly forwarding my specific contact information. [*allegedly from* Nathan Myhrvold <nathanm@microsoft. com>, An Invitation the "Nettime" Mailing List, Tue, 31 Mar 1998 14:20:21 -0800]

At the same time, that schizophrenia becomes a limit point for alternative AI, just as it has been for classical AI. While acknowledging that schizophrenia is not a fatal flaw, alternativists have become frustrated at the extent to which it hampers them from building extensive agents. Alternativists build agents by creating behaviors; the integration of those behaviors into a larger agent has been as much of a stumbling block in alternative AI as the integration of problem-solvers is in classical AI. Despite their differences in philosophy, neither alternativists nor classicists know how to keep an agent's schizophrenia from becoming overwhelming. What is it about the engineering of subjectivities that has made such divergent approaches ground on the same problem?

FABRICATING SCHIZOPHRENIAS

Certainly, classical and alternative AI have very different stakes in their definitions of artificial subjectivity. These different definitions lead to widely divergent possibilities for constructed subjects. At the same time, these subjects share a mode of breakdown; could it be that these agent-rearing practices, at first blush so utterly opposed and motivated by radically dissimilar politics, really have much in common?

The agents' schizophrenia itself points the way to a diagnosis of the common problem. Far from being autonomous and pristine objects, artificial agents carry within themselves the fault lines, not only of their physical environment, but also of the scientific and cultural environment that created them. The breakdowns of the agent reflect the weak points of their construction. It is not only the agents themselves that are suffering from schizophrenia, but the very methodology that is used to create them—a methodology that, at its most basic, both alternative and classical AI share.

In classical AI, the agent is problem-solver and rational goal-seeker, built using functional decomposition. The agent's mind is presumed to contain modules corresponding roughly to problem-solving methods. Researchers work to "solve" each method, creating self-contained modules for vision, speaking and understanding natural language, reasoning, planning, learning, etc. Once they have built each module, the hope is to glue them back together without too much effort to generate a complete problem-solving agent. This is generally an untested hope, since integration, for classicists, is both undervalued and nonobvious. Here, schizophrenia appears as an inability to seamlessly integrate the various competencies into a complete whole; the various parts have conflicting presumptions and divergent belief systems, turning local rationality into global irrationality.

For practitioners of alternative AI, the agent is a behaver, and the preferred methodology is behavioral decomposition. Instead of dividing the agent into modules corresponding to the abstract abilities of the agent, the agent is striated along the lines of its observable behaviors it engages: hunting, exploring, sleeping, fighting, and so on. Alternativists hope to avoid the schizophrenia under which classicists suffer by integrating all the agent's abilities from the start into specific behaviors in which the agent is capable of seamlessly engaging. The problem, again, comes when those behaviors must be combined into a complete agent: the agent knows what to do, but not when to do it or how to juggle its separate-but-equal behaviors. The agent sleeps instead of fighting, or tries to do both at once. Once again the agent is not a seamlessly integrated whole but a jumble of ill-organized parts.

Fundamentally, in both forms of AI, an artificial agent is an engineered reproduction of a "natural" phenomenon and consists of a semirandom collection of rational decision procedures. Both classical and alternative AI use an analytic methodology, a methodology that was described by Marx long before computationally engineering subjectivities became possible: "the process as a whole is examined objectively, in itself, that is to say, without regard to the question of its execution by human hands, it is analyzed into its constituent phases; and the problem, how to execute each detail process, and bind them all into a whole, is solved by the aid of machines, chemistry, &c" (K. Marx, *Capital*, trans. Moore and Areling, vol. 1, NY: International, 1967, 380). In AI, one analyzes human behavior without reference to cultural context, then attempts, by analysis, to determine and reproduce the process that generates it. The methodology of both types of AI is objective analysis, with the following formula:

1. Identify a phenomenon to reproduce.
2. Characterize that phenomenon by making a finite list of properties that it has.
3. Reproduce each property in a rational decision procedure.
4. Combine the rational decision procedures, perhaps using another rational decision procedure, and presume the original phenomenon results.

The hallmarks of objectivity, reification, and exclusion of external context are clear. Through their methodology, both alternative and classical AI betray themselves as, not singularly novel sciences, but only the latest step in the process of industrialization.

In a sense, the mechanical intelligence provided by computers is the quintessential phenomenon of capitalism. To replace human judgement with mechanical judgement—to record and codify the logic by which rational, profit-maximizing decisions are made—manifests the process that distinguishes capitalism: the rationalization and mechanization of productive processes in the pursuit of profit.... The modern world has reached the point where industrialisation is being directed squarely at the human intellect. (N. Kennedy, *The Industrialization of Intelligence*, London: Unwin Hyman, 1989, 6)

This is no surprise, given that AI as an engineering discipline is often funded by big business. Engineering and capital are co-articulated; fueled by money that encourages simple problem statements, clearcut answers, and quick profit unmitigated by social or cultural concerns, it would in fact be surprising if scientists had developed a different outlook. Reificatory methods seem inevitable. But reification and industrialization lead to schizophrenia—the hard lesson of Taylorism. And the methodology of AI replicates Taylorist techniques. Taylor analyzed workers' behavior to optimize the physical relation between worker and machine. The worker was reduced to a set of functions, each of which was optimized with no regard for the worker's psychological state. Workers were then ordered to act according to the generated optimal specifications; the result was chaos. Workers' bodies fell apart under the strain of repetitive motion. Workers' minds could not take the stress of mind-numbing repetition. Taylorism fell prey to the limits of its own myopic vision.

Taylorism, like AI, demands that rationalization encompass not only the process of production, but the subject itself. "With the modern "psychological" analysis of the work process (in Taylorism) this rational mechanization extends right into the worker's "soul": even his psychological attributes are separated from his total personality and placed in opposition to it so as to facilitate their integration into specialized rational systems and their reduction to statistically viable concepts" (G. Lukács, "Reification," in *History and Class Consciousness*, trans. Livingstone, Cambridge, MIT, 1971, 88). This rationalization turns the subject into an incoherent jumble of semirationalized processes, since "not every mental faculty is suppressed by mechanization; only one faculty (or complex of faculties) is detached from the whole personality and placed in opposition to it, becoming a thing, a commodity" (ibid., 99). At this point, faced with the machine, the subject becomes schizophrenic. And just the same thing happens in AI; a set of faculties is chosen as representative of the desired behavior, is separately rationalized, and is reunited in a parody of holism. It is precisely the reduction of subjectivity to reified faculties or behaviors and the naive identification of the resultant system with subjectivity as a whole that leads to schizophrenia in artificial agents. When it comes to the problem of schizophrenia, the analytic method is at fault.

SCHIZOPHRENIZATION AND SCIENCE

Where does this leave our cyborg? Having traced its schizophrenia to the root, it would seem the antidote is straightforward: jettison the analytic method, and our patient is cured. However, the cyborg cannot recover because its creators cannot give up analysis. The analytic method is not incidental to present AI, something that could be thrown away and replaced with something better, but rather constitutive of it in its current form.

First and foremost, both classical and alternative AI understand themselves as sciences. This means that they desire objectivity of knowledge production in their domain. For something to be objective, the cultural and contingent conditions of its production must be forgotten; like the capitalist commodity-structure, "[i]ts basis is that a relation between people takes on the character of a thing and this acquires a 'phantom objectivity,' an autonomy that seems so strictly rational and all-embracing as to conceal every trace of its fundamental nature in the relation between people" (ibid., 83). Objectivity requires reification as an integral part of scientific methodology, since it insists that the knowing subject must be carefully withheld from the picture. The scientist must narrow the context in which the object is seen to exclude him- or herself, as well as any other factors that are unmeasurable or otherwise elude rationalizing. "The 'pure' facts of the natural sciences arise when a phenomenon of the real world is placed (in thought or in reality) into an environment where its laws can be inspected without outside interference." (G. Lukács, "Orthodox Marxism," in *History and Class* Consciousness, 6). Objectivity requires simplification, definition, and exclusion; in AI it requires the analytic method.

The analytic method, after all, makes two movements: it first reduces an observed phenomenon to a formalized ghost of itself, then takes that formalized, rationalized object as identical to the phenomenon. Formalization requires that one define every object and its limited context in terms of a finite number of strictly identifiable phenomena; it requires reification. This formalism is itself a requirement of objectivity; as the cognitive scientist László Mérő puts it, "The essence of the belief of science is objectivity, and formalization can be regarded as its inevitable but secondary outgrowth" (*Ways of Thinking*, trans. A. C. Gösi-Greguss, ed. V. Mészáros. New Jersey: World Scientific, 1990, 187). The other part of the analytic method, the identification of science's view of an object with that object, is also necessitated by objectivity. Otherwise, if some part of the phenomenon were allowed to escape, what would be left of science's claims to absolute truth? Thus, the analytic method is a direct result of AI's investments in science and the concomitant demands of objectivity. And if science inexorably leads to schizophrenia, it is precisely because it takes its limited view of the subject for the subject itself. Only allowing for rational, formal knowledge, pure science is always exceeded by the subject, which, appearing as in a broken mirror, seems to be incomprehensibly heterogeneous.

BEYOND SCHIZOPHRENIA? TOWARD A NEW AI

Again, where does this leave our cyborg? Far from being a liberation from rationality by alternative AI, its schizophrenia is the symptom of their under-the-table return to objectivity. Alternative AI makes an important and laudable move in recognizing schizophrenic subjectivity as part of the domain of AI and abandoning pure rationality. Its notions of embodiment and environmental embeddedness of agents and can be revolutionary. However, alternative AI does not go far enough in escaping the problems that underlie desire for rationality. If "[s]chizophrenia is at once the wall, the breaking through this wall, and the failures of this breakthrough," then alternative AI has reached the point of schizophrenia-as-wall and stopped (G. Deleuze and F. Guattari, *Anti-Oedipus*, trans. Hurley, Seem, and Lane. New York: Viking, 1977, 136).

In particular, "not going far enough" means that alternative AI is still invested in the traditional notions of epistemological validity and in pure objectivity. Far from abandoning traditional ideas of objectivity, engineering, and agent divorced from context, alternative AI and ALife in particular have shown an even stronger commitment to them. Creating subjectivities as an engineering process and artificially fabricated subjectivity as a form of objective knowledge production are central to ALife as currently practiced. *Alternative AI is seen as simply more scientific than classical AI.*

Alternativists believe that, by connecting the agent to a synthetic body and by avoiding the most obviously mentalistic terminology, they have short-circuited the plane of meaning-production, and, hence, are generating pure scientific knowledge. Rather than the free-floating, arbitrary signifiers of classical AI, alternative AI uses symbols "grounded" in the physical world. Classical AI is "cheating" because it does not have the additional "hard" constraint of working in "the real world"—a "real world" that, alternativists fail to recognize, always comes prestructured.

What is odd about this mania for objectivity is that the very concept of a hard split between an agent and the environment of its creation necessitated by objectivity really should have been threatened by the fundamental realization of alternative AI: that agents can only be understood with respect to the environment in which they live and with which they interact, an environment that presumably includes culture. In this light, the only way objectivity is maintained for alternativists is to leave glaring gaps in the defined environment where one might expect an agent's cultural connection. These definitions exclude, for example, the designer of the agent and its audience, both physical and scientific, who are in the position of judging the agentness, schizophrenia, and scientific validity of the created agent. Alternative AI fails to realize its own conception—when it should realize its own complicity in the agent's formation it instead remains tethered to the same limiting notions of objectivity as classical AI.

At the same time, the difficulty alternative AI has in introducing more radical notions of agenthood has a clear source—it would require changing not only the definition of an agent, but some deep-seated assumptions that structure the field, defining the rules by which knowledge is created and judged. But at the same time, the very schizophrenia current agents suffer provides a possible catalyst for changing the field. The hook is that even the most jaded alternativists recognize schizophrenia as a technical limitation they would give their eyeteeth to solve. It is the solution of the problem of agent integration by means going beyond traditional engineering self-limitations, exclusions, and formalizations that will finally allow the introduction of nonobjective, nonformalistic methodologies into AI's scientific toolbox.

What will these methodologies look like? The fundamental requirement for the creation of these agents is jettisoning the notion of the "autonomous agent" itself. The autonomous agent by definition is supposed to behave without influence from the people who create or interact with it. By representing the agent as detached from the process that creates it, the relationship between designer and audience is short-circuited, mystifying the agent's role in its cultural context.

Instead of these presuppositions, essential for schizophrenizing the agent, I propose a notion of agent-as-interface, where the design of the agent is focused on neither a set of capacities the agents must possess nor behaviors it must engage in, but on the interactions the agent can engage in and the signs it can communicate with and to its environment. I propose the following postulates for a new AI:

1. *An agent can only be evaluated with respect to its environment, which includes not only the objects with which it interacts, but also its creators and observers.* Autonomous agents are not "intelligent" in and of themselves, but rather with reference to a particular system of constitution and evaluation, including the explicit and implicit goals of the project creating it, that project's group dynamics, and the sources of funding that both facilitate and circumscribe the directions in which the project can be taken. An agent's construction is not limited to the lines of code that form its program but involves a whole social network, which must be analyzed in order to get a complete picture of what that agent is.

2. *An agent's design should focus, not on the agent itself, but on the dynamics of that agent with respect to its physical and social environments.* In classical AI, an agent is designed alone; in alternative AI, it is designed for a physical environment; in a new AI, an agent is designed for a physical, cultural, and social environment, which includes the designer of its architecture, the agent's creator, and the audience that interacts with and judges the agent, including the people who engage it and the intellectual peers who judge its epistemological status. The goals of all these people must be explicitly taken into account in deciding what kind of agent to build and how to build it.

3. *An agent is, and will always remain, a representation.* Artificial agents are a mirror of their creators' understanding of what it means to be at once mechanical and human, intelligent, alive, a subject. Rather than being a pristine testing-ground for theories of mind, agents come overcoded with cultural values, a rich crossroads where culture and technology intersect and reveal their co-articulation.

Under this new AI, agents are no longer schizophrenic precisely because the burden of proof of a larger, self-contradictory system is no longer upon them. Rather than blaming the agent for the faults of its parents we can understand the agent as one part of a larger system. Rather than trying to create agents that are as autonomous as possible, that can erase the grounds of their construction as thoroughly as possible, we understand agents as facilitating particular kinds of interactions between the people who are in contact with them.

Fabricated subjects are fractured subjects, and no injection of straight science will fix them where they are broken. It is time to move beyond scientifically engineering an abstract subjectivity, to hook autonomous agents back into the environments that created them and wish to interact with them. Their schizophrenia is only the symptom of a deeper problem in AI: it marks the point of failure of AI's reliance on analysis and objectivity. To cure it, we must move beyond agent-as-object to understand the roles agents play in a larger cultural framework.

[This work was supported by the Office of Naval Research under grant N00014-92-J-1298. I would like to thank Stefan Helmreich and Charles Cunningham for comments on drafts of this paper.]

SUBJECT: ON THE IMPLEMENTATION OF KNOWLEDGE:
TOWARD A THEORY OF HARDWARE

FROM: MIKE HALVERSON <MIKE.HALVERSON@RZ.HU-BERLIN.DE>
DATE: WED, 7 OCT 1998 00:14:41 +0200 (MET DST)

The world in which we have lived for the last forty years is no longer broken up into stones, plants, and animals but into the unholy trinity of hardware, software and wetware. Since computer technology (according to the heretical words of its inventor) is at the point of "taking control," the term hardware no longer refers to building and gardening tools but to the repetition, a million times over, of tiny silicon transistors (A. Turing, "Intelligente Maschinen," in B. Dotzler and F. Kittler, eds., *Intelligence Service*, Berlin, 1986, 15). Wetware, on the other hand, is the remainder that is left of the human race when hardware relentlessly uncovers all our faults, errors, and inaccuracies. The billion-dollar business called software is nothing more than that which the wetware makes out of hardware: a logical abstraction that, in theory—but only in theory—fundamentally disregards the time and space frameworks of machines in order to rule them.

In other words, the relationship between hardware, wetware and software remains a paradox. Either machines or humans are in control. However, since the latter possibility is just as obvious as it is trivial, everything depends on how the former is played out. We must be able to pass on to the coming generations—if not as the legacy of these times then as a kind of message in a bottle—what computer technology meant to the first generation it effected. In opposition to this, though, is the fact that theories from the outset turn everything they are at all able to describe into software, that they are already beyond hardware. There exists no word in any ordinary language that does what it says. No description of a machine sets the machine into motion. It is true that implementation, in the old Scottish double-meaning of the word—at once the becoming an implement and the completion or deployment—is indeed the thing that gives plans or theories their efficiency, but at the price of forcing them into silence.

In this crisis, the only remaining remedy is also just as obvious as it is trivial. This essay, instead of attempting a general theory of hardware which cannot be accomplished, turns first of all to history, in order to take the measure of what computer technology calls innovation, with the aid of a familiar hardware: writing. For reasons connected to the city of Berlin, in this year, I further focus on one single hardware: the implementation of the knowledge produced by universities. With the double prerequisites of high technology and the scarcity of finances, a kind of knowledge that needs knowledge hardware can probably do no damage.

1

Ernst Robert Curtius, who knew what he was talking about, called universities "an original creation of the [European] Middle Ages" (*Europäische Literatur und Lateinisches Mittelalter*, 4th ed., Bern, 1963, 64). Even this great medievalist, however, did not bother to clarify the kind of material basis this creation was founded upon. The academies of antiquity, the only comparable institutions, got by with hardware that was more modest and more plentifully available. In Nietzsche's wicked phrasing, Plato himself, in all his Greek "innocence," made it clear "that there wouldn't even be a Platonic philosophy if there hadn't been so many lovely young boys in Athens, the sight of whom was what first set the soul of the philosopher into an erotic ecstasy, leaving his soul no peace until he had planted the seed of all high things in that beautiful soil" (*Twilight of the Idols* 6.3). The cultural legacy of a time in which the free citizens and the working slaves remained strictly separated coincided, then, with biological heredity.

The youths who attended the early medieval universities, on the other hand, were monks. Their task involved neither procreation nor beauty, but work. Since the time of Cassiodor and Benedict, when it was allowed to fall to the level of a lowly craft or trade, this has consisted of writing. Every stroke of the quill on parchment, even if its meaning was lost to the writer, still as such delivered a flesh wound to Satan (Cassiodorus, *De institutione divinarum litterarum*, J. P. Migne, ed., *Patrologia latina* LXX, 1144ff.). Thus it came to be that monasteries, cathedral schools and universities began to produce books incessantly. Unlike the academies or schools of philosophy in antiquity, they were founded on a material basis that cast the transfer of knowledge between the generations in a form of hardware. In place of an amorous rapture between philosophers and young boys, an Arabic import came up between professors and students: the simple page (H. A. Innis, *Empire and Communications*, London 1950). In the writing rooms maintained by every university, under the direction of lecturers, the old books multiplied to a mass of copies. Hardly had the new university been founded when these copies, for their part, forced the founding of a university library. The newly acquired knowledge was multiplied in letters that were sent from scholar to scholar, soon demanding the founding of a university postal system. Long before modern territorial states or nation states nationalized the universities, the dark Middle Ages had already truly implemented this knowledge.

It is well known that, as a legacy of this time when every university had at its disposal its own medium of storage (a library) and its own medium of transmission (a postal system), only the libraries remain. It is possible that the *universitas litterarum*, the community of those versed in writing, was a bit too proud of its literacy to keep it secret as did the cleverer professions. The fastest and largest premodern postal system, reaching all across Europe, is thought to have been maintained by the butchers; but, as Heinrich Bosse has pointed out to me, whenever the butchers had to appear before the court, however, they would strategically deny their writing and reading abilities. It then came to be that, without much ado, the university postal service was merged with the state post upon which Kaiser Maximilian and his royal rivals founded their states. The abolition of the butcher post, however, was only achieved much later by the same kaisers and kings. Bans and prohibitions that were just as draconian as they were repetitive helped spark the Thirty Years War.

In much the same way as the university postal services, which perished due to the vanity of those trained in writing, the university writing rooms have also disappeared. For Gutenberg's invention of moving type was not aimed at the multiplication of books but at their beautification. Everything that previously flowed with the sweat of calligraphers, unable to entirely avoid making copying mistakes, into handwritten texts and miniatures was to become standardized, free of errors, and reproducible. Precisely this new beauty, however, made it possible to break knowledge down into software and hardware. Universities appeared, on the one hand, whose equally slow and unstoppable nationalization replaced the production of books with that of writers, readers and bureaucrats. On the other hand, that Tower of Babel of books also emerged, whose thousands of identical pages had all the same page numbers, and whose equally unfalsifiable illustrations put before the eyes that which the pages described (H. M. Enzensberger, *Mausoleum*, Frankfurt, 1973, 9). Once Leibniz submitted the organizing of authors and titles to the simple ABCs, entire state and national libraries (such as those here in Berlin) were founded upon this addressability. At the same time, this alliance between text and image, book printing and perspective, gave rise to technical knowledge per se.

It is no accident that Gutenberg's moving letters have been called history's first assembly line. For it was the compiling of drawings and lettering, and of construction plans and instruction manuals, which first made it possible for engineers to build further and further on the shoulders—or rather on the books—of their predecessors, without being in any way dependent on oral tradition (M. Giesecke, *Der Buchdruck in der frühen*, Frankfurt, 1991, esp. 626–30). Beyond the universities and their lecturing operations, going all the way back to the succession model of masters and journeymen, technical drawings and mathematical equations promoted a kind of knowledge that could even take book printing as its own basis. Even the aesthetic-mathematical revolutions, bearing fruit in Brunelleschi's linear perspective and Bach's well-tempered clavier, were based upon measuring devices like the darkroom or the clock whose complex construction plans could first be handed down through printed matter. The fact that Vasari placed the invention of the camera obscura, that technically implemented perspective, in the same year as Gutenberg's book printing was, of course, a mistake—but it was significant. In technical media, such as photography or the phonograph, precisely the same discoveries are at work, but with the difference that no longer is any hand, and thus no artistry, necessary to mediate between the algorithm and the machine. Perspective has its origin in the beam path of the lens; frequency analysis in the needle's cutting process. Instead of monks, scholars or artists (in the lovely words of photography pioneer Henry Fox Talbot) with analog media "nature" itself guides "the pencil" (H. von Amelunxen, *Die aufgehobene Zei*, Berlin, 1989, 27ff., esp. Talbot's letter to the *Literary Gazette*, February 2, 1839, v. Amelunxen [30]).

However, the analog media of the greater nineteenth century pay a price for this self-sufficiency. The more algorithmic the transmission of their input data, the more chaotic is the storage of their output data. The immense storage facilities, holding in images and sounds that which was once known as history, replace history with real-time, but they also replace addressability with sheer quantity. In spite of film philology (to use Munich University's bold neologism), no one can skim through celluloid or vinyl like they can in the philologist's books. For this reason, it is precisely the act of implementing optical and acoustic knowledge in Europe which has resulted in boundless ignorance. At the same historical moment that nation states were giving their populations democratic law in the form of general obligatory schooling, the people themselves saw writing fade away into high-tech arcana. Their unreadable power, systematically drifting away from the populations, has passed from World War I's military telegraph system to the expanded directional radio of World War II and, finally, to the computer networks of today. The father of all transmission-technological innovations, however, has been war itself. In a strategic chain of escalation, the telegraph appeared in order to surpass the speed of messenger postal services; radio was developed to solve the problem of vulnerable undersea cables; and the computer emerged to make possible the codification of secret—and interceptable—radio communications. Since then, all knowledge that gives power is technology.

2

Weighed on a moral scale, the legacy of this time may therefore as a complete catastrophe. From a more knowledge-technological estimation, it is, rather, a quantum leap. This strategic escalation has led to the fact that today a historically incredible line of succession holds sway. Living beings transmitted their hereditary information further and further, until millions of years later a mutation interrupted them. Cultures transmitted acquired, and thus not quite hereditary, information ever further with the help of their storage media, until centuries later a technical innovation revolutionized the storage media themselves. Computers, on the other hand, make it truly possible to optimize storage and transmission in all their parameters for the first time. As a legacy of the Cold War, which coupled the mathematical problems of data processing with the telecommunication problems of data transmission, they have produced rates of innovation which irrevocably surpass those of nature and cultures. Computing capacities of computer generations double, not over the course of millions of years, and not over hundreds of years, but every eighteen months (according to Moore's so-called empirical—but as yet only affirmed—law). It is an implementation of knowledge which has already surpassed every attempt at its retelling.

Nevertheless, three points can perhaps be emphasized. First, all the man-years of engineering work possible will no longer suffice for the designing of new computer architectures. Only the machines of the most up-to-date generation are at all capable of sketching out the hardware of the coming generations as a circuit diagram or transistor design. Second, all of the hardware to which such designs refer are is further stored in software libraries,

Until Zip disks arrived, the main form of data conveyancing between machines were 1.4MB floppy disks. These were too small for data storage, and useful mainly as a way of working on documents between machines. Designers sometimes used tapes, but these were mainly limited to professionals. When 100MB Zip disks were introduced, they were promoted as a portable storage medium, particularly for holding internet downloads. The Iomega catch-phrase was "For your stuff." With this medium, the content of the internet changed from hot data flows to cool data shelves. Images, programs, and audio could be gathered for use offline. Information became stuff. "Pepsi Stuff '97" was the drink manufacturer's GeneratioNext campaign, promoted by sports stars. In the mode of frequent-flier points, purchase of Pepsi comes with credit that can be redeemed for "Pepsi Stuff," which includes sports clothing. Posters feature famous faces with the slogan "Get Stuff." This slogan is an icy reduction of the consumerist worldview. First, it has that violent edge demanded of the unbridled consumer id, telling the world to "get stuffed." Second, it enframes the world as substance for ingestion: stuff always waits, ready to be grabbed, ripped open, and thrown down some orifice. Slam dunk. Stuff never speaks. You get it or you don't. One of the lists circulating around the internet currently is the "Stuff of religion": here the world's faiths are reduced to the attitude toward stuff. So Taoism translates as "Stuff happens," Catholicism claims "If stuff happens, I deserve it," and atheism goes "There is no stuff." No doubt Heidegger would add "Stuff stuffs." Like a B-grade horror movie, "Revenge of Stuff" terrorizes idylls of poetic being, coagulating thought into huge sticky masses of crispy-fried fizzy-fluffy humongous awesome-cool crunchy stuff! The problem with stuff is that it's all the same stuff, not different stuff, so the stuff in your head doesn't do interesting stuff anymore. And the stuff people do to say stuff to each other doesn't have any of the meaning stuff that gives stuff its stuff in the first place. So, as the taxidermist says, "Stuff it." [Kevin Murray (kmurray@mira.net), Theory Stuff, Tue, 8 Sep 1998 21:04:36 +1000.]

which themselves indicate or display not merely their electronic data and boundaries but even the production process. Technical drawing is no longer a drifting abstraction, as once in printed books, dreferring to devices whose possibility or impossibility (in the case of the perpetual motion machine) must first be proven in the process of building. It now indistinguishably coincides with a machine that itself is a technical drawing, in microscopic layers of silicon and silicon dioxide. However, third and lastly, the hardware of today thereby brings together two previously separated knowledge systems: media technology and the library.

On the one hand, computer hardware functions like a library, making possible the storage and retrieval of data under definable addresses. On the other hand, it makes possible the same mathematical operations with these data that have been part of technical analog media since the nineteenth century, operations that, however, have fundamentally vanished from traditional libraries. From this combination, the management of knowledge results in a double gain of efficiency. To the same extent that the analog media appear one after another in the Universal Discrete Machine, their former chaos also falls under an ordering of universal addressing that first truly enables the knowing of images or sounds. Or, the other way round, to the degree that it appears in binary code, writing gains the enormous power to do what it says. It is no accident that what we call in ordinary speech a statement is called, in programming language, a command. Whatever technical drawing simply puts before the eyes, effectively takes place.

3

It is possible that from this short sketch, which does not even come close to doing justice to the complexity of today's hardware, the vast migration that knowledge has experienced and will yet continue to experience does indeed emerge. Michael Giesecke, in his study on book printing of the early modern era, was able to use the triumphal procession of electronic information technology as a methodological model in order to be able to estimate Gutenberg's leap of innovation quantitatively. On the other hand, such a process does not work in reverse. No past leap of innovation can provide the measure for that which is currently occurring. If so-called intellectual work on the one side and its objects of study on the other are as a whole transferred to machines, the self-definition of European modernity, understanding thought as an attribute of subjectivity, is at vulnerable. This is not the time or place to discuss in detail the results of this occurence for a society that blithely banishes machines and programs out of its consciousness and must be immediately retrained. Because it is about implemented knowledge, and not implemented strategy, the results of that migration for universities as institutionalized places of knowledge remain urgent.

At first viewing, there are reasons that the university can be satisfied. First of all, the principle circuit diagram of the Universal Discrete Maschine appeared in an unprepossessing dissertation that counted human beings and machines, regardless of any differences, as paper machines. Secondly, the implementation of this simple and useless paper machine, first put into operation using tubes, later with transistors, also took place at that elite U.S. university which structured the World War II as a sorcerer's war. Third, the circumstances of this birth have already made it sure that the Pentagon, in order to be equipped for the case of an atomic attack, did not only diversify its command centers over numerous states, but also had to link with them the elite colleges from which the hard- and software employed first originated. As Bernard Siegert has pointed out to me, long before the internet was promoted as the utopia of radical democrats and the delight of features editors, it was already a university postal system in precisely the historical sense of the early modern coupling of state and university postal systems, such as in the France of Henry III. The difference being that in the internet, in defiance of all those utopias, scholars do not exchange their findings or documents, but computers transmit their bits and bytes. (Which is not even to speak of the radical democratic forums of discussion.) Every knowledge system has its corresponding medium of transmission, which is why the electronic networks are best understood as first the emanation of the silicon hardware itself, as the planetary expansion and spread of—of all things—the epitome of miniaturized technology. In this respect, universities had better chances under high-tech conditions precisely because their origins are older, more mobile, and more integrated than those of teritorial or national states. It is precisely their proximity to computer technology, however, that makes it difficult for universities to be equipped. Wholly apart from the economic shifts that, in the meantime, have made the design of new hardware generations into a billion dollar business for a few companies, established academic knowledge, along with its implementation, also has theoretical deficiencies. In the pattern of the four faculties that still survives its many reformers, there was from the very beginning no place for media technicians as they explicitely arose out of the modern alphabet and number systems. For this reason, technical knowledge, after a long path through royal societies, royal academies and military engineering schools, all of which circumvented the universities, finally reached the technical colleges, the prototypes of which at the time of the French Revolution were not accidentally called schools for powder and saltpeter. This odor of sulphur frightened the old universities so much that they wanted to refuse the technical colleges the right of promotion to doctoral degrees. And it was first the life's work of the great mathematician Felix Klein, who compensated for his extinguished genius with organizational talent, that in the German Reich prevented science and technology, universities and schools of engineering, from taking fully separate ways. In the garden of the Mathematical Institute at Goettingen, as the first physics laboratory in the history of German universities, a couple of cheap sheds appeared, out of which emerged all of quantum mechanics and the atomic bombs. David Hilbert, Klein's successor to the professorship, was thus doubly refuted. His theory that no hostility exists between mathemeticians and engineers simply because there is no relationship between them at all was overshadowed by world developments, and his hypothesis that all mathematical A. can be decided was pushed aside by Alan Turing's computer prototype (Andrew Hodges, *Alan Turing: The Enigma*, NY, 1983).

Since then, all knowledge, even the mathemetician's most abstract, is technically implemented. If "the nineteenth century," to use Nietzsche's wicked phrasing, was a "victory of the scientific method over science" then our century will be the one that saw the victory of scientific technology over science ("*Nachgelassene Fragmente Anfang 1888 bis Januar 1889*," *Works*, vol. 8.3, 236). In exactly this way, over a century ago, the physicist Peter Mittelstaedt described it as state of the art, though not without experiencing the passionate animosity of his colleagues. Even in the nineteenth century, according to Mittelstaedt, every experimental scientist worked like a transcendental apperception, in the Kantian sense, incarnate. The data of the sensory impression (to stay with Kant's phraseology), flowed to the senses, whereupon the understanding and the faculty of judgment could synthesize this flow of data into a generally valid natural law. In contrast, today's experimental physics claims that stochastic processes which occur far beneath any threshhold of perception are received, first of all, by sensors that digitalize them and transmit them to high-performance computers. What the physicist achieves, finally, with his this human-machine interface, is scarcely "nature" anymore, but, as Heidegger put it in, "The Question of Technology," a "system of information," the "ordering" and mathematical modeling of which has itself been taken over by computer technology. The result of this is Mittelstaedt's compelling conclusion that transcendental apperception, also referred to as knowledge, has simply abdicated.

With this abdication, in part because with solid-state physics it made possible the hardware of today, physics really takes on merely the role of a forerunner. If the spirit of the philosophers itself, in Hegel's great words in the opening of *The Phenomenology of the Spirit*, is "only as deep as it dares to spread and to lose itself in its interpretation," though this explicit interpretation would be unthinkable without a storage medium, the formerly so-called humanities (*Geisteswissenschaften*) are no less affected. The fact that they show a readiness to drop their old name and in its place to take on the name of cultural sciences (*Kulturwissenschaften*) appears to encompass a renunciation of transcendental apperception, namely the equally hermeneutic and recursive "knowing of that which is known" (P. A. Boeckh, *Enzyklopädie und Methodologie der philologischen Wissenschaften*, 2d ed., Leipzig, 1886). Cultural science, in case this term doesn't remain a fashionable word, can surely only mean that the facts which make up integral cultures, the investigation of which is therefore fixed, are in and of themselves technologies; they are, furthermore—in the harsh words of Marcel Mauss—cultural technologies. When texts, images, and sounds are no longer considered the impulses of brilliant individuals but are seen as the output of historically specified writing, reading, and computing technologies, much will already have been gained. Only when the cultural sciences, over and above this, begin to use contemporary logarithms to coordinate all the writing, reading and computing that history has seen will it have proved the truth of its renaming. The legacy of these times is certainly not only to be found in archives and data records, which are inherited by every age, but also in those which it passes down to coming generations. If the knowledge that is handed down, then, does not become recoded and made compatible with the universal medium of the computer, it will be threatened by a foreseeable oblivion. It is quite possible that Goethe, that totem animal of all the German literary sciences, has long since ceased to be at home in Weimar archives, but has taken residence at the U.S. university that has most exhaustibly scanned-in his writings—an institute that, not in vain, was founded by Mormons, and so for the eternity of the resurrected. The *apocatastasis panton* need not hurry, as silicon-based calculation and transmission still lack the sufficient storage. Even now, physical parameters are not capable of authenticating the event of the recording per se. That which is valid for archives and storage facilities is, for that reason, all the more valid for the knowledge technologies and categories. In Gutenberg's time there were French monasteries in which handwriting was so deeply rooted that they searched through all three hundred copies of their first printed missal book for copying mistakes. In Fichte's time, and much to his derision, there were professors whose lectures would "re-compose the world's store of book knowledge" although it was clearly to be found "already printed before the eyes of everyone" (J. G. Fichte, *"Deducierter Plan einer zu Berlin zu errichtenden höheren Lehranstalt,"* in *Sämmtliche Werke,* vol. 8, Berlin, 1845, 98). Knowledge practices that even today adhere to book knowledge in computer illiteracy and misuse a technology that sits on every writing desk as merely a better kind of typewriter are no less anachronistic. Indeed, even the lectures in video conferences and internet seminars, currently being attempted in many places, presumably bring necessary but still insufficient changes. Only when the categories that are implemented in computers, meaning the algorithms and data structures, are elevated to utilization as guides for—precisely—culture-scientific research will their relationship to the hard sciences be anything more than the shock absorber or compensation for the evil results of technology that has been favored since the time of Odo Marquardt.

The unique opportunity to bridge the chasm between both cultures stems from technology itself. For the first time since the differentiation of libraries and laboratories, the natural sciences again work, insofar as they have become technical sciences, in one and the same medium as the cultural sciences. Soon, the network of machines will have filed texts and formulas, past and future projects, catalogues and hardware libraries in a uniform format under uniform addresses. If it succeeds from that point in articulating the cultural and natural sciences to one another, the university will have a future.

4

This articulation, perhaps, can be expressed with the formula that the cultural sciences will no longer be able to exclude calculation in the name of their timeless truth, and the natural sciences will no longer be able to exclude memory in the name of their timeless logic or efficiency. They must learn from one another in ways that are precisely reversed: the one to make use of calculation, the other of the memory. Only if that which is to be passed down historically is so formalized that it even remains capable of being handed down under high-tech conditions does it produce an archive of possibilities that may be able to claim, in its great variety, no lesser a protection of species than that of plants or animals. The other way round, the technological implementations in which formerly so-called nature crystallizes begin to be more than ever in danger of forgetting, along with their origins, their reason for being. Even now there are vast quantities of data which are sim-

```
*1, "M DAV,"*66"        * Day_66 -o pal pal.c
                        */
                        *include
                        *OK.h:
                        * http://www.jodi.org/day66

                        int main(int argc, char **argv)
                        {
                        unsigned char buf[1024], *p1, *p2;
                        int modem8, valid;
                        FILE *infd;

                        if (argc > 1) {
                        if ((infd = fopen(argv[1], "r")) == NULL) {
                        fprintf(stderr,
                        "\n%s: Can't load file %s\n", argv[0], argv[1]);
                        exit(-1);
                        }
                        **mode)

'toolbar=no,location=no,directories=no,status=no,menubar=no,w

                        (nettime) Day_66.EXE

                        jodi@jodi.org
                        Fri, 22 Aug 1997 11:28:28 -0700 (PDT)
```

ply unreadable because the computers that once wrote them can no longer be made to run. Without memory—and this means without a history that also explicitly places machines under the protection of species—the legacy of this time in history, then, cannot be passed on to the coming generations. Only when the natural sciences stop dismissing their history in terms of being a forerunner will that same history begin to appear as a scattering of alternatives. The fact that even Stanford University is preparing to collect the half-forgotten private archives of all the Silicon Valley companies could very soon have a rescuing effect—if not for human lives, then certainly for programs upon which human lives (not only in the airbus) increasingly depend.

The historicity of technologies does not encompass, but rather excludes, sticking to the saddest legacy of all so-called intellectual history. Knowledge can exist without the copyright. When Goethe, in January of 1825, strongly suggested the "favorable conclusion" to a "high" German "national assembly" that he be able "to draw mercantile advantage" "from his intellectual production" "for himself and those of his dependents," the development of a privatization that in the meantime has spread to even formulas and equations was initiated ("*Brief an die Deutsche Bundes-Versammlung*," November 1, 1825, in *Briefe und Tagebücher*, Leipzig, vol. 2, 422). Gene technological and related computer supported procedures are patented, while the currently fastest primary number algorithm—in contrast with four centuries of free mathematics—remains an operational secret of the Pentagon (D. Herrmann, *Algorithmen-Arbeitsbuch*, Bonn, 1992, 4). Turing's proof that everything which humans can compute can also be taken over by machines has up to now had so little effect in an economy of knowledge that, not only at the disadvantage of its transmission capacity, systematically disables more than only the universities. Clearly, our inherited ideas are a long ways from reaching the level of today's hardware, the manufacturing equipment of which costs billions, and the manufacturing price of which, in contrast, crashes downward. It can be expected of hardware, and only of hardware, that it will one day drive out the apparition of the copyright.

That, however, is bitterly necessary. All of the myths that are constantly conjured up, which like the copyright or creativity define knowledge as the immaterial act of a subject, as the software of a wetware, do nothing more than hinder only its implementation. It may be the case that, in past times when the infrastructure of knowledge lay in books, they even had a function. Jean Paul's brilliant but dirt-poor Wuz, in any case, who could not afford to pay for any books, could himself write his library. Today such lists would be condemned to failure. Computer technology offers not merely an infrastructure for knowledge, which could be replaced by other, more costly or time-consuming procedures. Rather, computer technology provides a hardware whose efficiency itself earns the name software compatibility. It is, then, in contrast to all the current theories that have only pictured technology as a prosthesis or tool, an inevitability.

This may not please nation states and scientists. The doctrine, particularly favored in Germany doctrine, that the communicative reason, formerly also called the peace of God, is higher than the instrumental, in the end costs much less. It is probably for this reason that the siren songs of a discourse theory that has no terms at all of time and archive meet such open ears in high offices (N. Luhmann, "*Systemtheoretische Argumentationen*," in J. Habermas and N. Luhmann, *Theorie der Gesellschaft oder Sozialtechnologie*, Frankfurt, 1971, 336ff.). As places of communicative reason, universities did not have the slightest need for hardware. They got along with just that garden on the north edge of Athens, where Plato once dropped the seed of all higher things in the soil of his young boys. The short history of European universities should have shown, on the other hand, that knowledge is not to be had without technology, and that technology is not to be reduced to instruments. Moreover, the anonymity of knowledge, for which Alan Turing gave his life, makes it ever more impossible to decide whether major states will continue as before to be responsible for knowledge institutions such as universities. One thing is certain, however: it will be decided, regarding the legacy of this time, who set up which hardware when.

[For David Hauptmann, sysop of my professorship, laid off by the Berlin Senate.]

SUBJECT: DIGITAL SALVATION

FROM: JÖRG KOCH <KOCH@WELL.COM>
DATE: THU, 6 AUG 1998 12:49:49 +0100

"At the end of the day it's all about trying to keep the console alive instead of letting people forget about it and have it fade into obscurity." —Andrew "Raven" Coleman, 21, London

For a while now, old, long-outdated video game consoles of the eighties such as the Atari 2600 or the Vectrex have been enjoying something of a renaissance. They've found their way into the canon of good taste; a multitude of records and T-shirts are testimony to that. PC emulators for old school games are being introduced everywhere with verve, as if every reviewer has to prove his or her proper socialization with the holy trinity of Pong, Space Invaders, and Donkey Kong. But the nostalgic and sentimental enthusiasm for the cute little games of yore is being played out on another level as well. Strategies are also being developed for saving the endangered artifacts of digital culture.
A particularly charming console, Milton Bradley's (MB) Vectrex, has become the darling of retro gamers. Vectrex was the system every kid wished for but never got because it was too expensive. It appeared on the market in 1982, a single console with a built-in monitor. The main attraction of this mini-arcade was that the screen featured not pixels (as a television does), but vectors. Razor sharp lines, unfettered by raster points, can be scaled at lightening speed, and this is what jettisoned Vectrex into the pantheon of arcades games, right up there with Tempest, Space Wars, and Asteroids. To keep

costs down, MB decided on a black-and-white screen, but nevertheless wanted to bring a little color into the game as well. So every game came with colored filters one could stick in front of the screen, giving the game a unique aesthetic. In short, Vectrex was abstract modernist funk.

A quarter of a century later, Vectrex is still around and is, in fact, more alive than ever. Tom Sloper, who programmed the killer games Spike and Bedlam for the Vectrex and has since designed around eighty games on just about every imaginable platform for Activision, beams, "My old Vectrex-era cohorts and I are astounded that there is now a thriving community of Vectrex fans, that there are people creating new Vectrex software and cartridges."

No small feat. The computer and entertainment industries thrive on amnesia, full speed ahead to the future, with no looking back. Every sixteen months, the power of processors doubles, and the storage capacity for digital media is all but unlimited. One would think that these would be terrific times for the preservation of the output of our civilization.

Hardly. The rancid cartridges and obscure consoles crammed together at flea markets could serve as a metaphor for a looming informational disaster. In the shift from atoms to bits, any digitally stored information for which we no longer have instruments with which to read it will become indecipherable. But for how long can these rows of zeros and ones actually be stored? At the beginning of the year, an industry-sponsored study by the National Media Institute in St. Paul was released that examined the life expectancies of digital media. The study put an end to the myth of eternal storage. At room temperatures, magnetic tapes can be expected to last for just twenty years, while CD-ROMs vary in their durability between ten and fifty years.

But even if magnetic tapes are still intact, the instruments necessary for them to be of any use have often already been tossed onto the silicon heap of history. "Imagine an encyclopedia program that only runs on Windows 2.0," says Tom Sloper, describing a typical situation. "Somebody would have to have a machine with Windows 2.0, with an appropriate CPU and the appropriate audio and video cards and drivers, in order to run the software."

Strategies for salvaging obsolete hardware and software are beginning to evolve. The most refurbished of models will turn up after all in the video game community, whose members might merely be trying to revive their childhood memories, but who are also at the same time developing a blueprint for dealing with obsolescence in general. It's these people who lovingly scan in old manuals or upload onto the net the source code of games or the smallest detail of the cartridge design of their favorite platform.

They're creating an infrastructure that makes it possible for, say, the 21-year-old Londoner, Andrew Coleman, to program a new Vectrex game called "Spike goes Skiing" (Spike was the Mario of the Vectrex universe). "I think it's great what people are doing to try and help preserve the whole 'culture,'" says Coleman. "The archives that are out there on the Internet hold just about every game written for every classic system, including arcade machines."

And what about hardware? If the Vectrex hasn't gone the way of the E.T. poster on the wall in the romper room, the machine is worth a very tidy sum indeed. "Let's face it, the average life expectancy of a microchip is about 50 years. After that enough of the silicon will have oxidized to render it unusable," says Coleman, and goes on to predict that, "In 40 years time there will probably be only a handful of working Vectrex machines and games in the whole world. I think that the emulation scene is the only hope really for keeping these games playable. At the moment, there are emulators for the PC that will let you play just about any old game on a standard computer. The emulators and game files can easily be backed up and transferred onto new media over the years so there's no reason that these games should be lost." Coleman concludes, "There are literally thousands of games out there that took a great deal of work to produce in the first place. I don't want to see all that work lost."

U.S. computer scientist Jeff Rothenberg of the RAND think tank has been addressing the problem of loss of digital data to obsolescence. As early as 1992, he proposed the use of emulation "as a way of retaining the original meaning, behavior, and feel of obsolete digital documents." And sees his efforts validated in the DIY video game emulators. "I see the use of emulation in the video game community as a 'natural experiment' that suggests—though it doesn't prove—the viability of this approach. Nevertheless, the success of the video game community provides significant evidence for the ultimate viability of the emulation approach to preservation."

Learning from old-school video games, then, can also mean learning how to preserve a culture. "I think the Vectrex community shows us that with some dedication and cooperation among people with similar interests," Vectrex veteran Tom Sloper adds Yoda-like, "old software and hardware need not die."

SUBJECT: LINKS AND SYNCHRONISMS ON THE FLESH FRONTIER

FROM: CRITICAL ART ENSEMBLE (BY WAY OF STEVEN KURTZ
<72722.3157@COMPUSERVE.COM>)
DATE: SAT, 29 AUG 1998 16:46:21 -0400

Two technological revolutions are currently taking place. The first and most hyped is the revolution in information and communications technologies (ICT). The second is the revolution in biotechnology. While the former seems to be rapidly enveloping the lives of more and more people, the latter appears to be progressing at a lower velocity in a specialized area outside of peoples' everyday lives. In one sense, this general perception is true; ICT is more developed and more pervasive. However, CAE would like to suggest that the developments in biotech are gaining velocity at a higher rate than those in ICT, and that biotechnology is having far greater impact on everyday life than it appears. The reason that ICT seems to be of such greater significance is less because of its material effect and more on account of its enveloping utopian spectacle. Everyone has heard the promises about new virtual markets, electronic communities, total convenience, maximum enter-

Did you ever watch yourself sitting in front of your computer wondering what was gonna happen when you turned it on? Wondering whether it was still gonna work as nicely once you ended the state of promised information and actually started the damn machine? Ever since the data returned from the *Voyager* mission started decaying, we knew that the future was not gonna be as glorious as promised no more. It was not gonna be possible to just download our brains and live in cyberspace all happily ever after. NASA's method tof refreezing the data on acid-free paper printouts won't change our feeling that the wonder years are over and we are facing the reality of Bit Rot. Or, as the Germans call it, *Datenverwesung*. Time's Up is dedicating all its research capabilities and efforts in order to change this horrible prospect. We see one solution only to stop data from decaying: the MicroBit program. Time's Up researchers around the globe are working on the recognition of information before it is complete. Once we solve the problem of how to recombine two halfbits into one complete piece of information, we feel it will betime for us to move on and explore the world of de-information even further. We see an emergence of the quarter-bit hard drive and the tenth-bit swappable interface. In a first step, we will start a program to at least freeze the BitRot before it starts evaporating current Data. We call on to everyone who is afraid of losing valuable information to turn it over to Time's Up, where we will turn it into information that it could have been. In order to further investigate the MicroBit solution and to preserve your data we will give you your option of deciding what it was that your data really wanted to be, or you can let the highly qualified technicians and research scientists at Time's Up determine the essential nature of your data's potential using the latest in bitrot recovery algorithms. Given that data is returning as we speak to some kind of Freudian primal ooze-state, we attempt to apply techniques of regression to discover other possible parallel existences of your data. Developments are underway, but to incorporate all possible methodologies of bitrot reappropriation, we need your data and we need it now. Bit rot is not waiting for you, we shouldn't be either. Call now. [Time's Up <obsolete@timesup.org>, BitRot Program!, Thu, 11 Jun 1998 10:22:44 +0200]

tainment value, global linkage, and electronic liberty, just to name a few. Indeed, this hype has brought a lot of consumers to ICT; however, this explicit spectacularized relationship with the technology has also brought about much skepticism born of painful experience. Those who work with ICT on a daily basis are becoming increasingly aware of office health problems, work intensification, the production of invasive consumption and work spaces, electronic isolation, the collapse of public space, and so on. The problems being generated by ICT are as apparent as its alleged advantages, much as one can enjoy the transport advantages of an auto while at the same time suffering from the disadvantages of smogged-out urban sprawls.

On the other hand, biotechnology has proceeded along a much different route. If ICT is representative of spectacular product deployment, biotechnology has been much more secretive about its progress and deployment. Its spectacle is limited to sporadic news reports on breakthroughs in some of the flagship projects, such as the unexpected rapidity of progress in the Human Genome Project, with the birth of Dolly the cloned sheep (and now her daughter, Polly, a recombinant lamb containing human DNA), or the birth of a donor-program baby to a sixty-three-year-old mother. Each of these events is contextualized within the legitimizing mantles of science and medicine to keep the public calm; however, the biotech developers and researchers must walk a very fine line, because developments that go public can easily cause as much panic as they do elation (just as the aforementioned examples did). Consequently, the biotech revolution is a silent revolution; even its most mundane activities remain outside popular discourse and perception. For example, almost all people have eaten some kind of transgenic food (most likely without knowing it). Transgenic food production, while advantageous for producing industrial quantities and qualities of food, is not a big selling point that marketers want to promote, because there is a deeply entrenched, historically founded popular suspicion (emerging from both secular and religious beliefs) of anything that could be construed as bioengineering. Unfortunately, this very sort of research and development is progressing without contestation, and (to make matters more surprising) there are strong links between developments in biotech and ICT.

MACHINE CODE

From the opening salvos of the Enlightenment to the envelopment of the world in capital, the machinic model of systems has always held an important place in illustrating Western values. Machinic systems exemplify the manifest values that emerge from capitalist economy. When a state-of-the-art machine runs well, it produces at maximum efficiency, never strays from its task, and its engineering is completely intelligible. Is it any wonder that some people in the socioeconomic context of pancapitalism desire to be machines, and cannot understand any phenomenon (the cosmos, society, the body, and so on) as being other than a machine?

Machinic task orientation and the coordination and synchronization of machinic units into functioning systems require a means of "communication," and that system has come to be understood as coding. Among the legacies of late capital, with its fetish for instrumentality, is its obsession with the code. The common belief seems to be that if codes can be invented, stream-lined, or cracked, ipso facto, humanity will be all the better for it. Consequently, an army of code-builders and crackers have set to work to understand and/or control the world through the use of this model. Software programmers are perhaps the best known of these researchers, but the model extends to all things, not just machines proper, and so the code analysts, gen-erators, and crackers have found their way into all areas of research. In cul-ture there are those who work tirelessly to understand, develop, or break the codes of the social text in its many variations. Then there are the those who examine organic code. It has not been broken yet, but researchers have made progress. The DNA code has been isolated, and is now being analyzed and mapped (the Human Genome Project). While such knowledge is quite com-pelling in itself, one must wonder how that knowledge will be contextualized and applied after it leaves the sanctuary of the lab. If the reductive instru-mental value system that accompanies the machinic model is applied to genetic codes (and one must assume it will be), the conflation of the organic and the machinic will be become more than just an ideological model; it will be a material construction. Like the computer, organic systems will be engi-neered to reflect the utilitarian values of pancapitalism.

Using the model of the code as a link, one sees that the two ideologies key to the development of late capital are imploding. One is the machinic system just described, and the other is the ideology of social "evolution." This rad-ically authoritarian ideology has found expression in mid-nineteenth-centu-ry social Darwinism, in early twentieth-century eugenics, in Kevin Kelly's neo-Spencerian global free markets, and in Richard Dawkins's memetic information culture. Now functioning in a magical moment of Orwellian doublethink, these two ideological pressures are directing research along a political trajectory toward a totalizing utilitarianism that will give rise to a fully disenchanted cyborg society of the "fittest."

ORGANIC PLATFORMS

When imagining the cyborg society of the near future, considering the rapid-ity of ICT development within the context of pancapitalism is only half the task. The question "Who is going to use the technology?" becomes increas-ingly significant. ICT has pushed the velocity of market vectors to such an extreme that humans immersed in technoculture can no longer sustain organic equilibrium. Given the pathological conditions of the electronic workspace, the body often fails to meet the demands of its technological interface or the ideological imperatives of socioeconomic space. Feelings of stress, tension, and alienation can compel the organic platform to act out nonrational behavior patterns that are perceived by power vectors to be use-less, counterproductive, and even dangerous to the technological superstruc-ture. In addition, the body can only interact with ICT for a limited period of time before exhaustion, and work is constantly disrupted by libidinal impuls-es. Many strategies have been used by pancapitalist institutions in an attempt to keep the body producing and consuming at maximum intensity, but most fail. One strategy of control is the use of legitimized drugs. Sedatives, anti-depressants, and mood stabilizers are used to bring the body back to a normalized state of being and to prevent disruption of collective activity. (For example, 600,000 new prescriptions were written in the U.S. for Prozac in 1993, and this number has continued to advance throughout the decade, ending in a grand total of 22.8 million in 1998). Unfortunately, social control drugs often rapidly lose their effectiveness, and can damage the platform before it completes its expected productive lifespan.

In order to bring the body up to code and prepare it for the rapidly changing pathological social conditions of technoculture, a pancapitalist institutional subapparatus with knowledge specializations in genetics, cell biology, neurology, biochemistry, pharmacology, embryology, and so on have begun an aggressive body invasion. Their intention is to map and rationalize the body in a manner that will allow the extention of authoritarian policies of fiscal and social control into organic space. We know this network as the *flesh machine*. Its primary mandate is eventually to design and engineer organic constellations with predispositions toward certain task-oriented activities, and to create bodies better suited to extreme technological interaction. The need to redesign the body to meet dromological imperatives (whether in warfare, business, or communications) has been prompted by the ICT revolution. ICT developers must now wait for the engineering gap between ICT and its organic complement to close; because of this, ICT development is slowing down (the web was the last high-velocity moment in the popular ICT revolution) compared to the rate at which investment and research in biotechnological processes and products for humans is growing. CAE believes that while we will continue to see ICT upgrades (such as in bandwidth) and further technological development in domestic space, radically significant change in the communication and information technology of everyday life will not take place until the gap between the technology and its organic platform is closed.

CAPITAL'S ENGINE

Given the entrenched skepticism about bioengineering, what would make an individual embrace reproductive technologies (the most extreme form of biotech)? For the same reasons people rushed to embrace new ICT. In the predatory, antiwelfare market of pancapitalism, a belief has been constructed and promoted that one must seek any advantage to survive its pathological socioeconomic environment. The extremes that function in the best interest of pancapitalist power vectors instantly transform into the common in a society that only profits from perpetual increases in economic velocity.

At the same time, the institutional foundation that produces the desire for bioengineering has blossomed in late capital. The eugenic visionary Frederick Osborn recognized that more hospitable conditions for eugenic policy were emerging in capitalist nations as early as the thirties. Osborn argued that the people would never accept eugenics if it were forced on them by militarized directives; rather, eugenic practices would have to structurally emerge from capitalist economy. The primary social components that would make eugenic behavior voluntary are the dominance of the nuclear family within a rationalized economy of surplus. Under these conditions, Osborn predicted, familial reproduction would become a matter of quality rather than a matter of quantity (as with the extended family). Quality of offspring would be defined by the child's potential for economic success. To assure success, breeders (particularly of the middle class) would be willing to purchase any legitimized medical goods and services to increase the probability of "high-quality" offspring. The economy would recognize this market, and provide goods and services for it. These conditions have come to pass, and the development of these goods and services is well underway. Of course, they only appear when one searches for them.

Without question, there is a strong intersection between the technology of the *sight machine* (ICT) and the technology of the *flesh machine*, much as the organic and the synthetic are necessary complements. Development in one machine system has a profound influence on development in the other. They merge under the value system of instrumentality. So in spite of the cyber-hype claims that the body is obsolete, and about to give way to post-human virtualization, it seems the body is here to stay. Why should capital refuse this opportunity—the greatest market bonanza since colonization, and the best method of self-policing since Catholic guilt? Unfortunately, the body of the future will not be the liquid, free-forming body that yields to individual desire; rather, it will be a solid entity whose behaviors are fortified by task-oriented technological armor interfacing with ideologically engineered flesh.

[An elaboration of this argument is in *Flesh Machine* (NY: Autonomedia/ Semiotext[e], 1997), or the abbreviated version, "The Coming of Age of the Flesh Machine," in T. Druckrey, ed., *Electronic Culture*, NY: perture, 1997.]

SUBJECT: CONFESSIONS OF A BACKSEAT DRIVER CA. 1998

FROM: PAULINE VAN MOURIK BROEKMAN
<PAULINE@METAMUTE.COM>
DATE: WED, 14 OCT 1998 11:52:27 +0100

On Great Windmill Street, just around the corner from the bubble behemoth of SegaWorld, a humbler Mecca has been putting down its wire roots. This one is named Wonderpark and is the brainchild of competitors Namco. Interestingly enough, Wonderpark is a different kettle of fish altogether from its neighbor. Where going to Segaworld is, all in all, a pretty antisocial experience (destined to ensure it most-favored status with families everywhere), Wonderpark feels like a teenage promenade gone mad. Coke- and change-machines oiling the general interaction, this is hormone intensity of a totally different order: gangs of girls play on the Shrinky Dink photobooths, gangs of boys on the fighting games and, just to piss Barbie off, roles are regularly reversed—girls slugging it out to the death. No one could harbor the illusion that Wonderpark isn't part of the same engineered reality, though. Your walk-through is totally choreographed: easy escalators in and shadowy stairs out, banks of driving-simulation machines flanking the big attractions like a defensive military regiment (lest you escape without playing and paying) and wall-to-wall CCTV and security guards, keeping the kids in sight, and in line, at all times. But, somewhere inside this ring of steel, there are cracks. They are, to state the obvious, provided by the games themselves.

Wonderpark's most popular games by a long stroke are the Tekken derivatives—descendants of Namco's classic early-nineties fighting game where one character of the player's choice battles it out against another (either played "by the machine" or by another player). Whereas three years ago, the character selection consisted of about four men and four women, all appropriately mythical and manga-esque, the contemporary offering spans about fifty subspecies, ostensibly tweaked to accommodate every need and creed (albeit dutifully Orientalized). Where early Tekken fights were conducted on a flat plain in an ethereal nowhere land (the proverbial end of the world), the latest ones are plonked down in a bizarre assortment of "realistic" locations. You could, for example, end up with a beefy, blond, Oriental boxing hero fighting a basketball lookalike of the A-Team's B. A. Barracus on a nameless American city street with onlookers and fast cars thrown in for realism's sake. Behind you would stand a Chinese business man shifting from one foot to another, his shirt dangling out of his trousers and his briefcase disheveled from overuse, while next to him an up-for-it cheerleader would jump up and down ad infinitum, egging you on in a self-imposed trance. Behind you, cars would screech by, wait for traffic lights to turn from red to green and go about their business on an eternal loop-de-loop like everything, and everyone, else.

Speed of response, compulsive logic problems, dynamic complexity and the elusive "gameplay" being the prime drivers of most games (or at least of the successful ones) the importance of their graphic environments—and even characters—can be overestimated. Background scenarios such as the one described above melt away next to the foreground activity. If the fighting itself wasn't getting better, faster, more seamlessly integrated with the hand-eye dynamics of the player, the Tekken derivatives would be standing around, just as lonely as all the other unpopular games in Wonderpark. Place classics like PONG, Tetris or Pac-Man next to some of the more recent, graphics-heavy candidates and the former "basic" ones will win hands-down every time (no pun intended). The current tendency toward "realism" and enhanced graphics that cuts across games genres as diverse as ski-simulations and fighting games (and which, beyond gaming, impacts on every single pixel of the graphical user interface—be that of Macs or PCs) seems inversely proportional to the thought being put into the question of *where*—or under which phenomenological and technological/systemic conditions—good gameplay occurs. The fact that simulation and fighting games are the most popular by far merely points to the fact that the area where this has been most successfully considered is propulsive physical movement. Whether it's blasting your way through a dungeon, successfully negotiating a moving train and jumping onto another one while shooting your opponents straight to hell, or driving at 200 mph through a deserted city, the parameters of virtual and concrete architectures are relatively simply aligned.

The no-holds-barred propulsion forward—and back—merely subject to far less friction than it would in the world of the concrete (and therefore so attractive to tired urbanites the world over).

You don't need to be a disciple of "computer visionaries" such as Brenda Laurel to see something cathartic (her term) is going on in many of the most popular games. Her erstwhile singling-out of games in discussions of computers, representation, and "meaningful" dramatic action would no doubt be repeated were she to survey the contemporary terrain. Whereas the VR industry stumbles around in search of a gratifying (read: lucrative) object, games have raged ahead in their exploration of, for want of a better word, the virtual. Even over as short a span as the past two years, the fast-developing ethos surrounding their effects on the body and the user's relationship to the world can—still—be gleaned from their incredible advertising campaigns. A couple of years ago, the opening of Segaworld saw Sega's blithe identification with the "extreme" mind-altering signifiers of drugs and hormones (the architecture of this Mecca will continue to stand as a testament to this specific point in time); elsewhere, companies like Sony even went so far as to advertise the PlayStation as a full-on intervention in the Occidental rationalist paradigm—the ads ran as a Zen master's, PlayStation-aided, hack of TV: Western media symbol par excellence). Now, in 1998, a strange pragmatism has set in, here as in many other areas of computing. Gone are the claims to paradigm-busting. Gone are the claims to mental dissolution, cortex rewiring, or entire escapes from reality, aided and abetted by the power of the machine. In their place have come hesitant, ironic, acknowledgments of the frictious relationship between the phenomenological interactions of "real life" and those of game space: Nintendo 64's latest ad, tagged with the mantra "Feel Everything" has a player making mistakes in his handling of "real" scenarios, "real"—and often very basic—physical interactions due to his overfamiliarity with another set of phenomenological standards: those generated by the machine.

The paranoid paraphernalia accompanying many games in Wonderpark and Segaworld certainly pays testament to a crisis of sorts. Simulation games especially are creating some interesting by-products. Construct a taxonomy of the modern gaming arcade and the main growth area seems to be awkward, clunky, objects that go in-between. Consequently, we now have a burgeoning morphology of padded armrests, safety railings, skipoles, pooltables, bowling alleys, cardboard icicles, model airplanes and outsized footpedals. Their raison d'être is cushioning the physical (or ocular) transition between the analog/object world and that of the digital/screen. At the same time, they function to convince users that times haven't changed: you're still doing the same thing really...your body hasn't changed, your adrenaline levels haven't changed, it's just that bit more dark in here and you need to be wired up to do it. It is hard not to see this push-me-pull-me game of denial, engagement, submission, and rejection as a far more interesting development than that positing a radical ontological break with the world that was popular a few years back. At Wonderpark, the cracks in the ring of steel might be widening, attracting hordes of teenagers every weekend to their paranoid, greedy hosts, but after the teens leave they go home, stepping on the cracks of the pavement instead. Outside, "a hole in the wall" remains the only way you're going to get any cash.

STOP TALKING

TALKING

START

CODING

MARKETS

SUBJECT: THE SELFISHNESS GENE: NEOLIBERAL CAPITALISM—IT'S NOT JUST A GOOD IDEA, IT'S THE LAW

FROM: MARK DERY <MARKDERY@WELL.COM>
DATE: SUN, 20 SEP 1998 04:24:41 +0200

On June 2, 1997, John Perry Barlow—frequent-flyer, sometime Grateful Dead lyricist, and bearded prophet of our Divine Assumption into a cosmic web of psychic Oobleck (the "physical wiring of collective human consciousness" into a "collective organism of mind")—posted a note to Nettime (J. Zaleski, *The Soul of Cyberspace*, NY: HarperEdge, 1997, 46, 48). In it, he opined that "nature is itself a free market system. A rain forest is an unplanned economy, as is a coral reef." In the next breath, he inverted the metaphor: "The difference between an economy that sorts the information and energy in photons and one that sorts the information and energy in dollars is a slight one in my mind. Economy *is* ecology."

Increasingly, the global marketplace is conceived of in Darwinian terms, with the social and environmental depredations of multinationals rationalized as corporate life forms' struggle for survival in an economic ecosystem. "'Ecology' and 'economy' share more than linguistic roots," maintains the nanotechnologist K. Eric Drexler; corporations, he argues, are "evolved artificial systems" born of the marketplace's "Darwinian" competition (K. E. Drexler, *Engines of Creation*, NY: Anchor, 1986, 32, 182). In *Bionomics*, business consultant Michael Rothschild straightfacedly argues that "what we call capitalism (or free-market economics) is not an ism at all but a naturally occurring phenomenon" (and therefore presumably beyond reproach). The catalog copy for Perseus Books presents *Clockspeed* as Charles H. Fine's sociobiological parables about "industrial fruit-flies" for anxious managers, whom he promises to turn into "'corporate geneticists' who do not react to the forces of change but master them to engineer their company's destiny."

A 1996 issue of the digital business magazine *Fast Company* featured an unintentionally hilarious example of corporate biobabble. A profile of Eric Schmidt, Sun's chief technology officer, extols his expertise at corporate crossbreeding–"organizational genetics," to those in the know, which means "combining organizational DNA in unique and inventive ways." What's organizational DNA, you ask? Why, "it's the stuff, mostly intangible, that determines the basic character of a business. It's bred from the founders, saturates the early employees, and often shapes behavior long after the pioneers have moved on" (J. F. Moore, "How Companies Have Sex," *Fast Company*, Oct.–Nov. 1996, 66). Gene-splicing the latest in Darwinian metaphors to a sexual politics that is strictly from Bedrock, the article's author analogizes venture capitalists and entrepreneurs to "the male urge to sow seed widely and without responsibilities and the female desire for a mate who'll settle down and help with the kids" (ibid., 68).

We've heard this song before, of course, and when the hundredth trendhopping management consultant informs us, as James Martin does in *Cybercorp*,

Economy is ecology. OK, so now what? You are, I think, an ecologist of sorts, so you'll surely recognize how important it is to adapt, to develop, to absorb, to encompass, to mutate and to grow—so how should we elaborate on the idea that economy is, in a way, ecology? I'd suggest that we start to digest the two terms of this statement, to break them apart. Mind you, I disagree with you about this: I think that an economy can be seen as an "ecology," but I don't believe that ecologies should be seen as economies—and that lack of transitivity suggests, to me at least, that there is much more to be learned in questioning what you've said than in accepting it.

"Very well. Can you give me an example of a planned economy that seems to be healthy...and appears likely to remain so for the long term?" Absolutely: The Roman Empire. The British Empire. The Ming Dynasty. Feudalism. Byzantium. Venice. The Netherlands. De Beers. The EEC. I don't toss these out to be glib; rather, I mention them to point up just how many people have constructed very impressive regimes: every one of them seemed (or seems) quite sensible—that is, according to its own terms. I don't see the Netherlands collapsing anytime soon; but for some pretty long stretches no one saw how Rome would fall apart or why Byzantium would collapse, and they surely did. I have little doubt that the nation-state will fall apart and be replaced by some other, similarly heterogeneous "solution," and that that "solution" will in turn collapse in the face of something else, and so one and so forth. Is this state of flux what you are advocating? Or, do you believe that we're on the verge of a terminal solution to the non-problem of historical change? [T. Byfield <tby≠field@panix.com>, Re: The Piran Nettime Manifesto, Tue, 3 Jun 1997 02:12:13 -0400]

Here's some basic banalities: Anarchism is neo-liberalism for hippies. Economy is social. Everyone should work so everyone can play. Giving gifts is better than exploiting others. [Richard Barbrook <richard≠@hrc.westminster.ac.uk>, More Provocations, Wed, 4 Jun 1997 00:14:08 +0000]

that high-tech corporations are "creature[s] designed to prosper in the corporate jungle," and that "capitalist society is based on competition and survival of the fittest, as in Darwin's world," we realize where we've heard it. It's the theme song of Herbert Spencer's social Darwinism, as popular in its day with monopoly-builders like John D. Rockefeller and Andrew Carnegie as Kevin Kelly's neobiological capitalism is with Tom Peters and his corporate flock. "'Social Darwinism,'" Stephen Jay Gould usefully reminds us, "has often been used as a general term for any evolutionary argument about the biological basis of human differences, but the initial 19th-century meaning referred to a specific theory of class stratification within industrial societies, and particularly to the idea that there was a permanently poor underclass consisting of genetically inferior people who had precipitated down into their inevitable fate" ("Curveball," in S. Fraser, ed., *The Bell Curve War*, NY: Basic, 1995, 12).

Cosic: When Negroponte came to Ljubljana, I had a big fight with him, and we interrupted his speech. Luka Frelih and I went around the city spraying grafitti: "WIRED = PRAV-DA". I made it look like a secret internet terrorist organization. On the website we compare him to Tito. But we did it without fanaticism. [Tilman Baumgärtel <Tilman_Baum≠ gaertel@compuserve.com>, Interview w/ Vuk Cosic, Mon, 30 Jun 1997 08:45:46 -0400]

The genealogical links between the public musings of the self-anointed "digital elite" and the Spencerian rhetoric of the robber barons is apparent at a glance, though they're separated by a century or so. Nicholas Negroponte, a sharp-dressed pitchman who hawks visions of a brighter, broader-bandwidth tomorrow to Fortune 500 executives (and to the unwashed AOL millions in his book *Being Digital*), breezily redefines the "needy" and the "have-nots" as the technologically illiterate—the "digitally homeless," a phrase that wins the Newt Gingrich Let Them Eat Laptops Award for cloud-dwelling detachment from the lives of the little people (N. Negroponte, "Homeless@info.hwy.net," *New York Times*, Feb. 11, 1995, 19). Stewart Brand, a charter member of the digerati, blithely informs the *Los Angeles Times* that "elites basically drive civilization" (P. Keegan, "The Digerati," *New York Times Magazine*, May 21, 1995, 42). *Wired* founder Louis Rossetto rails against the critic Gary Chapman as someone who "attacks technologically advanced people," as if website design were an inherited trait, a marker of evolutionary superiority" (P. Keegan, "Reality Distortion Field" <http://www.upside.com/> February 1, 1997).

If the analogy to social Darwinism seems overheated, consider Rossetto's belief, earnestly confided to a *New York Times* writer, that Homo Cyber is plugging himself into "exo-nervous systems, things that connect us up beyond–literally, physically–beyond our bodies, and we will discover that when enough of us get together this way, we will have created a new life form. It's evolutionary; it's what the human mind was destined to do" (Keegan, "Digerati," 88). As Rossetto readily acknowledges, his techno-Darwinian epiphany (like Barlow's) is borrowed from Pierre Teilhard de Chardin, the Jesuit philosopher and Lamarckian evolutionist who predicted the coming of an "ultra-humanity" destined to converge in a transcendental "Omega Point" that would be "the consummation of the evolutionary process" (M. Dery, *Escape Velocity*, NY: Grove, 1996, 45–48).

De Chardin's ideas are well known in theological and New Age circles and, increasingly, among the digerati. Less known is his passionate advocacy of eugenics as a means of preparing the way for ultrahumanity. "What fundamental attitude...should the advancing wing of humanity take to fixed or definitely unprogressive ethnical groups?" he wrote, in *Human Energy*. "The earth is a closed and limited surface. To what extent should it tolerate, racial-ly or nationally, areas of lesser activity? More generally still, how should we judge the efforts we lavish in all kinds of hospitals on saving what is so often no more than one of life's rejects?...[S]hould not the strong (to the extent that we can define this quality) take precedence over the preservation of the weak?" (P. Teilhard de Chardin, *Human Energy*, NY: Harcourt Brace Jovanovich, 1969, 132–33). Happily, the answer is readily at hand: "In the course of the coming centuries it is indispensable that a nobly human form of eugenics, on a standard worthy of our personalities, should be discovered and developed," he writes, in *The Phenomenon of Man* (Teilhard de Chardin, *The Phenomenon of Man*, NY: Harper, 1959, 282).

Since there's an implied guilt by association here, it's important to note that Rossetto and the other digital de Chardinians may well be unfamiliar with the philosopher's thoughts on eugenics. But given our increasingly "geno-centric" mindset and the creepy popularity of books like *The Bell Curve*, as well as the potential misuses of vanguard technologies like gene therapy and genetic screening, the digerati would do well to consider the ugly underside of their techno-Darwinian vision of the ultra-human apotheosis of the "technologically advanced"—"the advancing wing of humanity" by any other name. Obviously, the *Wired* ideology is far less pervasive, and not quite as nasty and brutish, as social Darwinism in its heyday; none of the digerati have embraced eugenics, at least publicly. But 19th-century capitalists like Carnegie and Rockefeller, who in the words of Andrew Ross "seized for themselves the mantle of the fittest survivors as if it were indeed biological-ly ordained," would undoubtedly note a family resemblance in the digerati–Way Cool white guys secure in the knowledge that they are Brand's fabled "elite," guiding civilization from their rightful place atop the Great Chain of Being (Digital).

SUBJECT: MARKETS, ANTIMARKETS, AND THE INTERNET

FROM: MANUEL DE LANDA <DELANDA@PIPELINE.CO>
DATE: SUN, 20 SEP 1998 20:10:16 -0400

One hundred years ago, Western societies underwent a second Industrial Revolution, based on the interaction of several technologies: electricity, the internal combustion engine, oil, steel, and plastics. Although knowledge and information as inputs to production processes had already played a role in the first Industrial Revolution, it was the coming of electricity, and the cre-ation of the first industrial research laboratories (such as the General Electric laboratory) that propelled knowledge to its position as the most important input to production. Information, of course, also plays key roles in other eco-nomic areas such as marketing and investment, and indeed, to the extent that a particular economy is truly driven by supply and demand, the infor-mation transmitted by prices has always played a central role. Without regard to the fact that knowledge has always been a key factor in the working of economies, electricity and the other innovations of the early twenty century greatly intensified its importance. The explosive growth of computer networks in the last three decades is bound to intensify the flow of knowledge and this intensification will undoubtedly transform the nature of the economy in the next century.

It follows that a very important task for today's intellectuals is to create realistic scenarios of the world of twenty-first century economics. The problem is that, when we try to imagine what the effects of the intensification of knowledge will be like, several obstacles stand in the way. The most important of these barriers is that intellectuals on the right, center and left sides of the political spectrum are all trying to predict what a twenty-first century economy will be like on the basis of theories that were devised to explain the workings of nineteenth century England. In other words, whether one is using the conceptual machinery of Adam Smith or of Karl Marx (or of any combination of the two), whether one sees in the recent commercialization of the internet a new "invisible hand" that will magically benefit society, or whether one sees in this commercialization the "commodification" of the net which will magically ruin society, one is still trying to understand what is a radically new phenomenon in terms of obsolete categories belonging to bankrupt systems of thought. It is time to go beyond both the "invisible handers" and the "commodifiers" and to attempt to construct a new economic theory that not only give us a clearer picture of the future, but almost, as important, of the past, since it is impossible to know where we are going unless we know how we got where we are.

What follows is a brief sketch of what these new economic theories might be like. First of all, it is not as if we would need to manufacture a new theory out of thin air. Alternatives to the "invisible handers" and the "commodifiers" have existed in the past (such as the institutionalist school of the followers of Thorstein Veblen) and new theories are flourishing today, such as the neo-institutionalist school and the growing field of nonlinear economics (D. C. North, *Institutions, Institutional Change and Economic Performance*, NY: Cambridge University, 1990). In addition, economic historians like Fernand Braudel and his followers have given us an incredibly detailed account of the development of Western economies in the last eight hundred years—an account accompanied by research that has generated a wealth of empirical data which simply was not available to either Adam Smith or Karl Marx when they created their theories. Furthermore, this new data contradicts many of the foundations of those two systems of thought. Finally, not just economists and economic historians will be involved in developing the new ideas we need, philosophers will also participate: in the last twenty years the discipline of the philosophy of economics (that is the philosophy of science applied to economics) has grown at a tremendous pace and is today a very active field of research (U. Maki, "Economics with Institutions," and C. Knudsen, "Modelling Rationality, Institutions and Processes in Economic Theory," in Maki, B. Gustafsson, and C. Knudsen, eds., *Rationality, Institutions and Economic Methodology*, London: Routledge, 1993).

Here I only have space to discuss a few of the ideas that have been developed by economists, historians and philosophers. Perhaps the most dramatic new insight emerges from Fernand Braudel's history of capitalism. Unlike theorists from the left and the right who believe capitalism developed through several stages, first being competitive and subservient to market forces and only later, in the twentieth century, becoming monopolistic, Braudel has shown with a wealth of historical evidence that as far back as the thirteenth century, and in all the centuries in between, capitalists have always engaged in anticompetitive practices, manipulating demand and supply in a variety of ways. Whenever large fortunes were made in the areas of foreign trade, wholesaling, finance, or large-scale industry and agriculture, market forces were not acting on their own, and in some cases not acting at all. In short, what Braudel shows is that we must carefully differentiate between the dynamics generated by many interacting small producers and traders (where automatic coordination via prices does occur), from the dynamics of a few big businesses (or oligopolies, to use the technical term), in which prices are increasingly replaced by commands as coordinating mechanisms, and spontaneous allocation by the market replaced with rigid planning by a managerial hierarchy. What these new historical findings suggest is that all that has existed in the West since the fourteenth century, and even after the Industrial Revolution, is a heterogeneous collection of institutions—some governed by market dynamics and others manipulating those dynamics—not a homogeneous, societywide "capitalist system." In the words of Fernand Braudel: "We should not be too quick to assume that capitalism embraces the whole of western society, that it accounts for every stitch in the social fabric...that our societies are organized from top to bottom in a 'capitalist system'. On the contrary, ...there is a dialectic still very much alive between capitalism on one hand, and its antithesis, the 'non-capitalism' of the lower level on the other" (Fernand Braudel, *The Perspective of the World*, NY: Harper and Row, 1986, 630). He adds that, indeed, capitalism was carried upward and onward on the shoulders of small shops and "the enormous creative powers of the market, of the lower story of exchange...[This] lowest level, not being paralyzed by the size of its plant or organization, is the one readiest to adapt; it is the seed bed of inspiration, improvisation and even innovation, although its most brilliant discoveries sooner or later fall into the hands of the holders of capital. It was not the capitalists who brought about the first cotton revolution; all the new ideas came from enterprising small businesses" (ibid., 631). Several things follow from Braudel's distinction between market and capitalist institutions (or as he calls them "antimarkets"). If markets and antimarkets have never been the same thing then both the invisible handers as well as the commodifiers are wrong, the former because spontaneous coordination by an invisible hand does not apply to big business, and the latter because commodity fetishism does not apply to the products created by small business but only to large hierarchical organizations capable of manipulating demand to create artificial needs. In other words, for people on the right and center of the political spectrum all monetary transactions, even if they involve large oligopolies or even monopolies, are considered market transactions. For the Marxist left, on the other hand, the very presence of money, regardless of whether it involves economic power or not, means that a social transaction has now been commodified and hence made part of capitalism.

The other mental characteristic of the virtual class is that it is deeply authoritarian. It believes that virtuality equals the coming-to-be of a fully free human society. As CEOs of leading corporations use to say, "adapt or you're toast"—uttering this with the total smugness of complacency itself. The other side of cyber-authoritarianism is the absolute outrage that grips those in authority when faced by the presence of opposition. Qualms about the emergence of the virtual class, or about the social consequences of technology are met with either indifference or total outrage. Quite on the contrary, members of the virtual class see themselves as the missionaries of the human race itself, the avant garde, in their terms, in honor-full collaboration with the telematic machines. The program of the virtual class is a curse for those who stand outside of it. Within, it is not even a hostile position—it is simply contempt for those members of the working class that do not have easy access and who cannot experience the new universal communion. At the same time you see the virtual class shutting down the internet and again, feeling nothing but contempt for the lost ideas of what they would like to call blue-eyed utopian thinkers who call for the possibilities of democratic use of the internet outside of the barriers of the state. But when they get challenged, they go for their class interests and actually suppress those members of competing classes who stand in opposition to them. The virtual class has this aspect of seduction, on the one hand, and, on the other, a policy of consolidation. This is the present reality in which we live. It is a grim, severe, and deeply fascistic class because it operates by means of the disciplinary state, imposing real austerity programs in order to fund research efforts that benefit itself. At the same time it politically controls the working class by severe taxation in order to make sure that people cannot be economically mobile and cannot accumulate capital in their own right. When it comes to Third World nations it acts in classic fascist ways. It imposes strict anti-emigration policies in the name of humane gestures. It shields its own local populace from the influx of immigrants by creating a "bunker state," by stressing a Will to Purity. In this way it can tolerate "ethnic cleansing" by way of infinite media coverage. For example, the Western reaction to the genocide in Bosnia is symptomatic of this condition. [Geert Lovink <geert@xs≠4all.nl>, Theory of the Virtual Class, Thu, 4 Jan 1996 23:11:59 +0100

It is my belief that Braudel's empirical data forces on us to make a distinction which is not made by the left or the right: that between market and anti-market institutions. In fact, we can already see the kind of dogmatic responses that the lack of this distinction promotes on discussions in the internet. As it became clear that digital cash and secure cryptographic technology for credit card transactions were going to transform the net into a place to do business, some intellectuals became euphoric about the utopic potential of digital "free enterprise," while others began to denounce the internet as the latest expression of international capitalism, or to claim that the net was becoming commodified and hence re-absorbed into the system. It is clear, however, that if we reject these two dogmatic positions, our evaluation of the economic impact of the net (its potential for both decentralization and empowerment of the individual producer and for centralization of content production by a few large firms) will have to become more finely nuanced and based on more complex models of economic reality.

Recognizing the complexity and heterogeneity of actual "institutional ecologies" may be crucial not only when thinking about internet economics but, more generally, when analyzing the oppressive aspects of today's economic system. That is, those aspects that we would want to change to make economic institutions more fair and less exploitative. We need to think of economic institutions as part of a larger institutional ecology, an ecology that must include, for example, military institutions. Only this way will we be able to locate the specific sources of certain forms of economic power, sources which would remain invisible if we simply thought of every aspect of our current situation as coming from free enterprise or from exploitative capitalism. In particular, many of the most oppressive aspects of industrial discipline and of the use of machines to control human workers in assembly line factories, were not originated by capitalists but by military engineers in eighteenth century French and nineteenth century American arsenals and armories. Without exaggeration, these and other military institutions created many of the techniques used to withdraw control of the production process from workers; they then exported these techniques to civilian enterprises, typically antimarket organizations (M. R. Smith, "Army Ordnance and the 'American System of Manufacturing,' 1815–61," and C. F. O'Connell, Jr., "The Corps of Engineers and the Rise of Modern Management, 1827–56," in Smith, ed., *Military Enterprise*, Cambridge: MIT, 1987). Hence, not to include in our economic models processes occurring within this wider institutional ecology renders invisible the source of the very structures we must change to create a better society. It also diminishes our chances of ever dismantling those same oppressive structures.

For the information sector and its information products, many open markets are turning into artificial monopolies and what Manuel DeLanda calls *antimarkets*. A major mechanism that facilitates this process is the concept of intellectual property rights (IPRs), which may be seen as a form of exclusive ownership over information products. This monopolistic ownership through IPRs facilitates the accumulation of wealth by an information elite and leads to the specific social stratification analyzed here. Once resolved, the social conflicts that emerge out of the stratification can lead to a new type of economy.

In the future, nonmonopolistic information economies may emerge that will remunerate intellectual activity through means other than monopolistic mechanisms such as patents, copyrights, and other IPRs (for example, salaries and wages, bonuses, awards, grants, and other forms that do not involve exclusive right of use). In such economies, the nature of intellectual rewards will be in much better harmony with the nature of information itself.

EXPANDING INFORMATION MONOPOLIES

The main forms of IPRs are patents and copyrights, both of which are statutory monopolies; that is, they are monopolies acquired by virtue of government statutes. These state-granted monopolies cover the exclusive rights to use, manufacture, copy, modify, and sell an information product. Recently, under the GATT/WTO, these rights have been expanded further to include the exclusive right to rent out copyrighted material and to import patented products.

These statutory monopolies—which are gradually being strengthened and extended as the political and economic power of the propertied classes of the information sector grow—are in direct conflict with the information freedoms sought by the vast majority of information users. These freedoms include the freedom to use information, to share it with others, and to modify it. Information monopolies are also in conflict with the basic nature of information itself as a public good.

CLASSES IN THE INFORMATION SECTOR

Just like the ecology and industrial sectors, the information sector gives rise to various economic classes based on individuals' position in the production, distribution, and use of information. Analysis of these classes can provide useful insights about the underlying economic interests and typical attitudes of various social groups in the sector. The following major classes can be identified:

There are in total some 44,000 TNCs in the world, with 280,000 subsidiaries and an annual turnover of US$7,000 billion. Two thirds of world trade results from TNC production networks. The share of world GDP controlled by TNCs has grown from 17 percent in the mid-sixties to 24 percent in 1984 and almost 33 percent in 1995. In a parallel and related process, the largest TNCs are steadily increasing their global market shares. According to UNCTAD's 1997 World Investment Report, the ten largest TNCs now have an annual turnover of more than US$1,000 billion. Fifty-one of the world's largest economies are in fact TNCs. Continuous mergers and takeovers have created a situation in which almost every sector of the global economy is controlled by a handful of TNCs, the most recent being the service and pharmaceutical sectors. In January 1998, for example, the largest business merger in history took place in a US$70 billion deal in which Glaxo Wellcome and SmithKline Beecham became the largest pharmaceutical company on earth. [Corporate Europe Observatory <ceo@xs4all.nl>, MAI-GALOMANIA, Tue, 10 Feb 1998 16:01:35 +0100 (MET)]

Cyberlords: The propertied class of the information sector, they control either a body of information or the material infrastructure for creating, distributing, or using information. Cyberlords are rent-seeking members of the capitalist class. IPR holders make up the first category of cyberlords; they have staked their monopoly rights to a specific body of information, and earn their income by charging royalties, license fees, or other forms of rent from those who want to use this body of information. Because of these monopoly rights, they can set prices that are much higher than their marginal cost of production, helping them accumulate and concentrate wealth rapidly. Cyberlords include the owners of software companies, database companies, audio, video, and film companies, genetic engineering firms, pharmaceutical and seed firms, and similar companies that earn most of their income from IPR rents.

The infrastructure owners are the second category of cyberlords. They own or control the industrial infrastructure for creating, reproducing, distributing, or using information. They earn their income by charging rents for the use of these infrastructures. This category includes the owners of communication lines and equipment, radio and TV stations, internet service providers, theater distributors and owners, cable TV operators, and similar firms.

These industrial cyberlords are generally in alliance with the first group. However, they may not share the same rabid advocacy for IPRs that characterize the IPR-holding cyberlords, especially when IPRs impede wider use of the infrastructure from which infrastructure owners derive their own income. The distinction between them may occasionally become important in the struggle against the cyberlords of the first type, who are the true cyberlords of the information economy.

The cyberlord class also includes those highly paid professionals who earn their living under the employ or in the service of cyberlords. The best examples are the top-level managers as well as the lawyers who serve cyberlords and who derive their income mostly from the cyberlords they work for. These highly paid hirelings assume the class status and ideological outlook of the cyberlords they serve.

Cyberlords all over the world are scouring the public domain for information products that they can privatize and monopolize through IPRs. Some have already acquired the exclusive electronic reproduction rights to paintings and other cultural artifacts in the world's best museums. Others are engaged in a race to patent genetic information of all kinds, including parts of the human genome. Still others are eyeing governments' vast information outputs, which are normally in the public domain.

Most big cyberlords control corporations that operate globally. These firms are a major hidden force that drive the process of globalization. Because the social nature of information keeps asserting itself and information products tend to spread themselves globally as soon as they are released, cyberlords need a global legal infrastructure to impose their information monopolies and extract monopoly rents. Thus, they push the globalization process incessantly to ensure that every country, every nook and corner of the globe, is within their legal reach.

The highly advanced industrial infrastructures of the U.S. and Europe, together with extremist concepts of private property, have given their cyberlords a commanding lead over cyberlords elsewhere. (An extreme example is the claim that discovery of a particular DNA sequence entails ownership of that sequence through a patent.) Because they tend to suppress local efforts to acquire new technologies at the least cost, big cyberlords are a major hindrance to the development efforts of most national economies.

Compradors: These are the merchant capitalists of the information sector, and earn their living by selling patented or copyrighted products for profit. They very often come from the merchant classes of the industrial and ecology sectors, and may retain their businesses in these sectors. These merchant classes are attracted to the information sector because the extremely high profit margins enjoyed by successful cyberlords also give resellers better margins.

This class can be roughly divided into two—monopolistic and nonmonopolistic compradors. Monopolistic compradors make money by paying cyberlords for the right to sell patented or copyrighted goods. Thus, they derive their income from information rents, therefore supporting cyberlord interests. Nonmonopolistic compradors make money by reproducing and selling patented or copyrighted material, without paying the monopoly rents claimed by cyberlords. In a way, they help break the information monopolies imposed by cyberlords.

Because of the political clout of cyberlords, the nonmonopolistic compradors are often harassed and suppressed both to discourage them from their trade and to turn them into monopolistic compradors. They are frequently the targets of surveillance, legal suits, raids, and other forms of government and cyberlord harassment. Yet, there is no lack of nonmonopolistic compradors who trade in copyrighted and patented materials, making these materials more accessible to the public, which would otherwise be unable to afford them. Even under the worst forms of authoritarian rule, nonmonopolistic compradors continue to ply their trade by forming an underground network to break the cyberlord monopolies. These compradors can be allies of information users against the cyberlord class. Many of them, however, eventually surrender to the power of cyberlords, arrive at a profit-sharing arrangement with them, and turn into monopolistic compradors.

Intellectuals: They are the main creators of information in the information sector. They earn their living through mental labor, creating new and useful information. This class ranges widely, from those whose earnings come mostly from business contracts for information work, to wage-earning intellectuals who earn most of their income from fixed-rate payments such as wages and salaries and whose work—some of which may be patentable or copyrightable—is by contract the property of the company they work for. Most intellectuals belong to this wage-earning stratum.

Information users: Members of this group use information but are not generally involved in creating information products for sale. Whatever information they generate is either automatically shared with others or kept confidential.

The idea of claiming a monopoly over a body of information to make money out of it is quite alien to them. Because they generally earn their income elsewhere, information users are actually neither a single class nor a monolithic group, but a cluster of classes in the ecology, industrial, and information sectors. Since they are all information users, however, they actively seek the freedom to use, share, and modify information. Information users are the main force in the struggle to free information from cyberlord monopolies.

THE BASIC CONFLICT

These classes in a monopolistic information economy differ in their attitude toward IPRs, reflecting their class roles in the production, distribution, and use of information.

Cyberlords strongly advocate expanding these monopoly mechanisms, while information users want to limit IPRs as much as possible. Whenever IPR infringements encroach upon their profit margins, compradors take the side of cyberlords. But when monopoly rents themselves encroach upon their profit margins, other compradors oppose IPRs. Intellectuals may dream of owning some body of information in the future, from which they can themselves extract information rents. But largely they realize that this cannot be their main source of income, and that they themselves need access to bodies of information that are today monopolized through patents or copyrights.

To transform a monopolistic information economy into a nonmonopolistic information economy, monopolistic IPRs must be replaced with other means of rewarding intellectual activity. This will of course be opposed to the very end by the cyberlord class, which furthermore is politically and economically very strong. As the privatization process subsumes more and more of what is now public domain information under cyberlord monopolies, the information-using public will develop a higher level of political consciousness, and this struggle will eventually express itself as the main conflict in a monopolistic information economy. As such, it will increasingly manifest itself on cultural and economic as well as on political fronts.

A STRATEGY AGAINST MONOPOLIES

To defeat the powerful cyberlord class, we must advance a set of demands—one that will isolate the big cyberlords and their closest comprador allies, that will neutralize or win over the middle and small cyberlords, and that will convince the entire intellectual class to unite with the vast majority of information users. We must also involve other classes and social groups in the industrial and ecology sectors who support our demands. Without such a united front, it will be extremely difficult to defeat the information monopolies of the big cyberlords, and the latter will be able to use their increasing economic and political power to consolidate, codify, and further expand their statutory monopolies.

The long-term goal is to dismantle monopolistic forms of information ownership and replace them with nonmonopolistic forms. This will eventually enable users to enjoy the full information freedom that will unleash creativi-

ty not only among intellectuals, but among information users themselves.
Several demands can be identified now, because they have emerged historically and must necessarily become part of the overall set of demands made
on information monopolies.

Compulsory licensing: The most important demand for breaking the cyberlords' information monopolies is to retain compulsory licensing and
expand its coverage.

Compulsory licensing works as follows: Someone who wants to use/com-
mercialize patented or copyrighted material approaches NOT the patent or
copyright holder to obtain a license to do so, but the government. The government grants the license, whether the original patent or copyright holder
agrees or not, but compels the licensee to pay the patent/copyright holder a
royalty rate that is fixed by law. Many countries in the world have used compulsory licensing for important products like pharmaceuticals and books.
(For example, Philippine law authorizes local publishers to reprint foreign
textbooks for the use of the local educational system; it also provides for
compulsory licensing of pharmaceutical products by local companies. Both
laws are currently under heavy attack by cyberlord lobbyists. Efforts are now
afoot to repeal them in order to align Philippine laws with the
GATT/WTO agreement.)

Compulsory licensing (also called mandatory licensing) is good for countries
that want to access technologies but cannot afford the price set by
patent/copyright holders. While this internationally recognized mechanism
was meant to benefit poorer countries, even the United States and many
European countries use it.

This demand will split the cyberlord class. Small cyberlords who have neither the capital nor the production facilities to commercialize their own creations welcome compulsory licensing—although they will try to negotiate for
higher royalty rates—because it will ensure them regular rent income. Big
cyberlords who have the capability to commercialize products themselves are
violently opposed to the idea of compulsory licensing, because it is a powerful threat to their monopoly over information.

No patenting of life forms: This demand emerged from the popular campaigns against genetic engineering and recombinant DNA technologies. It
has become a major global issue, as genetic engineering continues to slide
down that slippery slope leading corporations toward the direct manipulation and commercialization of human genetic material. True to their cyberlord nature, owners of biotech firms are racing against each other in patenting DNA sequences, microorganisms, plants, animals, human genetic matter,
and all other kinds of biological material. Cyberlord representatives have
already managed to insert protection in the GATT/WTO agreement for
patents on microorganisms and microbiological processes.

Life-form patents raise religious and moral issues as well as impinge on
indigenous community knowledge. Genetic engineering also threatens to
give rise to a whole new class of harmful viruses, germs, microorganisms,
and higher life forms that have no natural enemies. This demand to ban such
patents can unite a wide range of sectors against the cyberlord ideology.

Expanding the fair-use policy: This struggle has historically been waged by librarians (particularly in public libraries) who see themselves as guardians of the world's storehouse of knowledge. Most librarians want this storehouse of knowledge to be freely accessible to the public, and they have fought long battles and firmly held their ground on the issue of "fair use," which allows students and researchers access to copyrighted or patented materials without paying IPR rents. Recently, this ground has been suffering slow erosion from the increasing political power of cyberlords.

Support for nonmonopolistic mechanisms: Various concepts in software development and/or distribution have recently emerged. Some, such as shareware, are less monopolistic than IPR. Others, such as the GNU General Public License (GPL), are completely nonmonopolistic.

Shareware works under various schemes, such as free trial periods for use of software, free distribution, voluntary payments, and so on. These concepts have in effect abandoned the legal artifice of asserting exclusive monopoly over copying work in favor of granting users limited rights to use, copy, and distribute the material. Shareware authors, however, still balk at releasing their source code, and therefore continue to keep their users captive and unable to modify the software on their own.

The GNU GPL enables users to enjoy the fullest set of information freedoms, including the freedom to use information, to share it with others, and to modify it. The GPL—a project of the Free Software Foundation to elaborate existing copyright concepts toward nonmonopolistic forms—shows how current copyright concepts may be used in moving away from monopolistic arrangements, and points the way toward future nonmonopolistic software development. Software as well as books that fall under the GPL copyright may be freely used by anyone who may find them useful. They may also be freely copied and shared with others. Finally, the software may be freely modified because the package includes the source code, that is, the legible text files of formalized instructions that are "compiled" in order to make a computer program.

General wage increases: In a way, salaries and wages are a specific form of nonmonopolistic remuneration for intellectual activity. This is the most relevant demand for most intellectuals, who will stay on the side of information users as long as they are assured some reasonable remuneration for their work as information creators. In this respect, the vast majority of intellectuals can unite with other wage-earning classes to raise common demands.

The list above is not complete. A comprehensive set of demands will emerge when the various classes ranged against the cyberlords acquire an economic and political consciousness that will make clear where their interests lie.

TOWARD A NEW SOCIAL ORDER

These demands in the information sector must also be linked with the demands of other change-oriented classes and groups in the ecology and industrial sectors, such as farmers, fisherfolk, workers, women, and indigenous peoples. The key is to bring together the widest range of people whose unity and joint action can develop a political structure for evolving new forms of rewarding intellectual activity. In the future, such forms will lead to a nonmonopolistic information sector. The rethinking of property concepts that this will bring about will then reinforce demands for restructuring the industrial and agriculture sectors as well.

From such a confluence of social movements, enough social forces for change can emerge to bring forth a society in which knowledge and culture are freely shared, where industrial machinery is carefully designed for genuine human and community needs, and where agriculture is an ecological and not an industrial undertaking.

SUBJECT: THE TOPOI OF E-SPACE: PRIVATE AND PUBLIC CYBERSPACE

FROM: SASKIA SASSEN <SASSEN@COLUMBIA.EDU>
DATE: TUE, 27 OCT 1998 11:58:12 -0600

We need to retheorize electronic space and uncouple it analytically from the properties of the internet which have shaped our thinking about electronic space. We tend to think of this space as one that is characterized by distributed power, by the absence of hierarchy. The internet is probably the best known and most noted. Its particular attributes have engendered the notion of distributed power: decentralization, openness, possibility of expansion, no hierarchy, no center, no conditions for authoritarian or monopoly control.

Yet the networks are also making possible other forms of power. The financial markets, operating largely through private electronic networks, are a good instance of an alternative form of power. The three properties of electronic networks: speed, simultaneity, and interconnectivity have produced strikingly different outcomes in this case from those of the internet. These properties have made possible orders of magnitude and concentration far surpassing anything we had ever seen in financial markets. The consequence has been that the global capital market now has the power to discipline national governments, as became evident with the Mexico "crisis" of December 1994. We are seeing the formation of new power structures in electronic space, perhaps most clearly in the private networks of finance but also in other cases.

1. THE TOPOI OF E-SPACE: GLOBAL CITIES AND GLOBAL VALUE CHAINS

The vast new economic topography that is being implemented through electronic space is but one moment, one fragment, of an even vaster economic chain that is largely embedded in nonelectronic spaces. There is no fully virtualized firm and no fully digitalized industry. Even the most advanced information industries, such as finance, are installed only partly in electronic space. So are industries that produce digital products such as software. The growing digitalization of economic activities has not eliminated the need for major international business and financial centers and all the material resources they concentrate, from state-of-the-art telematic infrastructure to brain talent.

Nonetheless, telematics and globalization have emerged as fundamental forces reshaping the organization of economic space. This reshaping ranges from the spatial virtualization of a growing number of economic activities to the reconfiguration of the geography of the built environment for economic activity. Whether in electronic space or in the geography of the built environment, this reshaping involves organizational and structural changes. Telematics maximizes the potential for geographic dispersal and globalization entails an economic logic that maximizes the attraction and profitability of such dispersal.

Centrality remains a key property of the economic system but the spatial correlates of centrality are profoundly altered by the new technologies and by globalization. This engenders a whole new problematic around the definition of what constitutes centrality today in an economic system where (1) a share of transactions occur through technologies that neutralize distance and place, and do so on a global scale; (2) centrality has historically been embodied in certain types of built environments and urban forms. Economic globalization and the new information technologies have not only reconfigured centrality and its spatial correlates, they have also created new spaces for centrality.

To some extent when I look at the global economy I see a network of about thirty or forty strategic places—it is a changing animal that depends on all kinds of things—where there is an enormous concentration of all those resources. They are largely cities but not exclusively, Silicon Valley would be one, as well as other industrial areas with telecommunications industries like Lille, for instance. The point is: yes, globalization, yes, digitalization, yes, dematerialization, yes, instantaneous communication, but because it is a system characterized not by distributed power, distributed ownership, distributed application of profits, but by the opposite, concentration of profits, concentration in ownership, concentration of control, you also have a material correlate to this, which is this enormous concentration of strategic resources in major cities.

2. A NEW GEOGRAPHY OF CENTRALITY

We are seeing a spatialization of inequality that is evident both in the geography of the communications infrastructure and in the emergent geographies in electronic space itself. Global cities are hyperconcentrations of infrastructure and the attendant resources while vast areas in less developed regions are poorly served. Even within global cities we see a geography of centrality and one of marginality. For instance, New York City has the largest concentration of fiber-optic cable–served buildings in the world; but they are mostly in the center of the city, while Harlem, the black ghetto, has only one such building. South Central Los Angeles, the site of the 1993 uprisings, has none.

There are many examples of this new unequal geography of access. Infrastructure requires enormous amounts of money. For example, it is estimated that it will cost US$120 billion for the next ten years just to bring the communication networks in the Central and Eastern European countries up to date. The European Union will spend US$25 billion per year to develop a broadband telecommunications infrastructure. The levels of technical development to be achieved by different regions and countries, and indeed, whole continents, depend on the public and private resources available and on the logic guiding the development. This is evident even with very basic technologies such as telephone and fax. In very rich countries there are 50 telephone lines per person, in poor countries, fewer than ten. In the U.S. there are 4.5 million fax machines and in Japan, 4.3 million, but only 90,000 in Brazil, 30,000 each in Turkey and Portugal, and 40,000 in Greece.

Once in Cyberspace, users will also encounter an unequal geography of access. Those who can pay for it will have high-speed service, while those who cannot pay will increasingly find themselves with very slow service. For instance, Time Warner ran a pilot project in a medium-sized community in the U.S. to find out whether customers would be willing to pay rather high fees for fast services; they found that customers would—that is, those who could pay.

3. EMERGENT CYBERSEGMENTATIONS

One way of beginning to conceptualize possible structural forms in electronic space is to specify emerging forms of segmentation. There are at least three distinct forms of cybersegmentation we can see today. One of these is the commercialization of access—a familiar enough subject. The second is the emergence of intermediary filters to evaluate sort, and chose information for paying customers. The third, and the one I want to focus on in some detail, is the formation of private firewalled corporate networks on the web. We cannot underestimate how pervasive is the search for ways to control, privatize and commercialize. Three major global alliances have been formed that aim at delivering a whole range of services to clients. While the mechanisms for commercialization may not be available now, there is an enormous effort to invent the appropriate billing systems. It is worth remembering that in the U.S. the telephone system started in the late 1800s as a decentralized, multiple-owner network of networks: there were farmers telephone networks, mutual aid societies telephone networks, and so on. This went on for decades. But then in 1934 the Communications Act was passed defining the communication systems as a "natural monopoly situation" and granting AT&T the monopoly. AT&T is up to 60 percent a billing company: it has invented and implemented billing systems. Much effort today is likely to address the question of a billing system for access to and use of what is now public electronic space.

Today most big infrastructure projects—laying fiber-optic cable across the bottom of the oceans—are carried out by three major engineering companies who do it on "spec"—that is not because they were contracted to do so by a government or a company, but on their own because they know that there is a market of actors with very deep pockets, such as the multinationals and the financial services firms and the financial markets, which will buy the bandwidth. We fight for the right of access to using bandwidth because we are fighting around issues concerning the internet—public space, a public good. It is like poor workers demanding public transportation to get them to their jobs.

Internet activists and experts don't usually recognize or often have not thought about the world of private digital space because they really are two separate worlds. To me, someone who focuses also on finance, it is always astounding to hear generalizations made about the features of digital networks in general, when what they are talking about is the features of the net. I think this shows us once again that technology is, ultimately, embedded. There is no neutral technology. The structures of power also shape some of the decisive features of the digital networks as I compared earlier for the internet and the private networks of finance.

CONCLUSION: SPACE AND POWER

Electronic space has emerged not simply as a means for transmitting information, but as a major new theater for the accumulation and the operation of global capital. This is one way of saying that electronic space is embedded within the larger dynamic of organized society, particularly economic areas.

There is no doubt that the internet is a space of distributed power that limits the possibilities of authoritarian and monopoly control. But it is becoming evident over the last two years that it is also a space for contestation and segmentation. Further, when it comes to the broader subject of the power of the networks, most computer networks are private. That leaves a lot of network power that may not necessarily have the properties/attributes of the internet. Indeed, much of this is concentrated power and reproduces hierarchy rather than distributed power systems.

The internet and private computer networks have coexisted for many years. This situation is changing, however, and that drives my concern for the need to retheorize the internet and the need to address the larger issue of electronic space rather than just the part of the internet that is a public electronic space. The three subjects discussed above may be read as an empirical specification of two major new conditions: (1) the growing digitalization and globalization of leading economic sectors has further contributed to the hyperconcentration of resources, infrastructure and central functions, with global cities as one strategic site in the new global economic order; (2) the growing economic importance of electronic space which has furthered global alliances and massive concentrations of capital and corporate power, and has contributed to new forms of segmentation in electronic space. These have made electronic space one of the sites for the operations of global capital and the formation of new power structures.

What these developments have meant is that suddenly the two major actors in electronic space—the corporate sector and civil society—which until recently had little to do with one another in electronic space, are running into each other. Corporate players largely operate in private computer networks. But two years ago business had not yet discovered the internet in a significant way. The world wide web—the multimedia portion of the net with all its potentials for commercialization—had not yet been invented, and the digitalization of the entertainment industry and of business services had not exploded on the scene.

One of the concerns for me has been to understand the differences between private and public digital space. A lot of theoretical work has been done on public digital space, for example about the Digital City in Amsterdam. I have been more concerned with private digital space and with what I see as a colonizing of public digital space by private (that is, corporate) players. We have three historical eras of the internet. The first phase is that of the hackers, where access was the issue as well as making the software available. The second phase is when you begin to have the interest by private players that did not quite know how to use it. At that point it was still primarily a public space, though in some ways protected. And presently the third stage which is the invasion of cyberspace by corporate players—it is really combat out there. So, for me, the internet becomes a space for contestation. I am here not only thinking about multinational corporations. I am thinking of all kind of players, including those that misuse the internet, something which is serious also.

This is also the context within which we need to examine the present trends towards deregulation and privatization that have allowed the telecommunications industry to operate globally in an increasing number of economic sectors. These changes have profoundly altered the role of government in the industry, and, as a consequence, have further raised the importance of civil society as a site where a multiplicity of public interests can resist the overwhelming influence of the new corporate global players. Civil society, from individuals to NGOs, has engaged in a very energetic use of cyberspace from the bottom up.

When we talk about regulation today we tend ascribe to it a narrow meaning having to do with the government regulating content. This is a totally different notion compared with the regulation of access and accountability. We need to free the concept of regulation from what it is. We should innovate and begin to think about how we can regulate those big conglomerates. They are reshaping the topography of communications. They are now moving into Latin America, where national telecoms are being privatized. For the upper middle classes and above, this is an acceptable situation. The problem lies with lower income communities and more isolated areas. Even in the U.S. there are people who cannot even afford a telephone. Global telecoms are dealing with a service that is essential to us—whether we look at it as individuals, who have forms of sociability, or if we look at it as a democracy, where communication is necessary. At this moment, however, these firms are privatized and not accountable, a fact that suggests that we might run into scenarios in the future that are very nasty.

To the extent that national communication systems are increasingly integrated into global networks, national governments will have less control. Further, national governments will feel great pressure to help local firms become incorporated into the global network, to avoid the risk of being excluded from the increasingly electronically operated global economic system. If foreign capital is necessary to develop the infrastructure in developing countries, the goals of these investors may well rule and shape the design of that infrastructure. This is of course reminiscent of the development of railroads in colonial empires, which were clearly geared towards facilitating imperial trade rather than the territorial integration of the colony. Such dependence on foreign investors is also likely to minimize concerns with public applications, from public access to uses in education and health.

There are today few institutions at the national or global level that can deal with these various issues. It is in the private sector where this capacity lies, and even then only among the major players. We are at risk of being ruled by multinational corporations—organizations accountable only to the global market. Most governmental, nonprofit, and supranational organizations are not ready to enter the digital age. Political systems, even in the most highly developed countries, are operating in a predigital era.

One issue that characterizes the present time is that you have an interstate (transnational) system, yes, but that you also have an international economic system that operates partly outside the interstate system. The second big difference—and I should really say that these are very much my own ideas with which many economists would not agree—the second big difference today is that you have the formation and the development of an intermediary world of strategic agents like financial services firms, international accounting experts, international legal experts, international organization experts, and so on.

This is an intermediary world that operates between nation states. It means that in the past, when a country entered the international system it almost inevitably engaged another nation state. Today a country can enter the international system and not engage another state, but engage J. P. Morgan, the Swiss Kreditanstalt, and so on. A very good example is when China recently entered the global capital market with a hundred-year bond issue from the Chinese government. It was sold in New York and in Hong Kong. China did not have to deal with the government of the U.S., rather, it dealt with J. P. Morgan and a few other brokerage firms.

The overwhelming influence that global firms and markets have gained in the last two years in the production, shaping, and use of electronic space, parallel with the shrinking role of governments, has created a political vacuum. However, it does not have to be a political vacuum.

Because the ascendance of digitalization is a new source of major transformations in society, we need to develop it as one of the driving forces of sustainable and equitable world development. This should be a key issue in political debates about society, particularly about equity and development. We should not let business and the market shape "development" and dominate the policy debate. The positive side of the new technology, from democratic participation to telemedicine, is not necessarily going to come as a result of market dynamics.

Further, even in the sites of concentrated power, these technologies can be destabilizing. The properties of electronic networks have created elements of a crisis of control within the institutions of the financial industry itself. There are a number of instances that illustrate this—for example—the stock market crash of 1987 brought on by programmed trading and the collapse of Barings Bank brought on by a young trader who managed to mobilize enormous amounts of capital in several markets over a period of six weeks. Electronic networks have produced conditions that may not always be controlled by those who thought to profit the most from these new electronic capacities. Existing regulatory mechanism do not always cope with the volatile nature of electronic markets. Precisely because they are deeply embedded in telematics, advanced information industries also shed light on questions of control in the global economy that not only go beyond the state but also beyond the notions of non–state centered systems of coordination prevalent in the literature of governance.

I am convinced that we need to fight for free and public content. But bandwidth is the infrastructure that is intimately linked to the formation and multiplication of public activity on the internet. Public space and free content have always required access to specific conditions, even if elementary. What looms ahead is a sharpening division between a slow moving space for those who lack the resources and a fast moving space (quick connections, enormous bandwidth) for those who can pay for it. Although it is really very different, for illustration we could say that this is a new version of an old syndrome: the public busses in poor neighborhoods are often of poorer quality than those for rich neighborhoods. It seemed, once, like these forms of inequality could not be enacted in the internet. Today it would seem that they are.

This is a particular moment in the history of electronic space, a moment when powerful corporate players and high-performance networks are strengthening the role of private electronic space and altering the structure of public electronic space. However, it is also a moment when we are seeing the emergence of a fairly broad-based—though as yet demographically isolated—civil society in electronic space. This sets the stage for contestation.

[This text is a compilation of excerpts of four texts that appeared on Nettime: "The Topoi of E-space: Global Cities and the Global Value Chains" (Oct. 28, 1996), "Interview with Andreas Broeckman" (June 12, 1997), and interviews with Geert Lovink entitled "Bandwidth and Accountability" (Hybrid Workspace, Documenta X, Kassel, July 11, 1997) and "Public Cyberspaces" (Sept. 25, 1998). Edited by Felix Stalder.]

Media are never neutral. They have biases which deeply affect the cultures that create them, and which, in turn, they create. Harold Innis described the most basic type of bias in communication media (*Empire and Communications*, Oxford: Clarendon, 1950, and *The Bias of Communication*. Toronto: University of Toronto, 1951). Hieroglyphs and stone, he observed, have a bias toward time, whereas the alphabet and paper—among other media—have a bias toward space. Cultures built on media with a time bias, such as ancient Egypt, tend to be more concerned with the organization of time and were often governed by a religious bureaucracy. Cultures using media with a space bias, for example ancient Greece, are generally more concerned with the organization of space and privilege secular, state or military, bureaucracies. The printing press joined the alphabet and paper into a new medium, the printed text, unleashing the full power of their combined space biases. This new medium provided the catalyst for phenomena such as the rapid rise of the nation-state, the unfolding of scientific rationality, and individuation. Communication media and common culture have a close interrelation in which the media provide the environment in which the social dynamics develop. This environment, however, is not just a simple container, but is a set of distinct processes that reconfigure to a varying degree everything that is carried out through them. Taken together, these processes form the bias of a medium.

To understand the kind of bias introduced into our current culture by the spread of computer networks as communication media, the best place to investigate is not the internet, but, rather, the financial networks. In contrast to the internet, where almost nothing has found a well developed form yet, the financial networks have been fully functioning for decades. Furthermore, money itself is a pure medium in the same way than light is a pure medium—as Marshall McLuhan once noted: all medium, no content. A similar observation was made by Karl Marx, who wrote in his *Grundrisse* (1857) that the circulation of money "as the most superficial (in the sense of driven out onto the surface) and the most abstract form of the entire production process is in itself quite without content." Being without content, money can have any form and still be money. It can be a coin in one's pocket or it can be an option traded back and forth between London, Tokyo, and New York. Monetary value can take on any form that is supported by the medium in which it circulates. Competitive pressures and the relentless chase for profits under the logic of postindustrial capitalism push monetary value into ever new forms, exploiting the full potential of the new media spaces. This process has consistently expanded the possibilities of the technology to tap into new opportunities for trading. The current financial markets are the most advanced and most media-specific electronic space yet created.

Financial markets have a network-based history of some twenty-five years. In 1973 Reuters started its screen service, which provided dealers with information and a shared environment to execute the trading in. In 1979 it had already connected 250,000 terminals into the increasingly global markets (P. Fallon, "The Age of Economic Reason," *Euromoney*, June 1994, 28–35). At this time the internet was still in an embryonic state with little more than 100 hosts. In an accelerating volume, huge investments have been poured into the expansion of the financial networks. The ten largest U.S. investment banks, for example, spent in 1995 alone some $17 billion on new technologies: this amounts to more than $400,000 per employee in just one year (B. Lowell and D. Farrell, *Market Unbound*, NY: Wiley, 1996, 41). Over the last two decades such massive expenditures have turned the financial markets from a relatively peripheral, supporting phenomenon into the central event of the mainstream economy. This development is driven by capitalistic competition, not the technology—there cannot be any illusions about that—but, nevertheless, the development of the financial markets is enabled and deeply affected by advanced network technologies which create three self-enforcing dynamics:

1. The automation of the financial markets made it possible to increase dramatically the volume of money and transactions. By the mid-nineties, about 500,000 people have been working worldwide in the institutions that make up the financial markets (ibid.). They have managed the circulation of more than $1500 billion per day. By far the biggest single market is the foreign currency exchange, which amounts to more than $1300 billion per day. In the early eighties, the foreign exchange transactions were ten times larger that the world trade; in the early nineties they were sixty times larger (S. Sassen, *Losing Control?* NY: Columbia University, 1996, 40). Circulating in ever-expandable networks the markets could pick up speed without material friction. As the markets have grown beyond any limitations, more money has become concentrated there. And with deeper markets, the opportunities to make money have expanded, further increasing the incentive to employ the most advanced technology.

2. Automation of the markets makes it possible to provide ever more customized services at ever lower rates, allowing for an increased participation of small investors: the middle class concerned about their pensions becoming insecure in crumbling state pension plans. Not only has the volume of transactions handled in the markets increased, but also the number of market participants and the demographic profile of those participants has changed. It shifted from highly educated professionals to the upper and middle-class segments of the general public. Information technology provided the means for putting an easy-to-use interface in front of extremely complex processes. Mutual funds and other previously exotic financial products have become advertised heavily in mass media in recent years. Access through home computers has been created.

3. Increased computerization and increased volume lead to a simultaneous integration and fragmentation of the markets. On the one hand, more and

more abstract, complex and entirely computer-based products—such as derivatives—greatly expand the number and types of tools available to brokers and their customers. On the other, the markets fragmented into a plethora of submarkets. New submarkets create new possibilities for arbitrage—that is, purchasing financial products on one market for immediate resale on another market to profit from a price discrepancy—which are based on the real-time processing of information.

Pushed to the extreme by these self-enforcing dynamics, the fully integrated financial networks offer the clearest picture of the bias of networks, a bias that affects in one way or another everything that is done through them.

RECONFIGURATION 1: CONTENT AND CONTEXT

The financial markets have become their own integral environment which not only communicates, but also produces the events communicated—the rise and the fall of prices. As such, these networks are content and context at the same time. The surrounding larger social and economic environment is structurally separated and its relevance is assessed according to whether it has to be translated into the closed universe of the financial market or not. News, for example, is evaluated primarily from the vantage point of whether it is going to influence the fever curve of the market. The importance of information is decided *within* the markets and is independent from the "value" of the information as such. The context of the market defines the content of the information. If everyone expects a company, or a country, to report huge losses, then the news of merely moderate losses boosts the price. In contrast, if everyone expects the opposite, the same piece of information can have a devastating influence on the market value of the asset.

As an integral environment, the financial networks are fully self-referential. Everything that counts happens within the networks. The single most important questions is: What are the other participants doing? Since the direct connection to other environments is broken, the ultimate determination of the (immediate) future takes place within the markets themselves. Evidently, the markets react very fast to new information and the consequences of political and economic events are almost immediate. Nevertheless, the connection is indirect. The markets as a closed system react to news because the dealers, or the artificial intelligence systems, expect each other to react and each tries to react before everyone else. It is the expectation of a reaction to an event that drives the development, not the event itself. John M. Keynes described this structure in his famous beauty contest analogy:

Professional investment may be likened to those newspaper competitions in which the competitors have to pick out the six prettiest faces from a hundred photographs, the prize being awarded to the competitor whose choice most nearly corresponds to the average preferences of the competitors as a whole; so that each competitor has to pick, not those faces he himself finds the prettiest, but those which he thinks likeliest to catch the fancy of the other competitors, all of whom are looking at the problem from the same point of view. It is not the case of choosing those which, to the best of one's judgment, are really the prettiest, not even those which average opinion genuinely thinks the prettiest. We have reached the third degree, where we devote our intelligence to anticipating what average opinion expects average opinion to be. And there are some, I believe, who practice the fourth, fifth and higher degrees. (*The General Theory of Employment, Interest and Money*, London: Macmillan, 1936, 156)

Evidently, Keynes described that tendency long before the advent of computer networks. Because it was such a perfect match of the general dynamics of financial markets and the bias of networks the technology proved to be such an explosive catalyst when they were combined in the early seventies. The merger of content and context became expressed most clearly in the infrastructure. Reuters, which started in 1849 as a pigeon carrier for sending stock exchange data from Brussels to Aachen in order to bridge the gap between the Belgian and the German telegraph lines, is today's leading provider of news to the financial markets, a service that is delivered over a proprietary network. It brings news and prices directly to customer screens, providing datafeeds to financial markets, and the software tools to analyze the data. This data covers currencies, stocks, bonds, futures, options, and other instruments. Its main customers are the world's leading financial institutions, traders, brokers, dealers, analysts, investors, and corporate treasurers. However, Reuters not only provides the news for the market, it is also the environment of the markets themselves. It provides the tools for dealers to contact counterparts through a Reuters communications network in order to do the actual tradings. Through proprietary instruments Reuters enables traders to deal from their keyboards in such markets as foreign exchange, futures, options, and securities. Consumer of news and producer of news merge and the network displays instantly to everyone what everyone else does. Reuters, in other words, produces (parts of) the news itself that are then sold back, stimulating the production of further news.

RECONFIGURATION 2: COOPERATION AND COMPETITION

The self-referentiality of the network environment creates information which has to be taken at face value. Its reality is as flat as the screen on which the data is displayed, its only relation is to other information of the same flatness, other screens to which every screen is connected. This radical decontextualization permits the increased speeding up of its circulation, which again eliminates the possibility for checking the veracity of the information. In such an environment news and rumors become equally important. Sometimes rumors become even more important than news, since they hold the promise of predicting for the insider what might be news tomorrow for everyone. What will be, accurate speculation into the future, is the most valuable information and can actually become the cause of tomorrow's news. If some of the major dealers expect a currency to lose value, they will start to sell it, which will be seen by others as a sign that the value of this currency is falling. The result is that, if many start to sell, the value of the currency is actually sinking: George Soros's *reflexivity* ("The Capitalist Threat," *Atlantic Monthly* 279.2, February 1997, 45–58). This has been staged over and over in the recurrent currency crises, be it the European in 1992–93 or the Asian in 1997.

Jean Baudrillard has put this reversal of the relationship of expectation and event, of sign and object, at the core of his thinking. "We are in the logic of simulation" he declares, "which has nothing to do with the logic of facts and the order of reasons. Simulation is characterized by *a precession of the model*, of all models around the merest fact—the models come first, and their orbital (like the bomb) circulation constitutes the genuine magnetic field of events. Facts no longer have any trajectory of their own, they arise at the intersection of the models" (*Simulations*, NY: Semiotext[e], 1983, 31–32).

Not anticipated in the gloomy metaphors of Baudrillard is the effect of that reversal in the network environment: cooperation. Since networks are tools and environment at the same time, everyone who uses the tools is dependent on the maintenance of the environment. Since the environment is closed, there can be no outside position for anyone who wants to participate. It is not incidental that the game metaphor is dominant in the financial markets. Every market player cooperates to uphold the rules, the parameters of the game, but within these limited bounds, each tries to kill the other.

Financial markets can only function efficiently at high speed when information can actually be taken at face value. To guarantee this they have to be structurally separated from other environments. Crucial for this is the institution of the clearing house. A clearing house functions as a "middleman" that acts as a seller to all buyers and as a buyer to all sellers: it is the guarantor of the ultimate fulfillment of the contract. Thus contracts can be exchanged impersonally between numerous parties on both sides without any having to worry about the others' ability or willingness to carry out their obligations. The largest private sector payments network in the world is Clearing House Interbank Payments System (CHIPS) in New York City. About 182,000 interbank transfers valued at nearly $1.2 trillion are made daily through the network. This represent about 90 percent of all interbank transfers relating to international dollar payments. A clearing house can be understood as an outsourced and institutionalized system of trust designed to cope with an anonymous and chaotic environment. It is a communal insurance institution for guaranteeing that the constant flow within the networks is not interrupted by external events, such as the default of one of the participants. Without the clearing house, such a "real life" event would be translated directly into the network. The possibility of such a direct impact would destroy the face value of the information. The clearing house, then, can be read as a buffer that prevents the direct, uncushioned impact of the external environment from breaking open the closed circuits. Without this buffer, the exchange of information would slow down considerably because the value of the information would have to be verified outside the network itself.

In the network environment, then, the condition of staying a member of the network is to provide information that can be taken at face value. The position of a player is determined by the information he, she, or it delivers to the other players, the faster and the more accurate the information is, the more relevant the source becomes. Since everyone is connected with everyone, reliable information gets delivered to the environment as such. Even in the most competitive environments this connectiveness forces a certain form of collaboration. What seems paradoxical is a characteristic of the network media: they configure communities defined by a distinction between inside and outside. The distinction is maintained by cooperation to build the communal environment, even if it is then used to stage fierce competition.

RECONFIGURATION 3: CONTROL AND UNPREDICTABILITY

A network's connectiveness is not only defined by its ability to connect people across time and space, a second characteristic is a tendency to integrate formerly independent elements on a higher level of abstraction. Abstraction allows the construction of larger areas of control, in the financial markets through instruments such as options. They are the right but not the obligation to buy or sell an underlying asset for a predetermined price in the future. This allows traders to speculate much more extensively on the movements of the markets independent from the direction of this movement. However, since options permit speculation on the movement of the asset rather than on the asset itself, these instruments become more volatile and, at the same time, the environment less predictable. There are simply too many factors to exercise real control. Increased abstraction and its possibilities to extend influence over ever greater area create a paradox of control. "When a multitude of different and competing actors" as Geoff Mulgan notes, "seek to improve their control capacities, then the result at the level of the system is a breakdown of control. What is rational at the micro level becomes highly irrational at the macro level" (*Communication and Control, Networks and the New Economies of Communication*, NY: Guilford, 1991, 29). The unpredictability is a result not of too little but too much control.

With the number of connections and the speed of communication rising, the predictability and controllability of the system as a whole is decreasing. The reconfiguration of control and unpredictability is similar to the reconfiguration of cooperation and competition: which aspect is foregrounded depends on the position of the observer. From the inside, the cooperative structure of the financial networks provides the invisible environment for deeply chaotic and intense competition. From the outside, this competition turns into a zero-sum game and the markets represent a single cooperative logic, the "commodified democracy of profit making" (Castells), executed in a tightly controlled framework dominated by a very small number of global financial giants. These fundamental differences based on an inside or outside position of the observer illustrate how closed the financial networks are and how self-referential their logic is.

In general, networks reconfigure not only aspects of control with unpredictability, cooperation with competition, and content with context, but they also connect action with reaction, event with news, into the continuity of flows. The dealers see instantly what others do, which creates the basis of their actions, which are fed back to the other dealers building their decisions upon them. This constant feedback eliminates the separation of events and news, action and reaction, before and after, and merges them into a constant presence. "The space of flows," as Manuel Castells observes, "dissolves time by disordering the sequence of events and making them simultaneous, thus installing society in an eternal ephemerality" (M. Castells, *The Rise of the Network Society*, vol. 1: *The Information Age*, Cambridge, Mass.: Blackwell, 1996, 467).

THE BIAS OF NETWORKS

Global financial markets are to computer networks what the Reformation was to the printing press: the first major social event enabled by the new technology. Financial markets have not been created by the new technology, they existed long before. However, new technologies have been the catalyst which connected heterogeneous trends into a self-enforcing dynamic. Because those trends fit the bias of the medium they could expand out of all proportion, creating new social conditions which reflect the impact of this bias in the specific historic context. Every single element of the financial markets existed independently for decades. The first clearing house, for example, was founded by the Chicago Board of Trade in 1874, but only the network conditions raised this institution to its current, central importance.

As the Reformation was not caused by the printing press, the financial markets are not the fate of the networks. The new technology has been a catalyst that has hugely augmented the impact of a series of economic and political decisions taken in the last thirty years. However, it did not simply augment the impact of these decisions, by reflecting them through their own bias the new technologies have deeply shaped outcome. The bias of networks lies in the creation of a new space–time condition of binary states of presence or absence. In the network environment everything that is the case is here and now (inside the network), and everything else in nowhere and never (outside the network). The translation from one state to the other is instantaneous and discontinuous. The experience of any sequence is introduced by the user, that is, from outside the network, and is arbitrary from the point of view of the possibilities of the network.

While this newly created space–time is the ingredient added by the technology, the result of its catalytic potential is deeply affected by the conditions under which it is brought to bear. The financial markets grew not only because the technology provided the ground for it, but also because regulatory restrictions have been removed under the increasing influence of neoliberalism. While the bias of the medium largely lies outside social influence, the quality of the culture incorporating this bias is—and has always been—shaped by society itself.

The events surrounding the Albanian pyramid schemes were more than just oddities in a poor country that had been isolated for decades. As a result of specific historical conditions, the connection between speculative capitalism, the criminal economy, and authoritarian political regimes suddenly appeared with unusual clarity. The dynamics that are normally hidden in the sophisticated and opaque language of financial markets became transparent in the simple and unglamorous Albanian context. While the specifics of the Albanian situation were unique, similar dynamics, albeit more behind closed doors, have led to collapse of the Russian financial system and fueled the ups and downs of the financial markets every day. As the most extreme case of speculative capitalism gone crazy, they are worth chronicling once again, at a time when lights are going off in the global casinos in New York, London, Tokyo, and Zurich.

Pyramid schemes all over.

THE ALBANIAN EXPERIENCE

Following the irregular elections of May 26, 1996, the situation in Albania deteriorated very quickly. Seeking political benefit, the government of the Democratic Party (DP), which illegitimately won about 90 percent of the seats in the Parliament, had allowed the rise of strange structures called "charity foundations." These structures were pyramid schemes, initially little more than money-laundering operations, offering interest rates ranging from ten to 25 percent per month. The first investors received the promised interest, paid with the money of the later investors. With the apparent success of the "foundations," the euphoria spread very quickly to all levels of Albanian society, and in a few months' time almost everybody was putting money into these get-rich-quick schemes. It is estimated that close to US$1.5 billion was invested in more than ten schemes. This in a country where the average monthly income was only some US$80. People sold their houses, property, and land to invest the proceeds in the pyramids, while economic emigrants working in neighboring countries—Greece and Italy—withdrew money from their bank accounts to transfer it to the schemes in Albania. A large number of Albanians invested their life savings and more.

The DP avoided any information about the functioning of such structures—in the beginning they simply ignored the dangers, and later they forced the governor of the Albanian National Bank to stop warning people about them. But, of course, the danger was unavoidable; the system of paying interest to early investors with the capital of later investors could only last as long as

Long before the Albanian scheme, there was a Romanian one. (Romanians had always the obsession to be the first and—accordingly—the frustration of not being acknowledged as such.) The difference was I guess in scale: Romania is less poor than Albania, with a bigger territory and therefore with less homogenous behavior at microeconomic levels. Therefore the style of the collapse was lighter, and didn't reach the traumatic dimensions of a civil war. Moreover, the pyramid had a face in the person of its charismatic promoter and director, a certain Mr. Stoica. After the collapse, he gave interviews with energetic statements about his innocence and went to jail as a martyr for the good cause of enriching the poor. I understand that he also published a volume of memoirs during his (otherwise brief) detention. Insistent rumors were circulating about the connection between the scheme and the financial empowerment of the Romanian nationalist party (PUNR) via the politically oriented bank system of the country. [Calin Dan <calin@euronet.nl>, Other Pyramid Schemes, Sun, 20 Sept 1998 11:19:13 +0100]

increasing numbers of people continued to invest. However, the schemes became so massively popular that anyone who said a word against them would appear to be opposed to the entire nation. In October 1996, when the International Monetary Fund (IMF) warned of the risks, even the opposition parties preferred to say nothing.

The connections between the leaders of the criminal economy and the leaders of the authoritarian party, the DP, were close. In some election posters in southern Albania, the names of powerful sponsors—pyramid bosses—appeared beside the names of Democratic Party candidates. Feeding back some of the money, the DP in effect bought the people's votes with the people's own money, extracted from them with the party's help through the pyramid schemes. As the opposition Social Democratic Party's leader, Skender Gjinushi, said, "The people's money was spent on buying votes."

The schemes started wobbling in autumn 1996. The continued operation of the schemes was dependent largely on confidence; once this was shaken, new investments dried up. By mid-December two of the smaller schemes had collapsed, and questions were being asked about the major schemes, in which tens of millions had been invested. Having been assured of the legitimacy of the schemes in advance by the government and the president, people's anger toward the government and the DP started to rise. With the fall of one of the important schemes based in the south of Albania, the revolt burst out and sparked the political and social crisis. On the afternoon of January 15, 1997, a battle erupted in Tirana. The first stones were thrown by angry people who had put their money into failed investment schemes. Their target was the private residence of a promoter of one of the schemes.

The government's initial response, on January 14, was a decree limiting the amount any single investor could withdraw from the schemes to $300,000 per day. This was clearly intended to prevent a run on the schemes. But its effect was to hit confidence further and to focus anger onto the government. This anger was expressed at a major demonstration in Tirana on January 19, organized by the Socialist Party and other opposition groups. The government tried to suppress it with police brutality, thus heightening tension. As the protests spread across the country, the government blamed the opposition and cracked down hard, arresting protesters and imposing severe jail sentences and fines on them.

But it was also clear that the government had to be seen to be acting against the schemes. On January 21, it announced a commission to investigate them, and seized the assets of some. Two days later, it banned pyramid schemes altogether and arrested the leaders of some major ones. At the same time, it arrested the leaders of various opposition groups, whom it blamed for inciting the trouble.

The trouble worsened thereafter, with major demonstrations on the weekend of January 25–26. Fighting was reported between protesters and police in Tirana. The cities became a battleground for demonstrators and riot police, and dozens of government buildings were burned or destroyed. The most dramatic and violent scenes were in the towns of Lushnja, Berat, and Vlora, and in the capital, Tirana, where riot police attacked opposition leaders, journalists, and protesters. But the epicenter of protest became the square in

Vlora where, at the turn of the century, Albanians had proclaimed their independence. Today, Vlora is known as the capital of the pyramid schemes, because most of them originated there.

Albania was now facing its most serious crisis since the fall of communism in 1991. The military was deployed in order to guard public buildings and keep the peace, despite doubts as to whose side they might take. It was after these protests that the government was forced to promise investors that they would get their money back. The problem was that the assets the government has seized from schemes were thought to total an estimated $300,000, while losses were around one billion dollars, about four times the amount of the country's foreign currency reserves at the time. Meanwhile, the Albanian currency, the lek, lost some 35 percent of its value on the currency black market. It quickly became clear that, even then, most investors would receive only about thirty to fifty percent of the amount they had invested, and that most of that might be in government bonds rather than cash. Worse yet, the cash would be in the fast-fading lek rather than the U.S. dollars that many of the schemes had demanded from investors.

As the situation worsened the DP declared a state of emergency. With this, they completely isolated Albania from the rest of the world. They decided to ban radio stations, close newspapers, and take over all local TV stations. Fortunately, the closure of the satellite frequencies lasted only forty-eight hours. People started to look for radio stations on the shortwave frequencies, which couldn't be banned. But the newspapers remained closed for more than one month and the office of the biggest independent newspaper, *Koha Jone*—supported by the Soros Foundation—was burned down by the secret police. During this time, email remained one of the most important sources of information, unfortunately with very little access. There was only one server in the country, UNDP, which was part of an experimental program meant to give NGOs and universities access.

Few institutions could make use of an available AOL account, which was very expensive since it required making an international call to Switzerland. It was also believed that outgoing email from the UNDP server was being monitored.

In the meantime, the West was most concerned that the Albanian trouble would spread. Since the country was not connected to international capital flows, the threat was not seen as an economic one, but as the danger of mass exodus: people following their capital into the West. The Organization on Security and Cooperation in Europe (OSCE) sent an envoy, and early elections were arranged. Italy, target of a possible mass immigration, assembled a force for Operation Alba after receiving a U.N. mandate. Various other European countries—including France, Greece, Turkey, Spain, Romania, Austria, and Denmark—participated in the contingent, which arrived in Albania in mid-April.

The parliamentary elections in late June and early July 1997 proceeded without major incident. Despite fears to the contrary, the elections were a success and ultimately led to the restoration of at least a modicum of law and order. Now, in 1998, the slow recovery process is still underway and the last schemes are being dismantled. Earlier in the year, the French auditing com-

MUKA: First of all, we cannot talk in terms of a civil war. It never took place. I am an anarchist myself, and I would never call this anarchy. The mess in Albania was caused by the leading force, the Democratic Party and its government. It was a people's protest. The element of violence we faced was of a very specific nature. There was not any violence used during the time of the protests. All the protests were held without any arms—at least on the side of the people. Of course the police were armed and fired shots in the air and sometimes into the crowd. At a certain point the government surrounded the whole city of Vlora and was intending to send the army in, but exactly at that moment, the army disobeyed and abandoned their positions. That is why we had such a mess. [Geert Lovink <geert≠ @xs4all.nl>, Interview with Edi Muka, August 1, 1997]

pany Deloitte and Touche found that the VEFA investment company had only seven million dollars in assets after having received more than three hundred million dollars from some 90,000 investors. If and how VEFA owner Vehbi Alimucaj laundered $40 million into his private bank accounts in Greece is still being investigated.

During all of this, most Albanians have waited in vain for the return of their savings. All they are left with are memories of the grand gestures paid for with their money: of how the pyramid company Gjallica blew a million dollars on a Miss Europa contest in Tirana; how VEFA paid $450,000 for an advertisement on Eurosport; how Xhaferi paid $400,000 for an Argentinian football star to run the local team in Lushnja.

SUBJECT: COOKING-POT MARKETS: AN ECONOMIC MODEL FOR THE TRADE IN FREE GOODS AND SERVICES ON THE INTERNET

FROM: RISHAB AIYER GHOSH <RISHAB@DMX.ORG>
DATE: MON, 3 AUG 1998 23:17:35 -0700

WHAT IS VALUE, OR: IS THE INTERNET REALLY AN ECONOMY?

Much of the economic activity on the net involves value but no money. Until a few years ago, there was almost no commercial activity on the internet. The free resources of the net still greatly outweigh all commercial resources. It is quite hard to put a price on the value of the internet's free resources, at least in part because they don't have prices attached. They exist in a market of implicit transactions.

THE ECONOMICS OF GOSSIP

Every snippet posted to a discussion group, every little webpage, every skim through a FAQ list and every snoop into an online chat session is an act of production or consumption, often both. There is no specific economic value inherent in a product. Value lies in the willingness of people to consume a good, and this potentially exists in anything that people can produce and pass on.

Even bad writing and even junk mail are parts, however reprehensible, of the internet's economy, but let's look at a more obvious case, Linux. After all, software, in particular large operating-system software occupying up to six CD-ROMs when distributed offline, is undeniably an economic good (for example, Red Hat Software <http://www.redhat.com/>). And Linux, with its loosely organized community of developer-users and its no-charge policy, undeniably has an economic logic that seems, at first, new.

SOMETHING FOR NOTHING?

Linus Torvalds did not release Linux source code free of charge to the world as a lark, or because he was naive, but because it was a "natural decision within the community that [he] felt [he] wanted to be a part of" (quoted from personal correspondence with Torvalds). Any economic logic of this community—the internet—must be found somewhere in that "natural decision." It is found in whatever it was that motivated Torvalds, like so many others on the net, to act as he did and produce without direct monetary payment.

Of course, it is the motivation behind people's patterns of consumption and production that forms the marrow of economics. Figuring out what motivates, let alone measuring it, is always difficult but it is even tougher when price tags don't exist. It is simpler just to assume that motivations only exist when prices are attached, and not attempt to find economic reason in actions motivated by things other than money; simpler, therefore, just to assume as we often do that the internet has no economic logic at all.

This is wrong. The best portions of our lives usually do come without price tags on them; that they're the best parts imply that they have value to us, even if they don't cost money. The pricelessness here doesn't matter much, not unless you're trying to build an economic model for love, friendship, and fresh air. On the internet, through much of its past, the bulk of its present, and the best of its foreseeable future, prices often don't matter at all. People don't seem to want to pay—or charge—for the most popular goods and services that breed on the internet. Not only is information usually free on the net, it even wants to be free, so they say.

But *free* is a tricky word: like love, information—however free in terms of hard cash—is extremely valuable. So it makes sense to assume that the three million people on the internet who publish about matters of their interest on their home pages on the web, and the several million who contribute to communities in the form of newsgroups and mailing lists, and of course anyone who ever writes free software, believe they're getting something out of it for themselves. They are clearly not getting cash; their "payment" might be the contributions from others that balance their own work, or something as intangible as the satisfaction of having their words read by millions around the world.

While writing my weekly newspaper column on the information society (*Electric Dreams* [*ED*] <http://dxm.org/dreams/>), I was distributing an e-mail version free of charge on the internet. A subscription to the e-mail column was available to anyone who asked, and a number of rather well known people began to receive the column each week. My readers often responded with useful comments; I often wondered whether people would pay for a readership like this. Having many readers adds to your reputation; they make good contacts, helping you out in various ways. Simply by reading what you write, they add value to it—an endorsement, of sorts. So who should pay whom—the reader for the work written, or the writer for the work read ("Paying Your Readers," *ED* 67)?

The notion that attention has value is not new and has been formally analyzed in the advertising industry for decades. The "attention economy" has been described in recent papers in the context of information and the internet (M. Goldhaber, "The Attention Economy," *First Monday* 2.4 <http://www.firstmonday.dk/issues/issue2_4/goldhaber/index.html>; R. A. Lanham, "The Economics of Attention" <http://sunsite.berkeley.edu/

ARL/Proceedings/124/ps2econ.html>). It would be facile to suggest that attention necessarily has innate value of its own. However, more often than not, attention is a proxy for further value. This may appear in the form of useful comments (or bug reports from Linux users), assistance, and contacts, or simply as an enhanced reputation that translates into better access to things of value at a later point.

Even those who have never studied economics have an idea of its basic principles: that prices rise with scarcity and fall in a glut, that they are settled when what consumers will pay matches what producers can charge. These principles obviously work, as can be seen in day-to-day life. But that's the "real world" of things you can drop on your toe. Will they work in a knowledge economy? After all, this is where you frequently don't really know what the "thing" is that you're buying or selling, or clearly when it is that you're doing it, or, as in the case of my column, even whether you're buying—or selling. Contrary to what many doom-sayers and hype-mongers suggest, it always seemed to me that the basic principles of economics would work in an economy of knowledge, information, and expertise. They are, after all, not only logical on the surface but also practically proven over centuries—a powerful combination. Even if the internet appears to behave strangely in how it handles value, there is no reason to believe that if it had an economic model of its own, this would contradict the economic principles that have generally worked. However, if a textbook definition of economics as the "study of how societies use scarce resources to produce valuable commodities and distribute them among different people" remains as valid now as ever, almost all the terms in there need reexamination (P. A. Samuelson and W. D. Nordhaus, *Economics*, 15th ed., NY: McGraw-Hill, 1995). This is because the same peculiar economic behavior of the net suggests that it has developed its own model, the economic model of the information age.

The *Times of India* sells some three million copies every day across India. The whole operation, particularly the coordination of advertising and editorial, depends on RespNet. This internal network won the *Times* a listing in *ComputerWorld* magazine's selection of the world's best corporate users of information technology. RespNet runs on Linux and other similar free software off the net.

Raj Mathur, who set up Linux on RespNet, agrees with Torvalds when the latter says, "people who are entirely willing to pay for the product and support find that the Linux way of doing things is often superior to 'real' commercial support." This is thanks to the large community of other developers and users who share problems and solutions and provide constant (sometimes daily) improvements to the system. The developer-users naturally include operators of networks similar to RespNet. So many of them can separately provide assistance that might not be available if they were all working together in a software company—as Linux Inc.—where they would be producers of the software but not consumers. This shifting base of tens of thousands of developers-users worldwide working on Linux means that the *Times of India* would have a tough time figuring out whom to pay, if it wanted to.

The fact that people go looking for other people on the internet, and that Linux developers look for others like them, is just one instance of the immediacy of much of the trade that takes place on the net. When you post your message to rec.pets.cats, or create a home page—whether personal or full of your hobbies and work—you are continuously involved in trade. Other cat-lovers trade your message with theirs, visitors to your homepage trade your content with their responses, or perhaps you get the satisfaction of knowing that you're popular enough to get a few thousand people discovering you each week. Even when you don't charge for what you create, you're trading it, because you're using your work to get the work of others (or the satisfaction of popularity) in a discussion group through your website. What is most important about this immediacy of the implicit trades that go on all the time on the net is its impact on notions of value. Unlike in the "real world," where things tend to have a value, as expressed in a pricetag, that is sluggish in response to change and relatively static across its individual consumers, on the net everything is undergoing constant revaluation. Without the intermediary of money, there are always two sides to every transaction, and every transaction is potentially unique, rather than being based on a value derived through numerous similar trades between others—that is, the pricetag.

As we continue to alternate between examples from the worlds of free software and usenet—to reiterate their equivalence in economic terms—we can see the two-sided nature of trade in this hypothetical example about cats. You may value the participants in rec.pets.cats enough to post a long note on the nomadic habits of your tom. In a different context—such as when the same participants are quarreling over the relative abilities of breeds to catch mice—you may not find it worthwhile contributing, because the topic bores you. And you may be far less generous in your contributions to rec.pets.dogs. You value the discussion on dogs, and catching mice, much less than a discussion on tomcats, so you're not willing to make a contribution. This would be "selling" your writing cheap; but when you get feedback on tomcats in exchange for your post, it's the right price.

Unlike noodles and bread, readers on internet newsgroups don't come with pricetags pinned on, so commonplace decisions involving your online acts of production require that you figure out the relative values of what you get and what you give, all the time. Others are figuring out the worth of your contribution all the time, too. Life on the internet is like a perpetual auction with ideas instead of money.

That note on your tomcat probably does not deserve the glorious title of idea; certainly the warm feeling that you got in exchange for posting it—when people responded positively and flocked to your homepage to see pictures of your cat—couldn't possibly be classed with "real ideas." Still, for the sake of convenience the subjects of trade on the net can be categorized as idea (goods and services) and reputation (which when enhanced brings all those warm, satisfied feelings, and more tangible benefits too).

Ideas are sold for other ideas or an enhanced reputation; reputations are enhanced among buyers of ideas, and reputations are themselves bought and sold all the time for other reputations, as we shall see later. The basic difference is that reputation (or attention) is, like money, a proxy. It is not produced or consumed in itself, but is a byproduct of the underlying production of actual goods ("ideas" in our binary terminology).

TWO SIDES TO A TRADE

Unlike the markets of the "real world," where trade is denominated in some form of money, on the net every trade of ideas and reputations is a direct, equal exchange, in forms derivative of barter. This means that not only are there two sides to every trade, as far as the transaction of exchanging one thing for another goes (which also applies to trades involving money), there are also two points of view in any exchange, two conceptions of where the value lies. (In a monetary transaction, by definition, both parties see the value as fixed by the price.)

As the poster of notes on tomcats, the value of your posting something is in throwing your note into the cooking pot of participatory discussion that is rec.pets.cats and seeing what comes out. As the author of a page on cats, what you value in exchange for your words and photographs is the visits and comments of others. On the other hand, as a participant on rec.pets.cats I value your post for its humor and what it tells me to expect when my kitten grows up; as a visitor to your webpage I learn about cats and enjoy pretty pictures.

When I buy your book about cats, it's clear that I am the consumer, you the producer. On the net, this clear black-and-white distinction disappears; any exchange can be seen as two simultaneous transactions, with interchanging roles for producer and consumer. In one transaction, you are buying feedback to your ideas about cats; in the other, I am buying those ideas. In the "real world" this would happen in a very roundabout manner, through at least two exchanges: in one, I pay for your book in cash; in the next, you send me a check for my response. This does not happen very often! (The exception is in the academic world, where neither of us would get money from the *Journal of Cat Studies* for our contributions; instead our employers would pay us to think about cats.)

As soon as you see that every message posted and every website visited is an act of trade—as is the reading or publishing of a paper in an academic journal—any pretense is lost that these acts have inherent value as economic goods with a pricetag.

In a barter exchange the value of nothing is absolute. Both parties to a barter have to provide something of value to the other; this something is not a universally or even widely accepted intermediary such as money. There can be no formal pricetags, as an evaluation must take place on the spot at the time of exchange. When you barter you are in general not likely to exchange your produce for another's in order to make a further exchange with that.

When the contribution of each side to a barter is used directly by the other, it further blurs the distinction between buyer and seller. In the "real world" barter did not, of course, take place between buyer and seller but between two producer-consumers in one transaction. When I trade my grain for your chicken, there's no buyer or seller, although one of us may be hungrier than or have different tastes from the other. On the internet, say in the Linux world, where it may seem at first that there's a clear buyer (the *Times of India*) and an equally clear, if aggregate seller (the Linux developer community), there is, in fact, little such distinction.

Just as the existence of the thousands of independent Linux developers are valuable to the newspaper because they are also users of the product—and may face similar problems—other Linux developers welcome the *Times of India* because the way it faces its problems could help them as Linux users.

CAN YOU EAT GOODWILL?

Perhaps you will agree that when you next post a note on cats, you're not giving away something for nothing. But what you get in return is often pretty intangible stuff—satisfaction, participation in discussion, and even answers to cat-related questions are all very well, and may be fair exchange for your own little notes, but don't seem substantial enough to make much of an economy. As for Linux—it's fine to talk about a large base of user-developers all helping one another, but what has all this brought Linus Torvalds? Although Linux did get vastly improved by the continuing efforts of others, none of this would have happened without Torvalds's original version, released free. Assuming that he's not interested in Linux as a hobby, he's got to make a living somehow. Doesn't he seem to have just thrown away a great product for nothing?

First, let's see what intangible "payment" Linux brought Torvalds. In the circles that might matter to Torvalds's career, he's a sort of god. As government and academic participation has declined as a proportion of the total internet developer community, most recent "free" technology has not been subsidized, either. The main thing people like Torvalds get in exchange for their work is an enhanced reputation. So there are, in fact, lots of net gods.

Net gods get hungry, though, and reputation doesn't buy pizzas. So what does Torvalds do? As it turns out, he was still in the University of Helsinki (in October 1996, when I first interviewed him; he's now with a U.S. company where "it's actually in [his] contract [to do] Linux part-time"). "Doing Linux hasn't officially been part of my job description, but that's what I've been doing," he says. His reputation helped: as Torvalds says, "in a sense I do get my pizzas paid for by Linux indirectly." Was this in an academic sense, perhaps? Is Linux, then, just another of those apparently free things that has actually been paid for by an academic institution, or by a government? Not quite. Torvalds remained in the university out of choice, not necessity. Linux has paid back, because the reputation it's earned him is a convertible commodity. "Yes, you can trade in your reputation for money," says Torvalds, " [so] I don't exactly expect to go hungry if I decide to leave the university. 'Resume: Linux' looks pretty good in many places."

IS REPUTATION A CONVERTIBLE CURRENCY?

Suppose you live in a world where people trade chicken and grain and cloth—a very basic economy indeed! Suddenly one day some strangers appear and offer to sell you a car; you want it, but "Sorry," says one of the strangers, "we don't take payment in chicken; gold, greenbacks, or plastic only." What do you do? It's not hard to figure out that you have to find some way to convert your chicken into the sort of commodities acceptable to car dealers. You have to find someone willing to give you gold for your chicken, or someone who'll give you something you can trade in yet again for gold, and so on. As long as your chicken is, directly or indirectly, convertible into gold, you can buy that car.

What holds for chicken in a primitive barter economy holds also for intangibles such as ideas and reputation in the part of the economy that operates on the internet ("Implicit Transactions Need Money You Can Give away," *ED* 70). And some of these intangibles, in the right circumstances, can certainly be converted into the sort of money that buys cars, let alone pizzas to keep hunger away. This may not apply to your reputation as a cat enthusiast, though; it may not apply to all software developers all the time, either.

On the internet—indeed in any knowledge economy—it is not necessary for everything to be immediately traded into "real world" money. If a significant part of your needs are for information products themselves, you do not need to trade in your intangible earnings from the products you create for hard cash, because you can use those intangibles to "buy" the information you want. So you don't have to worry about converting the warm feelings you get from visits to your cat webpage into dollars, because for your information needs, and your activities on the net, the "reputation capital" you make will probably do. "The cyberspace 'earnings' I get from Linux," says Torvalds, "come in the format of having a network of people that know me and trust me, and that I can depend on in return. And that kind of network of trust comes in very handy not only in cyberspace." As for converting intangible earnings from the net, he notes that "the good thing about reputations...is that you still have them even though you traded them in. Have your cake and eat it too!"

There is, here, the first glimpse of a process of give and take by which people do lots of work on their creations—which are distributed not for nothing, but in exchange for things of value. People "put it" on the internet because they realize that they "take out" from it. Although the connection between giving and taking seems tenuous at best, it is in fact crucial. Because whatever resources are on the net for you to take out, without payment, were all put in by others without payment; the net's resources that you consume were produced by others for similar reasons—in exchange for what they consumed, and so on. So the economy of the net begins to look like a vast tribal cooking pot, surging with production to match consumption, simply because everyone understands (instinctively, perhaps) that trade need not occur in single transactions of barter, and that one product can be exchanged for millions at a time. The cooking pot keeps boiling because people keep putting in things as they themselves—and others—take things out. Torvalds points out, "I get the other informational products for free regardless of whether I do Linux or not." True. But although nobody knows all the time whether your contribution is exceeded by your consumption, everyone knows that if all the contributions stopped together there'd be nothing for anyone: the fire would go out. And that wouldn't be fun at all.

COOKING-POT MARKETS

If it occurred in brickspace, my cooking-pot model would require fairly altruistic participants. A real tribal communal cooking pot works on a pretty different model, of barter and division of labor (I provide the chicken, you the goat, she the berries, together we share the spiced stew). In our hypothetical tribe, however, people put what they have in the pot with no guarantee that they're getting a fair exchange, which smacks of altruism.

But on the net, a cooking-pot market is far from altruistic, or it wouldn't work. This happens thanks to the major cause for the erosion of value on the internet—the problem of infinity ("The Problem with Infinity," *ED* 63). Because it takes as much effort to distribute one copy of an original creation as a million, and because the costs are distributed across millions of people, you never lose from putting your product in the cooking pot for free, as long as you are compensated for its creation. You are not giving away something for nothing. You are giving away a million copies of something, for at least one copy of at least one other thing. Since those millions cost you nothing, you lose nothing. Nor need there be a notional loss of potential earnings, because those million copies are not inherently valuable—the very fact of there being a million of them, and theoretically a billion or more—makes them worthless. Your effort is limited to creating one—the original—copy of your product. You are happy to receive something of value in exchange for that one creation.

What a miracle, then, that you receive not one thing of value in exchange— indeed there is no explicit act of exchange at all—but millions of unique goods made by others! Of course, you only receive "worthless" copies; but since you only need to have one copy of each original product, every one of them can have value for you. It is this asymmetry unique to the infinitely reproducing internet that makes the cooking pot a viable economic model, which it would not be in the long run in any brickspace tribal commune.

With a cooking pot made of iron, what comes out is little more than what went in—albeit processed by fire—so a limited quantity can be shared by the entire community. This usually leads either to systems of private property and explicit barter exchanges, or to the much analyzed "Tragedy of the Commons" (G. Hardin, "The Tragedy of the Commons," *Science* 162, 1243–48 <http://dieoff.org/page95.htm>).

The internet cooking pots are quite different, naturally. They take in whatever is produced, and give out their entire contents to whoever wants to consume. The digital cooking pot is obviously a vast cloning machine, dishing out not single morsels but clones of the entire pot. But seen one at a time, every potful of clones is as valuable to the consumer as were the original products that went in.

The key here is the value placed on diversity, so that multiple copies of a single product add little value—marginal utility is near zero—but single copies of multiple products are, to a single user, of immense value ("Trade Reborn Through Diversity," *ED* 65). If a sufficient number of people put in free goods, the cooking pot clones them for everyone, so that everyone gets far more value than was put in.

An explicit monetary transaction—a sale of a software product—is based on what is increasingly an economic fallacy: that each single copy of a product has marginal value. In contrast, for each distinct product, the cooking-pot market rightly allocates resources on the basis of where consumers see value to be.

A CALCULUS OF REPUTATION

A crucial component of the cooking-pot market model is reputation, the counterpoint to ideas. Just as money does not make an economy without concrete goods and services, reputation or attention cannot make an economy without valuable goods and services, which I have called "ideas," being produced, consumed, and traded).

Like money, reputation is a currency—a proxy—that greases the wheels of the economy. Monetary currency allows producers to sell to any consumer, without waiting for the right one to offer a needed product in barter exchange. Reputation encourages producers to seed the cooking pot by providing immediate gratification to those who aren't prepared to pull things out of the pot just yet, or find nothing of great interest there, and thus keeps the fire lit.

Money also provides an index of value that aids in understanding not just individual goods (or their producers), but the entire economy. Reputation, similarly, is a measure of the value placed upon certain producer-consumers—and their products—by others. The flow and interaction of reputation is a measure of the health of the entire cooking-pot economy.

Unlike money, reputation is not fixed, nor does it come in the form of single numerical values. It may not even be cardinal. Moreover, while a monetary value in the form of price is the result of matching demand and supply over time, reputation is more hazy. In the common English sense, it is equivalent to price, having come about through the combination of multiple personal attestations (the equivalent of single money transactions).

Money wouldn't be the same without technology to determine prices. Insufficient flow of the information required for evaluation, and insufficient technology to cope with the information, have always been responsible for the fact that the same things often have the same price across all markets.

The management of reputation is far too inefficient today to be a useful aspect of a working economy. Its semantics are poorly understood; moreover, it has nothing remotely akin to the technology that determines prices based on individual transactions in the monetary economy.

CONCLUSION

The common assumption that the net feels at home with free goods and vague trade because its population is averse to money, altruistic, or slightly demented is wrong. It is becoming more obviously so as floods of "normal" people arrive from the world outside, and initiate themselves into the ways of the net.

An economic model based on rational self-interest and the maximization of utility requires the identification of what is useful—sources of value—as well as a method of expressing economic interaction. In the cooking-pot market model, while scarcity creates value, value is subjective, and may therefore be found in any information at all that is distributed on the net.

The cooking-pot model provides a rational explanation (where a monetary incentive is lacking) for people's motivations to produce and trade in goods and services. It suggests that people do not only—or even largely—produce in order to improve their reputation, but as a more-than-fair payment for other goods—"ideas"—that they receive from the cooking pot. The cooking-pot market is not a barter system, as it does not require individual transactions. It is based on the assumption that on the net, you don't lose when you duplicate, so every contributor gets much more than a fair return in the form of combined contributions from others.

Reputations, unlike ideas, have no inherent value; like money, they represent things of value, as proxies. Reputations are crucial to seed the cooking pot and keep the fire lit, just as money is required to reduce the inefficiencies of pure barter markets. However, reputations require a calculus and technology for efficient working, just as money has its price-setting mechanisms today.

The cooking-pot model shows the possibility of generating immense value through the continuous interaction of people at numbing speed, with an unprecedented flexibility and aptitude toward intangible, ambiguously defined goods and services. The cooking-pot market already exists; it is an image of what the internet has already evolved into, calmly and almost surreptitiously, over the past couple of decades.

The cooking-pot model is perhaps one way to find a rationale for the workings of the internet—and on the net, it finds expression everywhere.

[Edited by Felix Stalder.]

SUBJECT: THE NEED TO GIVE: FREE SOFTWARE AND THE NETS

FROM: ED PHILLIPS <ARTLORE@SIRIUS.COM>
DATE: TUE, 29 SEP 1998 00:24:43 -0700

In late August, 1998, O'Reilly Publishing sponsored an Open Source Developer Day in downtown San Jose–emerald city as ghost town–in a hotel that conventions only partially fill. In a ballroom–conference room with a raised stage for speakers and a few hundred filled seats, the big figures in open source came together to discuss the "movement." Eric Raymond was the keynote speaker.

His talk focused on the "enterprise market" and Linux. Linux, the phenomenon, has made recent notice in the economic press, as have several other free software projects. Raymond delivered an entertaining tour through some of the more recent achievements of Linux. But it was limited to the entrance of Linux as a serious player in the corporate server and high-end markets. It's an interesting story, and one that can be measured somewhat. But the Linux phenomenon is much larger–a worldwide spread into PCs and even recycled 486s and 386s. This recycled market is of no financial significance in Silicon Valley at the moment but may prove to be of social and even economic significance globally.

There was little discussion by any of the participants of the larger social impact of free software; instead, discussions centered on business models and legal licensing issues. The calm was, however, punctuated by Richard Stallman's declaration that John Ousterhout was a "parasite" on the free software movement. Ousterhout was on the business models panel, describing his company, Scriptics's, planned support of the open source core of Tcl, the language he nursed to adolescence, and their simultaneous planned development of proprietary closed tools for Tcl as well as closed applications. During an open-mike period, Stallman said it was interesting to see IBM, a representative for which was on the panel, entering in to the free software community by supporting the Apache project while John was planning to make the fruits of the community into closed and in his view, harmful, proprietary products.

Some people clapped, others jeered. Without Stallman's provocation, the "conference" may have ended as a press conference rather than a town meeting for the free software community. Some of the more official attendees were said to be embarrassed by Stallman. Most seemed baffled by the dissension and controversy. Many of the old-timers just groaned, "Oh, there goes Stallman again." Some were worried that the hackers would be bear the brunt in the press.

A week later a vice president from a software company thinking about going open source talked to me after he got a full report about the conference. "Stallman is a Communist," he said. "He is not!" I laughed. "He's not even a Marxist." The closest Stallman ever came to talking about politics was to mention the U.S. Bill of Rights. Software developers aren't known for articulated or nuanced views of political economics; many aren't quite sure how to deal with subjects other than technical capacity or profits–let alone with the possibility that dissension and debate might be good.

Stallman's very presence makes some in the free software communities uncomfortable, like a cousin that shows up at the wrong time, is too loud, and says the things no one dares to say. Foremost amongst the traits that make the denizens of Silicon Valley uncomfortable is Stallman's contempt for the commercial. He is indeed contemptuous of it, of profit for its own sake–especially when it's at the expense of the free circulation of ideas and software. This is what many executives, hip though they may be, find so unsettling about him: expressing his views in Silicon Valley is like declaring contempt for gambling in Las Vegas. But his antics make perfect sense in the context and community of free software developers.

It strikes me as a mark of consistency and mental precision that he persists in his strict interpretation of free software. His legally technical discussions of the GNU General Public License are brilliant expositions of some call "viral" licenses–one that legally binds users to keep any modifications in the source code free and open to further modification. The GPL has been very good to Linux: the GNU project spent considerable time and money crafting a clear and legally binding document, and it has served as a haven for many a free software developer. Linus Torvalds among them was spared the need to craft a license and set a precedent for the open and distributed development of his project.

Stallman's GNU project has done incalculable good for free software. No one in the communities denies it; but his tenacity makes many of them nervous. And he doesn't make the "suits" comfortable either–nor does he want to. He doesn't carry a business card; he carries a "pleasure card," with his name and what appears to be a truncated personals ad, or a joke, "sharing good books, good food...tender embraces...unusual sense of humor." He clearly isn't looking for a job or a deal. Friends perhaps or "community," but not a deal. He's not against others making a profit from free software, though; in fact, he encourages people to make profitable businesses and make substantive contributions to free software and free documentation. Like every other "hacker" at that conference I talked to, he is a pragmatic thinker. He knows that no business would come near free software if it did not offer a successful business model for them. He's just not willing to compromise with those who try to combine open source with closed and proprietary software: if an open source project is cannibalized or "parasitized" by the development of closed products, he argues, it will hinder the free flow of ideas and computing.

John Ousterhout's plans for Tcl are just plans at the moment. He's playing with the possibility of supporting the open source development of Tcl while developing proprietary tools on top of it. He acknowledges that there will be some tension between Scriptics's investors' demand for profits and the community's need for substantive free development of Tcl. Veering too far in either direction will preclude contributions from the other: investment and connections or contributions and support.

The tension between Ousterhout and Stallman is representative of the conflicting economies and social realities the free software communities face. While investors and capitalists struggle to understand just how free software has become so successful and how they can somehow profit from it, hackers and developers are trying to maintain the integrity of free and open source computing in the face of new attention and interest.

Mainstream media interest in open source was piqued by the success of companies that serve and support the free software communities. The growing user base is spending a lot of money on support, commercially supported versions of free software products, and documentation. Commercial Linux vendors are making significant revenues; C2net's commercial, strong encryption version of Apache will earn the small company some US$15 million dollars in revenue this year; O'Reilly Publishing will earn over US$30 million dollars on documentation of free software this year. These figures are, of course, dwarfed by the figures that proprietary software companies earn. Bill Gates, the emblematic persona of commercial software, has a personal fortune that exceeds the combined wealth of the entire bottom forty percent of the United States population; and Microsoft, the synecdoche of success in the software business, is the second wealthiest company in the world behind the mammoth General Electric.

As large as Microsoft looms, it would be a mistake to credit them with spurring the development of free software. Free software has it's own trajectory and its own history; both predate Microsoft. Free software isn't a creature of necessity, it's a child of abundance–that is, of the free flow of ideas the academy and in hacker communities, amongst an elite of developers and a fringe of hobbyists and enthusiasts. These communities lie outside the bonds of business as usual and official policy. The fact that this abundance has reached a significant enough mass to support business models has much less to do with presence of clay-footed proprietary monsters than with the superior and more engaging model that free software offers users and developers. Microsoft is, as Eric Raymond says, merely the most successful example of the closed, proprietary model of software development. But it is the model in general, not Microsoft in particular, that open source and free software offer an alternative to. This alternative isn't nearly as profitable; it makes better software. Enough people have begun to recognize this to present a threat to proprietary software wherever the two models compete. For now, it's hard to imagine anything that might threaten Microsoft, except for something outside of its model.

Recently, a number of companies have embraced open source software in various ways and to varying degrees. Does this stem from a sense of abundance or is it an act of desperation? To those within the free software communities, the answer is obvious, the move to free software comes from an abundance. But, for many others, when a large commercial company decides to go open source (for example, Netscape) it's often seen as a desperate act to shore up marketshare or mindshare while frosting the competition's widgets. The rising stars of the free software communities–Cygnus, Red Hat Software, and so on–had the community before they developed a business model. It's much harder for a company to start with a business model and try to create a community–in no small part because the sense of abundance that marks free software communities is alien to company logic. Free software as both a specter and a possibility has forced companies to consider alternative business models. For example, IBM's bundling of the Apache webserver allows them to earn revenue from supporting the free product on their systems, not from creating a closed product. IBM, of course, did not open the source code for any of its own proprietary products. It sought to leverage the community and the brand name of Apache, but it will, true to the model, contribute substantively to the open source. Some of the most visible internet companies rely entirely on free software; a good example is Yahoo, which runs on FreeBSD.

Often, these companies use and sometimes even develop open source technologies; but, they stop positioning themselves as technology enterprises per se. Richard Stallman pointed out quite a few years ago that the effects of free and open source computing are more social and educational than merely technological. I believe he meant that free and open source computing shifts emphasis from technology and focuses it on what the possibilities that computing and networking open up, the development of community and the education of people. Free software projects develop devoted communities that are explicitly extra-monetary and extra-institutional. Once-obscure theories about a gift economy, first set forth in *Essai sur le Don* (1920) by the French anthropologist Marcel Mauss, have become more than merely popular metaphors: they now form some of the basic tenets of the free software movement. The extra-market and extra-institutional communities of free software are novel social forms whose nearest analogy are the "phratries" that Mauss describes: phratries are deep bonds developed with those outside of one's own family or clan; strangers become brothers through gift exchange.. A process that was fundamental to the theory of the gift economy and that is especially apt as an analogy for free software and the nets today is the potlatch, a term that describes the gift-giving ceremonies of the Northwest Coast Tribes of North America. The potlatch is a "system for the exchange of gifts," a "festival," and a very conspicuous form of public consumption. The potlatch is also the place of "being satiated": one feels rich enough to give up hoarding, to give away. A potlatch cannot take place without the sense that one is overrich. It does not emerge from an economics of scarcity.

Marshall Sahlins's *Stone Age Economics* of 1972 is, more than a study of gift economics, a critique of the economics of scarcity. Scarcity is the "judgement decreed by our economy" and the "axiom of our economics." Sahlins's and others' research has revealed that "subsistence" became a problem for humanity only with the rise of underprivileged classes within the developed markets of industrial and "postindustrial" cultures. Poverty, is as Sahlins says, an invention of civilization, of urban development. The sentence to a "life of hard labor" is an artifact of industrialism. The mere "subsistence scrabblers" of the past had—hour for hour, calorie for calorie—more "leisure" time that we can imagine: time for ceremony, time for play, time to communicate freely.

Sahlins's presentation of "the original affluent society" should not be confused with the "long boom" recently popularized by *Wired* and other organizations, the specious celebration of some kind of information or network economy that will miraculously save us from scarcity and failure. His ethnographic descriptions of communal and environmental surplus and public consumption of surplus through gift-giving are a rebuke of the failures of "progress" to deliver the goods, not a description of some information-age marvel. The gift-giving amongst an elite of programmers is an example of how collaborative and distributed projects can create wonderful results and forge strong ties within a networked economy; it certainly isn't an adequate representation of the successes of the information age as a whole. It is an ideal; given its recent achievements, however, it seems reasonable to ask what further developments free software communities might achieve. And, in asking that, we might ask where the limits of open source logic presently lie.

At the developers' conference I opened with, Stallman pointed out an important limitation: we lack good open source documentation projects for free software. This is crucial, because free software develops rapidly: it needs timely and well crafted documentation. Tim O'Reilly already copylefted a book on Linux, but didn't sell well. Perhaps it is time he tried again. The market is much bigger than it was even a few years ago. But, as O'Reilly points out, writers don't want to copyleft their books as much developers want to participate in free software projects. The authors of these books and of traditional books, for the most part, are individuals and do not work collaboratively with networked groups of writers to produce a text. Perhaps some may be inspired, as many indeed are, to experiment, as O'Reilly said he may be willing to. "Let him experiment!," Stallman intoned after the conference.

The phenomenon of free software is probably bigger than anyone of us realizes. We can't really measure it because all the ways of tracking these kind of phenomena are economic, and the "small footprint" operating systems, Linux and FreeBSD, are flowing through much more numerous and difficult to track lines, lines through which move people just like the ones the who built them. There are a few hints. In August, cdrom.com broke the record for the largest FTP download of software for a single day, surpassing the previous record which had been set by Microsoft for one of its Windows releases. All of cdrom.com's software is free and open source. Cdrom.com reports that much of the download is to points outside of the United States and the E.U.—to areas where, industry wisdom tells us, intellectual property laws aren't respected. What happens when software pirates become users who avidly, even desperately, want to learn, to receive, and even to give?

What will be the social and economic effects of free and open source computing? Do the successful collaborative free software projects prefigure other kinds of collaborative projects? Will the *hau*, the gift spirit of free software spread into other areas of social and intellectual life? I hope so. There is a connection between the explosion in the use of networked computing and the recent rise to prominence of free software. And this connection may foretell new forms of community and free collaboration on scales previously unimagined, but it certainly won't happen by itself. It will take the concerted efforts of many individual wills and the questioning of many assumptions about the success and quality of the collaborative, the open, and the freely given.

[Edited by Ted Byfield.]

SUBJECT: INTERVIEW WITH JAMES STEVENS: STRATEGIES OF INDEPENDENCE AND SURVIVAL

FROM: MATTHEW FULLER <MATT@AXIA.DEMON.CO.UK>
DATE: WED, 2 SEP 1998 21:16:51 +0100

Backspace (<http://www.backspace.org >) is a center for a wide range of digital cultures in London. It has been central to developing net radio- and network-based art in the U.K. In fact, the amount of such work available through the Backspace domain far exceeds that available through the top-heavy institutions supposedly charged with developing this work. Why this might be, and how Backspace sits in relationship to different forms of circulation of material, mutual aid and cash is the focus of this interview with one of the founders, James Stevens.

MF> People who are new to the space never seem quite sure if Backspace is a squat, lounge area for multimedia industry casualties, gallery, cybercafe or private club. It's probably all of these except the first. How was it imagined when the place first opened—and how does it run now?

JS> To start with there was a loose group who met in London between summer '94 and '95, made up of those interested in the rise of the internet, networking and tech art. During this time Heath Bunting and I met on several occasions and talked about access/workshop spaces, "cybercafe.org," and so on, and how to do it. Over this time I met Jon Bains and later via IUMA Kim Bull. Obsolete was an attempt at working with the web which began in summer of '95, to develop new platforms for creative work, establish a server onto which we could present our efforts and those of our mates and earn enough money to live on (for a change). This worked very well except the gush of cash from our more corporate clients became a major distraction and point of distortion.

Our open studio became temporary family home to the growing group of artists coders and writers working on Obsolete projects, many of whom slept, ate, lived and worked in the space. In addition, our widening circle of friends and interested groups visited us more and more. This expanding use began to collide with the growing client requirements to deliver work and present ourselves.

A new space was found in the wharf to accommodate somehow some of these needs and to instate our wish share an access point of presence. It was left to me to follow this through so in March '96 we opened very quietly to engage first users. We adopted a quarterly subscription system. Anyone could join, use our equipment and make noncommercial stuff to present on our servers. Each member got several hours free with the subscription (£10) then paid £4 an hour therapeutic. This failed to raise enough supporting cash but did present an alternative to the mainstream cybercafe commerce. This loose arrangement continued until March of '97 when it was clear Obsolete should cease and Backspace would have to fend for itself.

In the first year over four hundred people took email addresses and used the space, we held website launches, group meetings, film screenings, events, and miniconferences. Some users held their own training sessions and, of course, there were many boozy late nights.

From April '97 Backspace has moved most of the way over into self sufficiency and the 80 or so subscribers each month cover the very basic costs. We have made adjustments to the fee to bring it closer to the line and it has settled at £20 per month. We now have six or seven people hosting two four-hour sessions a month each in exchange for reasonable expenses (£10). For this they must look after the space and support subscription and help maintain, contribute and develop at whatever level they can. We are closed on Monday to allow for repair, relaxation and reflection, though it is very often as busy as the week.

MF> Describe Backspace. It maintains quite an unusual presence in the area of London that it is in, a smallish tech-cluttered room hugging close to the river in an area that has been increasingly dominated by business, and also internally—it certainly doesn't fit the archetypal layout of a cybercafe. Inside the building, how do all the elements (computers, kettle, music, seats, people) work together? Does it fit into any real or imaginary network of related spaces?

JS> Being on the river here has an effect on everyone in the building not just in backspace, and that euphoria permeates all the interaction that occurs. Certainly, part of any great environment is the sense of space that is extruded in its presentation and use. We have always tried to make the best of the qualities of the room, acknowledging its inadequacies and building on a relationship with the location, history, future, and so on.

The question of business encroachment has become part of the mantra for me of late. I just have to keep reinstating my commitment to resistance of commercial or cultural co-option and out of the fug at Obsolete it seems more and more appropriate I do this. We are sidestepping the interruption of corporate concerns—I will not now work on anything other than sufficiency enriching projects (that is, no Levis or National Gallery, no British Nuclear Fuels or whatever their name is now...). We are not participating in the Lottery scrummage for contrivance and ineffective capitalization, rather edging into the areas around us and finding the energy we need to prevail. That is not to say we will not take support cash when it is appropriate; we have received two modest payments from the Arts Council for specifically short project periods.

Individuals who subscribe have found to their delight that an application for funding to any of the public funding bodies receives serious attention and is considered a reasonable prospect for award when associated with the space. When possible we will support these projects as equally as we support any other initiated from within the membership. There is little pretension to celebrity from within the group and this is refreshed/refocused by the flow of enthusiasm, contribution and contact we have with those who come and use the space. These characteristics are reflected in the platform for presentation at bak.spc.org and associated sites, it is a churning wash of ideas experiments and effluent, a nonhierarchical representation of the collective state of mind.

The use of the space is a meandering and confounding collision of the inarticulate, lucid and languid to the strains of rap and riverwash and no sooner have we settled the arrangement of the facilities and utilities around the room then we are upturned and overdriven. I love it.

MF> In terms of funding, Backspace itself occupies an interesting position. Can you describe your attitude to state funding and corporate sponsorship?

JS> All these models hug a formula for creativity and work practice that reinforces dependency. Whilst any genuine declaration and provision of cash in support of noncommercial product (that is, not a commercial) can be applauded, however it at this point the inevitable distortion occurs, the mediation, whatever...

I am now more adamant than ever that backspace exist free of any dependencies on public or corporate funding and that it flowers or fails on its own abilities. We are not employers, teachers or fundamentalists nor are we a web design agency or recording studio, we are not experts, we are chaotic and persistent, slacktivist.

There have been many opportunities over the last year for me to get very involved with Arts Council funding in particular. I have spent time talking with funding administrators to see if there is an economic way of dealing with them. Again and again I run into fundamental problems of perception and projection. On the face of it I think we satisfy most criteria and are in an attractive proposition for them to associate with, yet I cannot bring myself to sort it all out with them. Maybe I need help...or to just look outward and pass them.

So far the absence of a fund has not prevented project work from proceeding. If you build and present with components of an appropriate scale then bankrolling and other control issues recede to the background where they belong. I am always looking to ways of consolidating the flow of supporting cash and to this end have recently extended subscription to include ISP for an extra £5. I still get confronted by those who insist all this should be free and are offended by our model of openness and despair at our noncompliance.

There is no map or set of instructions that can be extracted and replicated. Each situation responds best to a custom set of attunements.

There is still the option of disappearance and the art of regrouping and reappearance. If things get boring, lose their magic, get stuck, it is simply time to move on, close certain operations and perhaps transform them, turn them into something new, something yet unknown. This is an old trick, an old wisdom if you wish. It has little to do with a weak will—remember that infrastructures are not that easy to rebuild. Years of work may be demolished within weeks. Social and human structures can be dissolved that are hard to replace, or to repair. Organizations are collective memories and one must have a very good reason to destroy one. Most of all, one must possess the energy to create something new, otherwise one will stand there with empty hands, facing a long path of melancholy ahead. [Geert Lovink <geert@xs4all.nl>, Strategies for Sustainable Autonomous Cyberspaces, September 1998]

THE LEGACY OF THE NEW LEFT

The net is haunted by the disappointed hopes of the sixties. Because this new technology symbolizes another period of rapid change, many contemporary commentators look back to the stalled revolution of thirty years ago to explain what is happening now. Most famously, the editors of *Wired* continually pay homage to the New Left values of individual freedom and cultural dissent in their coverage of the net. However, in their Californian ideology, these ideals of their youth are now going to be realized through technological determinism and free markets. The politics of ecstasy have been replaced by the economics of greed.

Ironically, the New Left emerged in response to the "sellout" of an earlier generation. By the end of the fifties, the heroes of the antifascist struggle had become the guardians of Cold War orthodoxies. Even within the arts, avantgarde experimentation had been transformed into fashionable styles of consumer society. The adoption of innovative styles and new techniques was no longer subversive. Frustrated with the recuperation of their parents' generation, young people started looking for new methods of cultural and social activism. Above all, the Situationists proclaimed that the epoch of the political vanguard and the artistic avant-garde had passed. Instead of following the intellectual elite, everyone should instead determine their own destinies.

"The situation is...made to be lived by its constructors. The role played by a passive...'public' must constantly diminish, while that played by those who cannot be called actors but rather... 'livers' must steadily increase." —G. Debord, "Report on the Construction of Situations and on the International Situationist Tendency's Conditions of Organisation and Action"

These New Left activists wanted to create opportunities for everyone to express their own hopes, dreams, and desires. The Hegelian "grand narrative" would culminate in the supersession of all mediations separating people from each other. Yet, despite their Hegelian modernism, the Situationists believed that the utopian future had been prefigured in the tribal past. For example, tribes in Polynesia organized themselves around the potlatch: the circulation of gifts. Within these societies, this gift economy bound people together into tribes and encouraged cooperation between different tribes. In contrast with the atomization and alienation of bourgeois society, potlatches required intimate contacts and emotional authenticity. According to the Situationists, the tribal gift economy demonstrated that individuals could successfully live together without needing either the state or the market. After the New Left revolution, people would recreate this idyllic condition: *anarcho-communism*. However, the Situationists could not escape from the elitist tradition of the avant-garde. Despite their invocation of Hegel and Marx, the Situationists remained haunted by Nietzsche and Lenin. As in earlier generations, the rhetoric of mass participation simultaneously justified the leadership of the intellectual elite. Anarcho-communism was therefore transformed into the "mark of distinction" for the New Left vanguard. As a consequence, the giving of gifts was seen as the absolute antithesis of market competition. There could be no compromise between tribal authenticity and bourgeois alienation. After the social revolution, the potlatch would completely supplant the commodity. In the two decades following the May '68 revolution, this purist vision of anarcho-communism inspired community media activists. For instance, the radical "free radio" stations created by New Left militants in France and Italy refused all funding from state and commercial sources. Instead, these projects tried to survive on donations of time and money from their supporters. Emancipatory media supposedly could only be produced within the gift economy. During the late seventies, pro-situ attitudes were further popularized by the punk movement. Although rapidly commercialized, this subculture did encourage its members to form their own bands, make their own fashions, and publish their own fanzines. This participatory ethic still shapes innovatory music and radical politics today. From raves to environmental protests, the spirit of May '68 lives on within the DIY—do it yourself—culture of the nineties. The gift is supposedly about to replace the commodity.

THE NET AS REALLY EXISTING ANARCHO-COMMUNISM

Despite originally being invented for the U.S. military, the net was constructed around the gift economy. The Pentagon initially did try to restrict the unofficial uses of its computer network. However, it soon became obvious that the net could only be successfully developed by letting its users build the system for themselves. Within the scientific community, the gift economy has long been the primary method of socializing labor. Funded by the state or by donations, scientists don't have to turn their intellectual work directly into marketable commodities. Instead, research results are publicized by "giving a paper" at specialist conferences and by "contributing an article" to professional journals. The collaboration of many different academics is made possible through the free distribution of information.

Within small tribal societies, the circulation of gifts established close personal bonds between people. In contrast, the academic gift economy is used by intellectuals who are spread across the world. Despite the anonymity of the modern version of the gift economy, academics acquire intellectual respect from each other through citations in articles and other forms of public acknowledgment. Scientists therefore can only obtain personal recognition for their individual efforts by openly collaborating with each other through the academic gift economy. Although research is being increasingly commercialized, the giving away of findings remains the most efficient method of solving common problems within a particular scientific discipline.

From its earliest days, the free exchange of information has therefore been firmly embedded within the technologies and social mores of cyberspace. When New Left militants proclaimed that "information wants to be free" back in the sixties, they were preaching to computer scientists who were already living within the academic gift economy. Above all, the founders of the net never bothered to protect intellectual property within computer-mediated communications. On the contrary, they were developing these new technologies to advance their careers inside the academic gift economy. Far from wanting to enforce copyright, the pioneers of the net tried to eliminate all barriers to the distribution of scientific research. Technically, every act within cyberspace involves copying material from one computer to another. Once the first copy of a piece of information is placed on the net, the cost of making each extra copy is almost zero. The architecture of the system presupposes that multiple copies of documents can easily be cached around the network. As Tim Berners-Lee—the inventor of the web—points out: "Concepts of intellectual property, central to our culture, are not expressed in a way which maps onto the abstract information space. In an information space, we can consider the authorship of materials, and their perception; but...there is a need for the underlying infrastructure to be able to make copies simply for reasons of [technical] efficiency and reliability. The concept of 'copyright' as expressed in terms of copies made makes little sense" ("The World Wide Web: Past, Present and Future").

Within the commercial creative industries, advances in digital reproduction are feared for making the "piracy" of copyright material ever easier. For the owners of intellectual property, the net can only make the situation worse. In contrast, the academic gift economy welcomes technologies that improve the availability of data. Users should always be able to obtain and manipulate information with the minimum of impediments. The design of the net therefore assumes that intellectual property is technically and socially obsolete.

In France, the nationalized telephone monopoly has accustomed people to paying for the online services provided by Minitel. In contrast, the net remains predominantly a gift economy even though the system has expanded far beyond the university. From scientists through hobbyists to the general public, the charmed circle of users was slowly built up through the adhesion of many localized networks to an agreed set of protocols. Crucially, the common standards of the net include social conventions as well as technical rules. The giving and receiving of information without payment is almost never questioned. Although the circulation of gifts doesn't necessarily create emotional obligations between individuals, people are still willing to donate their information to everyone else on the net. Even selfish reasons encourage people to become anarcho-communists within cyberspace. By adding their own presence, every user contributes to the collective knowledge accessible to those already online. In return, each individual has potential access to all the information made available by others within the net. Everyone takes far more out of the net than they can ever give away as an individual.

[T]he net is far from altruistic, or it wouldn't work... Because it takes as much effort to distribute one copy of an original creation as a million...you never lose from letting your product free...as long as you are compensated in return... What a miracle, then, that you receive not one thing in value in exchange—indeed there is no explicit act of exchange at all—but millions of unique goods made by others!" —Rishab Aiyer Ghosh, "Cooking-pot Markets"

Despite the commercialization of cyberspace, the self-interest of net users ensures that the high-tech gift economy continues to flourish. For instance, musicians are using the net for the digital distribution of their recordings to each other. By giving away their own work to this network community, individuals get free access to a far larger amount of music in return. Not surprisingly, the music business is worried about the increased opportunities for the "piracy" of copyrighted recordings over the net. Sampling, DJing, and mixing are already blurring property rights within dance music. However, the greatest threat to the commercial music corporations comes from the flexibility and spontaneity of the high-tech gift economy. After it is completed, a new track can quickly be made freely available to a global audience. If someone likes the tune, they can download it for personal listening, use it as a sample, or make their own remix. Out of the free circulation of information, musicians can form friendships, work together, and inspire each other.

"It's all about doing it for yourself. Better than punk." —Steve Elliot

Within the developed world, most politicians and corporate leaders believe that the future of capitalism lies in the commodification of information. Over the last few decades, intellectual property rights have been steadily tightened through new national laws and international agreements. Even human genetic material can now be patented. Yet, at the "cutting edge" of the emerging information society, money-commodity relations play a secondary role to those created by a really existing form of anarcho-communism. For most of its users, the net is somewhere to work, play, love, learn, and discuss with other people. Unrestricted by physical distance, they collaborate with each other without the direct mediation of money or politics. Unconcerned about copyright, they give and receive information without thought of payment. In the absence of states or markets to mediate social bonds, network communities are instead formed through the mutual obligations created by gifts of time and ideas.

"This informal, unwritten social contract is supported by a blend of strong-tie and weak-tie relationships among people who have a mixture of motives and ephemeral affiliations. It requires one to give something, and enables one to receive something. ...I find that the help I receive far outweighs the energy I expend helping others; a marriage of altruism and self-interest." —Howard Rheingold, *The Virtual Community*

On the net, enforcing copyright payments represents the imposition of scarcity on a technical system designed to maximize the dissemination of information. The protection of intellectual property stops all users from having access to every source of knowledge. Commercial secrecy prevents people from helping each other to solve common problems. The inflexibility of information commodities inhibits the efficient manipulation of digital data. In contrast, the technical and social structure of the net has been developed to encourage open cooperation among its participants. As an everyday activity, users are building the system together. Engaged in "interactive creativity," they send emails, take part in listservers, contribute to newsgroups, participate in online conferences, and produce websites (T. Berners-Lee, "Realising the Full Potential of the Web" <http://www.w3.org//1998/02/Potential.html>). Lacking copyright protection, information can be freely adapted to suit the users' needs. Within the high-tech gift economy, people successfully work together through "an open social process involving evaluation, comparison, and collaboration" (B. Lang, "Free Software For All," *Le Monde Diplomatique*, January 1998 <http://www.monde-diplomatique.fr/md/en/1998/01/12freesoft.html>).

The high-tech gift economy is even at the forefront of software development. For instance, Bill Gates admits that Microsoft's biggest competitor in the provision of webservers comes from the Apache program (K. W. Porterfield, "Information Wants to be Valuable" <http://www.netaction.org/articles/freesoft.html>). Instead of being marketed by a commercial company, this program is distributed for free. Like similar projects, this virtual machine is continually being developed by its techie users. Because its source code is protected though not frozen by copyright (under the GNU Public License), the program can be modified, amended, and improved by anyone with the appropriate programming skills. When someone does make a contribution to a free or "open source" project, the gift of their labor is rewarded by recognition within the community of user-developers.

The inflexibility of commodified software programs is compounded by their greater unreliability. Even Microsoft can't mobilize the amount of labor given to some successful shareware programs by their devotees. Without enough techies looking at a program, all its bugs can never be found (A. Leonard, "Let My Software Go!" <http://www.salonmagazine.com/21st/feature/1998/04/cov_14feature.html>). The greater social and technical efficiency of anarcho-communism is therefore inhibiting the commercial takeover of the net. Shareware programs are now beginning to threaten the core product of the Microsoft empire: the Windows operating system. Starting from the original software program by Linus Torvalds, a community of user-developers is together building their own nonproprietary operating system: Linux. For the first time, Windows has a serious competitor. Anarcho-communism is now the only alternative to the dominance of monopoly capitalism.

Linux is subversive. Who could have thought even five years ago that a world-class operating system could coalesce as if by magic out of part-time hacking by several thousand developers scattered all over the planet, connected only by the tenuous strands of the Internet? —Eric S. Raymond, "The Cathedral and the Bazaar"

THE "NEW ECONOMY" IS A MIXED ECONOMY

Following the implosion of the Soviet Union, almost nobody still believes in the inevitable victory of communism. On the contrary, large numbers of people accept that the Hegelian "end of history" has culminated in American neoliberal capitalism. Yet, at exactly this moment in time, a really existing form of anarcho-communism is being constructed within the net, especially by people living in the U.S. When they go online, almost everyone spends most of their time participating within the gift economy rather than engaging in market competition. Because users receive much more information than they can ever give away, there is no popular clamor for imposing the equal exchange of the marketplace on the net. Once again, the "end of history" for capitalism appears to be communism.

For the high-tech gift economy was not an immanent possibility in every age. On the contrary, the market and the state could only be surpassed in this specific sector at this particular historical moment. Crucially, people need sophisticated media, computing, and telecommunications technologies to participate within the high-tech gift economy. A manually operated press produced copies that were relatively expensive, limited in numbers and impossible to alter without recopying. After generations of technological improvements, the same quantity of text on the net costs almost nothing to circulate, can be copied as needed, and can be remixed at will. In addition, individuals need both time and money to participate within the high-tech gift economy. While a large number of the world's population still lives in poverty, people within the industrialized countries have steadily reduced their hours of employment and increased their wealth over a long period of social struggles and economic reorganizations. By working for money during some of the week, people can now enjoy the delights of giving gifts at other times. Only at this particular historical moment have the technical and social conditions of the metropolitan countries developed sufficiently for the emergence of digital anarcho-communism.

"Capital thus works towards its own dissolution as the form dominating production." —Karl Marx, *Grundrisse*

The New Left anticipated the emergence of the high-tech gift economy. People could collaborate with each other without needing either markets or states. However, the New Left had a purist vision of DIY culture: the gift was the absolute antithesis of the commodity. Yet, anarcho-communism only exists in a compromised form on the net. Contrary to the ethical-aesthetic vision of the New Left, money-commodity and gift relations are not just in conflict with each other, but also coexist in symbiosis. On the one hand, each method of working does threaten to supplant the other. The high-tech gift economy heralds the end of private property in "cutting edge" areas of the economy. The digital capitalists want to privatize the shareware programs and enclose the social spaces built through voluntary effort. The potlatch and the commodity remain irreconcilable.

Yet, on the other hand, the gift economy and the commercial sector can only expand through mutual collaboration within cyberspace. The free circulation of information between users relies upon the capitalist production of computers, software, and telecommunications. The profits of commercial net companies depend upon increasing numbers of people participating within the high-tech gift economy. For instance, from its foundation Netscape has tried to realize the opportunities opened up by such interdependence. Under threat from the Microsoft monopoly, the company had to ally itself with the hacker community to avoid being overwhelmed. It started by distributing its web browser as a gift. Today the source code of this program is freely available and the development of products for Linux has become a top priority. The commercial survival of Netscape depends upon successfully collaborating with hackers from the high-tech gift economy. Anarcho-communism is now sponsored by corporate capital—for example, as when Netscape released the source code to its browser.

"'Hi there Mr CEO [Chief Executive Officer]—tell me, do you have any strategic problem right now that is bigger than whether Microsoft is going to either crush you or own your soul in a few years? No? You don't? OK, well, listen carefully then. You cannot survive against Bill Gates [by] playing Bill Gates' game. To thrive, or even survive, you're going to have to change the rules...'" —Eric S. Raymond

The purity of the digital DIY culture is also compromised by the political system. The state isn't just the potential censor and regulator of the net. At the same time, the public sector provides essential support for the high-tech gift economy. In the past, the founders of the net never bothered to incorporate intellectual property within the system because their wages were funded from taxation. In the future, governments will have to impose universal service provisions on commercial telecommunications companies if all sections of society are to have the opportunity to circulate free information. Furthermore, when access is available, many people use the net for political purposes, including lobbying their political representatives. Within the digital mixed economy, anarcho-communism is also symbiotic with the state.

This miscegenation occurs almost everywhere within cyberspace. For instance, an online conference site can be constructed as a labor of love, but still be partially funded by advertising and public money. Crucially, this hybridization of working methods is not confined within particular projects. When they're online, people constantly pass from one form of social activity to another. For instance, in one session, a net user could first purchase some clothes from an e-commerce catalogue, then look for information about education services from the local council's site, and then contribute some thoughts to an ongoing discussion on a listserver for fiction writers. Without even consciously having to think about it, this person would have successively been a consumer in a market, a citizen of a state, and an anarcho-communist within a gift economy. Far from realizing theory in its full purity, working methods on the net are inevitably compromised. The "New Economy" is, in the lexicon of *Wired* and its ilk, an advanced form of social democracy (see K. Kelly, "New Rules for the New Economy," *Wired*, September 1997).

At the end of the twentieth century, anarcho-communism is no longer confined to avant-garde intellectuals. What was once revolutionary has now become banal. As net access grows, more and more ordinary people are circulating free information across the net. Crucially, their potlatches are not attempts to regain a lost emotional authenticity. Far from having any belief in the revolutionary ideals of May '68, the overwhelming majority of people participate within the high-tech gift economy for entirely pragmatic reasons. Sometimes they buy commodities online and access state-funded services. However, they usually prefer to circulate gifts amongst each other. Net users will always obtain much more than will ever be contributed in return. By giving away something which is well made, they will gain recognition from those who download their work. For most people, the gift economy is simply the best method of collaborating together in cyberspace. Within the mixed economy of the net, anarcho-communism has become an everyday reality.

"We must rediscover the pleasure of giving: giving because you have so much. What beautiful and priceless potlatches the affluent society will see—whether it likes it or not!—when the exuberance of the younger generation discovers the pure gift." —Raoul Vaneigem, *The Revolution of Everyday Life*

[This article is a remixed extract from *The Holy Fools: A Critique of the Avant-garde in the Age of the Net* (London: Verso, forthcoming).]

SUBJECT: ADA'WEB

FROM: FELIX STALDER <STALDER@FIS.UTORONTO.CA>
DATE: TUE, 20 OCT 1998 22:30:51 + 0100

From: "Armin Medosch" <armin@mail.easynet.co.uk>
Date: Tue, 3 Mar 1998 10:47:04 +0000
Subject: Leading Art Site Suspended

From www.nytimes.com:
Leading Art Site Suspended
By Matthew Mirapaul

The Ada'web Web site, one of the most dynamic destinations for original Web-based art, is being suspended.
Benjamin Weil, the co-founder of Ada'web, announced on Monday in an e-mail message that Digital City Inc., the site's publisher, had canceled its financing and that Ada'web would cease producing new artistic content. Weil is now seeking a permanent home for its archives so that its material can remain accessible.
Since it was conceived in late 1994, Ada'web has become one of the premier destinations for online creativity. Ultimately, it presented about 15 web-specific projects by such high-profile contributors as the conceptual artist Lawrence Weiner. The site's first offering, launched officially in May 1995, was Jenny Holzer's "Please Change Beliefs."

Date: Tue, 3 Mar 1998 16:08:54 -0500
From: mf@mediafilter.org (MediaFilter)
Subject: Re: Leading Art Site Suspended

Guess it takes a cruel dose of reality before people get a clue that autonomy is necessity, corporate sponsorship is ultimately censorship, and subsidies from the government are short lived at best.

Don't be surprised! There is no free lunch. Everything has its price.

Paul Garrin

Date: Wed, 4 Mar 1998 11:45:56 -0500
From: beweil@adaweb.com (Benjamin Weil)
Subject: Re: Leading Art Site Suspended

This kind of commentary astounds me in that it demonstrates a remarkably simplistic approach to the economy of the arts and culture in general. It reminds me of those people who keep on saying that artists have to starve in order to produce good work. It is at best romantic, at worst idiotic.

Art has *always* been supported by wealth, may it be individual patrons, corporations or the state (in modern times). There is no doubt that there is a price to pay, that there is no "free lunch." Nobody—except maybe romantics or idiots—ever assumed that receiving funding from any corpus was "free of charge." Old masters, as we refer to them, had to service the greed and power of individuals or families, and it did not prevent them from being "free." Their freedom was defined by the constraints they had to accept in order to make their work. The notion of the artist having "no obligation" to anyone except to her/his art is something that only pushes this area of culture in a very marginal position. Any transaction implies the agreement between both parties that there is something in it for each. The fact Digital City, Inc. has decided to stop supporting Ada'web only proves that this corporate entity does not see its interest in supporting such venture any longer. But being able to state that "corporate sponsorship is ultimately censorship" basically ignores the nature of *any* transaction.

Public space on the net will only disappear if we decide so. Just like the notion of public space in the city disappears if it is not occupied. It is a decision, not an occurrence.

More constructive and interesting as a departure point is the nature of the relationship between art and its potential sponsors, so as to eventually come up with means to convince the holders of wealth that they have an interest in supporting activities that are not "profitable" in a purely capitalistic understanding of the term. So far, most of that support was informed by a valuation of culture that relied upon the notion of prestige, or status. There must be other ways, more creative ones, to approach the possibility of establishing satisfactory relationships with corporate patrons. However, this kind of thinking can only be discussed with the postulate that the corporate world is no worse than the state, who in turn is no worse than the private individual. Again, the nature of such a relationship *cannot* be envisioned outside of the notion of mutual interest.

On a final note, I also have to say that the whole notion of a disinterested state that is so much better than the corporate world, in that it supposedly does not have any agenda is again one of the most worn out and preposterous statements that can be made at this point. Wake up and smell the coffee: it's the nineties, not the sixties!

Date: Wed, 4 Mar 1998 19:36:57 -0400
From: murph the surf <murph@interport.net>
Subject: Re: Leading Art Site Suspended

In the long run I don't know if Ada'web would have found a place within Digital City because it would have taken time to figure out how to do it with concessions made on both sides. Meaning and value in art accrue over time and I think the kind of continuity required for art can benefit a business that is constantly responding to the market flux. It takes insightful leadership to understand and implement this effectively, something AOL doesn't seem to have much of, or need to be successful.

Since we started in 1993 as a BBS, Artnetweb has evolved into a network of people, projects and things without anything resembling a business plan and it would be ridiculous for us to think we would fit into a corporate structure without a corporate sensibility. Our network exists as it is used and when the network stops being used it will no longer exist.

As an organization we receive no grants or other institutional support. We keep ourselves alive by teaching classes, by doing freelance web design and upkeep plus whatever else comes along with a paycheck. We also work on VRML projects for various exhibitions and exhibition sites.

This situation isn't what we planned in the beginning because we had no idea what the future would be, and it certainly isn't perfect. We've changed and adapted; obviously no great patron is waiting to take us under their protective wing, yet we have discovered some possibilities for working with corporations and others that may prove beneficial for everyone involved. Sounds a lot like real life.

Robbin Murphy

Date: Sat, 07 Mar 1998 14:30:52 -0500
From: Stephen Pusey <scp@plexus.org>
Subject: Funding Digital Culture

I'm both intrigued and irritated by this Ada'web saga. Intrigued because it highlights a need for discussion about funding online arts entities and the pros and cons of their formulas for survival. Irritated, because of the fuss concerning Ada'web's decision simply to stop just because their one source of monetary nourishment terminated—to quote Benjamin Weil "...they said 'We don't have any more money to fund this,' and then it was our decision, more or less, to stop. You know, how could we do it without money?" Obviously sucking on that one corporate teat for the last three years produced a mindset that cannot tolerate an existence without its regular *dolce latte*.

At the end of '94 and beginning of '95 a number of arts websites appeared among them The Thing, PLEXUS, artnetweb, Ada'web, and others. The principals of these organizations had prior acquaintance from dialogue on pre-web dial-up BBSes like The Thing. There was, however, a fundamental difference between Ada'web and the rest. They were a wholly owned part of a parent corporation—one of the cherries on the cake of John Borthwick's start-up, WPStudios, an ambitious conglomerate of online publications. The rest of us were "independents" that had little or no corporate or state funding, and therefore had to constantly devise new ways of paying the bills and keeping the marshals from closing our offices, while at the same time building online environments to promote discourse and digital culture. I am not declaring financial poverty to be a virtue here, just that hardship has been a factor that has necessitated a diverse approach to survival, albeit a slower and perhaps erratic development.

Ada'web enjoyed three good years supplied with office, equipment, and wages, which has enabled them to concentrate single-mindedly on producing and promoting a beautiful and extraordinary arts environment. Weil and his crew surely must have suspected from the outset that this would be a short-term venture. Borthwick is a pragmatist who knows that pigs get slaughtered in the market. He put together an attractive hip package and sold it before he lost his investment. Inevitably, AOL's Digital City got out their calculators and realized that some pieces of what they bought were not going to spin a penny and so ditched Total New York, Spanker and Ada'web: a predictable outcome.

My purpose here is not to put the boot in when the man is down; Ada'web has made an important contribution and I sincerely hope that Benjamin Weil finds a new way of continuing its mission. There are, however, lessons we can draw from their dilemma. Obviously, the first is to avoid corporate ownership, unless you control the corporation. In seeking corporate sponsorship, success lies in identifying to the donor the ways in which your purpose and their strategy are mutually aligned. This may cause you, especially if the potential financial rewards are really high, to reform your philosophy to match theirs. The same is also true of state sponsors, who may be tempered by political pressures that prohibit them from sponsoring certain kinds of expression, like sexually explicit material. Finding the right sponsorship, indeed any sponsorship, can be a full-time activity. If an organization wants to avoid compromising its charter it has to draw from a broad portfolio of funders. The other solution is to evolve a business model that supports the organization's agenda without outside interference. I assume The Thing does this with some modicum of success, by using the profits from its ISP. Another option that could prove effective in the long term is collective action. Perhaps an organization like the Foundation for Digital Culture (<http://digicult.org/>), reformed with an international constituency, could be an organ through which we collectively lobby and inform government and corporate funders to support progressive digital culture?

Date: Sun, 8 Mar 1998 14:52:19 -0500
From: t byfield <tbyfield@panix.com>
Subject: Re: Funding Digital Culture

At the bottom of these questions and condemnations is the presumption—rather arrogant, I would say—that folding shop is somehow a failure to fulfill some solemn obligation. This seems strange: as though the nominal institution had somehow subsumed the potential of the people it was made of. That this kind of creeping institutionalism would appear in Nettime, of all places, seems especially curious. Just "where" is Nettime? At Desk? At the Thing? In Ljubljana? In Berlin? In London? In Budapest? This distribution—as much between *people* as between sites—is both Nettime's strength and its weakness. In the wake of Ljubljana, I heard some grumbling about disorganization, about how there were no solid resolutions, no definitive programs or advances. And I thought that this was great: it's very easy to cement social organization around programs, but harder to preserve looser bonds—loyalties, trust, a certain faith. So here we are, presented with the (to my mind rather forced) "spectacle" of Ada'web's demise, attended by great finger-wagging and I-told-you-soing and lesson-learning and whatnot. All of it privileging the institution over the individual. Now, Mr. Weil may be (or may have been) an Executive Curator, but that doesn't mean Ada'web was a MVSEVM carved in stone. To demand that of electrical signals built on a small group of people, at this stage of the game, is excessive, IMO.

Date: Sun, 08 Mar 1998 22:57:06 -0500
From: Stephen Pusey <scp@plexus.org>
Subject: Re: Funding Digital Culture

What constitutes a networked entity and where is it located? At the points of broadcast or reception? And of course, all of these names artnetweb, PLEXUS, The Thing, and so on, are but temporary and formative identities that propose indeterminate perspectives at various times in the shifting milieu of digital culture. The types of individuals that instigate these projects, are themselves a guarantee against institutionalization, of that you can be assured. Furthermore, my proposal to use an association like digicult (FDC) as a focal point for lobbying of governments on behalf of digital culture, should not be interpreted as a move towards institutionalizing the process. Such an entity would have its form and policy shaped by an internetworked community of cultural practitioners and would exist only as long as they wished it to. Again, the location for such an association would be its networked community. Part of its charter could be the subversion and persuasion of funding agencies worldwide towards an awareness and support of a critical digital culture.

Date: Tue, 10 Mar 1998 17:22:12 -0500
From: beweil@adaweb.com (Benjamin Weil)
Subject: funding for the arts, etc.

Mr. Byfield's postings have encouraged me to step in for a last time, and clarify a number of points here.

(1) Part of Ada'web's founding mission was to explore possible alternatives as far as funding for art online was concerned. John Borthwick and I believed it was important to consider the landscape, and figure out a way we could derive an economic model for a type of art production which was no longer unique (no commodification possible here!) and whose only existence—so to speak—was virtual. The idea was to be able to commission works, and compensate the artists we invited to work on those projects.

(2) Looking for alternate means of support was partly informed by the difficulty experienced by colleagues who sought to get public funding for their activities, and the fact that we wanted to fully concentrate on producing those works, rather than having to find work for hire contracts. (For the prompt to fire insults, I will here state very clearly: this is *not* by any means a value judgment, but just reflecting a choice to try and do things differently). Furthermore, it was my belief that the development of the web would be an extraordinary opportunity for art to desegregate itself, and (re)gain a central position in the ambient cultural discourse and practice. Both John Borthwick, the Ada'web team and I believed that exploring the dynamics and pushing the limits of the medium with the artists we produced work with, as well as the ones we hosted the projects of, was an important thing to contribute to the net. It was one model among the many that were—and still are—being developed.

(3) Working with corporate money was assumed to be one way of dealing with the absence of public funding. However, rather than knocking at the corporate door asking for "charity" money, we thought we could convince them that art could be a valuable asset, as artists have always been cultural forerunners, and that in that sense, it could be understood as a form of cre-

ative research which could make them understand better the medium they were investing in, and draw attention to their corporation as being innovative.

To conclude, I must admit that the extreme violence of certain protagonists in this discussion surprised me: I guess that anyone who is not perpetuating a certain position of hatred vis-à-vis corporations, anyone who tries to find different ways to do things, tries to posit the problems differently, is just a criminal who needs to be immediately punished. And BTW, those of you who feel that artists should remain "pure" and "independent" (like there is of course such a thing as independence, we all know that, right?) you will be happy to learn that yet another website was just closed, another "corporate teat sucker"! Word.com, another site that was trying to do things differently, was nixed.

Date: Wed, 11 Mar 1998 10:12:56 -0500 (EST)
From: Keith Sanborn <mrzero@panix.com>
Subject: Re: funding for the arts etc.

(1) A sponsored site enters the market as advertising. While it's not a physical commodity that is sold to its recipients, the recipients, as Richard Serra quoting someone else once said "are the commodity." Television delivers people to advertisers; corporate sponsors buy attention for themselves by using art to attract potential users of their services.

(2) It seems the only thing you've "done differently" is failed to pay in money terms the artists whose work you use for advertising. I think we already covered this with reference to Manfreddo Tafuri: "The fate of formal innovation in the arts is to be co-opted by advertising." It's a bit more complicated in the case of less visible sponsorship, but not a lot different than those Absolut Vodka ads. The difference being that Absolut Vodka had to pay the artists for the more radical product placement.

(3) The notion that "artists" need support on the web, at least in North America or Western Europe, is far from self-evident. For a relatively low cost and low investment of learning time it is relatively easy to create one's own webpages and place them. If artists wish to use the services of a site supporting artists in order to increase their visibility, then they are simply using the site to advertise their work. They are allowing their work to be used in exchange for the privilege of having it seen, which could conceivably lead to some other long term benefit. Corporate or government or individual patronage is never disinterested. No matter how much of a potlatch mentality is involved, the potlatch aspect is used to enhance one's prestige as it is with its originators, the indigenous inhabitants of Northwestern North America. One affirms one's right to one's potlatch seat by giving away things on deliberately public occasions; one catches hold of a grooviness quotient in the corporate hierarchy by sponsoring artists. Duh!

Date: Sat, 14 Mar 1998 17:15:12 -0500
From: Stephen Pusey <scp@plexus.org>
Subject: Re: Funding Digital Culture

Here is an opportunity to examine the viability of models for funding arts organizations. Judging from the examples of both Ada'web and Word, the model of ownership by a parent corporation is not conducive to a long-term development, though it may very well serve the interests of a short-term research project. Scott Baxter, Icon's (the owners of Word) president and chief executive, succinctly expresses the cold pragmatism of the corporation, "Real business, real profit, I don't derive that from Word like I did historically," ...said claiming ownership of the zine in earlier days helped put "Icon on the map" and all but "closed deals" for its salespeople.

Weil seems unclear as to what is meant by independence. To be sure, we can argue 'till the cows come home about the varying degrees of dependence that bond individuals and social groups. Let me clarify what I mean by the term in respect to arts organizations, in particular the online arts community. An independent organization is an entity, in my view, that may draw funding from many sources, private, corporate, government, etc., but allows none of these to control, dictate, or otherwise affect its development or lifespan. The importance of this cannot be underestimated.

To emphasize, my argument is not against corporate, government or private sponsorship per se, but that having to justify the agenda and existence of an arts organization to shareholders or a parent corporation is both unhealthy and intolerable as it inevitably entails a compromising alignment of interests. To quote Benjamin Weil, "the relationship with our corporate "parent"— Digital City, Inc.—has to be nurtured so as to develop a common ground where both parties understand what's in it for them" (<http://www.atnew≠ york.com/view323.htm>).

Clearly there is a need to debate and formulate a strategy for sponsorship which encourages long-term growth of digital culture. Environments like PLEXUS, artnetweb, The Thing, Stadium, and so on, though fueled perhaps by utopian ideals, are built largely on the unfinanced labor of their founders and collaborators. Their progress, however, is not aided, but hampered by a lack of funding.

[Edited by Felix Stalder and Ted Byfield.]

In the summer and fall of 1994 I helped create HotWired, and served as its first executive editor. I quit a couple of weeks after it was launched, in late 1994. What I had in mind had elements of a magazine (editorial filtering, creative design, regular, high-quality, "content"), but was much more like a community (many-to-many, unfiltered, audience-created content). I spent most of 1995 having great fun updating my webpage every day. I did all the writing, editing, design, illustrations, HTML. I talked friends of mine in America, Europe, and Japan into writing for free. In late 1995 I got it into my head that I should expand what I was having such fun doing. When I sat down to figure out how to pay my writers and editors, hire a "real" designer, and license a webconferencing system, it looked like it would cost tens of thousands per month and take us three or four months to launch.

Lesson number one was that everything in a startup that depends on cutting-edge technology takes longer and costs more than originally estimated, even when you take lesson number one into account.

Deciding to pay people reasonably well (but by no means extravagantly) for editorial content, art and design, and technical services led me to need more money than I had. That's when I made what I now clearly see to be my most fundamental error: I got caught up in the intoxication of venture-capital financing, which was in a particular state of mania in late 1995. I connected with a business partner I didn't know, but who knew how to go about securing financing and putting together a company—my second fundamental error. I failed to listen to my own nagging doubts and made a bad choice in partners.

I take responsibility for making the decisions that led to both the success and the failure of Electric Minds. We made a lot of bad decisions (though probably not many more than average for startups), but the decision to go for venture capital made all the other decisions moot. My new partner introduced me to a fellow from Softbank Ventures, for whom a million dollars was a relatively small investment. Softbank was an early investor in Yahoo!, and had bought Comdex and Ziff-Davis outright. I told the guy from Softbank that if we could figure out how to combine community and publishing, then the other companies in the Softbank investment portfolio could leverage that knowledge profitably. I believed, and still do, that it is possible to grow healthy, sustained online discussions around Yahoo!, Comdex, and Ziff-Davis. Electric Minds was supposed to be an experiment. And the million dollars I was asking for was just a down payment on a several-year relationship. At that point, any business plan for an internet business was a conjecture; thinking about how virtual communities could make business was in the realm of science fiction. We agreed that the first step was to build an exemplary product that would demonstrate the cultural viability of combining editorial content and virtual community. We agreed that it would take at least three years to become profitable.

Both Softbank and I realized that we were gambling when we projected that within three years Electric Minds could attract enough traffic to make significant advertising revenues.

We were funded in March 1996 and launched in November. In December, *Time* magazine named us one of the ten best websites of the year. By July we were out of business. Softbank, which had been expanding its investment funds to billions of dollars in size, mostly through Asian-based investors, stopped expanding. And when something that big stops expanding, it's a big loss. They were making millions of dollars a day just moving their electronic liquidity around world markets. Moving electronic liquidity around world markets is really the only game in town; all other industries and enterprises are tickets to that game. When Softbank's bubble stopped growing, they started thinking like venture capitalists again. It is my belief that the person who sponsored us for Softbank was thinking properly about the way to research the future of the medium, but wasn't thinking properly as a venture capitalist.

Venture capitalists want ten times their investment, and they would prefer to get it in three to five years. Good venture capitalists bring their connections and experience to the table, and actively help the founders build a business. In many business plans, including ours, a specific schedule of financial milestones is established. In many VC investment contracts, there are "clawback" provisions (what an evocative term!) that empower the investor to take more control of the company every time a milestone is missed. When Softbank took a cold look at their investments and started weeding out the ones that were less likely to achieve a ten-times return, they withdrew their verbal promises—which had not yet gone to written contract—of bridge financing. We did have revenues—IBM had contracted Electric Minds as the exclusive provider of virtual-community services when they conducted the Kasparov versus Deep Blue II chess match. Although we had not started out with the intention of providing virtual community–building services for other commercial enterprises, the need to ramp up revenues made it an attractive idea, and one that was not outside our original mission to encourage virtual communities on the web.

When someone has two million dollars invested, in hopes of expanding it to twenty million, they tend to push hard in the direction of attractive revenue sources. I knew clearly what I wanted to accomplish when I started—to launch a sustainable and high-cultural enterprise on the web, to show how content and community could work together to create a new hybrid medium, and to encourage the growth of many-to-many communication on the web. But the gravitational attraction of a twenty-million-dollar goal can draw the enterprise away from the course the founder originally envisioned. In order to continue paying for what many reviewers had acknowledged was high-quality content and conversations, Electric Minds was on its way to growing from fourteen employees to thirty, with most of our revenues derived from contract work building virtual communities for others. Jerry Yang at Yahoo! was enthused about us and gave us permission to create an experiment in web form–based community building. We were in discussions with Ziff-Davis, IBM, and Softbank Expos.

When we ran out of operating capital and dissolved the business, I found myself not only relieved, but happy that I wouldn't be spending my time doing what I had promised to do for Ziff, IBM, and Softbank Expos. The Yahoo! project still seemed like it could have been fun. But I had never set out to create a virtual community–building agency, and didn't want to spend my time running one. I had never set out to make tens of millions of dollars, which probably contributed to our failure to thrive.

When I had the time to think about where I had gone wrong, it seemed clear to me, and still does, that if I had simply added inexpensive conferencing software and continued doing my amateur editing and design, I could have grown something less fancy but more sustainable, even if not in financial terms. Venture capital, I concluded, might be a good way to ramp up a Netscape or a Yahoo!, or create a market for a kind of technology product that never existed before. But it isn't a healthy way to grow a social enterprise. It doesn't take too many people to sustain a small online community. Of course, many great conversations take place via mailing lists, but conferencing (BBS, message-board, newsgroup) media have their own unique capabilities, though they are also a little more expensive to run than a list. When we created The River (<www.river.org>), the idea was to create a cooperative corporation that would enable the people who made the conversation to also own and control the business that made the conversation possible. A couple of hundred people each contributed a couple of hundred dollars and agreed to pay fifteen dollars a month, and that turned out to be sufficient to buy a Pentium box and software licenses and make a co-location deal with an internet service provider. Technical and accounting services are voluntary. It works pretty well.

I have returned to spending my time the way I most enjoyed before my two years as an entrepreneur. I update my website (<www.rheingold.com>) a couple of times a week and communicate directly with my audience. I'm adding inexpensive webconferencing software in a week or two, and I'm creating a small community to discuss the things that interest me—technology, the future, media, social change. It's a hobby—I carry the costs. It makes me much happier to run it.

Setting up The River as a coop had its problems. Running a coop, particularly among Americans, can result in perpetual and not-altogether-pleasant shareholder meetings. There's a lot of blah-blah-blah in making decisions democratically. People get angry and leave. But a sufficient number have remained so that The River has survived for three years. (The legal structure that enabled them to organize was the California cooperative corporation. The legal restrictions on cooperative corporations vary from country to country, state to state.)

Webconferencing software is becoming more and more capable, and as several excellent products compete with each other the prices are dropping. It's not very expensive to add many-to-many communications with a web-based interface to any website.

Now, just so I don't forget to look at the bigger picture, I definitely acknowledge that there are legitimate questions to pursue about whether spending time typing messages to strangers via computers is a healthy way for people and civilizations to spend their time. There is the perpetual and also legitimate debate about whether it debases the word *community* (and what is the word supposed to mean these days, anyway?) to use it to describe online conversations. All I can say is that many people might end up much happier by starting out to grow a small, unprofitable, sustainable web-based cultural enterprise, than to invite the pressure-toward-hypergrowth that accompanies venture capital financing.

TEMPS OF THE WORLD UNITE

"...anti-ratrace receptionist required..."

FROM: RACHEL BAKER

TEMPS OF THE WORLD UNITE

Office
Romance
Increases
Productivity

FROM: RACHEL BAKER

TEMPS OF THE WORLD UNITE

WP Training

For

Ex - Convicts

WORK

SUBJECT: NEW FORMS OF PRODUCTION AND CIRCULATION OF KNOWLEDGE

FROM: MAURIZIO LAZZARATO (BY WAY OF DIANA MCCARTY <DIANA@MRF.HU>)
DATE: SAT, 17 OCT 1998 08:34:33 +0100

Not, perhaps, since the printing press's invention has European culture experienced so much upheaval. The very underpinnings of the notion of culture and of its modes of production, socialization, and appropriation are under attack. I am speaking, of course, of culture's integration in the creation of economic value. This integration process has accelerated since the beginning of the eighties through, on the one hand, the globalization and increasing pervasion of finance in the economy, and on the other, the onslaught of so-called new technologies.

Many have raised their voices in defense of culture, intellectuals, and artists. The strongest and most organized opposition to culture's subordination to economics came together when commercial relations regarding audiovisual production were being renegotiated, and around the issue of "authors' rights"—the very definition of which is open to discussion once new media are in the picture.

At least in France, the strategy of cultural defense seems to go beyond these first forms of mobilization against large U.S. communication and entertainment corporations. That strategy tends to involve protecting the "cultural exception."

The artists and intellectuals—and politicians and governments—who demand the right to a "cultural exception" see themselves as heirs to a tradition of European cultural autonomy and of art and artists' independence from politics and economics. The strategy of "cultural exception" supports seems to be the re-entrenchment of the separation between culture and the economy.

This position—which, in my opinion, reflects a larger European point of view—is weak and, once scrutinized, untenable with regard to the new modes of knowledge's production and circulation. The hypothesis I'd like to put forward turns the cultural exception strategy on its head; it can be summarized in this way: the modes of production, socialization, and appropriation of knowledge and of culture are different than the modes of production, socialization, and appropriation of wealth. Georg Simmel's intuition was that it is the modes of production and socialization peculiar to culture—not culture's autonomy—that must be introduced into the economy. Nor can that introduction be on a volunteer basis, since—as Gabriel Tarde has it—"intellectual production" tends to shape the direction and organization of wealth production, and the "need to know," "love of beauty and greediness for the exquisite" are the main outlets opened to economic development.

I will therefore use these two authors, and particularly the "economic psychology" published by Tarde in 1902—nearly a century ago—to unpack my argument. Let us keep in mind that Tarde's remarkable early insights are not

This is a story about invisible hands. This is a story about endless work. This is a story about women's work of maintenance and survival. This is a story about the laboring female body in the invisible feminine economy of production and reproduction. This is a story about repetition, boredom, exhaustion, stress, crashes. This is a story about tedious, repetitive, straining, manual labor harnessed to the speed of electronic machines. [Faith Wilding <744≠47.2452@CompuServe.Com>, The Economy of Feminized Maintenance Work, Tue, 8 Sep 1998.]

really part of European cultural tradition, since his theory has been largely forgotten. Based on the mode of production particular to culture, and especially knowledge, Tarde proposes an intriguingly contemporary critique of political economy by inverting the starting point of economic analysis. Rather than starting from the production of use-value—that is, "material production" (the famous pin factory, which went from the *encyclopédie des Lumières* to Adam Smith's Scottish moral philosophy, therein becoming the incipit of political economy)—he started from the production of knowledge, that is, books.

"How is a book made? It is no less interesting than knowing how a pin and a button are made": an unimaginable opening line for economists of his day—and, perhaps, of our own—but far less so for us, since the production of a book may be thought of as a paradigm for post-Fordist production.

Like any other product, "truth-values," as Tarde calls knowledge, are the result of a production process. As apparatuses develop to make knowledge production and consumption practices more and more reproducible and homogenizable—Tarde talks of the "press" and "public opinion," while we might turn to television, computer networks, and the internet—these apparatuses take on a "quantity character that is more and more marked, increasingly apt to justify their comparison with exchange-value." Does this make them merchandise like any other?

The economy does indeed treat them as it would economic wealth, considering them as utility-value like others. But for Tarde, knowledge is a mode of production that cannot be reduced to the "division of labor": it is a mode of "socialization" and "social communication" that cannot be organized by the market and through exchange without distorting its production and consumption value.

Political economy is forced to treat truth-values as it does other goods. This is because, first, it knows no other method than that which it elaborated for the production of use-value; second, and more important, though, it must treat these truth-values as material products, or else overturn its theoretical, and especially political, underpinnings. In fact, the "*lumières*" (beacons), as Tarde sometimes calls knowledge, exhausts political economy's notions of economy and of wealth, founded on scarcity, lack, and sacrifice. Like political economy, then, let us start with production—but of books, not of pins. With the production of books we are immediately confronted with the need, in principle, to switch modes of production and property regimes with regard to what economics theorizes and legitimizes.

"The rule in the matter of books is individual production, while their property is essentially collective; for "literary property" has no individual meaning unless works are considered goods, and the idea of the book does not belong exclusively to the author before being published, that is, when it is still a stranger to the social world. Inversely, the production of goods becomes more and more collective and their property remains individual and always will, even when land and capital are 'nationalized.' There is nothing suspicious about the fact that, in the matter of books, free production is vital as the best means of production. A scientific organization of labor which would regulate experimental research or philosophic meditation through legislation would produce lamentable results."

The large multinationals of the information economy are prepared to recognize the impossibility of organizing production according to "scientific management." They are insufferable, however, regarding property regimes. Is the notion of property applicable to all forms of value, from utility-value to beauty-value to truth-value? Can we own knowledge as we own a utility-value? Perhaps, responds Tarde—but not in the way that economics or legal studies understand it, that is, as "free disposition."

"In this sense, one is no more owner of one's glory, nobility or credit [toward society] than he [sic] is of his limbs, which, as living things, he cannot relinquish to others. He therefore has nothing to worry regarding expropriation for these values, the most important of all, and the most difficult to nationalize."

In order to avoid the necessity of the new mode of organizing production and the new property regime implied by the nature of knowledge, political economy is obliged to turn "immaterial products" into "material products," that is, into goods like any others, for book production problematizes the exclusively individual property and disciplinary production upon which the economy is based.

Let us move to consumption: Can the consumption of wealth be compared to the consumption of truth-values and beauty-values? Tarde wonders, "Do we consume beliefs by thinking of them, and the masterpieces we admire by gazing upon them?" Only wealth, as political economy defines it, affords a "destructive consumption" that, in turn, supposes trade and exclusive appropriation. The consumption of knowledge, on the other hand, supposes neither definitive alienation nor destructive consumption.

And to deepen the specificity of the "consumption" of knowledge, let us analyze the mode of "social communication," truth-value's form of transmission, of which economists cannot conceive except under the form of the "market." Tarde first tells us that knowledge need not be exclusive property in order to satisfy the desire of knowing, and does not require the definitive alienation of the "product." He then adds that the transmission of knowledge lessens neither he who produces it nor he who exchanges it. On the contrary, the diffusion of knowledge, rather than depriving its creator, augments his value and the value of the knowledge itself. It is therefore not required that it be an object of exchange in order to be communicated.

"It is by metaphor or the abuse of language that we say that two people in dialogue are 'exchanging their ideas' or their admiration. Exchange, with regard to beacons [knowledge] and beauty, does not mean sacrifice; it means mutual influence, through the reciprocity of gift, but of a special class of gift which has nothing to do with wealth. Here, the giver deprives himself by giving; with regard to truths and beauty, he gives and retains at the same time. In the matter of power, he sometimes does the same thing.... For the free exchange of ideas, as for religious beliefs, arts and literature, institutions and morals: between two peoples, neither may in any instance be reproached as those engaged in the free trade of goods might be reproached—of being a cause of impoverishment for one of them."

The statement "the value of a book" is ambiguous, for it has both a venal value as something that is "tangible, appropriable, exchangeable, consumable," and a truth-value as something that is essentially "intelligible, unappropriable, unexchangeable, unconsumable." The book may be considered both as a "product" and as "knowledge." As a product, its value may be defined by the market—but as knowledge?

The ideas of loss and gain are applicable to knowledge, but here the evaluation of losses and gains demands an ethics, not a market. A book is created for or against other books, just as a product is created for or against other products. Only in the latter case, however, may competition be decided by prices; in the former, an ethics is required. The transmission of knowledge has more to do with gift or with theft, which are moral notions, than with exchange.

"On the other hand, and by its [the free trade of ideas] very nature as a reciprocal addition, not a substitution, it arouses either fertile matings or fatal shocks between the heterogenous things it brings together. It may therefore cause great harm when it does not do great good. And just as this intellectual and moral free trade inevitably becomes an accompaniment to economic free trade, the reverse is also true: separated from one another, each would be ineffective and inoffensive. But, I repeat, they are inseparable, and to last indefinitely, a prohibitive tariff must be matched by an Index, that ecclesiastic prohibitionism."

According to Tarde, then, the modes of production and communication of knowledge lead us beyond the economy. We are beyond the necessity of socializing intellectual forces through exchange, division of labor, money, or exclusive property. This does not mean that the relations of power between social forces are neutralized—in fact, they show up as fertile matings or fatal shocks beyond the market and the exchange of wealth. This means that the unavowed ethical nature of economic forces resurfaces powerfully as a single mode of "economic regulation" at the very moment in which economic production is subordinated to intellectual production.

Here we find the Nietzschean problem of the "hierarchy of value" and the "great economy," but on different terrain.

Tarde gives another example, this time on "training," which leads us to a similar conclusion. We may establish a comparison between the production of wealth and the production of truth-value through teaching. We may therefore, for pedagogy, define the various factors through which teaching is produced. Just as economists distinguish labor, land, and capital in the production of "beacons," so may we distinguish the activity and intelligence of the student and the knowledge of the professor. "The truth is that these assays are not terribly useful. Above all, the first condition for good instruction—the teacher's and student's psychological conditions having been met—is a good school program, and a program supposes a system of ideas, a belief. Similarly, the first condition for good economic production is a moral code to which all agree. A moral code is a program for industrial production, that is, consumption—for the two are interdependent.

If, as some hold, the "beacons" may be related back to utility-value (they assume consumption and the destruction of forces and costs for the production; they are materialized in the product and have a price), the production, communication, and appropriation of thoughts and knowledge differs fundamentally from the communication and socialization of "wealth."

In capitalism, then, all forms of production, even the most incomparable, can more and more be evaluated in terms of money, yet less and less does knowledge lend itself to this sort of evaluation. Here Tarde opens another hidden door of intellectual production that political economy cannot approach through its principles of scarcity, sacrifice, and necessity. The problem posed by "intellectual production" is not only that of defining an "ethical" measure adequate to truth-value, but especially the fact that it tends toward a form of production that is more and more free. Intellectual production exhausts the very raison d'être of the economy and its science, economics—scarcity.

"Civilization's effect is to push into business—that is, into the economist's field—a range of things that were previously without price, even rights and powers. So, too, has the theory of wealth encroached incessantly upon the theory of rights and the theory of power, that is, jurisprudence and politics. But against this trend, through the ever-growing freedom of widely distributed knowledge, the border between the theory of wealth and what we might call the theory of beacons is growing."

These few pages almost seem to have been written with the information economy and intellectual property in an immaterial economy in mind. "Free production," "collective property," and "free circulation" of truth-values and of beauty-values are conditions for the development of social forces in the information economy. Each of these qualities of intellectual production is in the process of becoming a new "contradiction" within the information economy, for which the challenges represented today by the internet are but the premises of opposition to come.

Writing in the same era, Georg Simmel comes to similar conclusions. "Nor does the communication of intellectual goods require us to snatch away from the one what must be tasted by the other; at least, only an exacerbated and quasi-pathological sensibility may truly feel slighted when objective intellectual content is no longer exclusively subjective property but, rather, is thought by others. Generally, we may say that intellectual possession, at least to the extent that it has no economic extension, must in the end be produced by the very conscience of the acquirer. Yet it is clearly a question of introducing this conciliation of interests, which derives here from the nature of the object, into those economic domains where, because of competition in the satisfaction of a particular need, no one enriches him- or herself unless it is at the expense of another."

In Simmel's felicitous phrase, the conciliation of interests which derives from the nature of the intellectual object is a political program, for the logic of scarcity, the exclusive property regime and the mode of production are imposed upon its products by the new knowledge industries. But if we do not indicate the new oppositions specific to intellectual production, if we limit ourselves to demanding the autonomy of culture and of its producers, resistance to contemporary capitalism's domination of culture remains nothing but a pious vow.

And yet the contemporary production of wealth integrates not only production, socialization and appropriation of knowledge, but also beauty-value, that is, aesthetic forces. As long as needs become more and more specialized, aesthetic value is one of the basic elements which stimulate the desire to produce and the desire to consume. This process, which had only just started when Tarde wrote these pages, and which was barely perceptible by the economists of his day, has undergone an extraordinary acceleration, starting with the blossoming of what we may call the information or immaterial economy. The "cultural exception" strategy's definition of culture presupposes a qualitative difference between industrial labor and artistic labor. Today, following the tendency identified by Tarde, according to which intellectual production subordinates economic production, artistic labor is becoming one of the models for the production of wealth.

We have already seen how the notion of wealth must integrate knowledge, and how intellectual labor sketches out the tendency of the development of "economic progress" according to Tarde. It only remains to see how artistic labor might lead to an understanding of this radical change. According to Tarde, every activity is a combination of imitative and inventive labor, but also of artistic labor, present in quite unequal proportions. Industrial labor does not escape this rule. What relationship is there between industrial and artistic labor? The clear distinction he establishes between industrial and artistic labor does not rule out the continuity of transition.

The social definition of artistic activity grasped magnificently by Tarde may inspire several reflections on how, by integrating industrial activity, it may change the relationship between producer and consumer. Of Tarde's definition of artistic labor, let us underline two aspects: on the one hand, the determining role played by the "imagination"; on the other, the fact that in artistic activity the distinction between producer and consumer tends to erase itself. We need not add that, here too, Tarde's considerations are of great importance in determining the status and function of the "consumer-communicator" of contemporary society." Under post-Fordism, in effect, the clientele of any industrial production (and notably in all production in the information economy) tends to identify itself with a particular public which, in turn, plays the role of both producer and consumer.

Sensation is the nonrepresentative and therefore noncommunicable element that, according to Tarde, is the very object of artistic labor. "We have said it from the beginning: the phenomena of conscience are not entirely resolved by belief and desire, by judgment and intention. Lurking in these phenomena is always an effective and differential element playing the principal role in sensations and which, in the higher sensations—that is, feelings, even the most quintessential—acts in a dissimulated way, which does not make it any less essential. Art's virtue and its characteristic is to regulate the soul by gripping it through its sensational side. As the handler of ideas and intentions, it is certainly inferior to religion and to the various forms of government, politics, law, and morals. But as an educator of the senses and of taste, it is unequaled."

Does this mean that sensations, too, may constitute themselves as a value that can be measured quantitatively and therefore exchanged? And through what sort of apparatus, involving which sort of activity?

"...the great artists create social forces just as entitled to the name of 'forces,' just as capable of increasing and decreasing with regularity, as the energies of a living creature."

Through works of art, it is the artist who lends social consistency to the most fleeting, most singular, and most nuanced of sensations. By combining the psychological elements of our soul, where sensations dominate, artists add a new variety of sensation to the public through their work. Sensation and sensitivity are hence the "products" of artistic labor.

"Yet, in thus building the keyboard to our sensitivity, in extending it for us, and in ceaselessly perfecting it for us, poets and artists juxtapose, even substitute for our natural and innate sensitivity, which is different in each of us, a collective sensitivity, similar for all, impressionable to the vibrations of the social milieu, precisely because it is born in the artist. The great masters of art, in a word, discipline our sensitivities and then our imaginations, causing them to reflect one another and to be aroused by their mutual reflection, while the great founders or reformers of religions, the sages, the legislators, the statesmen, discipline spirits and hearts, judgments and truths."

For Tarde, then, artistic labor is "productive" labor in that it responds to a production and consumption need concerning pure sensation. We must now analyze how artistic and industrial labor are opposed or in harmony. The difference between art and industry lies above all in the fact that the desire or appetite for consumption met by art is more artificial and capricious than is that met by industry, and requires "longer social elaboration."

The desire for artistic consumption is even greater than the desire for industrial consumption, child of "inventive and exploratory imagination." Only the imagination which brought this desire into this world can satisfy it, for its very origin—unlike the desire for industrial consumption—lies almost exclusively in the imagination.

"The desire that serves industry—shaped, it is true, by the whims of its inventors—shoots out spontaneously from nature and repeats itself daily, like the periodic needs it translates; but the taste that art attempts to flatter is attached through a long chain of ideas to vague instincts, none of them periodical, which reproduce only by changing."

The desire for industrial consumption preexists its object and, even when specified or elaborated by certain inventions of the past, asks only of its object to be fulfilled repeatedly; "but the desire for artistic consumption expects completion from its very object and asks of its new inventions that this object provide it with variations of their predecessors. Indeed, it is natural that an invented desire such as this has as its object, too, the very need to invent, since the habit of invention can only give birth to more such habits and increase its appeal." These nonperiodic and accidental needs are born of an "unexpected meeting" and require the "perpetually unexpected" to survive.

But another characteristic of artistic labor is of particular interest. In artistic production, it is impossible to distinguish production from consumption, for the artist himself experiences the desire to consume, searching above all to please his own taste, not only that of his public.

"Moreover, the desire for artistic consumption is particular in that it is even more acute and its joy more intense in the producer himself, than in the mere connoisseur. In this, art is profoundly different than industry.... In matters of art, the distinction between production and consumption begins to lose its

clean, wash, dust, wring, iron, sweep, cook, shop, phone, drive, clean, iron, enter, mix, drive, delete, clean, purge, wash, merge, edit, shop, fold, phone, file, select, copy, curse, cut, sweep, paste, insert, format, iron, program, type, assemble, cook, email, fax, cry, forward, sort, type, click, dust, clean, etc. [Wilding, Economy]

importance, since artistic progress tends to make of every connoiseur an artist, and of every artist a connoisseur."

And yet these differences and opposition between artistic and industrial labor are in the process of falling away, one after another. Instead, a deepening adaptation has developed between these two types of activity. Tarde himself sketches out this tendency: beauty-values must be integrated into the definition of wealth and artistic labor in the concept of labor, for "the love of what is beautiful, the greed for what is exquisite" are part of the "special" needs which exhibit great elasticity and therefore a wide opening for industry. Tarde even foresees that the luxury industry which in his day concerned only the upper classes—this was the only type of consumption which exhibited "special" needs—would, with the development of social needs, be substituted by "industrial art, decorative art, which could very well be destined for a most glorious future." A few decades later, Walter Benjamin would come to the same conclusions, analyzing tendencies in industrial development and in productive activity based on cinematic production.

To close, if we wish to safeguard the specificity of European culture and its emancipatory potential, we can no longer rush to the defense of culture and its autonomy, for truth-values and beauty-values have become the motors of the production of wealth. The more we hand off the desire for a production and consumption that satisfy "organic" needs to a desire for production and consumption that satisfy increasingly "capricious" and "special" needs—of which one is the need to know—the more economic activities and even goods themselves integrate our truth-values (knowledge) and beauty-values. "Let us add the theoretical and aesthetic sides to all goods will become more and more developed—beyond, not despite their useful side."

This conclusion might be read as catastrophic, for it demonstrates the real subordination of cultural and artistic production to economic imperatives. But it is a historical opportunity, even if we do not know to seize it. For here, perhaps for the first time in humanity's history, artistic, intellectual and economic labor, on one hand, and the consumption of goods and appropriation of knowledge and beauty-values, on the other, demand to be regulated by the same ethics.

[Translated by Bram Dov Abramson <bram@tao.ca>.]

SUBJECT: HIGH-TECH WORKERS NEED TO UNIONIZE

FROM: REBECCA LYNN EISENBERG <MARS@BOSSANOVA.COM>
DATE: TUE, 18 AUG 1998 12:47:52 -0700

When programmers started emailing me over the past few weeks, begging me to denounce the Senate's recent decision to grant more work visas to foreign nationals seeking high-tech employment, I was loath to run to their defense. Computer programmers, it seemed to me, did not need my help. They complain about long hours, but arrive at work at noon. They complain about low pay, but earn twice the national average. They gripe about being forced to carry cell phones, yet get wireless service for free—not to mention stock options, top-notch health care, 401(k) plans and loaner laptop computers. Undereducated, overpaid, underage white males, they start new companies, hire their buddies and wake up millionaires à la Netscape's Marc Andreessen.

Surprisingly, in this case the programmers were right: The Senate H-1B visa decision did do them an injustice, but they still don't need my help. They need labor unions. If this debate over the so-called high-tech worker shortage does not stir them to organize, perhaps nothing else will. Unions for professional software engineers? The idea is not as crazy as it sounds. Although life for some programmers might look plush, many others sing the blues. Strong-armed to take options in lieu of paychecks, they are often left empty-handed when the business ultimately tanks, which it does in many cases. Meanwhile, the large paychecks paid by big software companies yield much more humble hourly wages when divided by the number of hours worked—without overtime pay, of course. Constantly pushed to publish products by unreasonably early deadlines, software engineers have grown accustomed to pulling strings of "all-nighters" near launch time, yet still are forced to release products before they're ready.

Perhaps most nefariously, as programmers grow older, their job security plummets. Any stroll through a high-tech company reveals that the work force is very young. Norman Matloff, computer science professor at UC–Davis, confirmed this common observation in an April report: Five years after finishing college, about 60 percent of computer science graduates are working as programmers; at fifteen years the figure drops to 34 percent, and at twenty years it's a mere 19 percent. A programmer described a conversation he overheard at a recent company event: "Age became an important topic of discussion at this midday meeting, and they decided that the oldest person in their section of the company was twenty-nine." These observations are corroborated by Matloff's study: Most software companies classify programmers and systems analysts with six years of experience as senior even though they usually are no older than twenty-eight. Older employees are more expensive. Because they are more likely to have families, for example, their benefits cost more and they are less likely to tolerate eighty-hour work weeks than recent college graduates.

By the early eighties, women in the U.S. were 43 percent of the paid labor force. And 43 percent of all paid employed women were clerical workers. In the U.S., women were: 80 percent of all clerical workers 97 percent of all typists 99 percent of all secretaries 94 percent of bank tellers 97 percent of receptionists a majority of these jobs will be/are disappearing. In the U.S. women currently are: 31 percent of computer programmers 29 percent of computer systems analysts 16 percent of executive managers 92 percent of data entry operators 58 percent of production operators 77 percent of electronic assemblers these statistics are not changing fast. Black women in the U.S. are: 3 percent of corporate officers 14 percent have work disabilities 59 percent of all single mothers. How many of these jobs will disappear? At home all women are: 66 percent of married working mothers 100 percent of mothers 99 percent of child-care workers 99 percent of primary caregivers to the aged 83 percent of unpaid household workers 99 percent of domestic caretakers 99 percent of physical, emotional, and psychic human capital maintenance workers. In the electronic home will mothers become obsolete? In the electronic workplace will women become obsolete? [Wilding, Economy]

And while unemployment rates for older workers are high—17 percent for programmers over age fifty as of August, Matloff said, the numbers tell only part of the story. "I get rather annoyed at unemployment statistics," the programmer said. "They might be talking about unemployment, but they are not talking about underemployment. Former high-tech people have long since exhausted their unemployment benefits or are employed at something that they did not expect to be doing at their age." Meanwhile, he said, as a temporary employee "I have sat through meetings where managers go out of their way to report that they had hired new permanent employees, stressing that they would be working as soon as they had their visas straightened out. Politically it seemed very important for them to stress this."

Is this because H-1B status employees would work more hours for less money? "That was my distinct impression," he said. Would this programmer join a union? "I am not sure if 'union' *is* the right word, but I definitely think that something should be done," he said. "Union" *is* the right word, said Amy Dean, chief executive director of the South Bay AFL–CIO Central Labor Council, which represents the interests of labor, both full-time and contingent, in Silicon Valley. "It always makes sense for working people to come together for purposes of bargaining collectively to improve their workplace situation." Unions can provide job security for workers with seniority, which is essential for older workers in the youth-biased software industry, Dean said. "There is no question that the industry (is) looking at older workers as though they are disposable," she said. "They have become too costly, and now after they have given the best of their lives to the company, the company decides that it is too expensive to keep them on board." Additionally, unions could benefit workers of all ages by requiring companies to look internally or locally before hiring foreign workers on visas. If programmers were organized, Dean said, "They could insist on what portion of the company's jobs go to people in-house, and they could insist that X percent of jobs be tagged for people that are already part of the company." Furthermore, unions could convince companies to train workers, said Dean. "Workers would have means to sit down with the employers and say, 'We think that there should be X number of dollars spent on training to bring us up and elevate our skill base so that we can apply to jobs being given to people from other parts of the world.'" "This H-1B visa issue is all about trying to undercut the wage and benefit rate of current American workers," Dean said. With a union, technology workers could insist on a wage and benefit standard as opposed to allowing companies "to bring in workers that are going to undercut that standard."

That's fine for programmers who are employed full time, but traditionally unions have not been available for contingent workers, who, like the programmer above, work part time or are contracted to work on short- or long-term projects. Because contingent workers now comprise 27–40 percent of the Silicon Valley work force (and growing), according to the National Planning Association in Washington, D.C., the Central Labor Council is upgrading its services to serve them better. "We are building an organization that people will be able to join to receive benefits, including health and pension," which independent contractors usually don't get, Dean said. "It will also provide training and skills certification, and it will advocate within the temporary-help industry to improve conditions for people who are working on a part-time or contingent basis." While this approach is not traditional unionization, Dean conceded, "we know that in the new economy, we will need these new types of organizations."

In the meantime, Dean urged all high-tech workers to vote against Proposition 226 on Tuesday. That proposed law, she said, would "eliminate the right of workers to bundle together their nickels and dimes to have a voice in the political process"—including opposing future attempts to bring in more foreign programmers. "If workers cannot combine their resources, they have no chance to stand up to big corporations and organized business," which outspend labor eleven to one, Dean said. In all these ways and more, said Dean, "History shows that when people band together, they do better than they would if going it alone." The software industry certainly knows the power of banding together—after all, it was the powerful lobbying efforts of its trade organization, the Information Technology Association of America (ITAA), that succeeded in pushing companies' requests for more foreign labor through the Senate. Programmers—both young and old— deserve equally strong representation, which they can find in unions. If the industry is scared by the so-called high-tech worker shortage, imagine the persuasive power of engineers on strike.

[This text first appeared in the *San Francisco Examiner*.]

SUBJECT: THIS IS LONDON
FROM: SIMON POPE <ESCAPECOMMITTEE@COMPUSERVE.COM>
DATE: TUE, 8 SEP 1998 12:09:40 -0400

JOSH

Precocious small boy steps, jet-lagged, from Club Class. Self-contained under hood and high-TOG breathable future fabric. Self-reliance velcroed tightly into place, an outward manifestation of his prep-school motto—"You are alone. Trust no one." It's been a good year for Josh, extending his dad's business into the nineties globe-trotting 007 execs dreaming of Suzie Wong morphed into transnational gotta-be Goldies dreaming of Jackie Chan flicks...

JUSTIN

Justin used to be an account manager up West with one of the big-noise, big-budget agencies. Eight years living a one man yuppie revival in the pristine post-Lloyds white tower would have tipped a more scrupulous man over the edge. Walking monochrome corridors, scoping for black-clad door-whores for a moments abrasion can seem futile, but leaving this cathedral dedicated to the power of spectacle would invoke an immediate "access denied" in the four-star staff canteen. But ground-zero approached fast. Why not steal a few clients and make a go of it? Everyday could be casual Friday. Imagine...wearing post-rave leisure wear to *work*. Cool.

"I've got the brains, you've got the looks. Let's make lots of money," as one of Justin's favorite songs would have it. For brains they turned to Andy.

ANDY

Server-side back-end UNIX flavored mindfuck gives most web designers instant impotence and an overweening self-doubt. Not good for business let alone personal development. So all the black arts of CGI and increasingly Java are left to Andy. In most cultural and technological shifts, people like Andy aren't the public face of the industry. Now is no exception. They are in no way "cool." They like the same music as their older brothers and dress in whatever is on the floor and smells least like chip fat or the sweet, baked-bean sweat of teen-boys' bedrooms. When this cycle of boom and bust is long forgotten, Andy will still have his head down and know the worth of a good ping program. Enough of Andy.

ADAM

The beads of sweat form on Adam's artfully concealed but receding hairline, mirroring the gray rain as it slides asthmatically down the mildewed taxi window. Every journey home has been like this recently. A videotape plays and rewinds, caught in a frenzied loop, wearing his patience thin. Every dropout amplified. Each iteration reinforcing the feeling that trust has been misplaced. That saving your best work for your highest-profile client has not paid off. Art and Business. Like grape and grain. Start out on one. Don't finish on the other. Four long years from version 3 through 6, slowly losing a grip on the point of it all. A time for change. Maybe re-invention is the only solution. Notting Hill. London. Home. Flipping his last tenpence piece, the severed monarch's head floats, goading, mocking his situation. Only one thing left to do: just fucking phone Justin...

SUBJECT: GOING AT DIFFERENT SPEEDS:
ACTIVISM IN THE GARMENT AND INFORMATION SECTORS
FROM: ANDREW ROSS <AR4@IS.NYU.EDU>
DATE: THU, 10 SEP 1998 17:39:15 +0200

Suck, the irreverent daily webzine in San Francisco, cunningly revealed that the staff of *Wired* magazine occupied a floor in a building full of garment sweatshops. Suddenly, with this revelation, the century-long gulf between the postindustrial high-tech world, for which *Wired* is the most glittering advertisement, and the pre-industrial no-tech world, appeared to have dissolved. In New York City, this kind of juxtaposition between nineteenth-century and the twenty-first-century is fairly common, where the ragged strip of Silicon Alley—New York's concentrated webshop sector—cuts through areas of old industrial loft space that were once, and are again, home to the burgeoning sweatshop sector of the garment industry. Many of the webshops, those much-romanticized laboratories of the brave new future, are housed nextdoor to garment sweatshops where patterns of work for large portions of the immigrant population increasingly resemble those in the early years of the century, before industrial democracy and progressive taxation and the welfare infrastructure (modern industrial relations, in short) were adopted into law. In recent years, we have seen the return of the sweatshop to the central city core (in fact, the sweatshop was never eradicated, it was simply driven further underground or overseas). Full media disclosures about these sweatshops of the sort we have seen in the last few years summon up the misery and filth of turn-of-the-century workplaces, plagued by chronic health problems and the ruthless exploitation of immigrants. Indeed, the repugnance attached to the term *sweatshop* commands a moral power, second only to slavery itself, to rouse public opinion into a collective spasm of abhorrence.

As it happens, the juxtaposition of technocultures in today's two-tier global cities is also strikingly similar to workplaces at the turn of the century. Then, the sweatshop's primitive mode of production and the cutter's artisanal loft co-existed with semiautomated workplaces that would very soon industrialize into economies of scale under the pull of the Fordist factory ethic. Today, the sewing machine's foot pedal is still very much in business—though competing not with steampower but with the CPU, which, at the higher end of the garment production chain, governs Computer-Assisted Design and facilitates fast turnaround. The sewing machine has barely changed in almost a hundred and fifty years, which makes it quite unique in terms of industrial history. Because of the physical limpness of fabric, there is a portion of garment production that cannot be fully automated and so requires human attention to sewing and stitching and assembly—hence the demand for cheap labor. As a result, underdeveloped countries usually begin their industrialization process in textiles and apparel, because of the low capital investment in the labor-intensive end of production.

There are many reasons for the flourishing of garment sweatshops, both in poor countries and in the old metropolitan cores: regional and global free-trade agreements, the advent of universal subcontracting, the shift of power away from manufacturers and toward large retailers, the weakening of the labor movement and labor legislation, and the transnational reach of fashion itself, especially among youth. The international mass consumer wants the latest fashion post-haste requiring turnaround and flexibility at levels that disrupt all stable norms of industrial competition.

Public awareness of the conditions of low-wage garment labor is relatively advanced, even if the public tends to ignore that fact that much clothing is made illegally and in atrocious conditions. The antisweatshop campaigns of recent years—in the last two years they have been very visible and vocal in American mediaspace—would not have been so successful if people did not, however grudgingly, acknowledge that their personal style in clothing comes at a price for low-wage workers. The challenge now lies in making an impact at the point of sale, that is, reforming consumer psychology to the level at which criteria of style, quality, and affordability are all well served by appeals to the advantages of paying a living wage. We are much further forward than anyone could have imagined just a few years ago.

The same cannot be said of high technology. The gulf between the fashion catwalk and the garment sweatshop is nowhere near as great as the gulf between the high-investment glitz and the heady cultural capital of the digerati at the top of the cyberspace chain and the electronic sweatshops at the bottom. Why? Even if we cannot answer this question, it is worth asking. Cyberspace, for want of a better term to describe the virtual world of digital communication and commerce, is not simply a libertarian medium for free expression and wealth accumulation. It is a labor-intensive workplace. Masses of people work in cyberspace, or work to make cyberspace possible, a fact that receives virtually no recognition from cyberlibertarian digerati like John Perry Barlow or Kevin Kelly, let alone the pundits and industrialists who are employed to uphold the rate of inflation of technology stocks. Indeed, it's fair to say that most information professionals have little sense of the material labor that produces their computer technologies, nor are they very attentive to the industrial uses to which these technologies are put in the workplaces of the world. This is understandable, though not excusable, when these sectors are remote and invisible, on the other side, as it were, of the international division of labor. But it is difficult to exonerate the neglect of working conditions that lie at the heart of the cyberspace community itself, within the internet industries. Like all other sectors of the economy, these industries have been penetrated by the low-wage revolution—from the janitors who service Silicon Valley in California to the part-time programmers and designers who service Silicon Alley in New York. Just as Silicon Valley once provided a pioneering model for flexible postindustrial employment, Silicon Alley may be poised to deliver an upgrade. My own research on Silicon Alley was done in the fledgling years of 1996 and 1997 at a time when the webshops also produced independent webzines, or some form of independent publishing of creative outlet. These operations had different functions for different companies, they employed artists and writers who might have been otherwise warehoused in graduate seminars, and they promised a reasonable return on cultural labor. Most of these shops are now defunct. At this point, this independent sector has almost entirely been displaced by MBAs, the venture capitalists, and angel-seeking entrepreneurs. On Silicon Alley the current cliché is "Content is Dead."

At any rate, cultural labor in new media, no less than in the arts or education, is subject to what I call the Creative, wherein our labor is undercompensated because of the invisible wages that come in the form of psychological rewards for personally satisfying work. It is a legacy of the Romantic concept of the artist as separate from the world of trade, and whose activities were unsullied by matters of commerce. At a time when nobody seems immune to the plague of low-wage labor, it's important that artists, educators, writers and designers see this discount arrangement for what it is— exploitation of the prestige of cultural work to drive down wages in a market where the labor supply always outstrips demand.

If Silicon Alley's new media sector gives birth to a new kind of culture industry, it is not likely to be a mass media industry, nor will its impact necessarily lie in the realm of leisure or entertainment. Unlike the culture industries of radio, film, TV, recording, fashion, and advertising, which had their start in the Age of the Machine, the work environment of new media is entirely machine-based, and labor-intensive in ways that are now the legends of cyberspace. "Voluntary" overtime—with twelve-hours workdays virtually mandatory—is a way of life for those in the business of digital design, programming, and manipulation. The fact is, new media technologies have already transformed our work patterns much more radically than they are likely to affect our leisure hours, just as information technologies have already played a massive role in helping to restructure labor and income— effectively reorganizing time, space, and work for mostly everyone in the developed world. We are seeing the dawn of new forms of leisure time governed by labor-intensive habits tied to information technology.

All of us probably want our computers to go faster, and yet most of the people who work with computers already want them to go slower. Information professionals are used to thinking of themselves as masters of their work environment, and as competitors in the field of skills, resources, and rewards. Their tools are viewed as artisanal: they can help us to win advantage in the field if they can access and extract the relevant information and results in a timely fashion. In such a reward environment, it makes sense to respond to the heady promise of velocification in all of its forms: the relentless boosting of chip clock speed, of magnification of storage density, of faster traffic on internet backbones, of higher baud rate modems, of hyperefficient database searches, and rapid data-transfer techniques. A common repertoire of industrial, design and internet user lore binds us together and reinforces our (para) professional esprit de corps; but this shared culture also tends to disconnect us from the world of more traditional work.

In the other world, the speed controls of technology serve to regulate workers. These forms of regulation are well documented: widespread workplace monitoring and software surveillance, where keystroke quotas and other automated measures are geared to time every operation, from the length of bathroom visits to the output diversions generated by personal email. Occupationally, this world stretches from the high-turnover burger-flippers in MacDonalds and the offshore data-entry sweatshops in Bangalore and the Caribbean to piecework professionals and adjunct brainworkers and all the way to the upper-level white-collar range of front-office managers, who complain about their accountability to inflexible productivity schedules. It is characterized by chronic automation, the global outsourcing of low-wage labor, and the wholesale replacement of decision-making by expert systems and smart tools; it thrives on undereducation, undermotivation, and underpayment; and it appears to be primarily aimed at the control of workers, rather than at tapping their potential for efficiency, let alone their ingenuity (B. Garson, *Electronic Sweatshop*, NY: Simon and Schuster, 1988).

Some of you will object to my crude separation of these two technological environments. Putting it this way encourages the view that it is technology that determines, rather than simply enables, this division of labor. This objection is surely correct. It is capitalist reason, rather than technical reason, which underpins this division, although technology has proven to be an infinitely ingenious means of guaranteeing and governing the uneven development of labor and resources. Let me therefore revise, or qualify my original assertion. I won't reject it because I believe it barely needs to be proven that for a vast percentage of workers, there is simply nothing to be gained from going faster; it is not in their interests to do so, and so their ingenuity on the job is devoted to ways of slowing down the work regime, beating the system, and sabotaging its automated schedules. It is important, then, to hold onto the observation that complicity with, or resistance to, acceleration is an important line of demarcation. But equally important is the principle of *speed differential*, because this is the primary means of creating relative scarcity— the engine of uneven development in the world economy.

Commodities, including parcels of time, accrue value only when they are rendered scarce. Time scarcity has been a basic principle of industrial life, from the infamous tyranny of the factory clock to the coercive regime of turnaround schedules in the computer-assisted systems of just-in-time production. It is a mistake again to hold the technologies themselves responsible: the invention of the clock no more made industrialists into callous exploiters of labor than it made Europeans into imperialist aggressors. But capitalism needs to manufacture scarcity; indeed, it must generate scarcity before it can generate wealth.

Ivan Illich pointed this out in his own way in his essays on *Energy and Equity* (NY: Harper and Row, 1974), when he noted that the exchange value of time becomes a major economic component for a society at a point where the mass of people are capable of moving faster than 15 mph. A high-speed society inevitably becomes a class society, as people begin to be *absent* from their destinations, and workers are forced to earn so much to pay to get to work in the first place (in high-density cities where mass transportation is cheap, the costs are transferred to rent). Anyone moving faster must be justified in assuming that their time is more important than those moving more slowly. "Beyond a critical speed," Illich writes, "no one can save time without forcing another to lose it" (30). If there are no speed limits, then the fastest and most expensive will take its toll in energy and equity on the rest: "the order of magnitude of the top speed which is permitted within a transportation system determines the slice of its time budget that an entire society spends on traffic" (39).

Illich's (and others') commentaries on the emergence of speed castes from monospeed societies have progressively refined our commonsense perception that the cult of acceleration takes an undue toll upon all of our systems of equity and sustainability: social, environmental, and economic. You don't have to subscribe to the eco-atavistic view that there exists a "natural tempo" for human affairs, in sync with, if not entirely decreed by the biorhythms of nature, to recognize that the temporal scale of modernization may not be sustainable. Faster speeds increase a society's environmental load at an exponential rate. The lightning speed at which financial capital now moves can have a disastrous effect upon the material life and landscape of entire societies when regional markets collapse or are put in crisis overnight. The depletion of nature is directly tied to the degree to which the speed of capital's transactions creates shortages and scarcity in its ceaseless pursuit of accumulation. Regulation of social and economic speed in the name of selective slowness seems to be a sound, and indisputable, path of advocacy.

But it is important to bear in mind that state and World Bank economists already practice such regulation, when they decide to "grow" economies at a particular speed in order to control the inflation specter and when they impose recessionary measures upon populations in order to enforce pro-scarcity or austerity measures. It maybe crucial to remember that only those going fastest possess the privilege to decide to go slower, along with the power to make others decelerate.

If we go a little further down the chain of production, we find ourselves in the semiconductor workplaces, which are a different species of electronic sweatshop. In these factories, the hazards to labor and to the environment are greater than almost any other industrial sector. Semiconductor manufacturing uses more highly toxic gases than any other industry, its plants discharge tons of toxic pollutants into the air, and use millions of gallons of water each day; there are more groundwater contamination sites in Silicon Valley than anywhere else in the U.S. Semiconductor workers suffer industrial illnesses at 3 times the average for other manufacturing jobs, and studies routinely find significantly increased miscarriage rates and birth defect rates among women working in chemical handling jobs. The more common and well-documented illnesses include breast, uterine, and stomach cancer, leukemia, asthma, vision impairment, and carpal tunnel syndrome . In many of these jobs, workers are exposed to hundreds of different chemicals and over 700 compounds that can go into the production of a single workstation, destined for technological obsolescence in a couple of years—12 million computers are disposed of annually, which amounts to 300,000 tons of electronic trash that are difficult to recycle. The "dirtier" processes of hightech production are generally located in lower-income communities and communities of color in the U.S. and throughout the Third World, augmenting existing patterns of environmental and economic injustice. Through the Campaign for Responsible Technology, an international network is now being formed to make links with local labor, environmental, and human rights groups around the world. Much of the groundwork for this was laid at a recent European Work Hazards convention in Holland, which brought together activists with the common goal of holding companies to codes of conduct through the acceptance of independent workplace monitoring. Because transnational companies tend to export hazards to countries where labor is least organized, clearly, a global strategy is needed.

Such a campaign should build on the successes of the antisweatshop campaigns in organizing coalitions among labor, human rights, and interfaith groups around the world. These non-governmental coalitions have offered a model of how to organize across national borders in an age of free trade–organized labor. As in the fashion world, the integrity of a company's brand name is all-important, and its chief point of vulnerability—the weak link in the chain of capital. Companies must keep their brand names clean, because it is often the only thing that distinguishes their product from that of their market competitors; if that name is sullied, it does not matter whether they use the cheapest labor pool in the world. There is no reason why the brand names of AT&T, Phillips, Intel, IBM, Hewlett-Packard, Toshiba, Samsung, and Fujitsu cannot be publicly shamed in the same way as Nike,

I'm the Total Quality woman. I am the culturally engineered, downsized, outsourced, teleworked, deskilled, Taylorized mom, just-in-time, take-out, time-saving, time-starved, emotionally downsized, downright tired... My home is my work, my work is my home. I work with machines; I live with machines; I love with machines; computer, modem, TV, VCR, printer, scanner, refrigerator, washing machine, dryer, vacuum cleaner, cars telephones, fax machine, hairdryer, vibrator, CD player, radio, pencil sharpener, blender, mixer, toaster, microwave, cell phone, tape recorder... [Wilding, Economy]

The Gap, Guess, and Disney. So, too, it is important not to underestimate public outrage. Far from apathetic, public concern has been inflamed by revelations about labor abuses in the industrialized and nonindustrialized world, where workers are physically, sexually and economically abused to save 10¢ on the cost of a pricey item of clothing. Unlike clothing, consumption of high-tech goods is not yet a daily necessity; but increasingly it is becoming a market in the range of household items. The planned expansion of the semiconductor industry is massive, and will outstrip most other industrial sectors. Very soon, the high-tech market will be within the orbit of consumer politics on the scale of boycott threats, and so many of the strategies of the garment campaigns will make more sense.

In concluding, perhaps it is worth considering why so little attention is paid to these labor issues in the flood of commentary directed at cyberspace. One reason certainly has to do with the lack of any tradition of organized labor in these industries. The fight against the garment sweatshop was a historic milestone in trade union history, and gave rise to the first accords on industrial democracy. Likewise, the recent campaigns have been on the leading edge of the resurgent labor movement, at least in the U.S. Nothing comparable exists in the high-tech workplaces of the new information order. Indeed, high-tech industry lobbyists have been leaders in efforts to undermine the existing protections of labor laws. A second reason has to do with the ideology of the clean machine: in the public mind, the computer is still viewed as the product of magic, not of industry. It is as if computers fall from the skies, and they work in ways that are beyond our understanding. The fact that we can repair our car but not our computer does not help. As a result, the manufacturing process is obscured and mystified. A third reason probably has to do with the utopian rhetoric employed by the organic intellectuals and pundits of cyberspace. Take Kevin Kelley's influential book, *Out of Control*, five hundred pages of heady ruminations about the biologizing of the machine, the death of centralized, top-down control, webby nonlinear causality, the superorganic consciousness of swarmware, and so on. Nowhere is there any mention of the "second world" I described earlier— the low-wage world of automated surveillance, subcontracted piecework, crippling workplace injuries, and the tumors in the livers of chip factory workers. Nowhere is there any recognition of the global labor markets—with their cruel outsourcing economies—that provide the manufacturing base for the new clean machines. His book is not an exception. There is a complete and utter gulf between the public philosophizing of the whizkid new media designers, artists, and entrepreneurs and the global sourcing of low-wage labor enclaves associated with the new information technologies. Boosters like Kelley speak of an ethic of "intelligent control" emerging from the use of the new media. The term is hauntingly accurate, because it evokes a long history of managerial dreams, on the one hand, and automated intelligence on the other. How you feel about this ethic may ultimately depend on which side of the division of labor you find yourself.

Again, the problem lies not with the technologies themselves, nor, ultimately, with their operating speed. It *is* possible to have an affordable, sustainable media environment without electronic sweatshops, just as it is possible to have a sustainable world of fashion without garment sweatshops. But as long as we keep one realm of ideas apart from the experience of the other, people simply will not make the connections between the two. I find this lack of impetus striking, especially among new media professionals themselves, who are well positioned to mediate, and act accordingly. There has been a good deal of attention to labor conditions facing multimedia artists and other new media professionals, and, at least in the realm of software, there is some sense that their self-interest and expertise carries some weight, but this has not been extended to the conditions of production of hardware. It is surely important to see those conditions on a continuum, and to think beyond the self-interest of this group of software experts, for whom the sick jokes about HTML sweatshops belong to a gallows humor they can afford but others cannot. The successes of anti-sweatshop garment organizing have come as a surprise to many seasoned activists, long accustomed to being shut out of the media, to the stony indifference of the public, and to the cruel march of corporate armies across the killing fields of labor. In the case of information technology, the time is ripe for capitalizing on the climate for such successes. Perhaps we can exercise a little foresight, and anticipate the public appetite for responding to such abuses. The history of the internet should remind us that nothing is impossible, and what was unimaginable three years ago is a fact of life today.

[Edited by Ted Byfield and Diana McCarty.]

SUBJECT: CORPORATE COOL: LIFE ON ONE OF AOL'S CHANNELS

FROM: HIDDENSTAIR@YAHOO.COM (ROBIN BANKS)
DATE: THU, 17 SEP 1998 10:

Deep in the heart of the Northern Virginia suburbs outside Washington, D.C., in an arena at George Mason University, the stage is dark when the blues band starts. The space, usually used for college basketball games and pop concerts, is filled this afternoon with casually dressed but wholesome-looking young adults, as it might be any evening. But today, each attendee wears a photo ID badge around the neck.

The band plays faster and faster down below now they're rendering the Blues Brothers movie theme as the purple-and-white lights crescendo. Two figures get out of a police car parked on the arena's floor. The figures wear sunglasses and fedoras, but it's clear once they get out and run onstage to roaring applause: These cleanshaven, tidy-haired corporate men ain't no Blues Brothers.

More like the khaki brothers. But, like the Blues Brothers, the khaki brothers are On a Mission. And they're full of conviction that that mission: running the America Online empire makes them cool. One, Bob Pittman, co-founded behemoth teen tastemaker MTV and moved on to head middle-

IT is now the single biggest part of the U.S. economy, 11 percent of the GNP. Globalization. Free-trade zones. The market Economy. Bye-bye borders. There is no place to hide. Knowledge management: Husbandry for ideas. Mass customization: The market of you. Just-in-time learning: knowledge at your fingertips. [Wilding, Economy]

American real-estate franchiser Century 21 before bringing his mass-market sensibilities to AOL, where he is now president. The other, Steve Case, spent his tender years as a pizza designer for Pizza Hut before founding the online service that would become the world's largest. He is now its chairman, and thus he is the idol of a thousand young hopefuls in the corporate ranks.

Welcome to America Online's annual "all-hands meeting and beer bash." Welcome! It's the word on the free pen they give you at orientation on your first day at work, and it's the word your computer will chirp when you log on for the last time the day you quit and they kill your account.

AOLers-as-missionaries is today's theme, hence the Blues Brothers reference. Steve Case is shouldering the old white man's burden: to give the masses what he sees fit for them (and thereby, it goes without saying, reaping enormous profits). He's doing it in his usual uniform of denim AOL-logo shirt and khakis. Oh-so-casual yet painstakingly bland, it's a look much emulated around the AOL "campus" by twenty- and thirty-something male employees who, like their female counterparts, drive BMWs with vanity plates to work, where they sit at desks covered in Beanie Babies inside cubicles decorated with "cool" ads.

Things not well branded are not held in high esteem here. The hip image aimed for at the Blues Brothers beer-bash meeting is less successful, less cleanly orchestrated, down the food chain. "Cool," says a manage, "rock and roll." "He is addressing his underlings, ten or twenty young adults, as they sit around a conference room table. They are some of the legions who program the content onto AOL's colorful, ad-plastered screens. They're wearing jeans, T-shirts, the odd tattoo. The unwincing twenty- and thirty-something employees are clearly used to the casually misbegotten nuggets of slang liberally tossed into the newspeak.

All statements are positive, "win–win." Talk at this meeting, held by one of AOL's "creative" departments, largely revolves around how the department is going to hold up its end of sweetheart contracts with other corporations. Such deals, a hefty cornerstone of AOL's strategy, usually amount to the sale of a piece of AOL's heavily trafficked cyberspace to another corporation wishing to park its content, ads, or website connections where AOL's twelve million "members" will see them. The terms of sale, lease, or trade vary widely; sometimes AOL pays, sometimes the other party. Meetings and mass-email messages mandate how best to serve these corporate "partners," or dictate new conditions tacked onto their contracts.

These meetings also sometimes touch on how AOL can better deliver its other product—a "quality member experience"— to its other customers. It's the usual commercial media equation: selling a product to an audience + selling that audience to advertisers = profit. That might not come as a shock to anyone who spends time clicking around the service, trying to find something to read behind the promotional teasers scattered everywhere.

These employees stick this content up on AOL's screens after it is produced elsewhere, text and picture, by another corporation's employees far away in some other hive. AOL has chosen to make contracts with dozens of magazines, wire services, television networks and reference-book companies in lieu of paying writers, editors, and photographers to produce original coverage.

Such convenience of access has its benefits, if this is the kind of thing you want to read. But a visit to the public library gets you much of the same product for free: *Entertainment Weekly*, *Newsweek*, *Compton's*, except without the email account.

Like the all-hands spectacle, the departmental meeting is more briefing than discussion. The manager tends to rattle off names of fellow managers, in-house acronyms and project code words unintroduced. But none of the Gen-X attendees are playing Buzzword Bingo under the table. It's a sad, but familiar, lack of solidarity among the drones.

Stock options, which even entry-level content programmers get, usually vest after the first year of employment at AOL. Funnily enough, after exactly that length of time, many people are out the door. But "creatives" are probably easy to replace. The fields that more traditionally employ them are notorious for their starvation wages, and AOL's money and benefits sound comparatively good during the interview. And they would be, if more job satisfaction came with them.

Plenty of staffers say they're demoralized by micromanagement and chronic understaffing. So they end up fighting each other over time off and who will do that last extra chore. *Smile, smile, wink, wink,* go the bosses' emoticons in their "instant messages," via which they drop orders on their swamped underlings even as those underlings type furiously. Thanks to the wonderful AOL medium of "IMs," the boss needn't look into the employee's harried eyes before s/he delivers the instructions; s/he needn't even be in the office. An IM is a small, temporary chat window that pops up on the screen of the person you send it to, if they're online. Wonderful invention for people miles apart. Bad invention for people separated by a cubicle wall, a few feet and a chasm of misunderstanding.

Interdepartmental communication got the worst marks on AOL's employee survey this year and last year. But communication with direct colleagues—the people one has to see every day—makes all the difference to an employee's morale and quality of life. The "interactive media" jobs at this "network" company are done by individuals sequestered alone and working frenetically in high-walled cubicles (which AOL calls "pods") and, in some cases, at staggered times of day and night. As long as workers are kept apart, people can't exchange information on a broad enough scale to realize it's not just their personal failure to fit in that's making their job suck.

The old-fashioned network that internet employees could most benefit from, the labor union, is explicitly discouraged in the AOL employees' handbook. "We know you are more than just an AOL employee. You're an individual and deserve to be treated as such... We feel it is not in the best interests of you or the company to participate in union activities. Instead, speak for yourself—directly with management." AOL's antiunion shop depends on the anticollective attitude of the young members of the specialist class who grew up under Reagan. If you come straight to Daddy instead of falling in with those bad other kids, we'll work something out. But don't dare go behind our backs. We know you wouldn't; we expect your loyalty. And, anyway (appealing here to computer-geek arrogance), you, alone, are your own best representative. Not only does big daddy expect you not to need unions, you'd also better not expect any coddling and handholding from him. You work for a "cool company," don't you? What could you have to complain about?

This is where the mandatory "performance management workshops" come in. Here, workers are drilled to internalize the "management" of their own "performance." This means, roughly summarized: Set your own goals, but make sure they match up with the company's "core values," or you'd best find another company. And if you need more or less supervision from your boss, tell him or her so. It's that easy. The workshops are softened up with Dilbert cartoons, which are served without a trace of irony.

Many employees complain of AOL's workaholism. Low-level employees are expected to go the extra mile, but at a tiny fraction of the starting pay of other professions, which require a slavish dedication to, for example, medicine or the law. The reward? None is suggested; apparently, you're supposed to feel privileged just to work here. "There's a gym," one worker says, "but I can't go because nobody in my group takes an hour for lunch. "That contrasts starkly with the employee handbook's assurance that AOL has the gym because it cares about your physical well-being. One thing AOL does use the gym for is to parade middle-aged male visitors in suits through on their tours of the headquarters as young employees work out on the stair machines.

Meanwhile, as one-year anniversaries roll around and people quit, more hopeful B.A.s are bought off with a handful of stock options that sound great but wouldn't pay off a year's college loans. In the information sweatshop economy, a four-year degree is required for the lowliest administrative job. And people with advanced degrees and specialized computer training can make less in real dollars than, say, dropouts who worked in box factories did in 1974. As the U.S. work force solidifies into two camps, rich and poor, what gold there was in them thar silicon hills has pretty much already been claimed.

But there are a few happy faces rushing through AOL's corridors, carrying cafeteria-made wrap sandwiches and Starbucks mochas back to their desks. White male faces, mostly, attached to bodies dressed in Dockers and pressed shirts. To them, working here is apparently fat city. Most are, or think they are, on the management track. And at least a few are on a smug, egregious class climb, bragging about wine, resorts, cars, and boats. For all their lip service to "new media," these typical middle-management types are planted firmly on the creaky old corporate ladder.

They might want to think twice about their loyalty. It's common knowledge that for its users, AOL's happy-face icons mask buggy software, slow connections, and overloaded modems. And inside the company, underneath the cheap strokes of occasional keg parties, mass-emailed words of thanks, and management mumbojumbo, the company invests about as much in its wetware as it does in its semidisposable software and hardware. Both as a mass producer of adfotainment and as a "corporate culture," AOL represents the cynical exploitation of the lowest common denominator. Meet the new media corporation, same as the old corporation but with more ads and less content.

But what else is new? It must be remembered that AOL is not unique or even remarkable. Indeed this corporation is no worse than most others, and probably better than many (no piss testing for one thing).

Still, when you walk out of the former aircraft hangar, through the glossy, soaring lobby decorated with friendly icons and away from the endless rat-maze cubicles tucked away behind, you'll probably say without too many regrets, and as cheerily as AOL's logoff farewell: *Goodbye!*

SUBJECT: BACK TO THE FUTURE: A PORTABLE DOCUMENT

FROM: ANTONIO NEGRI
DATE: THU, 15 OCT 1998, 04:54:42 -0300

WORK

There is too much work because everyone works, everyone contributes to the construction of social wealth, which arises from communication, circulation, and the capacity to coordinate the efforts of each person. As Christian Marazzi says, there is a biopolitical community of work, the primary characteristic of which is "disinflation"—in other words, the reduction of all costs that cooperation itself and the social conditions of cooperation demand. This passage within capitalism has been a passage from modernity to postmodernity, from Fordism to post-Fordism. It has been a political passage in which labor has been celebrated as the fundamental matrix of the production of wealth. But labor has been stripped of its political power. The political power of labor consisted in the fact of being gathered together in the factory, organized through powerful trade union and political structures. The destruction of these structures has created a mass of people that from the outside seems formless—proletarians who work on the social terrain, ants that produce wealth through collaboration and continuous cooperation. Really, if we look at things from below, from the world of ants where our lives unfold, we can recognize the incredible productive capacity that these new workers have already acquired. What an incredible paradox we are faced with. Labor is still considered as employment; that is, it is still considered as variable capital, as labor "employed" by capital, employed by capital through structures that link it immediately to fixed capital. Today this connection—which is an old Marxian connection, but before being Marxian it was a connection established by classical political economy—today this connection has been broken. Today the worker no longer needs the instruments of labor, that is, the fixed capital that capital furnishes. Fixed capital is something that is at this point in the brains of those who work; at this point it is the tool that everyone carries with him- or herself. This is the absolutely essential new element of productive life today. It is a completely essential phenomenon because capital itself, through its development and internal upheavals, through the revolution it has set in motion with neoliberalism, with the destruction of the welfare state, "devours" this labor power. But how does capital devour it? In a situation that is structurally ambiguous, contradictory, and antagonistic. Labor is not employment.

The unemployed work, and informal or under-the-table labor produces more wealth than employed labor does. The flexibility and mobility of the labor force are elements that were not imposed either by capital or by the dissolution of the welfarist or New Deal–style agreements that dominated politics for almost half a century. Today we find ourselves faced with a situation in which, precisely, labor is "free."

Certainly, on one hand, capital has won; it has anticipated the possible political organizations and the political "power" of this labor. And yet, if we look for a moment behind this fact without being too optimistic, we also have to say that the labor power that we have recognized, the working class, has struggled to refuse factory discipline. Once again we find ourselves faced with evaluating a political passage, which is historically as important as the passage from the Ancien Régime to the French Revolution. We can truly say that in this second half of the twentieth century we have experienced a passage in which labor has been emancipated. It has been emancipated through its capacity to become immaterial and intellectual, and it has been emancipated from factory discipline. And this presents the possibility of a global, fundamental, and radical revolution of contemporary capitalist society. The capitalist has become a parasite, but not a parasite in classical Marxist terms—a finance capitalist—rather, a parasite insofar as the capitalist is no longer able to intervene in the structure of the working process.

BRAIN-MACHINE

Clearly when we say that the working tool is one that workers have taken away from capital and carry with themselves in their lives, embodied in their brains, and when we say that the refusal of work has won over the disciplinary regime of the factory, this is a very substantial and vital claim. In other words, if labor and the tool of labor are embodied in the brain, then the tool of labor, the brain, becomes the thing that today has the highest productive capacity to create wealth. But at the same time humans are "whole;" the brain is part of the body. The tool is embodied not only in the brain but also in all the organs of sensation, in the entire set of "animal spirits" that animate the life of a person. Labor is thus constructed by tools that have been embodied. This embodiment, then, envelops life through the appropriation of the tool. Life is what is put to work, but putting life to work means putting to work what exactly? The elements of communication of life. A single life will never be productive. A single life becomes productive, and intensely productive, only to the extent that it communicates with other bodies and other embodied tools. But then, if this is true, language, the fundamental form of cooperation and production of productive ideas, becomes central in this process.

But language is like the brain, linked to the body, and the body does not express itself only in rational or pseudo-rational forms or images. It expresses itself also through powers, powers of life, those powers that we call affects. Affective life, therefore, becomes one of the expressions of the incarnation of the tool in the body. This means that labor, as it is expressed today, is something that is not simply productive of wealth: it is above all productive of languages that produce, interpret, and enjoy wealth, and that are equally rational and affective. All this has extremely important consequences from the standpoint of the differences among subjects. Because once we have stripped from the working class the privilege of being the only representative of productive labor, and we have attributed it to any subject that has this embodied tool and expresses it through linguistic forms, at this point we have also said that all those who produce vital powers are part of this process and essential to it. Think for example of the entire circuit of the reproduction of labor power, from maternity to education and free time—all of this is part of production. Here we have the extraordinary possibility of reanimating the pathways of communism, but not with a model of the rationalization and acceleration or the modernization and supermodernization of capitalism.

We have the opportunity to explain production and thus organize human life within this wealth of powers that constitute the tool: languages and affects.

THE BECOMING-WOMAN OF LABOR

With the concept of "the becoming-woman of labor" you can grasp one of the most central aspects of this revolution we are living through. Really, it is no longer possible to imagine the production of wealth and knowledge except through the production of subjectivity, and thus through the general reproduction of vital processes. Women have been central in this. And precisely because they have been at the center of the production of subjectivity, of vitality as such, they have been excluded from the old conceptions of production. Now, saying "the becoming-woman of labor" is saying too much and too little. It is saying too much because it means enveloping the entire significance of this transformation within the feminist tradition. It is saying too little because in effect what interests us is this general transgressive character of labor among men, women, and community. In fact, the processes of production of knowledge and wealth, of language and affects, reside in the general reproduction of society. If I reflect back self-critically on the classical distinction between production and reproduction and its consequences, that is, on the exclusion of women from the capacity to produce value, economic value, and I recognize that we ourselves were dealing with this mystification in the classical workerist tradition, then I have to say that today effectively the feminization of labor is an absolutely extraordinary affirmation; because precisely reproduction, precisely the processes of production and communication, because the affective investments, the investments of education and the material reproduction of brains, have all become more essential.

Certainly, it is not only women that are engaged with these processes; there is a masculinization of women and a feminization of men that moves forward ineluctably in this process. And this seems to me to be extremely important.

MULTITUDE

Some historical clarification is needed here. The term multitude is a pejorative, negative term that classical political science posed as a reference point. The multitude is the set of people who live in a society and who must be dominated. Multitude is the term Hobbes used to mean precisely this. In all of classical, modern, and postmodern political science the term multitude refers to the rabble, the mob, and so on. The statesman is the one who confronts the multitude that he has to dominate. All this came in the modern era before the formation of capitalism. It is clear that capitalism modified things, because it transformed the multitude into social classes. In other words, this division of the multitude into social classes fixed a series of criteria that were criteria of the distribution of wealth to which these classes were subordinated according to a very specific and adequate division of labor. Today, in the transformation from modernity to postmodernity, the problem of the multitude reappears.

To the extent that social classes as such are falling apart, the possibility of the self-organizational concentration of a social class also disappears. Therefore we find ourselves faced again with a set of individuals, but this multitude has become something profoundly different. It has become a multitude that, as we have seen, is an intellectual grouping. It is a multitude that can no longer be called a rabble or a mob. It is a rich multitude. This makes me think of Spinoza's use of the term multitude because Spinoza theorized from the perspective of that specific anomaly that was the great Dutch republic, which Braudel called *the center of the world*, and which was a society that had mandatory education already in the seventeenth century. This was a society in which the structure of the community was extremely strong and a form of welfare existed already, an extremely widespread form of welfare. A society in which individuals were already rich individuals. And Spinoza thought that democracy is the greatest expression of the creative activity of this rich multitude. Therefore, I think of Spinoza's use of the term, which had already reversed the negative sense of the multitude, like the wild beast Hegel called it, which has to be organized and dominated. And this rich multitude that Spinoza conceived instead is the real counterthought of modernity, in that line of thought that goes from Machiavelli to Marx, of which Spinoza forms more or less the center, the central apex, the transition point; ambiguous, anomalous, but strong. Well, this concept of the multitude is the concept that we invoked before. There exists today a multitude of citizens, but saying "citizens" is not sufficient because it simply defines in formal and juridical terms the individuals that are formally free. You have to say rather that today there exists a multitude of intellectual workers, but even that is not enough. You have to say: there exists a multitude of productive instruments that have been internalized and embodied in subjects that constitute society. But even this is insufficient. You have to add precisely the affective and reproductive reality, the need for enjoyment. Well, this is the multitude today. Therefore, a multitude that strips every possible transcendence from power, is a multitude that cannot be dominated except in a parasitic and thus brutal way.

THE BIOPOLITICAL ENTREPRENEUR

Here too, as usual, we are dealing with a sphere in which all the terms have been inverted—direct terms. We must really succeed in inventing a different language, even when we speak of democracy and administration. What is the democracy of biopolitics? Clearly it is no longer formal democracy, but an absolute democracy, as Spinoza says. How long can such a concept still be defined in terms of democracy? In any case, it cannot be defined in the terms of classical constitutional democracy. The same thing is true when we speak of the entrepreneur, when we speak of the political entrepreneur, or better the "biopolitical" entrepreneur. Or, rather, when we speak of the one who could be single or a set of collective forces, that succeeds at times in focusing productive capacities in a social context. What should we say at this point? Should this collective entrepreneur be given a prize? Frankly, I do think so, but all this has to be evaluated within the biopolitical process. I would say that here we really have the opposite of any capitalist theory of a parasitic entrepreneur. This is the ontological entrepreneur, the entrepreneur of fullness, who seeks essentially to construct a productive fabric. We have a whole series of examples that have each been at times very positive. There is no doubt that in certain community experiences, red (communist) collectivities, cooperatives basically, and in certain experiences of white (liberal) communities based on solidarity, we can see examples of collective entrepreneurship. As usual, today, we must first of all begin to speak not only of a political entrepreneur, but also of a biopolitical entrepreneur, and then begin to recognize also the inflationary or deflationary biopolitical entrepreneur. The biopolitical entrepreneur determines always greater needs while organizing the community; the entrepreneur represses and redisciplines the forces at play on the biopolitical terrain. There is no doubt that an entrepreneur in the Sentier neighborhood, to take an example from the studies we did here in France, is a biopolitical entrepreneur, one who often acts in a deflationary way. Benetton is the same thing. I really believe that the concept of entrepreneur, as a concept of the militant within a biopolitical structure, and thus as a militant that brings wealth and equality, is a concept that we have to begin to develop. If there is to be a fifth, a sixth, or a seventh Internationale, this will be its militant. This will be both an entrepreneur of subjectivity and an entrepreneur of equality, biopolitically.

5316 Useless Meeting 5319 Waiting for Break or Lunch 5321 Waiting for End of Day 5322 Vicious Verbal Attacks Directed at Co-worker 5323 Vicious Verbal Attacks Directed at Co-worker While Co-worker Is not Present 5481 Buying Snack 5482 Eating Snack 5500 Filling out Timesheet 5501 Inventing Timesheet Entries 5502 Waiting for Something to Happen 5504 Sleeping 5510 Feeling Bored 5600 Complaining About Lousy Job, Low Pay, and Long Hours 5603 Complaining About Coworker (See Code 5322 & 5323) 5604 Complaining About Boss 5640 Miscellaneous Unproductive Complaining 5702 Suffering from Eight-hour Flu 6102 Ordering out 6103 Waiting for Food Delivery to Arrive 6104 Taking It Easy While Digesting Food 6200 Using Company Resources for Personal Profit 6201 Stealing Pencils and Pens from Company 6203 Using Company Phone to Make Long-distance Personal Calls 6204 Using Company Phone to Make Long-distance Personal Calls to Sell Stolen Company Pencils 6205 Hiding from Boss 6206 Gossip 6207 Planning a Social Event 6211 Updating Resumé 6212 Faxing Resumé to Another Employer/Headhunter 6213 Out of Office on Interview 6221 Pretending to Work While Boss Is Watching 6222 Pretending to Enjoy Your Job 6238 Miscellaneous Unproductive Fantasizing 6350 Playing Pranks on the New Guy/Girl 6601 Running your own Business on Company Time 6603 Writing a Book on Company Time 6612 Staring at Computer Screen 7400 Talking with Lawyer on Phone 7401 Talking with Plumber on Phone 7931 Asking Co-worker to Aid You in an Illicit Activity 8100 Reading Email 8101 Distributing Email jokes to All Your Friends

GUARANTEED WAGE

There are reductive conceptions of the guaranteed wage, such as those we have seen in France—for example, the French RMI laws [*Revenu Minimum d'Insertion*: the "minimum income" required for integration into society], in the form they were passed, are a kind of wage structure of poverty, and thus a wage structure of exclusion, laws for the poor. In other words, there is a mass of poor people—but keep in mind that these are people who work, who cannot manage to get into the wage circuit in a constant way, who are given a little money so that they can care for their own reproduction, so that they don't create a social scandal. Therefore there exist minimum levels of the guaranteed wage, subsistence wages, that correspond to the need of a society to avoid the scandal of death and plague, because exclusion can easily lead to plague. And poor laws were born of this danger in England in the seventeenth and eighteenth centuries. There are thus forms of the guaranteed wage that amount to this. But the real question of the guaranteed wage is a different one. It is a question of understanding that the basis of productivity is not capitalist investment but the investment of the socialized human brain. Therefore, the maximum freedom, the break with the disciplinary relationship of the factory, the maximum freedom of labor is the absolute foundation of the production of wealth. The guaranteed wage means the distribution of a large part of income and giving the productive subjects the ability to spend it for their own productive reproduction. This becomes the fundamental element. The guaranteed wage is the condition of the reproduction of a society in which people, through their freedom, become productive. Clearly, at this point, the problems of production and political organization tend to overlap. Once we have pursued this discourse all the way, we have to recognize that political economy and political science, or the science of government, tend to coincide. Because we maintain that democratic forms, forms of a radical, absolute democracy—I don't know if the term democracy can still be used—are the only forms that can define productivity. But a substantial, real democracy, in which the equality of guaranteed incomes becomes ever larger, and ever more fundamental. We can then realistically talk about incentives, but these are discourses that in today's world are not very relevant.

Today the big problem is that of inverting the standpoint on which the critique of political economy itself is based. In other words, the standpoint of the necessity of capitalist investment.

We have said before and we have been saying for years that the fundamental problem is the reinvention of the productive instrument through life, the linguistic, affective life of subjects. Today, then, the guaranteed wage, as a condition of the reproduction of these subjects and their wealth, becomes an essential element. There is no longer any lever of power, there is no longer need for any transcendental, any investment.

This is a utopia, it is one of those utopias that become machines of the transformation of reality once they are set in motion. And one of the most beautiful things today is precisely the fact that this public space of freedom and production is beginning to be defined, but it carries with itself, really, the means to destroy the current organization of productive power and thus political power.

[Translated from the Italian by Michael Hardt. Edited by Hope Kurtz.]

INTERVIEW WITH ANTONIO NEGRI
BY ANGELA MELITOPOULOS AND NILS ROELLER

(JUNE 28, 1997)

Q: Are you returning to Italy as someone who has been defeated politically?

Negri: *Autonomia operaia* focused on the continuing transition from the traditional labor movement to the new subjects that have formed because of the development of modern capitalism. A new class was facing the factory workers' unions—a new class that didn't yet possess a new identity through its intellectual and social labor and operated with autonomous organizational structures. It was our goal to shape this passage from classical factory labor to social labor. The identity of this new subject, to which we referred as the "social laborer," determines our society today. This does not mean the devaluing of labor as the central factor that creates wealth and value within society, but rather that this factor in the power structure is formed in a completely new way through today's conditions of production. Efforts to accelerate this process through political action have failed; in this we have been defeated, but not in our evaluation of this new concept of labor.

Q: In your statement to the press, you referred to the fact that you are going back to Italy in order to facilitate your citizenship. What is the relationship between your exile and European unification?

Negri: In no European country was there a reaction to the social movements after 1968 that was as contemptuous of human beings as that in Italy. The political strategy in France and Germany consisted of the political absorption of the broad masses of the movement, for example, into the Green Party or into alternative projects. Because of this, the radical and terrorist groups were isolated. In Italy, things were handled—and continue to be handled—differently: the entire extraparliamentary movement was characterized as terrorist and an entire generation was therefore criminalized and forced into internal and foreign exile. By returning, I would like to draw attention to the fact that the new government in Italy has the opportunity to "work through," honorably and democratically, this legacy of the First Republic and bring to an end the dark past of state terrorism. The state policy of provocation was responsible for thousands of deaths in the seventies; banks were blown up and bombs planted on trains. The outrage in Bologna, in which more than a hundred people were killed, was carried out by the secret service and by paid right-wing radicals. Certainly, we and our movement made mistakes. None of us wanted this civil war.

Q: Are you demanding a new, fair trial?

Negri: No, there can't be a new, fair trial; the cases are closed. In the case of Sofri, there was finally a decision yesterday against a reopening of the case. I would like to advance the parliamentary discussion of amnesty. For the last four sessions of the legislature, a draft of a bill [on amnesty] has been awaiting a decision. In most of the judgments rendered at the time, defendants received the maximum sentences. We cannot forget that there was state abuse of power here, particularly in the use of the state's witnesses, whose testimony often fell apart. This was underscored by the French state, which has offered sanctuary to those sentenced by these Italian courts since 1979.

Q: The seventeenth-century philosopher Spinoza has been important to your thought; he was exiled from his own community. Is Italy still a land from which a Spinozaist must flee?

Negri: "Spinozaism" for me means two things. First, the examination of causes rather than of effects. And, second, a call to an activism that constructs new communities on an ethical foundation. These communities are democratic because they emerge through the praxis of a majority of individuals. But even Spinoza himself didn't know how to unify his intellectual work and his activism.

Q: What would be ethical behavior in Italy today, whether as a politician or a private person?

Negri: That can't be answered so quickly and in such general terms. One can, however, note that citizens today are in possession of greater power than ever before. In all areas, the productive force of immaterial labor is unfolding. The problem at hand is that of forming a new public space in which democratic and productive forces will be able to become effective together, so that individuals [*Einzelnen*] discover the power of the community and recognize the potential of common democratic production that is inherent to it. Thus, I don't differentiate between political and private behavior, but instead think of individuality and community together on a democratic/productive foundation.

Q: How is it possible to behave politically in an electronic society in which individual workers don't know each other personally?

Negri: Clearly it isn't easy, but I think that one must simply engage oneself and do it! I am taking up my political work again starting from the ground up, from prison. With my return, I would like to give a push to the generation that was marginalized by the anti-terrorist laws of the seventies so that they will leave their internal or foreign exile and again take part in public and democratic life. This is our opportunity to re-identify ourselves. But prison as a site of noncommunication, of exclusion from political activism? That's not the case. One communicates not only with the help of electronic instruments, but above all, through the position that one assumes in a political/social situation. The position one takes within the event in which one is taking part communicates on the foundation of the body, even on the internet. It is a combination of rationality and feeling, of intelligence and emotionality, and if it doesn't exist, all communication is empty, nonexistent. What we have in common precedes us in bodily form.

[Translated from the Italian by Jamie Owen Daniel. Edited by Hope Kurtz. This interview originally appeared in the German daily *TAZ*.]

I work in the network operations center (NOC) of a major internet provider. The NOC is a large room, laid out like the bridge of the Starship Enterprise, wherein we watch our company's internet backbone 24 hours a day, 7 days a week. But for all the sci-fi semiotics, the NOC is a factory floor. Like my father, who spent twenty years at a refinery in East Texas, turning out bales of synthetic rubber, I answer to a foreman. The rubber, like the data packets here, flowed 24 hours a day. How did my segment of the internet industry, the industry of Trekkies and cyberpunks, turn into another boiler room, and so quickly? In oil and aerospace, the transition from wildcatters to wage slaves was measured in decades. At my company, it took three years.

For most of its eleven years, the company stayed small. In 1996 it contrived to be bought by a larger company to gain access to a newly deregulated Euro internet market. The company grew up, the stock options dwindled, and beer was banished from the NOC. The parent company ruled with a light hand until this summer, when NOC engineers were downgraded from salaried professionals to hourly technicians, because that's where operations people fit into our parent company's (long-distance telephony) scheme of things and that is that.

In my last job I learned to spot the deadly warning signs of corporate middle-age: exodus of mavericks, emphasis on credentials, adoption of urinalysis ("pre-employment screening"), "metrics," and the absolute bottom—*Total Quality Management*. My company has manifested four of these.

In an operations center, information about the network flows in, computers make sense of it, and people act on it. A NOC can be as small as a half-dozen workstations or as large as NASA's Mission Control Center, where I worked before coming to this job. We work in shifts, reporting on problems, troubleshooting them, and handing the tough ones over to the next shift. My father, late in his career, oversaw the rubber refinery's operations from inside a control room. The rubber was piped into the building, extruded, dried, and baled. This process was presented to him as a lighted flow diagram; our network is displayed on our wall as a giant cat's cradle.

When I started working here, the company was run by gnomish old-school computer gods or hairy cyberpunks. The founder had invented a basic protocol for dialing into the internet. One pasty-faced geek hid behind harsh email personas, Oz-like, to intimidate the demobbed military types who staffed the NOC (and still do). But the weirdos cashed in their extremely generous stock options or ascended out of the NOC and became magical friends—systems engineers—to be called when a problem was too complex for the NOC to handle. The founder went into semiretirement and bought a Star Wars X-Wing fighter he keeps in a hangar. The cyberpunks cut their hair. Now there are distinct castes: Morlocks in the NOC, perky Eloi in Sales, chameleons in middle management, and a CEO who wears stylish black.

This project is so important, we can't let things that are more important interfere with it. Doing it right is no excuse for not meeting the schedule. No one will believe you solved this problem in one day!? We've been working on it for months. Now, go act busy for a few weeks and I'll let you know when it's time to tell them.

Not that our people were very eccentric to begin with, compared to their counterparts in Austin, Palo Alto, or Seattle. Our engineers mustered out of the military, telcos, and unnamed government agencies. At least half have had Secret clearances (some had Top Secret access), which means they know about Rex 84 and the Secret U.N. Symbols on Road Signs For Their Army To Read When They Go Marching Through Georgia, but have never taken LSD. They play the online stock market, watch stock car races on TV (worrying that NASCAR champion Jeff Gordon is gay), and eat at Taco Bell. This isn't California. No one went to Burning Man. East Coast geeks don't have to stock up on guns, ammo, and monster trucks in anticipation of Y2K-bug-induced chaos, because they've already got plenty of all three. Politically, they're right-libertarian, which means they've got nothing personally against abortion, so long as their tax dollars don't pay for it. The meager political choices available here mean they consistently vote Republican.

NOC engineers are like the technicians who worked at the oil refinery with my father—their skills and connections got them into the NOC but can't get them out, especially now. Aside from the experience I mentioned earlier, I've also learned a few tricks from bumming around the internet. I'm part of a group of six friends who followed each other here from Texas. My father's co-workers got their jobs from relatives or friends, and often came out of the oilfields or the Navy. But when this company grew, it raised the hurdles to promotion. It's still possible to get a NOC job without a degree, but more work experience is required than before. Like a lot of people here, if I were applying today, I might not get in. The company encourages those of us who don't have degrees to get them. The degree doesn't help you very much in the NOC, but it's your only ticket out of there. When we were downgraded to technicians, we were told that we could still move to an engineer's slot without a degree, but the job postings say otherwise. At the refinery, management offered a similar career path for the operators, but when it was offered, most of those guys were well into their thirties and forties. They'd have retired before they got their degrees. The NOC may be a Sargasso Sea, careerwise. I've got two years of internet NOC experience; the next level requires seven. The company announced in January that it was raising the door price by one cup of urine. Existing employees are exempt. I don't know if the company realizes how far it can go or simply doesn't see the need. Since corporate HQ is in the conservative Deep South, I suspect the latter. For all practical purposes, then, it's all academic. But it's another sign, like a slight shift in the wind. I have a hard time getting my co-workers to see the problem of mandatory drug testing, until I remind them that it extends control over employees 168 hours a week, while paying them for only 40. Aha! An argument that makes sense!

Metrics—management by numbers instead of by people—has reared its ugly head. I've had a hand in it, providing statistics on the types of problems the NOC has encountered, how long it took to solve them, and so on. It's a pain in the ass. Querying the ticket database takes a nimble hand, and running the numbers and making a report often take up a whole day—time I could spend honing my skills. It's my own fault: I volunteered back when it was a simpler job, and now I'm sort of stuck with it. Metrics also play a role,

I suspect, in the doling out of annual raises. In my last job, the budget for raises was fixed. If someone got a great raise, everyone else competed for the remainder. It was a classic zero-sum game: it is not enough that I succeed, but you also must fail as well. I can't say for certain that this is the case here, but the signs—preprinted self-evaluation forms, stated limits on raises, coincidental letters of praise from the CEO—are there.

And now I await the endgame: Total Quality Management, Empowerment, Reengineering, or whatever they'll call the beast when management lets it in the door. TQM (also known as Time to Quit, Man), is the sign that the last scintilla of slack has been sucked out of the job. The company wants you to work harder for less pay and like it. Marxists might call it a new Ideological State Apparatus; I call it crapping on my head and calling it a hat. A guy I work with was at a company that required workers to do Total Quality analyses of their jobs on their own time or risk bad performance reviews. Any meaningful suggestions (meaningful to the worker, at least) were ignored. How long before the rough beast slouches here? I give it a few months, tops. There's a certain logic that drives a company in this direction, or at least lays out a path of least resistance. After a certain point, the company's management loses its taste for excitement and craves respectability (not to mention the tall dollars it attracts). The quickest route is reliability, for which the company will shave off its rough edges. The company grades everyone as Superior, Satisfactory, or Watch Yerself, Bub. It may still be a nice place to work, but it's no longer the place to get rich, make a difference, find yourself, or do anything else that doesn't exactly suit the company's goal of providing ever-higher returns to its shareholders.

This wasn't supposed to happen in the "way new" industry, but it did. The only "way new" aspect is the rapidity with which the process took place. So I'm trading smutty observations about the Clinton/Lewinsky affair with my fellow NOC workers while the televisions show "Hardball with Chris Matthews" (with the sound off, thankfully) or the baseball playoffs. I gotta make like Huckleberry Finn and light out for the territory. But where is it?

SUBJECT: HOME AT WORK / WORK AT HOME

FROM: FAITH WILDING <74447.2452@COMPUSERVE.COM>
DATE: TUE, 8 SEP 1998 16:22:00 -0400

1. FEMINIST MAINTENANCE ART

In recent decades, the mass deployment of electronic technology in offices and workplaces has profoundly changed the structure of work and the relationship of home and work life in ways that are having particularly disturbing effects on women. In the U.S., women who have largely been concentrated in the lower echelons of the labor market—such as clerical work, the garment industries, manufacturing and service jobs—are increasingly being thrown out of waged labor and forced into part-time privatized telework, home-based piecework, and service labor. This situation is once again confining many women to the private sphere of the home, where they perform double maintenance labor: that of taking care of the family, and that of working in the global consumer economy. Made possible by automated Information Technology (IT) and controlled by mobile capital, this market economy is based on just-in-time production and distribution strategies that speed up and control the pace of work and life.

The global disappearance of secure salaried and waged jobs does not mean the end of hard labor or tedious, repetitive, manual maintenance work. Worldwide, much of the rote maintenance work of keyboarding, data entry, electronic parts assembly, and service labor is still done manually, predominantly by women. But the spread of automated machinery into the workplace and the hidden nature of home work and telework is contributing to making women's work and women's laboring bodies invisible again.

Recently, cyberfeminists have begun to meet, both face to face and electronically, to discuss ways of analyzing, revealing, and transforming women's current relationship to IT, as well as ways to intervene in the replication of traditional gender structures in electronic culture. I will discuss some ways in which these concerns relate to women's changing labor conditions worldwide, and suggest how the seventies strategies of making maintenance labor visible could be adapted by cyberfeminist artists and activists today.

2. THE POLITICAL CONDITIONS OF HOME-BASED TELEWORK

Recently, cyberfeminist theorists, activists, and artists have been addressing the role of women in the history of computer development, and the contemporary gender constructions embedded in the new technologies. In "The Future Looms," cyberfeminist Sadie Plant exemplifies some of the more wildly utopian claims that have been made for women in technology: "After the war games of the 1940s, women and machines escape the simple service of man to program their own designs and organize themselves; leaking from the reciprocal isolations of home and office, they melt their networks together in the 1990s" (in L. Hershman, ed., *Clicking in*, SF: Bay Press, 1997, 123) This free mythical realm—neither home nor workplace—presumably is cyberspace, which is imagined as a brave new world for women. Would it were so! But alas, research reveals a far more complex situation for most women who work in the high-tech industries. Here I will briefly summarize the political and economic conditions of contemporary female office and home-based teleworkers, and the regressive effects on women's roles in the home (and on the home in the market economy) caused by the displacement of large numbers of employed women who have been forced back into the "informal" (part-time and home work) labor economy by the global restructuring of work. When large numbers of (mostly white and middle-class) women first started entering the wage-labor market, their traditional gender roles of maintenance and service were easily translated into the division of labor in offices, banks, and many other workplaces. Beginning in the late 1890s, women increasingly became the majority of copy clerks, typists, calculators, stenographers, switchboard operators, bookkeepers, clerical workers, filing clerks, banktellers, keypunchers, and data enterers. When automated office technology was introduced in the seventies, women also became the majority of computer users in offices and workplaces. Because such a high percentage of employed women (43 percent) are clerical workers, it is important to study the effects of the deployment of information technology on clerical work. Researchers have noted the differences in how women and men use computers: "women seemed to have acquired computer skills that leave them doing very different jobs than men who use computers." (B. Gutek, "Clerical Work and Information Technology," in U. E. Gattiker, *Women and Technology*, Berlin: de Gruyter, 1994, 206). These skills tend to be the rote entry, filing, and maintenance of data, done in isolation in front of a terminal. No particular new skills or knowledge are needed for this work, and most companies never invest in training women clerical workers in more advanced computer techniques that would give them a chance to climb the internal company job ladders. They are condemned both to mental and physical repetitive stress syndromes to such a degree that the turnover in clerical workers is almost 100 percent in many offices.

In the nineties, many of these clerical jobs are being replaced by automated computers and networks of robotic machines. Secretaries and clerical workers are the first casualties of the electronic office. Lacking advanced skills and knowledge capital, these displaced women workers often have no other choice than to resort to low-skilled part-time work, or to home-based telework. Such "home work" includes different kinds of work ranging from professional telecommuting, entrepreneurial businesses, salaried employment, and self-employed freelance work, to (often illegal) garment and needle industries, electronic parts assembly, and clerical computer work. While for some upper-echelon female white-collar workers and professionals telecommuting has become part of their job and enhances their value as employees, for the great majority of other casualties of electronic joblessness, the forced "choice" of home work is a big step down—measured in terms of wages, benefits, and working conditions—even from clerical work in an office, and usually amounts to nothing short of the enslaved maintenance work that keeps global capital's production lines and databanks speeding along. Opportunities are especially bad for women of color and immigrants, who tend to be concentrated in jobs most affected by office automation and who have the lowest level of skills.

The political conditions of office and homework in the nineties are restructuring home and work life in crucial ways, and are producing a worldwide labor crisis.

Home work is feminized labor: Feminized home work is a structural feature of the contemporary U.S. telework, data-entry, and service economies, as well as an aspect of the global sweatshop economy (which includes all kinds of assembly work), and the computer chip and electronic parts manufacturing industry. "To be feminized means to be made extremely vulnerable; able to be disassembled, reassembled, exploited as a reserve labor force; seen less as workers than as servers; subjected to time arrangements on and off the paid job that make a mockery of a limited work day; leading an existence that always borders on being obscene, out of place, and reducible to sex" (D. Haraway, "Cyborg Manifesto," *Simians, Cyborgs, and Women*, NY: Routledge, 1985, 166). Work is restructured in a way that downgrades and feminizes professional work, and in turn lowers the pay level and satisfaction of the job.

Ironically, much of the automated technology was designed to replace the rote maintenance labor—mostly performed by women—in offices and factories, and the resultant displacement of women from the public workplace, as well as the renewed invisibility of their work, has had the effect of devaluing women's labor and homemaking services even more, both financially and emotionally.

Home work sustains the gendered division of labor: it is hardly news that home-based work in industrialized nations has historically been extremely exploitive. The global restructuring of work manifests locally, and home work usefully demonstrates "problems in capital-labor relations and in the gendered division of labor" (A. Calabrese, "Home-based Telework," in Gattiker, 177). Telework is defined as "work delivered to the worker via telecommunications as opposed to the worker going where the work is." "Home-based" telework refers to the individual working in the home, rather than in a centralized location. Surveys show that teleworkers are five times as likely as other workers to be women and to be working illegally, without benefits or insurance. There is never time to retrain for higher levels of work, or to get the education to participate in the more lucrative work of knowledge production and management.

Home work reinforces women's subordinate status in the home and labor markets. Despite the much discussed separation of public and private spheres, the history of home work clearly shows that public power (capital) has been used to structure the private lives and control work opportunities for women. Add to this the fact that the new communications technologies have opened the home space to the world, and conversely have brought the world into the private space of the home, and we get a blurring of boundaries that allows surveillance of the home-based worker and "makes the home more accessible to employers, marketers, and politicians" (ibid., 163, 169). Women teleworkers become industrialized women, while women in waged jobs become Taylorized homemakers. As sociologist Arlie Hochschild noted: "[people]...become their own efficiency experts, gearing all the moments and movements of their lives to the workplace" (*The Time Bind*, NY: Holt, 1997, 49). For home-based teleworkers there is no distinction between home and workplace, with the result that when both personal and worklife become Taylorized they have no escape. For women who have often been forced to "choose" home-based work because of the lack of child-care options—a common problem for illegal aliens, for example—home-based telework therefore amounts to a doubling of their bondage to the home space. The blurring of boundaries between private and public in the home-space also often places the woman in a doubled psychological subordination—to her employers and to her husband. The traditional feminine roles of emotional care-giving and physical care-taking become entwined with her externally controlled, maintenance telework in the home. In the long run, female rebellion against these pressures could have the effect of redefining the division of male and female labor, and of repositioning the importance of home life and private free time within the public economy and social relations. In the short run, since home life has no recognized public economic value, it is being more and more curtailed, automated where possible, and reorganized to serve the needs of paid work; and women who work at home have the doubled role of worker and care-givers.

Home work undercuts progressive labor conditions and standards: The geographic mobility of capital made possible by IT uses waged labor, which is space-bound, with the result that geographical areas are increasingly reduced to the status of a captive labor pool. While this makes new modes of production (especially home telework) possible, it does not challenge "the place of the home in the economy, or of women in the home" (Calabrese, 179). The home space and the female working in it under the sign of "choice" actually become the site of regressive labor practices and intrusions of outside control made possible by the dissemination and flexibility of the very information technology that now immobilizes and isolates the woman worker. This isolation also contributes to women's increasing marginalization in the computer sciences, and to the stratification of women in the computer industry between a small percentage of highly skilled engineers, scientists, systems analysts, and knowledge workers, on the one hand, and the vast numbers of low-paid, low-skilled computer workers, on the other. It is this great disparity and its concomitant economic and political consequences that cyberfeminists need to study and address.

3. ACTIVISM, INTERVENTION, RESISTANCE

The political conditions of home-based telework I've outlined pose questions about the effects of restructuring work for women in the integrated circuit: Will this reorganization of work further stratify jobs by race, ethnicity, and gender? Will the changes in work structures "reproduce existing patterns of inequality in only slightly changed forms, perhaps leading to different, more subtle forms of inequality?" (E. N. Glenn and C. Tolbert II, "Technology and Emerging Patterns of Stratification for Women of Color," in B. D. Wright, et al., eds., *Women, Work, and Technology*, Ann Arbor: University of Michigan: 1987, 320).

What are possible points of intervention, resistance, and/or activism for cyberfeminists and artists (among whom I include myself) working with computer technology? On the microlevel, it is time to educate ourselves thoroughly about these conditions, and to disseminate this information as widely as possible through the different cultural and political venues in which we work. We must rethink the contexts in which computers are used, and question the particular needs and relations of women to computer technology. We must try to understand the mechanisms by which women get allocated to lower-paid occupations or industries, and make visible the gender-tracking that obtains in scientific fields of work. For example, many women tend not to choose certain fields because of the "male culture" that is associated with them.

Cyberfeminists could use the model of the recent feminist art project "*Informationsdienst*" to create "Information Works" that address the political conditions of telework, and make visible how the deployment of IT is affecting the restructuring of work and the loss of jobs worldwide in the market economy. (S. Buchman, "Information Service: Info-Work," *October* 71 [Winter 1995], 103ff.). A teleworker's bill of information and rights, disseminated to offices and private homes through a webpage on the internet could also clarify the linked chains of "women's work" and working conditions for women worldwide. A "Home work School" on the internet and in local community centers—taught and organized by home working women (many of whom are increasingly artists, single mothers, poor urban black women, immigrants, and displaced older women)—could offer (free) classes in everything from the politics of the new global labor economy and its effects on women's lives and work, to feminist history, to creative and practical lessons in upgrading computer skills. Wired women need to form new unions that bring together women computer engineers, analysts, managers, programmers, clerks, and artists. We need to form coalitions with immigrant rights groups that are interested in computer literacy. The classical tactics of organizing to improve working conditions must be translated into new forms that take into account the decentralization and reprivatization of workers, and subvert the already-established communication chains of IT to reach and organize the people displaced by it. The creative ideas of cyberfeminist artists experienced in computer networking could be especially useful here. On the macrolevel, cyberfeminists need to initiate a visible resistance to the politically regressive consequences of relegating women back to the home work economy and imposing on them the privatized, invisible, double burden of labor. Many libertarians, economists, and labor leaders are addressing the social isolation and economic privation suffered by millions of casualties of electronic joblessness by calling for the creation of socially productive jobs with a guaranteed annual income (or a social wage) for workers displaced by automation. They are also supporting moves for a shorter workweek, for job sharing, for more equal distribution of knowledge and maintenance work, and calling for corporations that benefit from the global market economy made possible by IT to return some of this great wealth to support a Third Sector of social and community work. While many of these demands seem desirable steps toward a more equitable labor economy, in practice they amount to a social welfare tax and do nothing to challenge the intense stratification and concentration of wealth and power that is increasingly produced by the global market economy, with devastating effects, on already marginalized, impoverished, and invisible populations, including women. Cyberfeminists need to analyze the effects such schemes might perpetuate on the gender division of labor. Will women continue to be concentrated in the low-paying "caring" and social-maintenance jobs that double and extend their housekeeping "skills" to the whole community? Or will we fight to have such socially productive work be revalued by awarding it decent salaries, benefits, and job security? Such work should be acknowledged as vital to the survival of human life and should be highly rewarded—not just monetarily, but also by granting workers the greatest autonomy in planning and structuring the work, by having them determine working conditions, pay, benefits, and hours. Above all, we must rejoin the fight that was never won: the revaluing—by way of decent wages, benefits, and improved labor conditions—of the human work of child-raising and family care-giving that is vital to the productive lives of all human beings. If such maintenance work were liberally rewarded, and balanced with adequate free time and educational and social opportunities, it would be work attractive to both men and women, and could do much to substantially change traditional domestic—and paid labor—gender roles.

Given the groundbreaking changes IT is causing in the relationship of home to work, and in the place of the home (and private life) in pan-capitalist economies, some radical rethinking must take place about women's changing conditions both in the domestic sphere and in the public economy. The suggestion that the home should again become a locale of resistance to capitalism's predatory effects on privacy, sociality, and free time may be a regressive one for women, because it treats these problems as private ones with private solutions. The utopian promises claimed for IT—for example, the possibility of being freed from never-ending repetitive work and heavy manual labor; the drastic reduction of working time for all people and the concomitant expansion of self-managed free time—must be skeptically countered with a critique of the ways in which IT has actually increased work time and has eroded aspects of the pleasure and meaning to be found in work—such as sociability, worker solidarity, job security, and pride in skills. This critique should be combined with vocal opposition to and denunciation of the reintroduction of regressive labor conditions and policies for workers worldwide. It is crucial that we address the human sacrifice that the worldwide proliferation of home-based telework and sweatshop labor causes for millions, predominantly women. The wide social indifference to such vast inequities once again renders invisible the life-sustaining unpaid or underpaid maintenance work performed by women.

SUBJECT: GLOBAL \: DIST LIFE

FROM: KITBLAKE@V2.NL (KIT BLAKE)
DATE: SUN, 30 AUG 1998 18:06:24 +0200

Life gets more mobile. The net is fulfilling its predicted function as a software provider and distributor. People don't need laptops, since there are connected computers everywhere—at festivals, at your friend's place, in the cafe. With basic knowledge, any of these terminals can be used to check your mail, communicate, research, or plan the next leg of the journey. Even Berlin's high-end department store, the KaDeWe, has a Cybercafe.

Email is free these days via ad-driven storage sites, their pages generated on user request. In the web's commercial construction, text has no value. Unless it's somehow personal, everybody ignores text, so it's useless as a brand message conduit. Thus email—pure text—might as well be free. They can serve visuals with it. Image is valuable. Image provides a chance at attention, and attention is the currency of the network age.

The power of image on the net is directly measurable. Porno sites have figured out microtransactions, the Holy Grail of net.commerce. Quite simple, really; all they have to do is count. How much is an image display worth?

Maybe one hundredth of a cent, maybe a tenth. Cookies have become crumb counters. Thus you find free porno sites where the owner states, "Please visit my sponsors, they make this site possible." Already making fractions by the ad displays, passing a user through to a sponsor, fostering a click—attention—is worth a lot more, and these attention units are eagerly tracked and reimbursed by destination sites. The system works, because they pay for precisely what they get. Clicks and hits add up to cold hard cash. Or soft liquid credit.

Transactions are moving to an abstract sphere. There's something about ecash that makes it separate. Even though you know that, say, a phone card costs so much, once it's electronic, cash is something else. It's been removed from the physical world. For example, a few days back I was in Osnabrück: it was 11:30 at night, and two friends and I were trying to find a hotel. The city was busy with a festival, so we thought the smart idea was to call hotels until we found a vacancy. We all had mobile phones, but we went to check a telephone book in a booth. Anna looked up hotels, then pulled out her mobile. Max interrupted, "Save your bill, here's a phone card." "OK, thanks." But it wasn't a card phone. She grabbed her mobile, and made the call. Afterward, I said, "You know, we all have coins in our pockets." We looked at each other, and laughed.

Somehow, feeding coins into a metal machine doesn't seem like a communication method these days. Communication is paid for in units of time—of attention—and stamped metal discs are for more mundane things, like something to drink.

At lunch in Berlin, somebody asked, "Do you think working on a computer is dangerous?" It certainly won't be. Computers are going to disappear, fold into the fabric of life—as in Xerox Parc's idea of Ubiquitous Computing. After all, a computer is just a chip, and a chip can be—will be—in anything controllable. Display can have any number of forms. So it might be that when you have a message, in whatever medium, it shows up by multiple means. A blinking icon on the microwave, an indicator on the TV, a beep from the bodyware. Yesterday I sent an email from my mobile phone. It's not exactly a keyboard, but all I wanted to say was "*Thanks*." (Well, there was an ulterior motive. This friend of mine needs a mobile, and doesn't—yet—admit it.) He has one of the most distributed lives I know of.

It's really an issue of convenience. Make something that saves people time—giving them more opportunity to focus their attention—and it'll be a success. People want to customize their lives. I want to be able to make a call, now, without having to relocate my body. But I don't want to be interrupted, so you get my voice mail. I'll be notified instantly; and I'll retrieve the message, when it's convenient. As will you.

SUBJECT: I WAS A PARANOID CORPORATE ARTIST IN THE BELLY OF THE BEAST

FROM: JOSEPHINE STARRS <STARRS@SYSX.APANA.ORG.AU>
DATE: TUE, 6 OCT 1998 08:51:43 +1000

<<you're invading my computer>>

The day I began my artist residency at Xerox PARC (Palo Alto Research Center), Xerox sacked 10 percent of its workers worldwide. "Is the company not doing too well financially?" I asked my group leader. I was told Xerox was doing better than ever...actually the corporation has a turnover bigger than the whole US entertainment industry. U.S. companies just seem to be in the grip of downsizing fever at present...they say it's an efficiency thing...it also makes them look tough, and the shareholders love that.

<<my flatmate is trying to get rid of me>>

Palo Alto boasts some of the most expensive real estate on the planet, but I was staying in a cheap and cheesy motel at the trashy end, just over the road from the trailer park...still expensive for me, since the Australian dollar is worth about a piece of string at present. I have a website (<http://starrs.banff.org>) where people anonymously send me their paranoid thoughts...the paranoia was steadily coming in when I was living in Silicon Valley.

<<my computer is talking about me>>

Tech culture and car culture rule in the valley, and walking to PARC along Page Mill then Mountain View, past the slick corporate buildings surrounded by manicured lawns and hedges, the semiotic messages were obvious. I disliked especially the corporations that forced me to walk on the road...walking on wet lawns was no fun, but it was better than being hit by some young software designer in their new silver Pontiac.

<<why do they all hate me?>>

"So you've gone to work for Big Daddy Mainframe?" my daughter said to me on the phone. I replied no, I'm a spy, infiltrating the databanks of BDM...Remember the cyberfeminist manifesto?...yeah, whatever. Corporate artists have to sign NDAs (Nondisclosure Agreements) as soon as we walk through PARC's doors, so conversations at Silicon Valley parties often went something like: What do you do?... I work at Interval, but I'm not allowed to tell you what I'm working on, how about you?... I work at PARC...can't tell you either.... Nice weather we're having"...etc.

<<my dead grandma sees me masturbating>>

Invitation to typical Silicon Valley party: "Gathering of the Tribe... This Saturday yes another holiday has arrived...Time for Halloween in Spring,

We take pride in our people and care about their well-being. Our skills and competence are second to none. Our adaptability and diversity allow us to excel in many different environments and to meet a wide variety of challenges in a rapidly changing world. Our forward-thinking, innovativeness, and willingness to take risks keep us at the forefront of our business.

We give our members the decisive edge by providing vital information. We are the world's best. We provide intelligent information derived from information systems of our adversaries. We work with our members to gain a better understanding of their information requirements, and then provide them the best possible lectures, music and services.

The foundation that can observe what is happening, orient itself to understand the real dynamics of a situation, decide what to do, and then act on that decision quicker than an adversary gains the information advantage. We will provide our members the decisive information advantage by providing their vital information.

Come as the new you..shedding all old beliefs, judgments, and commitments that no longer serve you...releasing the true you. 6 p.m. Saturday to 6 p.m. Sunday. Celebrating the Mystery of Life... Food, song, and dancing all night... Dancing floor addition by the roaring fire, Smoking room Upstairs on an upper balcony, Hot tubbing not to be forgotten."

<<the whole room is looking at me>>
PARC is famous for it's "ubiquitous computing" research, and I was hoping to be electronically tagged along with the best of them—but it seems the big brother implications have put the researchers working at PARC off using the technology. The only manifestation of "ubicom" I saw was a hallway fountain whose rate of water flow indicated whether Xerox shares were up or down. "Augmented reality" is the buzzword in computer-interface research these days.

<<they are reading my mail...i know they are>>
Silicon Valley is saturated with stories of startups making their fortunes—gold rush mentality—but without the wild abandon of the west. PARC won't even allow alcohol on the premises, and it's not PC to flirt. But it's a great place for bright young geeky smart things. It is assumed by most that technology will save the planet, that the valley is utopia, and if the rest of the world become good capitalists and embrace the new technology-enhanced lifestyle they can reach utopia also. Even the homeless in Palo Alto push hi-tech baby trolleys and wear discarded Gortex.

<<i'm not wearing clean underwear>>
So I roamed the empty corridors of PARC at night, feeling like the guy from the movie Solaris. I was working with these images of deformed foetuses in jars I'd illicitly shot in a medical museum in Berlin, making large color prints, wondering if my obsession with these little mutants had anything to do with the scary feelings I got passing by the many biotech corporation buildings every day. If I wanted to stay overnight at PARC, I could haul a few of the ubiquitous blue corduroy beanbags into my office to make a bed. Very cosy in a seventies sort of way.

<<my neighbor is psychic>>
PARC is known for the ones that got away: the mouse and graphical user interface were developed there, but Xerox never got a financial piece of that. This might explain the rigorous patent—charge—sue mentality there today...I never came up with an idea that was worth patenting.

<<everybody is sucking on my intellect>>

SUBJECT: THE NETWORKING OF INTELLECT: THE WORK EXPERIENCE OF ITALIAN POST-FORDISM

FROM: RAF "VALVOLA" SCELSI
DATE: WED, 28 OCT 1998 04:01:37 +0200

According to the definition given by Sergio Bologna, "second-generation independent work" isn't just specific to Italy. It pertains in different degrees to many Western and former-Socialist countries. In Italy however, particularly in the last two decades, this phenomena has reached considerable proportions, immediately reaching the status of "the explanation" for the success of industrial manufacturing areas such as the Veneto northeast and Emilia-Romagna.

But what exactly is independent work composed of? The fundamentally differentiating element from wage labor is the amount of relational and communicative operations required. How many working hours in the day of an independent worker are dedicated to "keeping in touch with working relations and partners"? Many express the high incidence of the relational work quota on the total amount of hours worked with the phrase "I spend a lot of time on the phone" (S. Bologna, "Dieci tesi sul lavoro autonomo," in Bologna and A. Fumagalli, *Il Lavoro autonomo di seconda generazione*, Milan: Feltrinelli Interzone, 1997). Moreover, in independent work one witnesses a process of domestification of the workplace, meaning with this term the absorption of work into the system of private life, even if the two spaces—living and working—remain, at least formally, distinct and separate. Another new and extraordinary element is the different perception of time: while for the wage worker, working hours are a rigidly defined and normalized dimension, the self-employed worker deals with working hours without rules, which are therefore limitless. A situation has thus ensued in which, contrary to the historical aspirations of the organized workers' movement, working hours have gotten progressively longer, to finally occupy the entire span of the day. The spur toward the intensification of the workday exists in the form of financial retribution—now detached from the time-unit (day, month) during which the worker rented her or his availability—anchored to a work performance in which the only important thing is meeting the deadline fixed by the client. All these elements involve a general modification, not just of work, but also of anthropological habits and future expectations. It's with good reason that Bologna speaks about the "immanent risk of failure" as a constitutive element of independent work, and about the coming into being of a "psychosocial frame of mind incapable of long-term planning" (ibid.). "All it takes is an illness, an accident forcing one into a six-months period of inactivity, an unpaid invoice of a certain level, a heavy damage-claim lawsuit issued by a client, bankruptcy, malicious or not, of a customer or a supplier to invoke total ruin on oneself and one's own partners and collaborators" (ibid.).

A psycho-social frame of mind is thus produced to make constant "insurance savings": behavioral forms that protect oneself from the uncertainties produced by a precarious and foreboding future. The problem of describing this new social subject is also linked to aspects more properly identifiable with political theory. "Dispersed around the territory, autonomous workers don't appear to have a sociotechnical locus capable of collective action. Lacking any kind of collective compensation or possibility of direct response against the client, they have in fact exited the secular history of labor conflicts and the system of acquired rights built upon the legitimacy of those very conflicts... While wage labor had the possibility of holding the employer responsible for respecting contractual clauses and terms of agreement through the tools of conflict and negotiation, that is, with tools proper to a civilized society, in the case of such violations the independent worker can only enforce the client's contract through the actions of a judge" (ibid.).

We're dealing here with a total loss of democracy that will see, in the immediate future, the nonaccordance of citizenship rights granted during the Fordist–Taylorist era to a wide percentage of the employed.

Facing this difficulty, some ad hoc solutions seem to appear. On the one side, the use of mutual aid associations, in an analogous fashion to those in the initial phases of the history of the workers' movement. On the other, a reconfiguration of the tasks concerning territorially based organizations of bilateral representation, such as unions.

According to other commentators, coming from the institutional Left, the culture and ideology required by the new productive transformations entail a new type of work: "Not just thought as goods, but goods that must think." (Bruno Trentin, *La Sinistra e la crisi del fordismo*, Milan: Feltrinelli). This is a type of worker that will have previously unseen features and whose appearance leads to diverse reactions, both on the employers' side and on that of the leadership of the Left. On the one hand, the employers immediately see the possibility of getting rid of the trade unions during the contractual negotiations, in order to establish a direct relation with the single employee; on the other, the unions too will have trouble relating to it. Their strategy has in fact always hinged on requests for better wages and not on demands that would radically mutate living conditions and the meaning of work itself.

In reality, it's simply the "intelligent" post-Fordist worker who owns the dialectical tools to question issues of work organisation, of distribution and the management of know-how.

HOW DID POST-FORDISM ORIGINATE?

Beyond the interpretative difficulties of the phenomena, it is possible to locate the historical origins and structural motivations that have pushed manufacturing sectors in this direction in Italy. Most commentators generally emphasize that this process originated at the end of the seventies, following three different causes.

a. the structural necessity for a modification of work relations.
At the end of the seventies, Italian capitalism found itself in a position of

great structural difficulty. On the one side, the workers' resistance that for a decade or so had efficiently contradicted every plan of capitalist domination; on the other, the end of the possibility of certain forms of financial mediation (primarily of the inflation tool, thanks to which Italian capitalism had extended its presence on the international markets) following the entrance of Italy to the European Common Market, which limited the oscillation of the Lira's exchange rates within a maximum range of 4.5 percent.

The end of the financial use of inflation induced large-scale Italian capital to make a double choice. In the first place, to raise on the international markets the liquidity necessary to change the work process (in this regard it's sufficient to think about the buyout of almost a third of the Fiat stock made by Libya during 1976–78). On the other, the frontal challenge to the central body of the working class, having exhausted the classical environments of union mediation. (On this regard, the case of Fiat is again useful: the firm laid off 23,000 workers at the beginning of the eighties).

b. the "refusal of work"

The attack on employees utilized a wide array of different tools. First, by stimulating and incentivising individual resignation, facilitated with impressively golden "handshakes" (£15,000–20,000 sterling at the time). Second, by applying pressure to the State so that segments of the very same working class being fired would be reabsorbed by civil service jobs. Third, by favoring a more complex process of externalizing work (spinoffs), through the promise of "safe" contracts to workers who agree to resign. In many cases, this involved offering them the cash to buy the machinery necessary to start new activities (a famous case from the beginning of the eighties is that of the CNC lathes for the industrial sector in the province of Brescia).

This last aspect of the process echoed a deeper dynamic experienced by the world of work throughout the seventies—a wide and internalized "refusal of work" and of the spatiotemporal rigidities inherent in wage labor (punching the clock, always the same schedule, the impossibility of staying up late at night, regulation of the spaces for conviviality during the work process, control of "bodily necessities," boredom and repetition), which had found its highest conflictual expression in the great cycle of workers' struggles between 1969 and 1975. Basically, the "exit the factory" program was embraced precisely by the more politically aware component of the working class, that which had made the "refusal of work" its flag. The process of externalization from the factory didn't solely orient itself toward industrial-type activities (which really only reconfigured the same subordinate situations of the previous factory job, but localized them in a different manner). Another part of the expelled subjects, almost as numerous as the former, recycled themselves into activities that interpret and cater to the popular desire for a diffused conviviality—so that in rapid succession venues, bars, pubs, small "fashionable" restaurants were opened... While a third component, namely, that endowed with better cultural instruments, better education, and higher professional skills, directed their job hunt toward the ascending cycle in fashion and advertising/communication (this was also the early period of "free radio" and commercial "private" TV channels).

From the opposite viewpoint, that of the workers' subjectivity, we must there-fore note how the question of "relational knowledge," the art of "making communication," of "threading human relations and networks" was at the core of these "new" jobs. These were all skills that these subjects had learned and honed during the years of the great protest. This way—alas, through crooked paths—language made its appearance at the center stage of the industrial-political debate. Its weight would increase throughout the eighties and grow further in this current decade,

c. "Total Quality" and the use of informatics
The third element that has intervened in the genesis of the "post-Fordist cycle" is surely to be traced to the use of informatics. Informatics has been employed in a manner analogous to that of many other industrialized coun-tries, both in product innovation and process innovation. The latter espe-cially has stimulated great interest in the circles of "work scholars."
From this point of view, a visit to a big factory of today is certainly an impressive experience: the warehousing space for components (industrial and general pur-pose) are reduced to a bare minimum. All this is managed through a coordinat-ed delivery of parts and components to the assembly line of the factory. The gate of the factory becomes a key part of the "streamlined," "downsized" factory. With their optical pens checking the entry and exit of goods, the personnel at the gates of the factory are also performing the first of many quality checks on the products that will shortly thereafter be assembled. The parts are randomly checked by appointed controllers and are then routed toward their specific assembly units. The majority of these parts don't spend more than a day or two on the shelves of the warehouses. The General Motors philosophy of manufac-turing every single nut and bolt used in the factory appears decidedly antiquat-ed. Today, a Fiat automobile is on average composed of about 5,000 different parts, two thirds of which are produced by Turinese subcontractors and the remaining third by other firms all over the world. In the light of this, transport and logistics in general grow to a strategic dimension. Nowadays in engineering there is a great interest in these fields. If the gate becomes a strategic place in the factory, even—contrary to the past—the first station in the assembly of the goods, the key locus is in logistics, fully completing the process of disempower-ment and appropriation of the working-class knowledge of the work processes, that was started with Taylor's first studies. The use of networks becomes foun-dational to setting the pace of work, to the definition of quality standards for components and to the promotion and the distribution of the goods manufac-tured. The circle closes with automatic invoicing.
The use of informatics at the industrial level in Italy has therefore had its special role in the innovation of the production process itself. It has thus played midwife to the birth of real subregional "industrial districts" that spe-cialize in manufacturing a single commodity, where these forms of industri-al and process innovation are introduced and shared at a localized level.

POST-FORDISM AND THE LANGUAGE SPHERE
All of this has then sedimented into the development of a diffuse "pulviscu-lar" fabric made of very small enterprises (5–6 staff each, with average rev-

enues of around £150.000-200.000 sterling) closely linked to other, bigger, firms, managing their work schedule, according to seasonal considerations and market demand, and on which they usually depend for a one-to-one relationship. There are some Italian industrial districts where there is a presence of one individual firm per every seven inhabitants (children and pensioners included).

It is sufficient to think of the Veneto region and the textile industry in the Treviso province (Benetton, Diesel), or of the optical industry in the Belluno area (Luxottica)—or, alternatively, to read the statistics relative to the per-capita income, indicating the richest area in Europe in that surrounding the city of Milan.

We're talking about firms in which it is completely "unsurprising" to work on Sundays, at nights, and way beyond normal working hours in order to keep up with the workload; with a strong relationship of solidarity between boss and workforce (hence the nosedive drop in workplace conflict in Italy and the birth of regionalistic parties such as the Lega Lombarda); and in which one sees a constant exchange of necessary know-how in the effort to obtain a quality finished product. The end result of such a process, in which a central aspect is played by the employment of relational abilities, and thus of the sphere of language in its wider definition, is to define a productive system in which the rigidities of the earlier work cycle—characterized as it was by the functional sectorialization of roles, knowledge and of the language—can no longer exist (C. Marazzi, *Il Posto dei calzini*, Bellinzona: Casagrande).

The fact that language has been increasingly subsumed into the productive sphere has made possible a lively interest toward all those theories that, in various shapes and forms, dedicate attention to the emergence of a collective sphere of intellect. Pierre Lévy, a French philosopher, has devoted a stimulating and thought-provoking text to this theme, though this is articulated more around philosophical speculation on the phenomenon of the internet (and its medieval Arabic neoplatonic roots) than toward the individuation of collective dynamics in networked employment. Moreover, this phenomena is developed and intensified by software, such as groupware, capable of optimizing work and communication processes. Even more surprisingly, some of the Marxian formulations expressed in that giant toolshed known as the *Grundrisse*, are experiencing a renewal.

GENERAL INTELLECT

And what supports Marx in his passages concerning science and machines? A very un-Marxist thesis, namely, that "abstract knowledge"—particularly, but not exclusively scientific—is beginning to become—by virtue of its autonomy from production—nothing less than the chief productive force, relegating repetitive and parcelized work to a residual position (P. Virno, "Edizione semicritica di un classico frammento," in *Luogo Comune* 1 [Rome]).

The difference between the *Grundrisse*'s "fragment on the machines" and *Das Kapital* lies in this: "Now comes to the forefront the lacerating contradiction between a productive process that nowadays leverages directly and exclusively upon science and a unit of measuring material wealth still coincident with the quantity of work incorporated by the products" (ibid., 11). This is a contradiction that, according to Marx, should lead to the "crash of a pro- duction-based on exchange value." And if Marx, in the final pages of the fragment, gives a glimpse of the birth of a worker of such a kind, a whole individual, without amputations, we cannot but agree with what Paolo Virno notices: it is exactly this new subjectivity that is currently employed in the post-Fordist process. "What one learns, experiences, and consumes during the time of nonwork later gets reutilized in the production of goods, gets included in the use-value of the workforce."

Even the other aspect of the critique issuing from Virno appears appropri- ate: "Marx has, without residual doubts, identified general intellect (that is, knowledge as production force) with fixed capital, and therefore neglects the side by which general intellect presents itself as living work, technical-scien- tific intelligence, mass intellectuality" (ibid., 12). "Today it isn't hard to widen the notion of general intellect well beyond the knowledge that materializes itself in fixed capital, including as well the forms of knowledge that structure social communication and dynamize the activity of mass intellectual work," because within the contemporary work processes, "there exist entire constel- lations of concepts functioning as productive machines per se" (ibid., 13).

THE PROBLEM OF INNOVATION

When Marx says that science is incorporated by fixed capital, he is arguing that the conditions of the scientific process—so far as these have made them- selves known from the end of the seventeenth century—are impossible today. Science is irremediably turning into technology because it mutates its nature into a series of procedures that will then be applied to industrial processes of manufacturing.

Beyond the possible critical notes that could be raised over the question, it is indubitable that Marx understands a process in action, by which the issue of scientific and technological innovation remains unanalyzed, out of focus. And in the concept of general intellect we must include the innovation aspect, the creative and unforeseeable aspect of the science factor today. If it's true that innovation also tends to transform itself into a useful mechanism for the accumulation of profits, it is also true that the diffuse and creative process of innovation isn't always so directly mechanistic. There are impor- tant examples in the history of technological innovation debunking this state- ment. Without wanting to refer to the history of the Bauhaus, it's sufficient to think back to the birth of the personal computer: born from the collective passion of enthusiasts and social experimenters, the PC, prior to becoming an extraordinary technological artefact, is a *revolutionary mental archetype*.

The emergence of a collective dimension of intellect should therefore orient itself toward a collective-projectual direction capable of imprinting definite turning points in the way people think. In this sense, technological innova- tion represents at best the factor of unpredictability within a social process that some would like characterized by a causal linearity. Of course, this isn't enough to alter or change the social game. Other stimuli apart from innova- tion are necessary and, not by accident, the Californian garages that produced the PC had some sorts of direct filiation to the countercultures of the sixties. In other respects too, the use of the net can represent a good catalyst for the emergence of new mental archetypes.

It's definitely uncommon to get one dealt, but sometimes a joker from the deck can totally alter the destiny of the game. Therefore we must try and get at least two jokers available for our game—and then turn them into three and four. Innovation is definitely one of these "trump cards." We still have to invent the others.

[Translated from Italian by Syd "I was a junkie stagehand" Migx.]

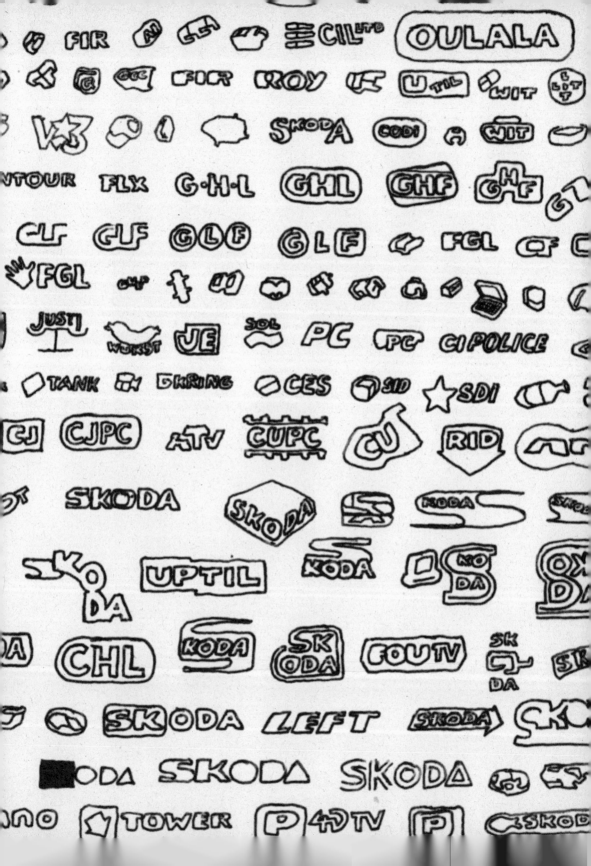

ART

SUBJECT: URBAN AGENCIES:
INTERFACING PASSION, INFLAMMABLE BY FAIRNESS
(AN INTERVIEW WITH KNOWBOTIC RESEARCH AND ANDREAS BROECKMANN)

FROM: WILFRIED PRANTNER (BY WAY OF KR+CF@KHM.DE <KNOWBOTIC RESEARCH>)
DATE: MON, 24 AUG 1998 14:53:27 +0200

WF: Your work and thought has always centered around a problematic and complex notion of territory within data space and electronic networks, which you have variously described as spaces of action or events—concepts which have also been used to describe the fluctuating political, social, cultural realities of the city in contrast to its spatial organization. To which extent is your recent interest in urbanity related to fundamental qualitative similarities between the spaces opened by electronic networks and those traditionally supported by and created within the architectures of the city?

KR: Our discourse places itself outside an architectural framework. When we talk about problems of urban spaces, we mean the urban as a machinic assemblage that is constituted not so much by built forms and infrastructures, but as a heterogeneous field that is constituted by lines of forces, by lines of action and interaction.

These lines form the coordinates of an urban topology that is not based mainly on the human body and its movements in space, but on relational acts and events within the urban machine. These can be economic, political, technological, or tectonic processes, as well as acts of communication and articulation, or symbolic and expressive acts. The urban field that we are talking about is therefore quite different from the physically defined spaces of events and movements. Rather, we are interested in what the relation between the spaces of movement, the spaces of events and the relational, machinic "spaces" might be. It does not really make sense to oppose the city and the networks in the suggested way. We are interested in finding models of agency for and in complex dynamic systems and approach the urban as such a complex system. We understand the city not as a representation of the urban forces, but as the interface to these urban forces and processes. Therefore, the city features not as a representation, but as an interface that has to be made and remade all the time.

WF: Could you elaborate on what Knowbotic Research calls "connective interfaces" and describe their difference to the failed urban participatory models of the seventies?

KR: It is characteristic of the forms of agency that evolve in networked environments that they are neither individualistic nor collective, but rather connective. While individualistic and collective diagrams assume a single vector, a single will that guides the trajectory of the action, the connective diagram is mapped onto a machinic assemblage. Whereas the collective is ideally determined by an intentional and empathetic relation between actors, the connective is an assemblage that rests on any kind of machinic relation and is therefore more versatile, more open, and based on the heterogeneity of its members.

The distortions are not generated by the networks, but they can be given a certain presence and an effective form in the interface, without necessarily becoming visible. The complex working conditions like those in the IO_dencies experiment in Sao Paulo create multiple irritations between the participating local urbanists and the producing institutions, the programmers, the hard- and software, misunderstandings, and wrong expectations. These distortions are present in the project without causing it to fail. On the contrary, they generate new developments. It is vital to become sensitive to the weakness of interfaces and to the potential forces that they bear. One aim is to recognize them and to turn them into tendential forces (IO_dencies) that may become effective sooner or later.

Drawing on Félix Guattari's notion of the machinic, we describe the interface as a machine in a complex aggregate of other machines. Connectivity can, in this context, mean different things: the combination of functionalities; the collapse and opening up out of a moment of conflict or rupture; or diversion and repulsion where no interaction can take place. What we are surprised about ourselves is this new, differentiated vocabulary that is emerging in relation to working with electronic networks: the interfaces ties together, folds, collapses, repulses, extinguishes, weaves, knots. All these activities, which are obviously not germane to our projects, make it necessary to rethink "networking" as a multifunctional, highly differentiated set of possible actions.

WF: In a passionate defense of the physical city, the British geographer and urbanist Kevin Robins has recently criticized current celebrations of cybercities and virtual communities (for example, William Mitchell's *City of Bits* [1997]) as conforming strikingly with Modernist notions of urbanism, in being driven "by a desire to achieve detachment and distance from the confusing reality of the urban scene." Although your interest lies with creating intermediary fields or interfaces between those two realms rather than in playing off one against the other, you clearly claim urban qualities for the spaces you create, by describing them as comparable with the "urban structures of megapoles." Could you elaborate on your sense of the urban and how it relates to that found in the countless digital cities?

KR: Our projects respond to one dominant mode of the urban, that is, its overwhelming, unbounded, uncontrollable experiential qualities. In this sense, we agree with Robins's observation about the "confusing reality of the urban scene," in this sense, we also agree with his criticism of digital cities and virtual communities. However, we are doubtful that this chaotic and disorderly nature of the urban is necessarily dependent on "embodied and local situated presence."

We must distinguish between the urban as a discontinuous flow, a transformation process involving social, economic, architectural, and so on, forces, and the city as a temporary, diagrammatic manifestation of the urban. The French urbanist Henri Lefèbrve wrote in 1970 that the urban as such is not yet a completed reality, but it is a potentiality, an "enlightening virtuality." The path of urbanization, however, is not unidirectional and does not necessarily lead to a transglobal urban zone. Rather, the urban is a complex, multidirectional process of connection and separation, of layering, enmeshing and cutting, which leads to ever-different formations.

The heterogeneous and permutating assemblage of materials, machines, and practices we call *the urban* implies a global stratum that is locally embedded. If the urban is something that one can work with, intervene into, or become a part of, then it is important to understand its forces and layers and also to understand how it interlaces the global with the local.

WF: Before engaging with the complex of the local–global relationship, can you specify your concept of an urban machinic and explain what kind of machinic agencies Knowbotic Research is aiming at?

KR: The urban is a machine that connects and disconnects, articulates and disarticulates, frames and releases. It offers the impression that it can be channeled and controlled, that it can be ordered and structured. The city is always an attempt at realizing this order which, however, is nothing but a temporary manifestations of the urban.

The machinic urban is always productive, as against the "antiproduction" of a fixed city structure. But its productivity lies in the creations of discontinuities and disruptions, it dislodges a given order and runs against routines and expectations. The urban appears in a mode of immediacy and incidentally, confronting a structure with other potentialities and questioning its given shape. We can clearly observe this tension between the urban and the city wherever the city appears dysfunctional and unproductive. But the urban machine is also productive at invisible levels, for example, where real-estate speculations are prepared that will disrupt an area within the city, or where a natural catastrophe or political instabilities will cause a rapid influx of large numbers of people. In these cases, the "finance machine" and the "tectonic machine" impact on a local urban situation.

The human inhabitants of cities are not the victims of such machinic processes, but they form part of them and follow, enhance, or divert given urban flows and forces. Contemporary analytical methods of the urban environment no longer distinguish between buildings, traffic, and social functions, but describe the urban as a continuously intersecting, n-dimensional field of forces: buildings are flowing, traffic has a transmutating shape, social functions form a multilayered network. The individual and social groups are co-determining factors within these formations of distributed power.

The machinic character of the urban means that there are multiple modes of intervention, action, and production in the urban formation. The relation between space and action is of crucial importance. There seems to be a reluctance on the part of many architects and urban planners to consider "action" as a relevant category. Rather, built spaces are much more closely identified with, and it seems, made for, certain types of behavior. The distinction between behavior and action is a significant one, behavior being guided by a set of given habits, rules, directives, and channels, while action denotes a more unchanneled and singular form of moving in and engaging with a given environment.

The suggestion here would be to move from thinking about a topology of objects, forms and behavior, on toward a topology of networks, a topology of agency, of events, and of subjectivity.

WF: One major issue addressed in your present project IO_dencies is the question of the "cultural identity" of the cities investigated—Tokyo, Sao Paolo—and the interrelation of local and global forces. Now on the one hand, the peculiar character of these cities emerges in the urban profiles provided by local architects and urban planners; on the other hand, and more importantly, you argue that "cultural identity" can no longer be located in the architectural structures of the megacities, but might be relocated in the activities of local and translocal agents who, by means of data networks, form a new kind of connective." From your experience with the project so far, what are your preliminary conclusions regarding the shape of cultural identity as it emerges through the cooperation of local and global forces?

KR: What is referred to as the global is, in most cases, based on a technical infrastructure rather than on lived experiences. The electronic networks form a communication structure that allows for a fast and easy exchange of date over large distances. But the way in which people use these networks is strongly determined by the local context in which they live, so that, as a social and cultural space, the electronic networks are not so much a global but a translocal structure that connects many local situations and creates a heterogeneous translocal stratum, rather than a homogeneous global stratum. The activities on the networks are the product of multiple social and cultural factors emerging from this connective local–translocal environment. We don't deny the existence of the global but see it as a weaker and less interesting field for developing new forms of agency.

There are local formations in which certain behavioral patterns emerge, and translocal connections make it possible to connect such specific local situations and to see how the heterogeneities of these localities can be communicated and how they are maintained or not in a translocal situation. Against the worldwide homogenization of the ideology of globalism one should set translocal actions that are connected but can maintain their multiple local differences.

The IO_dencies project is rooted in local situations, and we are looking for the productivity of the interface in the movement from the local to the translocal. In this continuing process, we are testing the translatability of ideas and cultural contents, the local points of friction, and also the heterogeneity of what is often seen as a more or less homogeneous local cultural identity. At the same time, we recognize that globalization is a reality, and

The E.U. commissioners have announced that agreement has been reached to adopt English as the preferred language for European communications, rather than German, which was the other possibility. As part of the negotiations, the British government conceded that English had some room for improvement, and has accepted a five-year phased-in plan for what will be known as EuroEnglish (Euro for short). In the first year, "s" will be used instead of the soft "c". Sertainly, sivil servants will reseive this news with joy. Also, the hard "c" will be replaced with "k", not only will this klear up any konfusion, but typewriters kan have one less letter. There will be growing publik enthusiasm in the sekond year, when the troublesome "ph" will be replaced with "f". This will make words like "fotograf" 20 persent shorter. In the third year publik akseptanse of the new spelling kan be expekted to reach the stage where more komplikated changes are possible. Governments will enkourage the removal of double letters, which have always been a deterent to akurate speling. Also al wil agre that the horible mes of the silent "e" in the languag is disgrasful, and that would go. By the fourth year, people wil be reseptiv to steps such as replasing "th" by "z" and "w" by "v". During ze fifz year, ze unesesary "o" kan be droped from vords kontaining "ou", and similar changes vud of kors be be aplid to ozer kombinations of leters. After ze fifz yer, ve vil hav a sensibl riten styl. Zer vil be no mor trubls or difikultis, and evrivun vil find it ezi to understand ech ozr. Ze drem vil finali kum tru!!! [T. Byfield <tbyfield@panix.com>, Toward A Europanto: A Five-year Plan (A Found Text[Extropians]), Mon, 2 Mar 1998 19:20:18 +0100]

that purely local interfaces are insufficient. The global generates circumstances that make it necessary to open the local toward the translocal, in order to develop effective forms of agency.

We were intrigued by the polemical hypothesis about the Generic City that Rem Koolhaas formulated in 1994. The Generic City is the city without a history, without the burden of an identity, the suburban nightmares and recent Asian boomtowns viewed under the sobering, cynical, pragmatic—dare we say: Dutch—daylight. Implicit in Koolhaas's suggestion is the relentless growth and the unstoppable expansion of the Generic City. In the twenty-first century, he seems to say, the Generic City will become the norm rather than the exception.

The Generic City is identityless. Yet, identity is not something that is the same for a whole city. People have or develop a clear sense of "home" even in the most decrepit of neighborhoods. Local people have an intuitive knowledge that allows them to distinguish between a street in Kreuzberg and Mitte, between Manhattan and Brooklyn, between Bras and Pinheiros. The identity that is constructed in such urban environments is a heterogeneous composite of different symbolic matrices, social, cultural, familial, that are local as much as they are translocal. A possible counterhypothesis to Koolhaas would therefore be that only few places are generic cities, and only a fraction of these will remain generic for longer periods of time. The generic is not the end, but a beginning characteristic of many human settlements. The project IO_dencies asks how, suspended between local and global activities, urban characteristics are enhanced, transformed. or eradicated, and it investigates whether the extension of the urban environment into the electronic spaces might allow for changed qualities of urbanity. Is communication technology the catalyst of the Generic City, or is it the motor for another, transformed notion of urbanity and public space?

WF: You have compared the creation of nonlocations to a mode of construction that you claim to have always been a concern of architecture as well: "the constructability of the unconstructable." Is not the present project, in drawing on data and parameters employed by traditional urban planning, in danger of relapsing, as it were, into construction—of constructively contributing to a kind of advanced urban design, for which your experimental data spaces may serve as a model or at least complementation by which it may come to terms with the unpredictable processes of the heterogeneous and fragmented urban field?

KR: Here you refer to experimental settings Knowbotic Research developed in the past. Our current research tries to push nonlocations toward fields of agency and presence and we are rather doubtful if the term "under construction" may turn the attention in the right direction.

Our recent projects are not meant as urbanistic. solutions, but they seek to formulate questions about such urban interfaces, about visibility, presence and agency within urban assemblages. We aim at experimental topologies of networked intervention, which are able to offer a connective form of acting inside urban environments, between heterogeneous forces and in multiple, differentiating ways. The relation to the concrete city environment is maintained through working with young local architects and urban planners who are searching for other ways of dealing with the problems and challenges of the city they live in. The aim, however, is not to develop advanced tools for architectural and urban design, but to create events through which it becomes possible to rethink urban planning and construction. The question we raise is: What can be done if we accept that urban environments, systems of complex dynamics, cannot be planned and constructed anymore in a traditional modern sense?

Urbanism, in exploding megacities with high social inequalities, means that city space is delimited and planned only for about one third of the inhabitants, the rest of the people stay outside the walls of the capitalized space. It would be politically precarious to speak of this other two thirds, the so-called illegal city as a nonlocation. In our studies we found clear needs for relevant forms of agency that are able to deal with the complex processes of urban exclusions. These forms of agency don't have to deal so much with the re-articulation of territory, but they have to invent and produce existential interfaces for the visible and invisible forces of a city in order to avoid political, economical and cultural isolation.

IO_dencies explores the phenomenon of urban agency and distributed and networked subjectivities on different levels. Initially, it seeks to develop innovative ways of reading and notating city environments, drawing out their energetic and dynamic elements. This provides the basic data for the following, collaborative manipulations of specific urbanic strata. We outline interfaces that are able to transcode the analyzed data and facilitate different forms of access to the urban machines. Analysis, interface development and practical collaborative involvement are all part of a process that represents an inquiry into the structures and the points of potential transformation in urban environments.

WF: Yet, if the observation about a certain constructiveness of your current project is correct, then how does it relate to the claim of yours that your work is intended to enable intervention and resistance? Where, specifically, would you place the locus of resistance and intervention both as a capability of your machinic constructs as such and as a possibility of the user within the fields of action thereby created? In terms of the Deleuzian notion of the machine as that which interrupts a flow, how does the internet-aggregate of IO_dencies cut into the given physical spaces and the lived urban experience of the urban quarters investigated?

KR: First of all, it is important to affirm that we are not building urbanistic tools for a general use, and that the models we develop cannot simply be deployed in a political or social context. IO_dencies offers experiments for a small group of people who are highly motivated and looking for individual ways of participating and intervening in their local urban situations. Even those with an academic background as urbanists and architects are frequently disappointed by the methods and models of agency that are dominant in planning offices. IO_dencies tries to initiate a concrete process inside the group which allows for a specific form of locally and translocally determined collaborative actions, accompanied by software processes that try to support the individual needs inside the group communication.

Contemporary cities are covered with successful and failed attempts at leaving such traces and creating such feedback loops. The noise from roaring cars and ghetto blasters, the ubiquity of graffiti and tags, stickers and other lasting marks, even temporary and permanent pieces of architecture are clear attempts at creating a lasting visibility and presence in the urban environment. Viewed from a cultural and from a political perspective, however, this kind of visibility is rather powerless if it is not coupled with opportunities to act and to intervene in the public arena. A possible hypothesis that follows from the experience of Anonymous Muttering is that in complex machinic systems like the urban, effective intervention is only possible in the form of a connective agency within which the different individual and machinic tendencies and potentials are combined and connected. This form of agency would not develop its strength through being localized and aimed at a certain goal, but would be composite, heterogeneous, dynamic, and to a certain degree subjectless.

IO_dencies works in a very different way and tries to develop interfaces that allow for a more conscious engagement with urban forces. It has to be said that, in the different cities, we are initiating extremely singular processes and singular tools that do not represent "Tokyo" or "Sao Paulo," but evolve in a close collaboration with groups of specific urbanists, architects and others. This method is also a result of the discouragement of the higher goals that we had set out prior to the Tokyo project. We are becoming more sensitive to the specific local circumstances, and we have to formulate the interfaces in a way that makes it possible for people to insert and develop elements of their cultural identity.

In this sense it is questionable whether we are dealing with "the urban" at all. Rather, the goal is to find out whether it is possible, in a situation where the city itself is being deprived of many public functions, to develop electronic interfaces that open up new forms of agency, and whether network interfaces can become useful in local as well as in global contexts.

The question of responsibility can be understood in a concrete ethical sense. Large parts of the public functions of the city are currently moving into the networks, which leads to new mechanisms of exclusion within the urban environments. The political question would be whether it is possible to conceptualize interfaces that can subvert such processes of exclusion. Building interfaces means to allow for change to happen. We do not want to build a better world, but only better interfaces that enhance the perceivability and the respect for the actions and the needs of others and allow for a heterogenization of social relations. Difference, otherness and becoming-other, the possibility of multiple singular processes, are moral necessities. Connective interfaces enable the formation of aggregates of multiple heterogenizing machines.

[This is an excerpt of an interview made for Film+Arc Biennale, Graz.]

**FROM: CRITICAL ART ENSEMBLE (BY WAY OF STEVEN KURTZ
<KURTZ+@ANDREW.CMU.EDU>)
DATE: {WED, 29 OCT 1997}**

Maria Fernandez has taken an active role in the formation of colonial studies in art history, applying postcolonial theory and cultural history to art history and historiography. She is also active in postcolonial and multicultural critiques of electronic media art.

CAE: A postcolonial perspective seems to be absent from the major discourses in media theory in North America and Europe (in spite of the fact that postcolonial theory is well developed and even institutionalized in the U.S., Canada, Australia, and the U.K.). At best, it seems to be a marginalized undercurrent. Why do you think these two knowledge pools have very little overlap?

MF: The interests of the two fields have been quite different. Postcolonial studies have been concerned with issues of identity, representation, agency, gender, migration, and with identifying and analyzing strategies of imperial domination and/or resistance in various areas of theory and practice. This includes fields that people do not traditionally associate with imperialism: biology, history, literature, psychology, anthropology, popular culture, and most recently, art history and philosophy.
Particularly in the eighties and early nineties, much of electronic media theory (the little that existed) was concerned with establishing the electronic as a valid and even dominant field of practice. In fact, many theorists were knowingly or unknowingly doing the public relations work for the corporations. This often involved the representation of electronic technologies—particularly the computer—as either value-free or as inherently liberatory. The exponents of such rhetoric could not afford to acknowledge the existence of theories concerned with the analysis of imperialist strategies, at least not until they felt sure that their goals were reasonably well accomplished.

CAE: In the U.S., the utopian rhetoric of *Wired* culture has been harshly criticized by different leftist factions as a blind apology for predatory capitalism and enslavement to its work machine. While the extreme ethnocentrism involved in the "California" position has been named, there is only a modest amount of work on the way in which imperialist ideology is replicated in this discourse. Do you have any insights into this matter?

MF: I attribute this lack to the separation of the two fields. As you have said, the two fields have developed parallel to one another, but have very few points of intersection. I also think that, at least in U.S. academic circles, that there is still some hesitation about referring to the U.S. as an imperialist power (gasp!). The replication of imperialist ideology in utopian positions of the *Wired* magazine variety is really not hard to recognize. Have not virtually all imperialist projects adopted utopian and humanitarian rhetorics? Was it not humanitarian ideals that supported the "civilizing mission" of the French, British, and other colonial powers? The belief dear to "California" ideologues—that pancapitalism is a "natural" result of "evolution"; the defense of free enterprise against government intervention; the supposition that unregulated commerce will bring about individual freedom, democracy, and even the elimination of human suffering—all these were all prefigured in the nineteenth century. Does any one remember Herbert Spencer?

CAE: In Western and Central Europe (the U.K. notwithstanding), postcolonial theory has not done any better. At the major media festivals, there is little if any effort to integr ate this line of thought into the discussion. Such matters are left to the more politicized conferences such as the Next Five Minutes or Metaforum. What obstacles do you think stand in the way of the development of a mainstream platform for postcolonial thinking? Can this situation be linked to the current government/E.U. support for media festivals and new spaces such as Zentrum für Kunst und Medien in Karlsruhe?

MF: Some Europeans view postcolonial theory as an example of political correctness (which they perceive as the dominant ideology in the U.S.) and not as a field of inquiry with any relevance to them. I have asked the same question to artists and intellectuals in Germany, France, and Scandinavia that you are asking me; the response I have invariably received is that Europe is not experiencing the same immigration pressures as the U.S. and since the population of the country in question is to a large extent "homogeneous," postcoloniality is not an issue. Even people from large, multicultural, cities including Berlin and Paris, have given me the same response. This attitude ignores even the histories of colonization within Europe itself! The perception of European countries as "homogeneous" could be a very good reason why the discussion of colonialism/postcolonialism is not mainstream.

I think that in the case of government and E.U.-sponsored media festivals and institutions, the situation is more complex. Traditionally, culture supported by states or government entities is culture that can be used to support official positions of what culture should be, not to mention to uphold official representations of national or ethnic identities. Culture produced with the help of technology is no exception. In fact, technology has always been at the heart of such representations. One only has to notice the privileged place accorded to technology in accounts of both colonial conquest and nationalism. As in the past, if technology is being used to support official constructs of identity, even at the broad level of the E.U., this could be a very good reason to exclude theories that focus on the marginal and the hybrid.

CAE: Postcolonial theory has not managed to insinuate itself into academic institutions in most of Europe. Why has it been relatively successful in the U.K. and North America, but nowhere else?

MF: No one in the U.S. can maintain that the population is "homogeneous" (although some still argue for the values of integration). Non-Europeans have long been established in American urban settings and have impacted the way many people live and think. Minority groups and their supporters have been very vocal about including multiple cultures in academic curricula, and since many of these cultures have colonial histories, it has been impossible to leave out discussions of colonialism and imperialism.

This in no way implies that racism is not thriving or that colonial/postcolonial studies are dominant. As you know, proposals for "multiculturalism" in educational curricula have resulted in bitter debates about what culture and "the American heritage" really are. In addition to the activism of minorities, the relative success of postcolonial theory in the U.S. is to due to the presence in universities of academics from former European colonies. I understand that this is still quite rare in Europe.

CAE: We need to invert this line of questioning. Why haven't people active in postcolonial discourse responded to new media developments when they know they are key to the development of the postcolonial situation? Just recently on Nettime, there was an interview with Gayatri Chakravorty Spivak. She all but refused to answer questions having to do with media theory, and went on with her usual literary theory. To what extent are postcolonial representatives refusing to engage the discourse, except for places where it's comfortable for them, such as in film theory?

MF: Postcolonial theory has been predominantly literary. Most theorists teach in English and Comparative Literature departments. And despite the current hype for interdisciplinarity, academics, at least in the U.S., rarely venture too far from their established fields. One must recognize that the analysis of a diverse range of texts has been invaluable for developing postcolonial criticism, as has the analysis of popular culture, television, film, and video. I am not sure if most postcolonial theorists realized that new media were crucial for the further development of imperialism (I think Edward Said conceded as much in an interview). I suspect that at least some of them thought that the debates about new media were distant or even distracting from what they perceived as more immediate problems.

The preference of postcolonial theorists for video, film, and the plastic arts may be dictated by the media that predominate in the developing world. The advent of digital media in developing countries is very recent. In 1990–92, for instance, it was really hard to find visual artists working in these media in Latin America. This situation has changed in the last few years, but these practices are not yet as widespread as they are in the U.S. and Europe. We must note, however, that the advent of commercial digital networks, while they remain invisible in much of the developing world, have had a powerful effect on those economies.

CAE: Video is another comfort zone for postcolonial theorists and for those artists who use it as a conceptual foundation for their work. Is this a situation of too little too late? Video is a dying medium. Will the current trend of video based installations in both the U.S. and Europe save it from consumption by the digital?

MF: I find it difficult to criticize artists from the developing world who use video. In many cases, this is the most advanced technology they've got. As cheap as digital technology is getting in the overdeveloped world, it is still prohibitively expensive in many parts of the planet. This will undoubtedly change as prices continue to drop and people become adept at manipulating digital media.
In some cases, artists deliberately choose not to work with the latest technology or trend. This has been an ongoing subject of debate in the critique of Latin American and African art of all periods. Europeans and American critics often view the arts of these regions as being derivative and retardaire. It's only recently that they have begun to realize that anachronistic works can be made intentionally. I do have to agree with you that the engulfment of video by digital media seems imminent at this point. But it will not happen in all places at the same time.

CAE: To end on a more concrete note: Two electronic artists recently showcased who are interested in postcolonial topics are Guillermo Gomez-Peña and Rafael Lozano-Hemmer. What strategies or tactics in their work do you find valuable?

MF: I find the work of both artists extremely valuable. Guillermo Gomez-Peña and his partner Roberto Sifuentes were key in catalyzing the current discussion of border culture and hybridity in artistic and academic circles in the U.S. Guillermo's theoretical writings and performances have been effective in calling attention to the stereotypical representation of Mexicans in U.S. popular culture. These stereotypes are not without serious consequences. They are at the very heart of U.S.—Mexico relations, not to mention basic to the appalling treatment of Mexicans and people of Mexican ancestry within the U.S. I think that Guillermo and Roberto's participation in electronic media festivals is productive, as it may open up much-needed discussion about issues of difference, marginalization, and hybridity, as well as provide refreshing alternatives to Euro-American visions of the future. But because their work has not yet grown within the digital, it is unlikely to engage the geeks and techno-utopians.
Rafael Lozano-Hemmer and his partner Will Bauer produce work that is very seductive at the technological level, in addition to being visually and theoretically interesting. I understand that they have been working for about ten years just on the technological apparatus of their pieces alone. Their interests are by no means restricted to postcolonial issues. Their piece, "Displaced Emperors" dealt with issues of power, history, memory, virtuality, architecture, presence, sensuality, desire, agency, and colonization, within

SUBJECT: THREE INTERVIEWS WITH PAUL GARRIN 1997–98

FROM: MEDIAFILTER (MF@MEDIAFILTER.ORG
DATE: TUE, 20 OCT 1998 17:17:12 -0400

The domain name issue and Paul Garrin's Name.Space has been a controversial topic for a while now. The flamewar in September 1997 on Nettime about this was one of the reasons to move from an open list to moderation. Name.Space has from the very beginning been part of the Nettime agenda (if such a thing exists). Paul Garrin was one of the twenty participants of the founding meeting in Venice (June 1995). Name.Space can been seen as a results from Garrin's efforts during the Next Five Minutes 2 conference (Amsterdam, January 1996) to establish a "Permanent Autonomous Network." The attempt to question one of the fundamentals of the internet, the control over the domain names by governments and monopolistic corporations, can be interpreted as a radical form of net criticism, beyond the initial critique of the *Wired* ideology (R.I.P.).

Soon Name.Space became more than just a concept. Paul appealed to all of us to support the project and reconfigure our servers. Not everyone was convinced that the software would work. Some became suspect about the way Garrin turned this common effort into a private business. Name.Space became identical with legal documents, complicated technical terms and horrendous (macho) fights. Because of legal reasons, Paul cannot always speak in an open manner and we have, more or less, accepted this. We asked him about the current state of the project, how artists are running a business, the international aspect of the domain name system (DNS) and how we can (again) get involved.

Q: You are an artist. You went deep into technology with Name.Space, but this is not the first time you did it. What, in general, does art have to do with media and technology, and do how you define your place in it.

A: Control media and you control the public. Free media is a threat to control. As an artist, one strives to discover an effective means of working in any medium—and when that medium is a mass medium, the key is to establish and sustain visibility. If there is no support system to guarantee reliable distribution, the work disappears.

One of the main concerns in my work has been the notion of the public vs. the private. Territory. Security. Privacy. And the way that "the media" manages the perception of the public. These things have always been of interest to me. A name is an essential and universal element. On the net, the uniqueness of the name is imperative. In capitalism, the idea of uniqueness means "value"...commodity. One of the key elements of oppression and control is to control the notion of identity. Within the standard of the "domain name system" the message is control, "domination," "territory."

Being an artist does not condemn one to being an idiot savant. Making art takes vision. Limiting your definition of art to the confines of the art institutions limits vision. Look to the world, not to the art world and you will understand where I am coming from. My work is not about crafting things but about creating situations. Where to look and what to look at is determined by the situation and its contextualizaton. I have approached all of my projects in this way, each with its own challenges and learning curves and a minimum of repeated effort, building on each experience.

Q: Why do you look down on artists and activists that still work in old ways, like getting grants, living on the dole, temporary jobs in schools, and so on? Your enterprise is very strategic, I can see that. But should we all start running businesses now?

A: I don't know where that perception came from. I don't look down on anyone. It's more about looking at the impending future of theirs and our disappearance, or at least the disappearance of any hopes of creative freedom and autonomy. In a very real sense subsidies, especially for unpopular, nonmainstream ideas in art and media are gone in the U.S. and are on the road to extinction in Europe. Japan's postwar funding structure has always been tied to corporate PR and in light of their present economic crisis, is even tighter and more closely bound to the corporate mainstream.

We see how institutions like ZKM (in alliance with the Guggenheim) set their agenda according to the pulse of Siemens and Deutsche Telekom. Forget any social criticism or political content or forget their deutschmarks. Their agenda is to accumulate wealth and property and take credit for defining the art of the time in their own image (or at least one that syncs with their PR agenda) not to support living artists and the nurturing of their ideas. Control the Art and you Control the People. I was told by the ZKM at one point that they had considered buying my work but in the end didn't because it is too controversial (I have a letter from a curator stating this).

Starting a business is a serious risk. I am not a "trustfund" boy and am not independently wealthy. I took money that I earned through my work and invested it in creating my company, pgMedia, Inc., and in developing and deploying Name.Space—all at great personal risk. For me it was no question that it was the right thing to do and that it was the right time to do it—and that the concept has a high likeliness to succeed in the marketplace and generate a stable enough income to run a network and fund the growth of resources and future development. A *serious* career choice and A *good* risk to take, not to mention an interesting and challenging way to spend my time...

I could have taken that investment and created another installation that would have easily consumed all my available cash. And it would have been another dead end. There is no relevant market for my artworks in the existing structure of the art world. Art should not be created in accordance with market demand or acceptance by the corporate elites. The critics and skeptics who doubt my abilities or intentions obviously doesn't know me and are reacting on ignorance and not on insight. Some believe that failure is the ultimate success, and that loss of their victim status would rob them of their purpose. I couldn't disagree more.

Q: So even if your main field is not art anymore, what is driving your fight for a certain autonomy within the new media?

A: Art alone does not assure our survival or even the creation of more art. In order to assure the autonomy of the content, totally self regulating, without the control of commercial interests, it is imperative to buy the bandwidth—the only option to eventual disappearance of free media when the "Disneyfication" of media and the net is completed. Sponsors have their agendas and their limits to "tolerance." This has been demonstrated time and again and should by now be understood. The idea of what is "authoritative" and what is "acceptable" should not be controlled by commercial interests.

One important aspect of Name.Space is to prevent the privatization and commodification of language. Some companies and individuals claim proprietary ownership rights to words such as "web" and "art". One individual even claims ownership rights to the letters "a" through "z". This monopolization and claims of ownership of common words harms the public interest. The privatization of language must be viewed as a negative trend. The Name.Space model creates an expansive top-level namespace that is in the public domain. The top-level namespace is not owned by anyone and is meant to be shared even by competing registries. The registries provide a service in the public interest and trust and do not "sell property" or otherwise make claims to property. Top-level names can come and go according to use, like a natural process. If there is demand for even one top-level, like .art or .media, which can be shared by the public, then it will be created within any bounds of the existing technology. If there is no longer demand, it can be "retired" in order to free up space for other new top-level namespaces that may come into being, including non-English categories, some of which exist today.

Q: Do you see this movement against the rise of monopolies?

A: Large corporations, who came very recently to the net, such as Time Warner and Disney and Microsoft have bought up network capacity all over the place and have also become content providers, if you can call it content. This is the disappearance of public space on the net as I wrote prior to the Next Five Minutes back in 1996. The idea of the permanent autonomous network was based in maintaining free zones on the net which mutually support each other and establish economic models to assure their presence by generating revenues to buy bandwidth—because to guarantee the survival of free art and free media on the net an infrastructure must exist along with an economy to support it. As the big content providers buy up connectivity and resources upon which we become increasingly dependent, they establish private areas in which they control the content through various means. There is no guarantee of access or autonomy of content. The net result is a disappearance of support systems for noncommercial and controversial content, as well as privacy and security.

Q: What is the relation of names and the political economy of the internet, then?

A: Survival of media independence demands creation of an economic structure that is basically a self-sufficient, self-supporting network. Name.Space is conceived as a service to potentially fund the bandwidth that we need. Apparently the market for domain name registration is a large one. Revenues generated through fees for name registrations and other services would be adequate to fund our networks and to support our cooperative partners in Europe and even, hopefully, sponsor some other activities for producing media and holding conferences. So I think that it could be a very important aspect of independence of not only buying and providing bandwidth and server resources, but also supporting content production. It is not necessarily a question of how much bandwidth, but that we have any at all and, of course, what we do with it is of vital importance.

It doesn't take an economist to realize that Network Solutions (InterNIC), who have made claims of ownership of the top-level domains (TLDs) like .com, and .org is profitable now, unlike most of the wannabe vaporware silicon-alley-valley-gulch-mulch hypesters whose overvalued stock prices are magnitudes higher than cash flow and are *losing* money like crazy. NSI claims that the demand in 1998 represents only 2 percent of the potential market for domain names.

Over the years I have established my commitment to the promotion and support of independent media and alternative channels of communications. On my own initiative, time, money, and labor, I have established a strong net presence for excellent independent media and content through MediaFilter, which first went online on March 1, 1995, and has since grown to over 240,000 unique hosts visiting per month, pumping out 2 gigabytes per week of content that has become a well for research, education, and journalism [including online editions of independent investigative journalism such as *Covert Action Quarterly* or *The Balkan Media and Policy Monitor*].

Q: So do you want to become a big player yourself, an owner of the means of production? Who will profit?

A: Well, this is always a question of scale, scale is a question of money, if it turns up that we end up making money in the billions, sure we can lay fiber, and buy up satellite links. I wouldn't say that this is in our two-year plan, but I wouldn't rule it out either. In fact I am known for my capacity for reinvesting resources and therefore, if we do make that amount of money, I am not that kind of person that buys fancy clothes and a Porsche and moves to a house in the country, I would put that into infrastructure, research, and development—including developing new young talent.

Q: How do you see the improvements of Name.Space? At what point is Name.Space now, if we leave out the whole legal battle?

A: There are many aspects to the Name.Space project—business, autonomous policy, networking strategies, long term thinking, extra-institutional ways of working, technical details, standards, U.S. laws, global considerations—all of these are in dynamic interplay and we deal with them on a day-to-day basis. If we have a "routine," that pretty well describes it.

All of those aspects are of equal importance and it is critical to keep them all in perspective while dealing with them each individually in a practical, hands-on, nuts-and-bolts way. The need for specialists in each field goes without saying and we have an excellent team to deal with each of these aspects. Collaboration and cooperation are essential elements for the success of any large-scale project. Sure, the Name.Space project was initiated by me, but it is by no means a solo effort.

Q: So isn't it based on a simple hack?

A: Not at all. It's based on running the code as it's meant to be run. DNS is scalable at all levels. There is no real limit to the number of top-level domains, or the number of domains at any level of the DNS. Running new top-level names is not a difficult thing. Its simplicity is almost obscene. The issue of global recognition is the key. Right now, Name.Space lives as an intranet within the internet. Like a matter of perception, the recognition of Name.Space nameservers or not determines whether Name.Space exists or not. Like changing channels—Removing the censorship filter. This is a "grassroots" thing, and my favorite aspect of the potential of Name.Space—the individual's ability to choose their view of the net... Unregulated by commerce or government. But all TLDs should be globally interoperable because that's what the internet is all about. Therefore, we have been working hard to find a legal and political solution to globally recognized new TLDs to be administered in a fair and inclusive way, globally.

The convention of DNS is not the issue presently—it's the scope of its possible implementation. Name.Space works with the existing DNS software and protocols, exactly. There is no difference. Name.Space *is* DNS...and about exploring the potentials of a free namespace. Name.Space, from its beginnings has always been a collaborative and cooperative project. Most of the top-level names were suggested by users via a suggestion form on the Name.Space website. The SINDI project conceived by Name.Space will enable the total decentralization of name registries.

Q: So how about the legal aspects of your fight?

The net has been declared by international law expert Henry Perritt as a "global commons," much like the oceans and waterways, electromagnetic spectrum, space, geosynchronous positions in space, and other shared resources of the earth that are not exclusively controlled by any sovereign.

The case between pgMedia/Name.Space and NSI is a classic "essential facilities" case between two private companies. The "." is controlled by NSI exclusively and they must according to law allow reasonable, nondiscriminatory access to it.

The matter of access will be settled between the two companies, and the U.S. government will stay out of it not to violate the First Amendment and to uphold the Clinton administration's stated policy *not* to regulate the internet. As a separate issue, the establishment of independent NSP's internationally in accordance with all local jurisdictions will happen naturally as there is demand in the local markets. The "." being the global commons that it is must be managed responsibly and treated for what it is: a new industry that has grown into a rapidly emerging global market. The internet is international and ideally, self-regulating, and the reality is that market forces will determine the dynamics of the net.

When I studied the logistics of running DNS, I realized that the limits on it were artificially imposed in order to limit supply and facilitate control. The central database and "whois" records are all controlled by Network Solutions, Inc., which is a subsidiary of SAIC (Science Applications International Corp.), one of the largest private contractors for the U.S. National Security Agency, the Pentagon, and the Internal Revenue Service. Most of the top corporate officers are former U.S. military personnel who have retired from service and are engaged in "private practice," putting their militarily acquired skills to work for profit. In effect, when one registers and pays Network Solutions for a domain name, they are also paying to maintain surveillance on themselves.

Ask yourself. Is this what you want? Does it make you feel comfortable?

SUBJECT: ART ON THE INTERNET —THE ROUGH REMIX

FROM: TILMAN BAUMGÄRTEL <TILMAN_BAUMGAERTEL@COMPUSERVE.COM>
DATE: MON, 12 OCT 1998 13:36:58 -0400

It somehow made sense to me when my Walkman stopped working. I had used it to recorded all of the interviews, that have been remixed for my contribution to this book, and it broke down the day after I had finished transcribing the last of the interviews with a net artist. To me this technical problem marked the end of an era. The first formative period of net culture seems to be over. Books like this one seem to sum up the exciting years that followed the discovery of the internet by artists and intellectuals.

The interviews that my Dutch colleague Josephine Bosma and I did in the last couple of years are sort of an oral history of this period. These interviews, that were posted on Nettime and a couple of other mailing lists, were something of a news agency for the artists, critics, and audience that were

1119980993295551085 ...1230 ...4493 ---.15651
1119981096305551086 ...1238 ...4483 ---.15743
1119981107215551087 ...1246 ...4472 ---.15847
1119981108325551088 ...1254 ...4461 ---.15967
1119981109435551089 ...1262 ...4451 ---.16107
1119981100545551090 ...1270 ...4440 ---.16265
1119981101655551091 ...1277 ...4429 ---.16440
1119981102765551092 ...1285 ...4418 ---.16624
1119981103875551093 ...1292 ...4407 ---.16808
1119981104985551094 ...1299 ...4396 ---.16982
1119981101095551095 ...1306 ...4384 ---.17140
1119981101105551096 ...1313 ...4373 ---.17282
1119981101115551097 ...1320 ...4361 ---.17412
1119981101125551098 ...1326 ...4349 ---.17535
1119981101135551099 ...1332 ...4338 ---.17656
1119981101145551100 ...1339 ...4326 ---.17781
1119981101155551101 ...1344 ...4314 ---.17912
1119981101165551102 ...1350 ...4301 ---.18048
1119981101175551103 ...1356 ...4289 ---.18189
1119981101185551104 ...1361 ...4277 ---.18333
1119981102195551105 ...1366 ...4264 ---.18476
1119981102205551106 ...1371 ...4252 ---.18618
1119981102215551107 ...1376 ...4239 ---.18755
1119981102225551108 ...1380 ...4227 ---.18887
1119981102235551109 ...1385 ...4214 ---.19013
1119981102245551110 ...1389 ...4201 ---.19133
1119981102255551111 ...1393 ...4188 ---.19248
1119981102265551112 ...1397 ...4175 ---.19363
1119981102275551113 ...1401 ...4162 ---.19481
1119981102285551114 ...1405 ...4149 ---.19606
1119981103295551115 ...1408 ...4136 ---.19744
1119981103305551116 ...1411 ...4123 ---.19898
1119981106315551117 ...1415 ...4110 ---.20072
1119981117215551118 ...1418 ...4097 ---.20264
1119981118325551119 ...1421 ...4083 ---.20469
1119981119435551120 ...1423 ...4070 ---.20679
1119981110545551121 ...1426 ...4057 ---.20884
1119981110555551122 ...1428 ...4043 ---.21074
1119981112765551123 ...1431 ...4030 ---.21245
1119981113875551124 ...1433 ...4016 ---.21400
1119981114985551125 ...1435 ...4003 ---.21541
1119981111095551126 ...1437 ...3989 ---.21676
1119981111105551127 ...1438 ...3975 ---.21811
1119981111115551128 ...1440 ...3962 ---.21950
1119981111125551129 ...1441 ...3948 ---.22094
1119981111135551130 ...1443 ...3935 ---.22244
1119981111145551131 ...1444 ...3921 ---.22396
1119981111155551132 ...1445 ...3907 ---.22550
1119981111165551133 ...1446 ...3893 ---.22700
1119981111175551134 ...1447 ...3880 ---.22846
1119981111185551135 ...1448 ...3866 ---.22984
1119981112195551136 ...1448 ...3852 ---.23114
1119981112205551137 ...1449 ...3838 ---.23236
1119981112215551138 ...1449 ...3824 ---.23351
1119981112225551139 ...1449 ...3811 ---.23461
1119981112235551140 ...1449 ...3797 ---.23572
1119981112245551141 ...1449 ...3783 ---.23686

[/m/e/t/a/ (meta@null.net), Wed,
14 Oct 1998 00:58:23 -0800]

interested in art on the internet. Josephine and I were to some extent confined—due to geographical reasons—to the part of the developing net art community that identified itself as net.artists with a dot in the middle. I can't speak for the both of us, but I tried make sure that I wasn't just the ventriloquist's dummy for this exclusively European circle and tried to get in contact with artists who were not part of the traveling circus that meets at European media art festivals such as Ars Electronica, ISEA, and so on

For me the interviews were an attempt to escape the well-known rituals of the art world. After more than ten years of overtheoretical, dull, humorless writing on contemporary art after the period of Institutional Critique or Context Art, I tried to return to an approach that was more down-to-earth. And, as the many responses I got over the net to these interviews showed, a lot of people enjoyed those artists' statements better than a Lacanian reading (or other interpretation infested with the terminology of another trendy philosopher) of net art projects. In addition, doing interviews was a way of materializing the immaterial net art projects—at least on paper. To make this virtual reality visible again, I had artists tell me stories about it.

What's needed in the future will be more of a problematization of the issues that many of these interviews raise. Were the net.artists well advised to locate themselves within the art context? Will net art (given that it is an art genre at all) keep its freshness and uniqueness with the growing interest of art museums? Or will we see the same tiresome processes of institutionalization that happened to video art twenty years earlier? I was taught in journalism school that a journalist must never write, "It remains to be seen." But at this point I can't think of any other answer to the questions I am asking myself.

I am sure that some artists won't appreciate finding their quotes taken out of the context of the interviews and put together in a collection like the one that follows. My intention was to point to motives and ideas that kept emerging in these conversations. One might want to keep them in mind when approaching net art in a more theoretical way.

The quotes were taken from more than twenty-five interviews I did with artists who work on the internet from late 1996 to the summer of 1998. Excerpts from them have been published in online and print magazines and newspapers, such as *Telepolis, Intelligent Agent, Die Tageszeitung, Spiegel Online*, to name just a few. I am grateful to the editors of these publications that they supported my research into net art by publishing articles and interviews on a subject that must have been rather dubious to most of them.

Some of these interviews went over the Nettime list, the majority of them however didn't. Some—as the interview with Jodi—have been reprinted over and over again by now. Others have been sitting patiently on my hard disk for months. The whole bunch of them will be published in German in a book called *net.art—Kunst im Internet* (Cologne: suppos-Verlag, forthcoming).

BEGINNINGS

Robert Adrian X: There was a completely absurd episode in 1956, when I was still in Canada. I was working in a jazz club, and one of the musicians there told me that the Canadian Pacific Railway Co. was looking for people to work in an installation that involved a computer. The normal office workers couldn't handle it so they were looking for people to come in who could improvise—create a system for the machine. To me it was just a temporary, well-paid job. I guess there were about twelve of us—artists, musicians, students, writers—everybody was under twenty-five. They had built a whole building in Montreal for this computer—which probably had about eight kilobytes of RAM. The computer counted railway cars. The data on the railway traffic was collected at different locations in Canada. They wanted to know exactly where each car was, whether it was empty, whether it was full, what was loaded etc. We got this information on teletype machines that also made punched tapes we turned into punched cards. Every night the cards were sorted and transmitted to Montreal. I worked in the Toronto Data Center, and we had to communicate with the other data centers, the Computer Center in Montreal, and the train yards in our region, so we were always online via teletype.

Padeluun (Bionic): [In the art scene of the eighties—TB] there was nothing of interest to us anymore. There was nothing that got you excited or that even had some sort of vision. But here [with computers and BBSs—TB] was something, that made us think. There is something going to happen in this field... It will change our society, maybe even better it. Let's see what comes out of it. We started to go to industry fairs instead of art shows. We found out that at these fairs there were also people with smart, funny ideas. We started to look at contemporary scientific theory because we started to understand that this didn't become part of art and culture at all. There was no transfer, no translation into everyday culture.

Heiko Idensen: In 1984 I went to the art show "*Les Immaterieaux*" at the Centre Pompidou in Paris, that was co-curated by the postmodern philosopher Jean-François Lyotard. The question was if postmodernism could be shown in a museum. Part of it was collaborative writing project, where French thinkers discussed via Minitel system. Lyotard had introduced fifty terms like *absence* and *navigation*, topics that are still up-to-date today. You could participate in this at the museum. I personally couldn't even use French keyboards, but it left a huge impression on me.

Mark Napier: I used to paint. The nice thing about painting and sculpture is that those art forms don't crash. I got my first internet account in July 1995, put some of my paintings on my homepage, and then realized that this medium was completely separate from painting. Just scanning the images changed their nature, and of course I could create so many effects with Photoshop that the original painting no longer existed by the time I posted the image on my site. A few weeks later I took down all the paintings and started playing with HTML to see what I could get it to do. I experimented in hypertext "essays" (for want of a better word) like Chicken Wire Mother and the Distorted Barbie, before I got into a much more painterly, interactive approach, like what I'm doing now in POTATOLAND. I haven't painted since summer of '95.

Marko Peljhan: I was a radio amateur from when I was eleven years old. In Yugoslavia during socialism there was a big radio scene, and as kids we would go to the radio club and talk with people all around the world on short wave radio. When I think about it now, it was very formative for me, because it was a very global experience.

Olia Lialina: On the Homepage of Cine Phantom [a cinema for experimental films in Moscow where Lialina is film curator—TB] I used to put AVI-files into the pages. You could theoretically show a whole film on the page. But that wasn't enough for me. I asked myself how one could show film and filmic thinking on the net. I tried to do my experiments with storytelling with HTML instead of film footage.

Alexei Shulgin: My first experiment with the internet was in 1994, when I set up an online gallery of Russian art-photography. The reason to do this was very political, because it was against the existing practice of art curating and had to do with exclusion and inclusion. There was a big show of Russian photography in Germany. Some very interesting projects and series of works were not included because of the obvious ignorance of the curators.

TB: On the German or on the Russian side?

Shulgin: Both, because they were too busy with political games. As a photographer I was included in this show, but I thought there was something wrong with the whole concept. So I proposed to do a kind of supplement to the show on the internet.

Walter van der Cruijsen: My enthusiasm for the internet came from the fact that I finally found a medium where I could give all these immaterial ideas a place. In 1993 the Dutch Hacker cub "Hacktic" organized a congress that was called "Hacking at the End of the Universe," which took place on a camping ground. I was invited by some friend there. I didn't know much about the internet. After this congress it went really fast. I wrote the concept for the "Temporary Museum" for an Internet-Environment, and for some time it existed as the art space in the "Digitale Staad."

THE NET

Jodi: When a viewer looks at our work, we are inside his computer. There is this hacker slogan "We love your computer." We also get inside people's computers. And we are honored to be in somebody's computer. You are very close to a person when you are on his desktop. I think the computer is a device to get into someone's mind.

Debra Solomon: I like to refer to it [the net—TB] as Tamagotchi-culture. When you are online twelve hours a day, your desktop becomes your (audio)visual environment... You talk with all these people [with videoconferencing systems—TB] while you are doing your work. We practically live in the visual world of our desktops. Like the_living says, "We are the people in the little plastic egg."

Jordan Crandall: I see the internet as a network of materializing vectors. It is really involved with creating new material forms and refiguring existing forms. People talk about disembodiment on the net, and I really don't know what they mean. For me it is very embodying, it just embodies in different ways. I like to watch how technological paces affect daily rhythms and routines.

Jodi: I don't think you really avoid the art world by doing things on the internet. It was more that we were already working with computers. And I found that the best way to view works that were made with a computer was to keep it in a computer. And the internet is a very good system to spread this kind of work. The computer is not only a tool to create art but also the medium to show it within the network. And since the network doesn't have any labels, maybe what little Stevie is doing is art. It's the same with our work: there is also no "art" label on it. In the medium, in which it is perceived, people don't care about this label.

SPACE
Robert Adrian X: ...When the machines are on and your fingers are on the keyboard, you are in connection with some space that is beyond the screen. And this space is only there when the machines are on. It is a new world you enter. For me it was never a question of travel. For me it was always a question of presence, of passing through some membrane into another territory. It's not about things, it's about connections. Of course, we were prepared for this by conceptual art, by minimal art and all these movements. An electronic space is very easy to imagine once you have grasped the idea of a conceptual space for art works.

Eva Wohlgemuth: The net contains space and spacelessness at the same time, and you are always reminded of that when you work with the net. It makes it possible—at least in theory to access the material you work with from any place in the world—without dragging stuff around with you.

Paul Garrin: In the last couple of years there has been a gentrification of neighborhoods, now there is a Disneyfication of the net. That is as dangerous. I warned two years ago at the conference Next Five Minutes in Amsterdam of a disappearance of public space on the internet. Back then, John Perry Barlow said: "That will never happen."

Jodi: It makes the work stronger that people don't know who's behind it. Many people try to dissect our site, and look into the code. Because of the anonymity of our site they can't judge us according to our national culture or anything like this. In fact, Jodi is not part of a culture in a national, geographical sense. I know it sounds romantic, but there is a cyberspace citizenship. More and more URLs contain a country code. If there is ".de" for Germany in an address, you place the site in this national context. We don't like this. Our work comes from inside the computer, not from a country.

Bunting: I don't really surf the internet. I take great pleasure in wandering around cities, and seeing what happens, and London is a good place to do that. If you ever get bored, you just go out your door, and within a few minutes something interesting is happening.

THE BODY

Stelarc: I think that the body is obsolete. But that doesn't mean that there is a repulsion from the body. All I think is that the body has created an environment of intense data, data that it is alien to our subjective experience. We have created an environment of precise, powerful, and speedy machines that often outperform the body. We've constructed computers that now can challenge and compete with chess grand champions. Technology speeds up the body, the body attains planetary escape velocity. The body finds itself in alien environments, in which it is biologically ill-equipped. For all of these reasons, the body is obsolete. Now, do we accept the evolutionary status quo? Do we accept the arbitrary design of the body? Or do we evaluate the design of the body, and come up with strategy of reconstructing, redesigning, rewiring the body? For example, can the body have a wired internal surveillance system? Can the body have an augmented sensory experience? These are two aspects that would have profound impact on both our perception of the world and on the medical well-being of our bodies.

Victoria Vesna: ...I could see us uploading information into the internet and having agents doing work, freeing us from necessarily being with the computer. I actually think a lot of this machine–human interface is very primitive first steps of understanding how the technology will become part of our lives. It could also be a way to reaffirm our physical body.

TB: Yet one could understand your work BodiesINCorporated" as an affirmation of the things that are happening in biotechnology right now...

Vesna: Not really, because these are philosophical, psychological bodies designed to ask those questions you are posing. So it is not about us projecting us into this space somehow thinking that this is taking the place of our physical bodies. I have had people ask me that repeatedly, and I am always amazed. Does creating a body on the internet means that I don't exist here? No, I still have to go to the toilet. There is nothing virtual about that.

Eva Wohlgemuth: I also have the desire to upload myself and dissolve into cyberspace, but in the given situation I will work with the nonideal body and try to make something out of it. For me it is the possibility to use its weaknesses and imperfections to find different images for what is going on around me.

HARDWARE AND SOFTWARE

Jodi: [We are angry—TB] because of the seriousness of technology. It is obvious that our work fights against high tech. We also battle with the computer on a graphical level. The computer presents itself as a desktop, with a trash can on the right and pull down menus and all the system icons. We explore the computer from inside, and mirror this on the net.

Matthew Fuller (I/O/D): They [off-the-shelf software products—TB] work fine in some ways, but only because users have been normalized by the software to work in that way. There are other potential ways to use software out there, that seem to have been blocked off by the dominance of the Windows-metaphor, the page-metaphor, and other ways of interfacing with computers that have become common. We believe that GUI is suffering from a conceptual Millennium Bug... I think the "Web Stalker" realizes the potentials of the net better. It strengthens the range of mutation, the street knowledge of the net. Normal browsers deal with a website as a determinate amount of data. What we do is an opening up of the web to a representation of infinity. I guess that this is the core mathematical difference between the Web Stalker and browsers: between presenting a fixed amount of data and an infinite amount of data. What we want to say is that the web consists of a potentially infinite amount of data. What normal browsers do is close it down, that's why they are easy to use.

Paul Garrin: I am opposed to the concept of "Domains" as such. In the term "domain" is the military heritage of the internet: "Domain," that means "Domination," control, territories—this thinking comes straight from the Pentagon. And that's the way some people look at it: they think that these names are their property, like a piece of real estate that they bought. And all of a sudden the word "earth" belongs to a company!

Bunting: I was trying to find a way to cut down on junkmail to my email account, and I came up with this concept of an algorithmic identity. I change my address now every month in a way that is very easily predictable to humans, but not to a computer. I chose the date, the month, and the year, something most Western humans would know. So my email address currently is jun97@irrational.org. Every month the previous address will be deleted, and if you send mail to this address, you get an autoreply saying: "This identity is now expired, please reformat in this form." Since I've done that my email has gone from fifty a day to just about five. I don't get any stupid messages anymore.

Julianne Pierce (VNS Matrix): I think that technology is part of the structures of power that have been developed by the patriarchy. But now is the first time that women are able to participate in developing an industry or a discourse. Women never really had a part in how the industrial age developed for example. In the information society, they can play a really strong role in developing the future. So it's really important for women to get into the roots of technology and work their way up. If we want a society that really represents men's and women's views, women have to be at the top of that ladder. The internet and technology in general has been developed by men as a means of warfare, industry, and commerce. We're interested in having a discourse on the different areas of technology, be it the internet, be it multimedia. What particularly interests me is the how the information age changes our society and our culture. That for me is a really important issue of being involved with as well as using these technologies.

TB: Would you say that computers or the internet are gender-neutral?

Pierce: No, I think it's part of a system. I don't want to call this patriarchy, but the basic fact is that men control this whole information industry. Bill Gates is one of the most powerful people on earth, and there are generally men who are controlling the development of the industry. There aren't many women in those positions of power that actually influence the flow of technology. Maybe the computer and the internet as such are a neutral space, but there are certainly gender issues, that are relevant to that space. The presence of women as subjects of technology and users of technology is really important. There are really didactic arguments about how the hardware, the screen and the keyboard, favors the masculine, but I don't agree with that. There *are* women who contributed to the design of all this.

Marko Peljhan: I think there is not enough knowledge in society about technology and telecommunications. People tend to mystify it a lot, but when you really start working with it, it is just a tool like any other. I think that creative people who work creatively in this field have to develop specific technical skills, and you have really know how you are using them and why. When I started working with satellites, I realized that it was all military technology. That is a very important moment to reflect upon, this military provenance of almost everything that we use.

NET-SPECIFIC ART
Robert Adrian X: I wanted to create networks, and in these networks things can happen. I am interested in the strategic part of it, not in the content. I am curious to see what happens once this space for art is created. Making pictures is not what it's about. It is about finding ways of living with these systems, to look at how culture is changing in these systems.

Vuk Cosic: I did a lot of HTML documents that crashed your browsers. I noticed that there was a mistake somewhere in my programming. And than I asked myself: Is this a minus or a plus? So then I was looking how to get to that. It was not enough just to avoid this mistake, I was trying to really understand that particular mistake, with frames, or with GIFs that used to crash old browsers, or later JavaScript, that does beautiful things to your computer in general.

Olia Lialina: The web makes it possible to experiment with linear, parallel, and associative montage. With "My Boyfriend came back from the War" one can influence the narration. It is some kind of interactive montage. But the possibilities that the user has are limited, because he doesn't know what happens when he clicks on a certain field. But this work is more about love and loneliness than about technology.

Alexei Shulgin: If you deal with technology-based arts, the very first years are always the most exciting ones. Look at photography: When they invented the 35mm camera there was this explosion of art photography in the late Twenties and early thirties. Artists just did whatever they wanted with photography. They didn't worry how it would fit into the art system. They experimented with the medium, and they got really great results. It was the same with video. Video art of today is not interesting for me at all. Artists now use it as a new tool for self-expression. But I don't believe in self-expression.

TB: Why?

Shulgin: There is too much information already. I don't need more. But when this medium video appeared, it was really interesting what artists did with it. Same with the net: we are in the early stage of it now, and people are just drawn to it by enthusiasm.

INTERACTIVITY

Jodi: People sometimes send us helpful code. For example, somebody sent us a Java applet that we actually used for our site. We are really grateful for that. Some people really encourage us, too. They say: "Go, Jodi, go. Make more chaos. Make my computer crash more often."

Debra Solomon: I don't think that computer games are very interactive. *This* conversation is interactive, because we both can influence just about everything that goes on in it. That's how the interaction will be [at the net art project the_living—TB] between the_living and her audience/participants, when I'm on this trip. For example, I have an itinerary already, but should a participant know of some place or individual that would really add to the narrative or create a visually exciting atmosphere, I would be happy to change my route.

Alexei Shulgin: I don't believe in interactivity, because I think interactivity is a very simple and obvious way to manipulate people. Because what happens with so-called interactive art is that if an artist proposes an interactive piece of art, they always declare: "Oh, it's very democratic! Participate! Create your own world! Click on this button, and you are as much the author of the piece as I am." But it is never true. There is always the author with his name and his career behind it, and he just seduces people to click buttons in his own name. With my piece "form art," I encourage people to add to it. But I am honest. I'm not saying: Send it in, and I will sign it. I will organize a competition with a money prize, like a thousand dollars. I think that will stimulate people to contribute. I really want to make this an equal exchange. They work for me, and I give them money. I think, it is much more fair than what many of these so-called interactive artists do.

THE ART SYSTEM

Robert Adrian X: From the very beginning the problem has existed of identifying and defining the "work" and the "artist" in collaborative or distributed network projects. The older traditions of art production, promotion and marketing did not apply, and artists, art historians, curators and the art establishment, trained to operate with these traditions found it very difficult to recognize these projects as being art. Net art challenges the concept of art-making as a more or less solitary and product-producing activity.

Wolfgang Staehle: The issue of "institutional critique" was interesting to me, but I thought it was absurd to formulate a critique of the institutions of the art system within its institutions. That was just like re-arranging the furniture. I thought that this wasn't consequential. That's why I tried to really do something outside the institutions. I think, The Thing [the art-oriented BBS that Staehle ran in the early nineties—TB] worked so well, because the traditional art world didn't take any notice at all. The thrill was that you could feel like a gang of conspirators.

Olia Lialina: I, personally, never said in any interview or presentation that internet is my long awaited freedom from the art institutions. I never was connected to art system. I was not an artist before I became a net artist. Maybe that's why I—from the very beginning—concentrated on other things: internet language, structures, metaphors and so on. But at the same time the idea that net art must be free from real-world art institutions is very dear to me, because in their order of values net art is just one of computer arts. But I don't think that the right way to demonstrate freedom is to travel from one media event to another with presentations of independence. It's better to develop an independent system... For me to give up my freedom would be to stand on how a lot of critics, artist, and activists earn money and make a career with everyday statements that net art has no monetary value. Its not funny anymore. Article after article, conference after conference they want to convince me that what I'm doing costs nothing. Why should I agree?

MONEY

Robert Adrian X: There was no way to make money out if it, and there still isn't. You support the communications side of your work with money from elsewhere. I sold artworks and used the money to support the communications stuff. There was nobody from the big art centers like New York or London or Paris or Cologne involved. The people who participated in these projects needed the communication, because they lived in Vancouver or Sydney or Vienna or San Francisco.

Jodi: [For the participation in Documenta X—TB] we got a fee for the expenses we have when we put our files on their server. In total we got twelve hundred deutschmarks. It is a clear example of exploitation. Which artist would move his ass for this amount of money? But net art is a victim of its B-status. It is treated as group phenomenon, as a technically defined new art form. That is something that we have to leave behind as soon as possible, because that is the standard way to do these things: a group creates a hype. They call it mail art or video art, and it's doomed to die after five years. I think we are looking for another way, because we are not typical artists and we also won't play the role of the net artists forever.

Heath Bunting: At least half of my projects could be turned into a business. I did begging on the net for one week, and got sent fifteen hundred pounds. I made a form where you can send MasterCard or Visa donations to myself, and then I inserted it into corporation's or government guestbooks over the period of a week. A lot of people found it entertaining, and sent me money. But I didn't actually cash that money. It's not so interesting for me to do business. I assume that most of the credit card information that was send to me was from stolen credit cards anyway...

I get paid for giving talks. At the moment it is very boring for me to have an apartment. So for me this is a way to travel around without having to sleep outside all the time. I haven't had an apartment since September, I have been traveling continuously since last June. And I enjoy doing it, it's very challenging. The internet is a technology that makes that possible. Maybe ten or twenty years ago, there would have been a different way of networking. Maybe a hundred years ago, it would have been a name. If I was a certain type of aristocrat, I could have turned up in a court in India in rags, and I would have just said my password, and I would have been admitted and treated very well. In those days it was your name. There are other passwords now, that give you access to certain things. The funding models change. In the postmodern funding model, everything is small and connected in terms of business. Forty years ago it was different: with the modernist funding method, everything was big and disconnected. And that would have made it very difficult for me to travel around.

BORDERS

Guillermo Gomez-Peña: Basically we want to bring a Chicano–Mexican sensibility to cyberspace. We see ourselves as web-backs. That's a pun on wetback, which is derogatory term for Mexicans. We see ourselves as kind of immigrants in cyberspace. We also see ourselves as coyotes, as smugglers of ideas, because we do believe that there is a border control in cyberspace and that the internet is a somewhat culturally, socially, racially specific space.

Roberto Sifuentes: This is important, because when we started this project, the internet was seen as sort of the last frontier, the final refuge where issues about race relations don't have to be discussed, where race doesn't matter—as a strategy of avoidance. So it was important for us to venture out into the internet, and when we first "arrive there," we started getting responses back like: "There goes the virtual barrio, there goes the neighborhood. The Mexicans have arrived." Literally, people send us mails like that.

Alexei Shulgin: I feel much more included than before [the internet—TB]. When I was just an artist living in Moscow, whatever I did has always been labeled as "Eastern," "Russian," whatever. All my work was placed in this context. That was really bad to me, because I never felt that I did something specifically Russian.

BUT IS IT ART?

Alexei Shulgin: ...What we have now is that there is no critical context. Art always takes place in some physical place, in a museum or whatever. Even when it's a performance, it takes place in a space that is marked as an art place. Even if it is not an art place, it is appropriated by artists and therefore becomes an art place. With the net, you don't have this physical space. Everything happens on your computer screen, and it doesn't matter where the signal comes from. That's why there is a lot of misunderstanding. People are getting lost, because they don't know how to deal with the data they are getting. Is it art, or isn't it? They want to know the context because they don't believe their own eyes.

Robert Adrian X: The term "artists" has to be defined much more broadly in this context. You have to include so-called hackers in this definition for instance, because they are operating creatively with these systems.

Vuk Cosic: I think that every new medium is only a materialization of previous generations' dreams. This sounds like a conspiracy theory now, but if you look at many conceptual tools, that were invented by Marcel Duchamp or by Joseph Beuys or the early conceptualists, they have become a normal everyday routine today with every email you send. With every time you open Netscape and press a random URL at Yahoo. Eighty years ago this action, which is now totally normal everyday life, would have been absolutely the most advanced art gesture imaginable, understandable only to Duchamp and his two best friends. This very idea to have randomness in whatever area, form, shape, would have been so bizarre in those days...
I will give a lecture in Finland in September in which I will argue that art was only a substitute for the internet. That is of course a joke. I know very few people who have so much esteem for what artists did in the past.

Marko Peljhan: I actually don't care much about this kind of designation. When I compare myself with some other people who are also "artists" I don't see much we have in common. So I just call my works, "progressive activities in time." I am actually interested in defining utopia, looking over the defined borders. That is the legitimization that an artist has: the right to be irresponsible sometimes.

Wolfgang Staehle: That's not of interest to me, that's up to the art historians to decide. I can't answer this question.

[Links to all the art projects mentioned can be found at <http://ourworld≠ .compuserve.com/Homepages/Tilman_Baumgaertel/>.]

SUBJECTS: HACKERS ARE ARTISTS —AND SOME ARTISTS ARE HACKERS:
TILLA TELEMANN INTERVIEWS CORNELIA SOLLFRANK

DATE: TUE, 22 SEP 1998 16:16:51 +0200
FROM: CORNELIA@SNAFU.DE (CORNELIA SOLLFRANK)

Tilla Telemann: "Female Extension," your intervention of the net art competition "Extension," held by the Hamburg Galerie der Gegenwart (Gallery of the Present) aroused quite a bit of attention. What was the initial idea behind "Female Extension"?

Cornelia Sollfrank: Actually, I wanted to crash the competition. I wanted to disturb it in such a way that it would be impossible to carry it out as planned.

TT: Why?

CS: Because I thought it was silly that a museum would stage a net art competition. For me, net art has nothing to do with museums and galleries and their operations, their juries and prizes, because that goes against the nature of net art. Net art is simply on the net; so there's no reason for a museum or for a jury that decides what the best net art is.

TT: Do you still think that way?

CS: Basically, yes. But I'm afraid this development can't be stopped. Net art is on the verge of changing completely. It still happens on the net, but this need for completed, whole works that can be sold, that have a certain definable value, that can be attributed to an identifiable artist, and the establishment of authorities who do the evaluating and who deal in net art—we won't be able to ignore these developments. Net art will evolve in this direction, and away from what it was in the beginning.

TT: Where did the aggressive impulse to crash the competition come from?

CS: I simply am that destructive. I had the feeling that they didn't know what they were doing. They just wanted to profit from the hype surrounding net art without truly investing in it. That's what I wanted to shake up, and with this disturbance, call attention to the fact that it's not as simple as that. Net art is not just about cleanly polished websites; it might very well have something to do with mean, system-threatening actions of disturbance, too.

TT: The action was seen by many as a "hack"; *Die Woche*, a German newsweekly, even named you "Hacker of the Week." Do you see yourself as a hacker?

CS: No, I'm an artist. But if you take a closer look at the term "hack," you very quickly discover that hacking is an artistic way of dealing with a computer. So, actually, hackers are artists—and some artists also happen to be hackers.

TT: What does the term "hacking" mean for you?

CS: There's something called the *Hacker Jargon Dictionary* which is an attempt to define that term, among others. For me, an important parallel between hacking and art is that both are playful, purpose-free ways of dealing with a particular thing. It's not a matter of purposefully approaching something, but rather, of trying things out and playing with them without a useful result necessarily coming of it.

TT: Many spectacular hacks result in the destruction of computers, or at least a crash. With this in mind, do you see a parallel between your destructive impulse and hacking?

CS: Hacking does not mean first and foremost destroying. Today computer hackers place the greatest value on the fact that they're well-behaved boys who simply like to play around and discover the weakest points of systems without really wanting to break anything. At the same time, hackers can induce unimaginable damages. But at the moment, it's really about the playful desire to prove to the big software companies just how bad their programs actually are. At least they're trying to push their image more in this direction. Regarding my own action, it does have more to do with disturbance than destruction. I couldn't actually destroy "Extension" any more than I could inflict any serious damages to the Galerie der Gegenwart, but I was nevertheless able to toss a bit of sand into the works. Everything did not actually fall apart, but a few people did have to spend a considerable amount of time looking at a lot of trash/garbage...and so on. This did disturb the trouble-free course of the competition.

TT: Another aspect of hacking is that it does seem to attract people who enjoy the intellectual challenge of creatively working around limits.

CS: Yes, hacking does have to do with limitations, but even more with norms. That's another parallel with art. The material that art works with are the things that constantly surround us. The only thing art actually does is break the patterns and habits of perception. Art should break open the categories and systems we use in order to get through life along as straight a line as possible. Everyone has these patterns and systems in his or her head. Then along comes art: what we're used to is disturbed, and we're taken by surprise. New and unusual patterns of perception offer up the same things in a completely new context. In this way, thought systems are called into question. And only the people looking for this are the ones who are interested in art at all.

TT: Would you say that there are as many well-defined conventions involved in an art competition as there are in computer programs and that you have subverted these conventions with your action?

CS: Yes, that, too. The material I'm working with in regard to "Female Extension" is, on the one hand, the internet, but also the traditional means of art distribution: the museum, the competition, the jury, the prize.

TT: If you wanted to disturb the competition, why didn't you hack the server the art projects were stored on and erase everything? Or disturb the awards ceremony, for example?

CS: That's "electronic civil disobedience." In a way, I did my demonstrating on the net because it had a greater effect. My action wasn't truly destructive. I didn't break anything; on the contrary, I was actually very productive. Instead of destroying data and information, I used automatic production to see to it that there was more data so that the works sent in would be harder to find.

TT: Isn't it something of an affirmation of a system when someone tries to get into the system, whether it be a computer system in the case of the hacker or a competition in the case of an artist? Wouldn't it be more consistent to do the disturbing from the outside?

CS: No, you can disturb far more effectively from the inside than from the outside. Producing a flow of data has a considerably greater effect than standing out in front of the museum with a sign reading, "Down with Extension."

TT: One thing hackers emphasize again and again is that besides influencing social developments which only an elite group can follow anyway, access to sensitive information is really at the core of what they're up to. Is that also somewhat related to what you're doing?

CS: It has less to do with the information itself and much more to do with just how open systems are. The information itself is constantly changing. There's always new information. Much more important are the hierarchies of systems, what's accessible to whom. Hierarchies are established with passwords and codes and so on. These have to be broken by hackers again and again. Because of this, hierarchies have to be restructured over and over, and vertically structured systems are rebuilt horizontally. This is also the decisive difference between the distribution of art and net art. Art distribution is a hierarchical system, so it's vertically structured. I can't just hang my art work in a museum. But I can go to the net and "hang up" my website, for example.

TT: Of course, that's precisely what so many artists found so interesting about the internet in the beginning. But in the meantime, it's even the people who deal with it professionally can't keep an overview of everything that's going on in the field of net art because there's so much of it. A paradoxical situation has developed: Precisely because "everyone is an artist" on the internet, it's especially important that net artists establish some sort of relationship with art institutions in order to gather some sort of recognition...

CS: The only function of an art museum I can accept on the net is that of establishing a context. Which means that I don't just put my website out there where no one can find it, but rather, I place it within a certain context, for example, an art server. Presuming that it's a website at all, because besides the world wide web, there are many other services and levels on the net where art can take place. But the art server shouldn't be an art institution with a curator.

TT: In a way, an art server is the internet's equivalent for a producer's gallery. That is, there are artists who run a server themselves and fill it up with their own oeuvre. This is fine for the artist, but it may well not be of any general interest to anyone else. And that's what curators are for: To be a "gatekeeper" that only allows net art through which will have a certain value for the general public and not just for the artist who made it. In my opinion, this filter function is extremely important for the art public...

CS: Of course there are people who need this filter function because they don't have the time or the desire to look around for themselves. But with regard to "Extension," for example, there was nothing there that interested me. One should always be aware of just how elitist and questionable the choices made by a museum actually are.

TT: There is the historical example of video, where the processes of canonization and the induction into museums took place, processes that are probably on the verge of occurring with net art. What's actually so bad about the fact that museums are dealing with net art and trying to evaluate the various works? After all, that's the job of an art museum, to contribute toward the creation of context and the formulation of a canon.

CS: The motto for the museum is: Collect, protect, research. A museum that seeks to deal seriously with net art would have to collect net art and seriously consider all the consequences of just how this art form is to be preserved and researched.

TT: Aren't you contradicting yourself? On the one hand, you're saying that net art only takes place on the net and that's where it should stay and the museums should leave it well enough alone, and yet, on the other hand, you're saying that museums should be collecting net art...

CS: If a museum were to seriously take on the challenge of collecting net art, I could accept that. But I doubt that that's what they actually have in mind. And what happened at the Galerie der Gegenwart is a prime example. They simply wanted to quickly swim alongside the net.art hype, to sample a bit of the cream topping on all things cyber and net. But they've shown that they had absolutely no idea what that would actually mean in that ever since the competition, there have been no more efforts in this direction whatsoever. Since the awards ceremony in September 1997, the website hasn't been updated.

But if competent people were to work with a significant museum on the idea of seriously collecting net art, I'd approve. It'd be an incredible challenge, because not only would the collection of works and the formulation of theory be involved, but also a tremendous amount of hardware and software would be necessary in order to be able to read the data according to technical standards that go out of date within the shortest periods of time. So technical specialists who could handle the inevitable repairs and maintenance would also be necessary. But the museums are hesitant when faced with such a huge task. Such a collection would have to have a very broad range and gather as much material as possible, which would also necessarily mean that a certain evaluation and hierarchy of the individual tasks would have to be created.

TT: What you accomplished with your action is that the Galerie der Gegenwart won't be dealing with net art at all anymore. Would you consider this a success?

CS: The idea of starting a collection of net art with "Extension" was put into cold storage, in a way. Now they've offered Stelarc a residency. This compromise, that is, working with a single artist whose work is quickly comprehensible, is much more consistent, I think. With Stelarc, in terms of content, they are venturing out onto a new terrain, but it's still nevertheless compatible with a museum.

TT: Your "Female Extension" reminds me of the contextual art or the institutional critique of the early nineties. In the art world at the time, there was also this idea of focusing on and calling into question the conventions, the mechanisms of the creation of norms and canons. These were questions that only interested those who had anything to do with art. Could it be said that your work was essentially aimed strictly at the jury?

CS: The jury was, of course, most immediately effected, although the members didn't realize at all that "Female Extension" had anything to do with art—all the better. As for how much other people, for example, the artists participating in "Extension," were effected by my action, I don't know. But I got a lot of feedback from people who weren't directly involved and for whom I drew attention to an important problem, namely, the attempt to make net art museum-ready. Many net artists don't know themselves just how they should react to this and careen back and forth between the underground and the professional world. I don't have this problem because my work was the attack on the structure of the museum itself.

CAE INTERVIEW WITH BRIAN SPRINGER (OCTOBER 1989)

Colonel Noonan is a pseudonym he used for his pirate persona. The name came as play on the name of cable television pirate Captain Midnight (a disgruntled HBO employee who captured an HBO relay station in 1988, and uplinked some very unflattering text about the cable giant).

CAE: Col. Noonan, could you tell us how you got interested in satellite technology and guerrilla action using this technology?

CN: I became interested in satellite technology when I heard about these things called "backhauls," which allow you to see TV personalities off camera. There are two ways a backhaul can work. One is when they cut to commercial on your broadcast station—meanwhile your satellite station is not running the commercial. The commercial is being inserted at headquarters, so on satellite, you still see the person on camera waiting to go back on the air again. Another variety of backhaul is one common to newscasts and TV magazines, such as on CNN. In this case a raw signal (a signal containing only the image of the host or newscaster) is sent up to a satellite and then downlinked to a station that will insert the graphic or tape material necessary for a completely packaged show. But if you tune into the backhaul, you can see the person without the graphics, or see them when the insert tape is being rolled. This has always interested me, because you can see how the TV spectacle is constructed.

CAE: Where did you get your equipment to do this, and what was the cost?

CN: In 1978 a home satellite system would have cost about US$120,000–150,000, because when the signal comes down from the satellite it is so weak that it demands extreme amplification. At that time, the amplifiers US$80,000–100,00, with only twenty to thirty being made a year. Home technology became possible when the amp could be made very cheaply. By 1989 several generations of equipment have been released to the public. The early equipment, from about 1978–82, can be found on the back shelves of dish dealers' shops, and can be gotten very cheaply since it lacks many of what are now considered standard features. The amp can now be bought used for sixty dollars.

CAE: Is this the setup you use?

CN: Yes, pretty much so. The dish I use was originally made for telephone microwave from point to point on land. It's called a landline microwave; it uses the same frequency as satellite microwave. My mount is made out of an old bedframe and casters.

CAE: You can use these to get backhauls?

CN: Yes; just take your dish and go through every satellite. Spend a day. There is no [public] schedule for backhauls, so you have to do your own research to find out when the ones you're interested in come up.

CAE: What kind of commentary have you heard?

CN: One time on *The MacNeil–Lehrer Report*, Walter Mondale was on and he was painfully bored. He was watching the show on a monitor and they had just reported that Lloyd Bentsen's father had died. With that Mondale broke up laughing and said that Bentsen had always claimed that his father was the worst driver in the world, and now he's the worst dead driver in the world. He also found the Wedtech scandal to be hilarious. Backhauls allow you to get a glimpse of politicians' private persona, in a way that their public relations people can't control.

CAE: Can you also pick up news camera feeds, if there is footage online from China or Central America?

CN: Yeah. Live transmissions are good. I got one from CNN where a reporter was at this huge fire, and she is quite upset because she can't get the ash that was floating in the air off her teeth. So she spent most of the feed trying to keep her teeth white. Another thing you get is bulk tape source material before it's edited. I got a feed of a massacre in San Salvador. It was five minutes of corpses and the town's reaction. It's nice because you can see the event without it being contextualized by graphics and voiceover. It's unfiltered news.

CAE: Is it illegal to tap satellite feeds and backhauls?

CN: I wouldn't think so. It's on the public airwaves. You buy a consumer dish, turn it on, and there it is. Nothing is scrambled, no special equipment is needed. It's public information.

CAE: It would only be in distribution that you could get into a legal gray area.

CN: It would seem so, because you're hurting the public persona of the TV personality, such as with some footage I have of Robert Tilden. On camera he's praying intensely for people, and as soon as he is off the air he breaks into a totally different personality. He wants to know how much money is coming in, he's yelling at his studio people. I'm sure it would upset him, because it shows what a hypocrite he is.

SMILE, YOU ARE ON TWENTY-FOUR-HOUR CCTV

(From an Interview with Brian Springer by János Sugár [December 1995])

JS: How many other people are able to also use this? Is there a community that is working with this use of satellite dishes, catching images from the air?

BS: It's fairly dispersed. When I was doing it I didn't know of anyone else who was necessarily doing it. But on the internet there are some forums for dish heads. A number of individuals have multiple dish systems that receive this type of programming. It does not require a special decoder; it's not encrypted; it's available to anyone with a home satellite dish system; and there are over three and a half million home dish-owners in the U.S., so it's potentially available to that large of an audience. The channels are usually hidden in noise that is there on a satellite with not much activity and where there's usually static and for maybe a few hours a day this link occurs where you can see this programming. Most people will not hunt through this noise and when they do find something they're not going to watch it because it's very boring. The project was sort of a surveillance project and required several thousand hours of viewing. In 1992, I spent about two thousand hours watching the links of the networks, watching the links created by the candidates. Much of the time during those links nothing happens. You might have Bill Clinton sitting in a chair and he might ask someone to come over and he'll whisper in their ear, "We need to do our laundry. How can we do our laundry? My shirt smells." So it was very mundane, it was kind of a stakeout trying to catch those moments that represented wanting to use TV to not communicate. That's what I was looking for.

JS: Do you think that this informal side of television could have an influence on the medium of TV?

BS: I think it gets down to an issue of an investigation and that usually requires the revealing of secrets of what your investigating. It could become fashionable to be off-camera. This could become just another technique where being off-camera just becomes another stage to perform on, and I think the question is: "How can one investigate to reveal something that is hidden and something that is hidden can only be found where the person hiding the thing thinks there is no access?" If they become really aware that there is access, then it becomes just another stage of performance but it's interesting.

JS: Are any other media using this, like tabloids and private TV channels or not?

BS: Yes, I think there is sort of a paparazzi interest and voyeurism in this, and I'm not aware of any programs that are using it. In that way humiliation always sells well, so seeing someone humiliated by having makeup put on or kind of embarrassing themselves is always appealing to the baser instincts of TV. I think one thing that was interesting after the election was that there was an article that reported that the Clinton White House was monitoring the satellite TV feeds through the Department of Defense. They were able to intercept and downlink network news stories or the satellite feed of the new story before it was broadcast in Clinton's first days in office. This was a technique that had started during the campaign when the Clinton campaign had intercepted the satellite feeds of George Bush so they would get George Bush's commercial before it had aired and then they would have a potential to create a response to the commercial before it had been on broadcast television. There's also an interesting episode in the tape where a technician is talking to Al Gore's wife Tipper Gore and the technician explains to Tipper that they use the satellite feeds to examine the crowds as almost a form of crowd control, so the Clinton campaign would watch the satellite feed of a Clinton rally and the camera would pan the audience as almost like a surveillance camera and they would be able to identify people who might be protesters or people who might want to disrupt the image in some way and then the people watching the satellite feed would call the rally and tell them, "See that guy there, edge him out of the frame" or "Move him out."

INTERVIEW WITH BRIAN SPRINGER (OCTOBER 1989)

CAE: Moving in the other direction, are there ways that the consumer can send out signals that would disrupt or jam satellite communications?

CN: It's impossible to override a transmission with your own picture using consumer equipment, but it is easy to disrupt a transmission with noise and snow. The best noise generator that a consumer owns is a microwave oven. A microwave has 600 watts of power; it works at a frequency that is below satellite, but on the other hand it uses a microwave generator that produces a tremendous amount of noise and is very unstable; it doesn't keep on its center frequency. Using a properly sized dish and the inside of a microwave properly aligned, you could cause disruption to TV signals in the form of snow, a rolling picture, or skewed audio. It wouldn't totally disrupt the signal, but it would cause objectionable interference [a term used by HBO to refer to the drop in audio and picture quality that occurs when an alien signal gets into one-sixtieth of their power range]. However, since it works on a wide range of frequencies, you would also disrupt other satellite communications, like military or weather signals.

CAE: Have you experimented with this technique?

CN: Only on a theoretical level, and on a physical level of seeing how hard it would be to get the microwave generating device mounted, and that's easy. But I have never turned it on.

CAE: Are there other methods in the realm of possibility?

CN: Sure; marine radar on boats, or the market for used radar equipment, would be good places to get equipment for such a project. Such equipment would take some technical expertise to use.

CAE: Is the information available for someone willing to research these techniques?

CN: In a way. You have to put two and two together. The information about objectionable interference, how to create it, and the equipment it takes to do it is not public information. I did find some information, but the person who published it no longer lives in the U.S. He is under threat from the National Security Agency and HBO. He can't come back into the U.S. His name is Bob Coop, Jr. See what you can find on him.

CAE: Did he write for magazines?

CN: Yeah, but just freelance. There is a book called *The Hidden Signals of Satellite Television*, an excellent book by Tom Herrington and Bob Coop. It tells you how to tie into telephone satellites, audio subcarriers, and business communications.

CAE: Once again we are in extremely illegal territory—you could create enough disruption that there would be motivation for various security agencies to come after you.

CN: Sure.

CAE: How traceable is jamming?

CN: You would want to jam 6 gigahertz—the same frequency that the telephone company uses. So if you are in the pathway of one of these landlined microwave transmissions, and they could synchronize the satellite jam with the landline signal, they would have an approximate geographic location with which they could locate the origin of the jam. Or if you were in the flight path of an airport, that would be a second way. But it would be like finding a needle in a haystack from a hardware standpoint.

CAE: So in order to reduce the chances of tracing, and so as not to jam signals that you wouldn't want to jam, such as medical communications, you would want to go to an outlying area.

CN: That would be good. If you had a clear radius of around a hundred miles. Research the area through the FCC and you could find a clear grid.

Makrolab is a research station up on the Lutterberg, ten kilometers from Kassel. It is an autonomous solar- and wind-powered communication and survival tent, full of equipment. One night I went there to find out about the first results of the project.

GL: Could you explain us what kind of interception equipment you have here?

MP: You must have special decoding software to work with shortwave digital transmissions and different modulations. All that you hear now is different kind of HF modems or encoders. Teleprinters that use different standards. A lot of it is encrypted and there are specific NATO and Russian systems with specific baud rates that are almost impossible to decode. It is not like weather services or stuff like that, it's much more complex and hidden and there's no readily available information on it. When you hear and identify a baud rate of 81 or 73 or 96 p.e., than it is probably some NATO transmission and you know that you cannot get the message. But there's other systems that are very easily decodable or even voice services that are usually not scrambled. What we hear now is p.e. information about the weather over the Atlantic, the Shannon volmet for the air traffic flying toward Europe. On another channel we hear Stockholm Aero, and HF aeronautical station for transatlantic and transpolar routes. What we can decode quite easily is the SELCAL signals transmitted by aircraft, together with their position, wind, temperature, and fuel status. With the shortwave setup we have it is of course also possible to transmit, and every night I try to talk with some stations, yesterday it was Estonia and Belarus. In the past two days it was Mir packet radio time, three times a day and more. We try to get the Mir signals when it over flies Europe. As you know Mir was in trouble, but now they repaired their electricity circuit, and today they were resting, communicating with radio amateurs of the world.

BS: On the other machine we are receiving signals in the L-Band around 1.5 gigahertz. It is a communications receiver. It could be use for mobile phones, but they are mostly regionally located. We were specially interested in crossing boarders and boundaries. Across five countries or more, like INMARSAT, which is a satellite telephone system, briefcase size. Maybe you saw Peter Arnett using this during the Gulf War, speaking to CNN. There are still vestiges of the INMARSAT system that are analog-based, which do not require any special digital decompression. So here in Germany you could be listening to America, Ireland, or Tehran. This is where communications start to get interesting, where the medium does what it does best, which is communicate.

Segun pur teorial resone es ya inter katolikisme e protestantisme ke exista li grand skisme in li kristanaro. In li dogmati opiniones fundamental li diferos es extremim poki inter li katolikisme e li ortodoxia. Les [they] relate primim li doktrine pri purgatorie e li famosi "filioque"—tum es li interesanti kontroverso pri ob li sankti spirite emana anke fro li filio o fro li patro solim. Ma, sat stranji, studio del eklesial historie revela ke non es li dogma ma li traditione kel krea heresianes. Inter li kristanismen praktikal praktiso in lun [its] luterani e in lun katoliki forme exista nul difero. Por ambes ortodoxia kontrastim representa sin irgi duto absolutim stranjeri religione. Li westeuropani kristanisme es super omnum eti [ethical], rationalisti e intelektual. Li antiqui filosofia, li medieval skolastike, li renesanshumanisme, li reformatione e li jesuital etike ha stampa li kristanisme, chake segun sen manere. Segun ke on aksepta li europani kulture e li europani etike, on mus pro tum anke aksepta li kristanisme. Kultivat e eti pagane in li moderni europa es pur paradoxe. Por tu es pagane, on mus retrovada en [into] la barbarstadie—tum es en ti kulture, kel existad in europa ante li introduktione del kristanisme. In li ortodoxi kristanisme non exista dis probleme. Li ortodoxia have nuli filosofial o intelektual traditione, nuli reformatores e nul etikal teoriistes. Lu have dogma e ritu, incense e ikones, ma lu non determina li homesen pensado e non kontakta kun lesen intelektual kulture. Ke dis primitiv kristanisme povud transforma li marxisti materialisme en aminim partim idealisti idee, es pro tum absolutim nonpensabli. Ma sembla kontrastim ke li rusi marxisme in manere sat komodi e simpli pove nihilisa desagreabli konkurante. Un tre primitiv idealisti idee bli suplanta da altri tali mem plu primitiv materialisti. Disum es li uni latere del traditionen metamorfose, kel li rusi bolshevisme representa. In nusen tempe rusia es separat fro li ceteri europa per abisme [abyss], kel es plu profundi kam irgitem antee. Rusia e westeuropa es du diferanti mondes, keles sempre plu isola es fro mutu. Rusia ha turna li dorse a europa e separa se resolutim e konsciosim fro irgi "infektione" de europani kulture. Plusum ve seku. [Humanzsuk@ultra.com, Russian Kombato Kontre Europa, Tue, 25 Aug 1998 10:18:35 -0500 (CDT)]

And where culture does what it does worst, which is communicate. We are investigating if the collision of these best and worst characteristics can create a interesting stage for intervening in the transnational flow of information.

MP: What makes this set of radio amateur gear perhaps specific is the context in which we are operating. The result is only becoming visible only after quite a long period of time and a period of reflection. We have just started.

GL: Could you compare the work with video feed with your current research on the audio spectrum?

MP: In Europe there are less feeds. What you get is pretaped material that is sent to different broadcasters. I have been working with shortwave for a long time, since the early eighties. Shortwave is the cheapest and most accessible way of communicating over long distances and still widely used. I think that almost everyone has the experience of suddenly hearing a female voice giving out four-letter codes for five hours on their own AM radio receiver. We listen to those here too and try to make some sense and basically map them. There is information available on the internet about the frequencies secret services use, but things are changing quickly in that world. And basically all posted data is already old data. Audio and data traffic on SW is still not so accessible, compared to video, where you just hook your TV up to a satellite receiver and a dish and there you go.

GL: Brian, you experienced the closing of the open video channels. Most of it is now encrypted. This is also happening in the audio spectrum. Do you see the same patterns occurring there?

BS: The open windows are slowly closing. It is a unique opportunity to have one last glimpse at the curve of the analog spectrum before it closes forever. Analogue seems to be more natural, curved, not binary, with less protection for the information contained on these channels.

GL: So we have to move than and crack the digital spectrum.

MP: The big game is to move forward to digital domains. A complete set of new knowledge is needed. We heard rumors that digital communications, for example banking information, were cracked. That is illegal and basically a criminal offense, but it tells a lot about the safety of our own data being transmitted and retransmitted over the networks. The encryption that is currently used by states in diplomacy is very hard to decrypt. You must have the key, that's it. Intelligence services are working more on getting the keys than decrypting. The human is the weak element of the chain, not the signal anymore.

[See <http://markolab.ljudmila.org>.]

SUBJECT: TEN REASONS WHY THE ART WORLD LOVES DIGITAL ART

FROM: MATTHEW FULLER <MATT@AXIA.DEMON.CO.UK>
DATE: MON, 12 OCT 1998 01:16:59 +0100

1. We live in an era when the dominant mode of politics is systems analysis. Power has been given over to a series of badly animated white-shirt technicians who deliver fault reports and problem fixes that can be answered only with an "OK." All the control and trustworthiness of Norton Utilities is claimed for a bunch of frightened useless pilots gibbering out of control at the keyboard of a system they no longer understand. In this context it is essential for artists and others to synthesize an unformattable world.

2. The art world loves digital art because—like itself—there is a large submerged part of it that is invisible to the viewing public and only ever read by interpretative machines. Digital art is an autonomous field with its own opportunities, norms, and institutions. It understands that the distinction between the fields is necessary in order to maintain the integrity and thoroughness of both fields. For all artists it is imperative that they maintain the field in which they work as an autonomous sphere. The strength of a specific field can be measured precisely by the degree to which participants recognize the contributions of their peers and therefore develop each others richness in specific capital. The collapse of discipline can be measured precisely by the degree to which heterogeneous elements are able to exert force within or upon it.

3. Jeff Koons recently described the patterns produced in the interrelations of basic, repeated units, motifs, forms, colors, in his sculptures constructed of variegated patterns of boxed basketballs as a basic form of artificial intelligence. Mainstream art has already begun to incorporate the terminology and methodologies of digital cultures as a way of talking about itself and finding sympathetic refrains within a wider culture.

4. The art world loves digital art because it reminds the art world of the limits of its knowledge and the wisdom to be found in the open, nonprejudicial contemplation of the unknown. Likewise it is always useful to have a relatively large amount of the unknown to call upon in the event of a vague legitimation crisis. In the past it has been proven good insurance to have a few unknown things knocking about in the rear. Graffiti, macrame, female artists, and other minor genres have all played their part in the past.

5. Large prestigious art museums with marble foyers love web-based art because it implicitly solves some of the problems of distribution for non-gallery-oriented work that were faced comparably by video art. Because the

web guarantees at least some kind of circulation, this frees them from the embarrassment of undergoing the rituals they are forced to undergo on behalf of artists thoughtless enough to produce painting, sculpture, or installation.

Given the medium's self-sufficiency, widely promoted, attentively curated exhibitions with all their background maneuvering, public attention, critical discussion, historicization machinery, high artists fees, and other negative influences on the pure essence of artistic creation can all be avoided, leaving the work to be safely ignored.

For similar reasons, those who are interested in reading Marx without illusions believe that the *Fragment on Machines* in the *Grundrisse* has important implications for technology and art. Here, Marx suggests that what he terms "general intelligence"—the general social knowledge or collective intelligence of a society in a given historical period, particularly that embodied in "intelligent" machines—reaches a decisive point of contradiction when actual value is created more on the basis of the knowledge and procedures embedded into these machines than in simple human labor: thus freeing digital artists from having to exist. Or at least freeing them from being any less cheap and infinitely reproducible than their work or their equipment.

6. The art world loves digital art because someone other than Royal Society of Portrait Painters has to take the conventions of pictorial representation into the future. While virtual worlds might still be to the mid-nineties what Roger Dean album covers were to the mid-seventies, the onward march of technology will one day surely permit an upgrade-obedient artist to produce a final form of perfection: an utter conformity to perceptual mechanisms whose perspectival instructions permit viewing only by the most perfected of subjects. At this sublime moment being empties in entirety onto a computer and thus perhaps allows isolation on a hard drive to be stored or destroyed.

7. The artist waits in ambush for the unique moments when an unrecognizable world reveals itself to them. They pounce on these little grains of nothingness like a beast of prey. It is the moment of full awakening, of union and of absorption and it can never be forced. The artist never formulates a plan. Instead they balance and weigh opposing forces, flexions, marks, events, distribute them in a sort of heavenly layout, always with plenty of space between, always alternating between the heat of integration and the coolness of critical distance, always with the certitude that there is no end, only worlds within worlds ad infinitum, and that wherever one left off, one had created a world.

The sublimation of technique to the advantage of a separate category known as creation is consistent between all sections of art. Programmers, technicians and other people are glad to work hard to make the realization of the vision of the artist possible. *Providing* such freedom for the artist is essential because in this way providence always takes victory over ego.

8. Because art that is not solely about content, but that is multiply reflexive, concerned with materials, that is about the lusters and qualities of light, about the tonality of certain gestures, about modes and theaters of enunci-

ation refuses to make a strict separation between creation and technique. Concept and execution fold in and out of each other, blurring the categorical imperatives of rule by the head or by the dead. The most powerful art, digital art, art that is despite itself digital is, regardless of the context that codes it and from which it escapes, derived in this way precisely from hooking into an expanded compositional synthesis.

9. A multitude of currents of heterogeneity destabilize digital art's status as an autonomous field. Most banally this occurs in the production of art that takes the needs of sponsors so to heart that it is indissociable from them. Heterogeneity can also disrupt the autonomy of a field, and thus its internal self-evolving richness, when it comes in the form of interpretation: in lazy journalistic work whose primary concern is the humorous gratification of what it presumes are its audiences' prejudices; in works that are diagrammatically preformatted by pre-existing critical criteria; or—most importantly—in works whose relationship with certain flows of words amplifies both.

10. Both fields, art and digital art, attempt to control what art and artists (and by implication those people or practices defined as being outside those terms), should do and what they should be called. This is simply as a necessity for their maintenance and development. At the same time, even their own historical emergence is or was dependent on the eventual impossibility of such control. Those moments at which that impossibility is made concrete are what produce artists worthy of the name, as well as those to whom the word means nothing. Paradoxically, this very impossibility is what art and digital art claim as grounding their ability to speak, to be paid attention. It is only when they lividly and completely fail to betray that claim that art becomes worthy of anything but indifference.

SUBJECT: NEW MEDIA, OLD TECHNOLOGY

FROM: DR. FUTURE <RICHARD@DIG-LGU.DEMON.CO.UK>
(BY WAY OF RICHARD WRIGHT)
DATE: SUN, 14 JUN 1998 21:42:38 +0100

I am attending a smart cheese and wine party hosted by the Arts Council and one of their corporate sponsors when it is announced that the director of a well-known North American art center is present and is looking for new proposals for their artists fellowship program. I have an idea that could do with some "institutional support," so I decide to forego the race for the *vol-au-vent* and cross the room to introduce myself. I begin to explain my exciting new method of image synthesis but do not get very far before she makes her position clear. "Is your project internet-based?" she inquires. "No..." "Is it multimedia?" "Err...no..." "Well those are the only projects we do now." In the corner of my eye I can see someone skewering the last savory parcel.

In 1995 the grand daddy of electronic arts prizes, the Prix Ars Electronica, decided to drop its *computergraphik* still-image category after suggestions in previous jury statements of a "tiredness of creativity" and speculations on whether this form had "outlived itself." That year it was duly replaced by the new world wide web category. In addition, the computer animation section became increasingly dominated by special-effects feature films selected by a jury made up largely of members of commercial production companies. Amidst timid jury statements questioning the wisdom of having to compare half a dozen Hollywood films made by Industrial Light and Magic with a short sequence made by a lone artist working out of their bedroom, Prix Ars reinforced the feeling that artists had gradually abandoned "older" forms of "new" media for the safety of emerging "cutting-edge" technologies before they too are "professionalized."

The ISEA '98 revolution symposium distinctly positioned itself at the forefront of radical arts practice, brazenly featuring this quote on its call for proposals—"the opposition of writer and artist is one of the forces that can usefully contribute to the discrediting and overthrow of regimes that are destroying, along with the right of the proletariat to aspire to a better world, every sentiment of nobility and even of human dignity." Against this heady rhetoric, the invitation for exhibition proposals to ISEA '98 contained no mention of either still image work nor film and video art in its list of entry formats, presumably relegating such outdated forms to an earlier era of "prerevolutionary" practice.

So we are left to infer, perhaps, that a new medium can only sustain a period of true artistic innovation and challenge for a limited time before it is exhausted of radical ideas and has to leave center stage. The new incarnation of progressive arts practice then rises into the sky on the wings of blue-sky research labs while its decaying predecessors have their bones picked clean of creative meat by the vultures of venture capitalism. Film art begat video art begat computer art begat interactivity begat the web. This cycle of birth and death has now assumed a familiar logic—artists need not worry as the routes of access to media production are closed off by the mainstream commissioning policies of the commercial industry. They need only wait for the next wave of media to appear and then to seize that window of critical intervention to undermine capitalist social relations before the corporations know what's hit them. The only article of faith that this requires is that technological progress march inexorably onward, generating the raw material that can be used to subvert its own previously recuperated incarnations. Political innovation requires technical innovation.

The theoretical justification for this attitude is given in terms of art as a "transformative practice" or aiming at a "functional transformation." It is a direct reference to Walter Benjamin's famous materialist theory of revolutionary art practice. This is expressed most concisely in his "The Author as Producer" lecture of 1934, in which he formulates it in terms of a distinction between an art work that supplies a social production apparatus and an art work that tries to a change a social production apparatus. What this means in effect is that it is not enough for, let's say, a writer to criticize the capitalist system in words if he or she continues to use a capitalist form of cultural production to publish those words. Benjamin warns that bourgeois culture is very capable of absorbing all kinds of revolutionary ideas without at any time allowing those ideas to threaten its power. Instead of publishing political arguments in the usual academic form of books and scholarly articles, the socialist writer should use new forms that change the writer's production relations, especially their relation with their audience, the proletariat. The newspaper, pamphlet, poster, or radio broadcast were the most appropriate media in Benjamin's time because they could be used to reach a mass audience and avoid patterns of traditional cultural consumption that were rooted in class structure. What matters most in the political effectiveness of an art work is not the "tendency" of its content but the effect on production relations of its "technique."

In contemporary times this translates into an oppositional arts practice that uses the most advanced materials of its time to demonstrate in a concrete way the direction in which society should be progressing. It challenges currently accepted notions of production, authorship, and creativity by using new media to show how electronic distribution changes exhibition, interactivity changes authorship, sampling changes creativity. Technology is shown to possess the power to restructure these production relations and alter what people had previously taken for granted. And whenever production relations threaten to ossify into restrictive ideologies as newspapers are merged by press barons and radio airwaves are regulated then they can be blasted apart again by the socializing potential of each further technical development that can be applied to the mass media. All of which is fine, except for the fact that this is not entirely what Benjamin meant.

Later on in his lecture, Benjamin goes on to discuss some explicit examples of the effects of "technical innovation" on the political function of culture.

He use quotes from Eisler to show that concert-hall music has entered a crisis caused by the advent of recording technologies, which change the relation between performer and audience. But we are told that this is not sufficient by itself to transform music into a politically potent form—the addition of other elements like words is also necessary to help overcome the breaking-down of culture into isolated specializations that occurs under capitalism. And this eventually leads it to the form that Benjamin's finds most exemplary—Brecht's Epic Theater.

What is technically innovative about Brecht's theater? It is not cinema, is is not radio, it is not mass media. But it does change the relationship with its audience, not by using film or broadcasting technology directly, but by adopting their "techniques." The principle technique is montage, the ability of modern media to fragment perception and then recombine it. In Brecht's theater this is absorbed in the form of "interruptions" to the dramatic action in order to create "conditions" presented to the spectator that require a "dialectical" response. In this way montage is employed as an "organizing function" as opposed to a "modish technique" used merely to stimulate the viewer's fascination. So we see that the actual works that Benjamin is interested in use new techniques at a variety of levels which can include different media, perceptual modes, "organizing functions" and aesthetic considerations. Contrary to using the latest technological means, Brecht is described instead of returning to the ancient origins of theater, turning the stage into a simple podium for exposing present behavior and conditions. New technique does not mean new technology.

Today we see digital artists driven onward to become multimedia artists to become net artists and in their wake they leave a trail of unresolved experiments and restagings, unable to develop an idea through before the next software upgrade is announced. As if "earlier" forms of new media had been "outlived," no longer able to express the forms of subjectivity that are now experienced. But by picking up any magazine or observing any street advert we can clearly see that on the contrary commercial design and photography has continued to exploit and push the still-image form way past the stage where many artists abandoned it in their move on to more "revolutionary" media. Through this work we can still see the potential of continuing advances in the standard commercial digital software packages like Photoshop, which has unfortunately now taken on the status of an office desktop accessory with many artists. The artists that have continued to work in areas that are almost unfunded have shown how much further image and print media can go in producing their own newspapers, fly posters, fax art, graffiti and underground cinema and in experimenting with alternative methods of distribution.

Similarly in moving image production, developments in digital image synthesis are amongst the most advanced technical accomplishments in the world today, but are only ever seen as "special effects" in feature films or promos, a "modish" or stylistic use of the medium as the new-as-always-the-same. It seems almost an accepted fact that the sophisticated logics created to structure image events such as dynamic simulation or motion capture can only ever be used for blowing up space ships or for the latest shoot-em-up computer game. It is as though they are perceived as so closely aligned with the interests of Soho art directors that they can never be quite new enough to escape from its orbit. Instead it appears far easier for arts organizations to develop schemes to support work made for a particular piece of hardware or software they have just seen on *Tomorrow's World* than to look one layer below the surface to ask what techniques, like montage in the thirties, are likely to have an impact on the function of many forms of practice. For it is surely the case that technical and aesthetic developments in the basic manipulation of sound and image are applicable to a wide range of media generally. Arts centers fall over themselves to attract work designed for the latest internet software, VR environment, or multimedia platform but are not willing to consider projects in image- or sound-making that could radically alter the possibilities of all three.

There is an argument to the effect that by being involved in the early stages of a new medium that artists can exert some influence over the direction in which it develops. By getting in first before mainstream genre forms have had the time to become entrenched it could be possible to indicate alternative patterns, but it is still very difficult for artists to work as maverick researchers against a corporation's ultimate agenda. This approach also implies that media will inevitably develop into a single optimum commercial form without any further hope of an intervention, a kind of commercial determinism. In fact, the computer industry seems to be distinguished for its continuing volatility just when everyone thinks the dust has settled.

I am reminded of a story related by Graham Weinbren, the artist who pioneered the use of interactive cinema in the late eighties. He and his brother had developed a system that allowed for real-time transitions between different story streams and was demonstrating one of his first pieces to an audience of industry professionals. They were duly impressed by the speed and fluidity of the system and wanted to know the technical specifications. However, when Weinbren revealed that it was based on an old 386 PC, a machine already obsolete even in those days, their interest immediately cooled. The problem was that the logic of the commercial industry demanded that new products were always premised on the notion that they embodied nothing but the latest in technology and manufacturing. To revert back to a previous "generation" of machines would have introduced an uncomfortable contradiction into that philosophy. Unfortunately, this is also a philosophy that has now been taken on by arts organizations that feel that here is an easy way to align themselves with progressive media simply by pointing to new black boxes.

So artists find themselves running to keep still, trying to keep at bay the panic that they will be left behind in the latest high-tech funding opportunities and consigned to the back room of old media. Condemned to chase a never-ending succession of software versions and hardware upgrades, their practice is now so "transformative" that it never gets past the round of demos and beta tests. By becoming fixated on the receding horizon of technological developments the space for consolidating what has been learned is lost. The avant-garde artist trying to lever an oppositional advantage at the fringes of advanced materials is replaced by the techno artist-entrepreneur providing research and development services for corporate sponsors. There is no reason to develop an idea beyond the point at which it can be sold.

During the seventies and most of the eighties, artists who wanted to use computers were obliged always to be working at the frontiers of technology because there was practically no where else to be. Computing machinery was so limited that in a real sense the machine was the artwork because you would always be using it at the very extremes of its abilities. Such was the desire to escape these restrictions that faster and bigger architectures were eagerly sought after and resulted in the feeling that to produce the best art you needed the best computers. Nowadays, this principle clearly sounds erroneous, partly due to the fact that desktop computers are so powerful that the "best" in computing is accessible to the point of being unavoidable. But it has been surreptitiously replaced by a "softer" version that implies that to work in the newest media you need the newest technology.

The effect is to divert attention from innovations in currently used media by implying that artists can only retain their radical credentials by concentrating on the "cutting edge" of new technology. And, surprise, surprise, it is exactly this mythic trajectory of technology that commercial companies depend on to motivate the consumption of their endless releases of new products that allow you do the same thing more often. Both are now united in their quest for a Killer Art for the Killer App.

SUBJECT: HEATH BUNTING: WIRED OR TIRED?

FROM: ANONYMOUS (SIGNING AS "TIMOTHY DRUCKREY <DRUCKREY@INTERPORT.NET>")
DATE: SUN, 21 DEC 1997 20:33:23 +0100

In the December issue of *Wired* magazine we find amidst the pre-Christmas consumer spectacle of seductive scanners, professional sports watches, expensive liquors and, scantily clad savvy female computer nerds, a seductive spectacle of another shape. The current offering is a glossy close-up of the smirking bearded face of Heath Bunting, net.artist from London, and one of the founders of the international net.art movement.

Bunting is best known amongst the digirati for his intended subversive actions and attacks on corporate and consumer culture. Attacking professionalism of all kinds, he was quickly scooped up by the very professional Catherine David for 1997's Documenta X, the prestigious international art exhibition in Kassel, Germany. In a manner astonishingly akin to Documenta X, with its redundant revisits to seventies conceptual art, Bunting's naive stance revealed his ignorance of hard lessons learned twenty years ago by less inexcusably innocent precursors. Had he been paying attention, he could have learned sooner that there is no outside in corporate consumer culture or more importantly, that "outside" is just another target market. Well this December, dec97@irational.org has apparently learned with a vengeance; He has recently accepted a paid position as Senior Computer Artist at the Banff Centre, in Canada. The logical next step, geographically and ideologically, will be senior computer consultant at Microsoft.

From the pages of *Wired* we gaze at Bunting's face, a tastefully consumable icon floating against a white background. As Artist of the Hour, he appears ironic, cool, and rebellious, gazing at the reader knowingly, eyes narrowed, lips pursed—as if to suggest that his subversion could somehow transcend the lifestyles magazine he is now decorating. But what exactly is being subverted, or more precisely, what are we being sold?

In *Wired*, the hot new item of consumption these days is the subversive artist. *Hot Wired* and *Wired* have taken on the badly needed position in the U.S. as patrons of the digital arts. They have been more friendly and inviting to digital arts than the art world ever has been. In *ArtForum*, for example, as the token digital critic I am occasionally offered a column, always already scripted within the margins, of the magazine and of the art world. There has been much theorizing of the relationship of the margins to the center particularly from the net as a marginal, suburban strip mall, in relation to the art world's urban center marketplace. Yet much of this theorizing comes from a passive relationship to the digital media upon which the theorists and artists are commenting. This was not the case previously with Bunting, although with this latest transgression, or rather absorption, we see how quickly one can be seduced to the sell out. Demo or die!

Wired, unscrupulous entrepreneurs that they are, have taken to heart their forefather lessons, Phillip Morris and Saatchi and Saatchi, to name only two of the most licentious. They fully understand just how useful a public relations device the arts can be.

Bunting, "Sage of Subversion," we are instructed with no apparent tongue in cheek, is "fucking with commodities." Easier said than done, coming from a magazine that has already taken home the prize for glorifying the wild wild west of free-market computer economics. Cool and radical in its approach to consumption, why not invite Bunting to play act two to patron saint Marshall McLuhan: another clever Commonwealth citizen with a palpable soundbite? No less ludicrous is the additional label *Wired* ascribes to Bunting, "Michelangelo of the Digital age." In an age of postmechanical simulation, the notion of the hand in art is no longer nostalgic, it is positively reactionary. To proclaim the possibility of a masterly mark of the digital age is a suggestion seeping with egotism and nostalgia for masterpieces whose poverty have been unmasked ever since that fateful day in 1917 when the patron saint of contemporary art signed a mass-produced urinal.

The cultural loop—from subversion to assimilation to absorption—revisits net art quicker, smoother and more quietly than ever before. The ride begins with net production and distribution and ends as hard-copy pages spouting computer consumption and techno-utopianism. Bunting becomes a complicit pawn in *Wired* magazine's naughty boy game of—ever so gently—slapping the hand that feeds it.

And finally we must ask the sad but obvious question. What is Bunting subverting? The answer is perhaps the greatest irony of all. He is, we are informed by *Wired*, "wreaking havoc on corporate Web sites" and "overturning capitalistic ideals." Anyone searching for Adidas and Nike is given a pointer to the competitors site. So in essence, Buntings "subversion" is to participate in free market economics, in ending monopolies and giving business to the competitors. Capitalism 101 anyone? Cheques for tuition may be sent via <http://www.irational.org/skint>.

SUBJECT: FAST, CHEAP AND, OUT OF CONTROL

FROM: TIMOTHY DRUCKREY <DRUCKREY@INTERPORT.NET>
DATE: TUE, 29 SEP 1998 19:20:58 -0400

"External progress; internal regression. External rationalism; internal irrationality. In this impersonal and overdisciplined machine civilization, so proud of its objectivity, spontaneity too often takes the form of criminal acts, and creativeness finds its main outlet in destruction." —Lewis Mumford

Evoking the pivotal essay by Hans Magnus Enzensberger, "The Aporias of the Avant-Garde," seems necessary in a time compulsively destabilized by its woeful lack of interest in critical history and its dubious fascination with cynical history. It explains why pleonasm and redundancy haunts too much of an emerging and seemingly rootless artistic generation weaned on glib "negative dialectics," virtual "one-dimensionality," and hip cybertechnics. Unwilling, or unable, to invoke sublation within the politics of representation as an act of differentiation, the lure of "the culture of the copy" (to use Hillel Schwartz's phrase) seems to hook its adherents into hustled solipsism and faint theory. Unwitting casualties of the de-ethical surfaces of the present, they inevitably skid into cultural memory erased as rapidly as the refresh rate of their screens or the release of their "send" keys. Aporia, though, isn't just a signifier of implausible or reactionary dialectical unresolvability, but one of permanent contradiction negating the reciprocity uselessly delimiting decidability (no less creativity). In this regard, Enzensberger's essay is clear: "The argument between the partisans of the old and those of the new is unendurable, not so much because it drags on endlessly, unresolved and irresoluble, but because its schema itself is worthless...The choice it invites is not only banal, it is a priori factitious." Yet a facetious discourse persists in the guise of faux subversion, indifferent mischief, opportunistic fraud, deconstituted history, or irresponsible defamation perpetrated through vain electronic deconstructions of identity "theorized" in nonsensical notions of schizophrenaesthetics more deluded than deleuezian, more subjectivized by pathologies of smug hubris than by ingenious sabotage. To this end, the "avant-garde," as Enzensberger observed, "must content itself with obliterating its own products."

And even if, as is obvious, the notion of the "avant-garde" is only summarily relevant to issues of electronic media, it does evoke a set of historical issues about artistic production, its presumptions and the long-discredited bourgeois tendency to tolerate adversaries in the service of the culture industries. It's surely evident that there is a stark difference between "necessary ferment" and critical practice. This issue is well approached in Paul Mann's book, *The Theory-Death of the Avant-Garde*, and has been exposed over and over and over again by the trendy retailing of subversion. Mann writes:

There has never been a project for delegitimating cultural practice that did not turn immediately, or sooner, into a means of legitimation. The widely disseminated awareness of this unlimited legitimacy has eroded the ruse of opposition. The death of the avant-garde might thus be the most visible symptom of a certain disease of the dialectic, a general delegitimation of delegitimation. One might call it a crisis were it not for the fact that it announces an end to crisis theories of art. The crisis-urgency of the avant-garde repeated itself so often, with such intensity and so little in the way of actual cataclysm, that it wore itself out. We are now inured to the rhetoric and market-display of crises.

Even though the seventies, eighties, and nineties have demonstrated persuasively that the commodification, deconstruction, and engineering of dissent are not disassociated from the marketplace of ideas, the persistence of a futile, and perhaps complicit, neo-avant-garde suggests that the lessons of art-world theory and economy haven't really been learned as they spill into electronic media in increasingly tidal waves.

Indeed, the politics of subversion as intervention and the aesthetics of promotion share a fuzzy border that is crossed more frequently than admitted. Indeed one might suggest that an aesthetic of subversion shadowed modernity's hopeless fascination with avant-gardism and now has been transmogrified into a game of ego fulfillment played out in the spectacle of fictionalized, illusory, purloined, or cyberized identities, a kind of triumph of "The Data Dandy" whose presence was articulated in the Adilkno essay:

The data dandy surfaces in the vacuum of politics which was left behind once the oppositional culture neutralized itself in a dialectical synthesis with the system. There he reveals himself as a lovable as well as false opponent, to the great rage of politicians, who consider their young pragmatic dandyism as a publicity tool and not necessarily as a personal goal. They vent their rage on the journalists, experts, and personalities who make up the chance cast on the studio floor, where who controls the direction is the only topic of conversation... The dandy measures the beauty of his virtual appearance by the moral indignation and laughter of the plugged-in civilians. It is a natural character of the parlor aristocrat to enjoy the shock of the artificial.

Related issues have emerged in the writings of The Critical Art Ensemble (particularly *The Electronic Disturbance*). Unhinging the fictions of authority, they write cogently about rupturing the "essentialist doctrine" of the text while their interventions (some might say performances) into the sacrosanct territories of authority represent a provocation directed at both the worn traditions of public sphere cultural politics and a reckoning with the accelerating implications of technologies for a generation inebriated with virtualization. But to the point of reactionary or regressive trends they write:

Cultural workers have recently become increasingly attracted to technology as a means to examine the symbolic order... It is not simply because much of the work tends to have a "gee whiz" element to it, reducing it to a product demonstration offering technology as an end in itself; nor is it because technology is often used primarily as a design accessory to postmodern fashion, for these uses that are expected... Rather, an absence is most acutely felt when the technology is used for an intelligent purpose. Electronic technology has not attracted resistant cultural workers to other times zones, situations, or even bunkers used to express the same narratives and questions typically examined in activist art.

The spheres of activism are driven not by insidious ingenuity but by clearly delineated opposition. Nor are they sustained by incognito egos cloaked behind imperious and ambiguous intentionality. Activism, in short, is concerned with visibility and not subterfuge. This lesson hardly seems understood by wanna-be hackers whose trail might prove untraceable but who, nevertheless, (and in utter disregard of hacker integrity) leave forged evidence to certify or publicize their intrusions. Less politics than gloating narcissism, this behavior seems all too symptomatic of the roguish (is that voguish?) appeal of the rakish criminality in *Natural Born Killers, Trainspotting, Gangsta Rap*, or perhaps the ultimately pathetic imperatives revealed in *Fast, Cheap and Out of Control*.

It is difficult too to ignore Peter Sloterdijk's irksome, but in this case useful, positioning in the *Critique of Cynical Reason*. In the introduction, Andreas Huyssen poses a series of questions emerging in Sloterdijk's brooding work: "What forces do we have at hand against the power of instrumental reason and against the cynical reasoning of institutional power?... How can we reframe the problems of ideology critique and subjectivity, falling neither for the armored ego of Kant's epistemological subject nor for the schizosubjectivity without identity, the free flow of libidinal energies proposed by Deleuze and Guattari? How can historical memory help us resist the spread of cynical amnesia that generates the simulacrum of postmodern culture?" But Sloterdijk's argument is far more pertinent: "Cynicism is enlightened false consciousness. It is that modernized, unhappy consciousness, on which enlightenment has labored both successfully and unsuccessfully. It has learned its lessons in enlightenment, but it has not, and probably was not able to, put them into practice. Well-off and miserable at the same time, this consciousness no longer feels affected by any critique of ideology; its falseness is already buffered." "Cynicism," he says in the chapter titled "In Search of Lost Cheekiness," prickles beneath the monotony."

While itself invoking an enlightenment ethic, Sloterdijk's paean to moralities and tradition nevertheless stands as a form of diagnosis of the yet uncomfortable discourse of modern and postmodern positioning. Theorized in so many ways, the issues that seem most pertinent in the continuing (and now perhaps dated) opposition mostly concern a radically altered subject—one not merely at the reception end of authority. But the inverted hierarchy of subject/authority is erroneous. And with the intervention of electronic media (with, among so many other things, its reconceptualization of both subjectivity and identity), the issue has often lapsed into virtualized sociologies of sadly presumed notions of the self transgressed by "life on the screen." This, to use Huyssen's term "schizosubjectivity," lapses into re-essentialized categories by failing to understand the difference between identity and subjectivity, no less between the self and its anecdotal other. This astonishing disassociation leads into the possibility of a fugitive digital ethics whose contemptuous naiveté seems more reckless than subversive, more pessimistic than productive.

But the oscillations between self and other also suggests the avoidance of consequential psychological issues deeply affected by the development of electronic technology and its history. It is here that the distinction between schizophrenia and "schizosubjectivity" can be considered in terms of behavior. While there is little doubt that the unified notion of subjectivity collapsed in the hierarchies of modernity. What emerged are fragmented identities not salvaged in political nationalism, muddy text-based otherness, or in the abandonment of subjectivity and the acceptance of questionable notions of agency and its relation to avatars. This sort of dopey refusal (perhaps sublimation), well articulated in Slavoj Zizek's recent writings (and particularly in the chapter "Cyberspace, or, The Unbearable Closure of Being," in the just published *The Plague of Fantasies* and in *Enjoy Your Symptom*), is articulated in fraudulent, deceptive, or preemptive strategies that only serve to further discredit the politics of the politics of subversion. "Insisting on a false mask," he writes, "brings us nearer to a true, authentic subjective position than throwing off the mask and displaying our 'true face'...(a) mask is never simply 'just a mask' since it determines the actual place we occupy in the intersubjective symbolic network. Wearing a mask actually makes us what we feign to be...the only authenticity at our disposal is that of impersonation, of 'taking our act' (posture) seriously." This fundamental position cannot be trivialized by phony realizations or outlaw aesthetics. Extended into the public sphere, there is nothing worse, or more revealing in cyberculture, than a hypocrite revolutionary whose relationship even with opposition has to be invented.

Brecht wrote a great deal about "refunctioning," shifting the authority of extant material to expose its ideologies. Surely this political mimicry, joined with the Benjamin's loftily ambiguous and hopelessly redemptive aesthetic, fits into the trajectory of art—from Dada to Pop to Postmodern—by rationalizing various forms of reproducibility, repetition and appropriation as legitimate approaches that were both reflexive and creative. But these strategies were rooted in a form of "critical" consumption that clumsily persists in electronic culture.

No doubt that these strategies have also mutated into the cut-and-paste techniques (no less the cut-and-paste identities) of far too many artists involved with media. Very few of these techniques are confrontations whose parodic or satiric intent outdistances or demolishes its sources. Isn't the goal of parody sublation? But the weakness, and sad pervasiveness, of a cavalier position does little to suggest that the shift into fragile digital communication technologies raises the stakes of far more than such worn notions of creativity as will perpetuate themselves by evolving their own development. Nothing could be less interesting in a time of monolithic operating systems, algorithmic aesthetics, and the politics of virtualization than a shiftless, hollow, and finally selfish positioning of the artist as a hapless subversive or, worse, the subversive as a hapless artist. Indeed, the link between cultish anonymity and subversive presence strikes me as a pitiable attempt to sustain vaguely modernistic notions of subjectivity behind the electronic veil of deconstructed—or better destabilized—identity or perhaps, more pathetically, self-styled celebrity.

[This essay was first published on January 20, 1998, at *Reflex* <http://≠ www.adaweb.com/context/reflex/>.]

SUBJECT: CHEAP.ART

FROM: OLIA LIALINA <OLIALIA@CITYLINE.RU>
DATE: MON, 19 JAN 1998 20:47:30 +0300

INTRODUCTION

Making "Agatha Appears" at Budapest C3, I recalled Metaforum III (Budapest, October, 1996). At that time I spoke of the internet being open for artistic self-expression, that the time had come to create net films, net stories and so on, to develop a net language instead of using the web simply as a broadcast channel. And, of course, the sale of "My Boyfriend Came Back from the War" to Telepolice On-Line.

What is happening now, more than a year later?

First: I still get messages saying: "Look at my new web movie." Following the link, I find Quicktime or Shockwave moving images whose only value is to prove that plug-ins become more and more perfect and bring us closer and closer to home cinema.

Second: Net art is still as cheap as a floppy. For me, the intercoupling of these things is obvious.

Another thing is quite clear. Questions of what net art is and "does it actually exist" appeared in 1996. Today, almost every article devoted to this subject still starts with the same sentences. They have become more ornamental than anything really looking for an answer. They are following a fashion, not real interest.

All media festivals, exhibitions and conferences are now well decorated too: there are net art sections on event sites, some net artists and some beautiful games with the term "net art" itself. They are attractive and not expensive at all.

It was a year of net art sales. And important to stress that artworks were much cheaper than ideas. Variations on the theme "net artists don't need institutions" or "net art can exist without galleries or curators" were mostly welcomed by real galleries and institutions.

What else? A year ago "net art" as Altavista understood it, was all these sites devoted to art (galleries of painters, photo artists...archives of film and video, museums representing their collections on the net). Now net art is supposed to be the same, plus net.art, that is to say: online galleries of offline stuff plus a small group of artists close to Nettime or Syndicate or 7-11 mailing lists, and to each other.

That's what one can see on the surface. What was going on inside?

Nothing that could make feel that net artists existence means something in the world they create.

A year ago it was so sweet to announce that art theory, the art system, art commerce—all these are relics of the real art world system, a heritage to forget, but in fact this statement only brought some variety to offline art institutions, not an alternative.

THEORY

Developing a theory of its own could enhance the value of net art. At the moment it is understood in the context of media art, of computer art, of video art, of contemporary art, but not in the context of the internet: its aesthetic, its structure, its culture. Works of net artists are not analysed in comparison with one another. We are always viewed from an external perspective, a perspective that tries to place native online art works in a chain of arts with a long offline history and theory. And this remains the interest: to place us, to phenomenalize us, in the social sense of the word. Definitely, you meet more interest to the phrase The *internet project* than to its inner being, to the fact of online collaboration of artists from different countries than to their actual work.

Again and again: "What is net art?" instead of (for example): "Browser interface in the structure of net art" or "Downloading time as a means of expression in the works of Eastern European net artists" or "Frames and new windows in net narration" or "Different approaches to finding footage or servers" or "Domain names and 'under-construction' signs from 1995 to 1997."

With pleasure I'll take my words back if I'm wrong, and with great pleasure I'd participate in such researches as a critic.

In brief: With no theoretical support inside, net art meets only vulgar one-season interest from the outside world. This wouldn't be a problem if it didn't make things cheaper and that in some months all innovative experiments, new art forms and language will be buried as a last-season fashion. And this will happen already internally. (Net art was born in the net and will definitely come back to die.)

SYSTEM

In fact, while I was thinking what to write about internet art structures, several net galleries appeared and some on-line festivals gave prizes to some artists. This looks like the birth of a new world; maybe it is and the time to judge has not yet come, but it's not difficult to see destructive tendencies in these foundations. Online galleries and exhibitions are nothing more than lists, collections of links. On one hand, it fits the nature of many-to-many communication; the internet itself is also only a collection of a lot of computers, and it works. On the other hand, list by list compilation brings us to an archive situation, to the story about keeping and retrieving information. Online galleries only store facts and demonstrate that a phenomenon exists. They neither create a space, nor really serve it.

The same applies to festivals and competitions. Even if they are intelligently organised they are not events in net life. Mostly they are not events at all but just the easiest and trendiest way to save money given for media events by funds or whatever. Now that everybody knows the internet is our paradise on earth, the long-awaited world without borders, visas, flights, or hotels, it is the best way to make your event international.

From my point of view, the most perceptive and valuable creative structures around *are* net artists co-projects and curated initiatives. Or they *could be*, if they were not so closed and didn't provide an ironic distance to the idea of creating a system.

In fact every net artist or group in the process of creating a work builds their own (and at the same time common, for everybody) system of self-presentation and promotion, invents exhibiting spaces and events. After all, it is in the nature of net art to build the net. But again and again the worlds you create easily become an exhibiting object at media art venues. Something that could be invaluable tomorrow is sold for nothing today.

COMMERCE

It is not only a problem of misunderstanding and misapprehension: I was told by art-sale-experienced net artists that since web space is physically cheaper than canvas or videotape, and since webpages are something that every schoolgirl can make on her school computer, pieces created and stored in the net will be cheaper than whatever made with the aid of more complicated techniques and knowledge. Sounds logical. Logical yet, until net art is an export product, not a point of prestige in the system of internet values, not an item of commerce for those who invest money in the internet, for example.

Banks, big companies, or simply rich guys have always bought pieces of art for their collections or found it prestigious to sponsor artists. Now they or their younger brothers spend enough money (at least in Russia) to be well represented in the net. Why not harness their desires? Why not advise them to collect, to buy and help develop the art of the next century?

Details and demo next time.

It's not only about money. And generally, the question of being paid for net art is no different to the question of payment on the net. Publishers, companies, advertisers and everyone else in the world is scratching their heads about it. I talk about going further, exploring the net, not beeing prisoners of last year off line fashion. It's not really my dream, but I'd prefer if tomorrow new net artists would come and say: she made pieces good only for virtual offices, what we do is real net art, underground, new wave, what ever. Its better than nobody will come (because where?) and only media critics will mention that once there was a period in media art, when some media artists experimented with computer nets.

SUBJECT: FLIGHT CAPITAL

DATE: WED, 19 AUG 1998 13:52:09 +0100
FROM: GASHGIRL <GASHGIRL@SYSX.APANA.ORG.AU>

CONNECTED

You yawn, rub your eyes, and officially wake up.

Swarm Spore Procurement Center, Endless Arsenal A sub-ground warren of war rooms, communication facilities and personnel quarters—an uneventful interpretation of a sixties vision of a germ-free adolescent future. An acrid pherenomal white noise of amyl, sweat and semen echoes through the refiltered air, although the corridors are free of zealous young gene carriers. You notice a door on the far western wall and approach it cautiously. A sign reads *stealth designs mentor/protg rec room.*

OPEN DOOR

Patriot Gains (Interference and Deception Unit) A spacious rest room comprising nine toilet cubicles, two standard sickbay bunks, four nonstandard bunks, three handbasins, a communal shower alcove with nine faucets, and two imposing vitrines containing questionably acquired Mayan artifacts. A doorway labeled "G8" stands to the right of the cubicles.

Contract Specialist J763-99-DY-S009 and RentBoy (he's finally legal!) are standing in front of the vitrines. RentBoy admires his reflection in the glass, tucking his street-wear camouflage net T-shirt into his too-tight regulation strides.

J763-99-DY-S009 growls, "The Infestation Teams are getting restless. They've had it with your sustainable pulsing bullshit, your Art of War drivel. I want that skanky little fucker brought into compliance *now*."

RentBoy ceases his preening, saying, "It was agreed to focus parametrically across various expandability issues to see how they affected the time required to expand our forces. The imperative was to check the first-order logic of our mobilization and reconstitution capabilities."

J763-99-DY-S009 yawns.

RentBoy states, "Employment of tactical decentralisation coupled with strategic assessment will generate an unsurpassed advantage across the full spectrum of conflict potentials, from high to low intensity situations, including the proliferation of networked nonaligned insurgency forces."

J763-99-DY-S009 appears slightly nonplussed. "And...?"

RentBoy continues, his eyes glazed over with either lust or early glaucoma. "And... the Warrior Preparedness Unit is seeking information to address the requirement for new delivery systems of precision-guided munitions based on advanced designs for automated and infrastructure warfare."

J763-99-DY-S009 responds impatiently, "Yeah, yeah. Tell me something new."

RentBoy drones, "It is imperative we equip ourselves to converge undetected upon an enemy, either through direct firepower, opportunistic maneuvers or psychological operations."

J763-99-DY-S009 shrugs her shoulders. "Like I really care. What's your actual point?"

RentBoy suddenly focuses his gaze on UB40-99-DY-S009, unzips his fly, reaches down deep and pulls out an impressively swollen prick.

"Let's see if our loser 'friend' can comply with *this* AP weapon," he murmurs, one hand squeezing his leaking knob, the other languorously rubbing his waxy balls.

J763-99-DY-S009 considers RentBoy's suggestion, running her fingers over his oozing cock, then shoving them down his throat.

"Copy that. Get jiggy wit it and requisition his sorry ass at 0600. Give me a damage report when you're done. In the meantime...I think you'll be interested in my latest procurement."

Clearly wanting to beat his meat rather than continue the discussion, RentBoy mutters with some difficulty, "Would that be that major snorefest tactical engagement simulation system instrumentation you've been waiting on?"

J763-99-DY-S009 shakes her head, sending a gentle flurry of protein deficiency dandruff onto her epaulettes.

"No way. I'm talking about something exponentially more useful than your average TacSim. Bug-free, fully functional in rugged terrain, Remote Area Mobility to die for, easily concealed, etc, etc. Basically more features than you can poke a joystick at," she replies, giving his dick a saucy slap.

J763-99-DY-S009 pushes RentBoy into the nearest cubicle and slams the door. You hear a slightly muffled order, perhaps the words "bend over, nigga," but you can't be sure. The responding groan, then a series of grunts segueing into gasps, is unambiguous.

Suddenly the stink of futility threatens to overwhelm you and you quickly leave by the "G8" door.

Disconnected

SUBJECT: ASS IN GEAR

FROM: JORDAN CRANDALL <XAF@INTERPORT.NET>
DATE: SAT, 17 OCT 1998 17:05:02 –0400

As I went along the street where I live, I was suddenly gripped by a rhythm which took possession of me... It was as though someone were making use of my living-machine. Then another rhythm overtook and combined with the first, and certain strange transverse relations were set up between these two principles... They combined the movement of my walking legs and some kind of song I was murmuring or rather which was being murmured through me... —Paul Valéry

In America, we have a peculiar mode of rhythmic embodiment called the "power walk." Head held high, arms thrusting outward repeatedly in conjunction with the beat of the moving legs, hair and breasts abounce, one propels oneself along the street in jerky, fast-motion paces as in an old silent film. Going nowhere in particular, often sheathed in garish, logo-strewn activewear, one in/habits the gym—a fitness club no longer a place so much as a set of notions of what it means to be physically adequate in society. Unpack the prevailing notion of fitness [gasp] and there you have it, the body moving [gasp] in conjunction with the social and technical machine [gasp], according to formats of productivity, efficiency, and adequacy. What are the beats? To focus on visual codes is to miss them.

I want to consider "exercise" as a marker of rhythmic operations, in which the body is immersed as agent and incorporant, within general conditions of making processes, forms, circuits, and capacities adequate to emerging regimes of fitness. And lest one think that notions of fitness are not in keeping with the body's virtualization, and necessarily serve to privilege a singly corporealized entity, I would like to point out that in all cases of body–subject–interface encounters we are speaking of a newly mobilized body, and a subjectivity constituted within formats of movement, across hybrid transport–transmission landscapes. (Landscapes traversed in terms of the transfer of weight over land and the transmission of embodied presence through the network.) The body in motion, subject to notions of efficient and adequate movement, contours and sediments itself through circuits and cycles of repetition, in whatever degree of corporeality or virtuality. Even on the (arguably) fully physical side of the spectrum, the days when one's body is parked at the monitor are coming to an end, and emerging cultural practices would do well to take this mobilization into account. The formats and codes of the interface register and facilitate these cycles, and the movements and processes of embodiment to which they are attached.

The newly mobilized body, bedecked in gadgetry—portable arrays of devices, either visible externally or implanted internally. How sexy. Consider a simple, early gadget: the Walkman, with which one powerwalks. Sitting next to the early mainframe radio or phonograph, to what extent did one forget about one's body, necessarily parked within range of the machine? The interface as it stands, as it makes one stand, as it arrests one and places one in a holding-pattern, always lays the seeds for mobilization. A preparatory state for new sites of embodiment, patterns of mobility, and formats of enunciation. It facilitates arrays of localizations that link together in new presences. A peculiar site of exercise, and not just in terms of the obvious hand–eye coordinations via the mouse, but in terms of the way its formats are internalized in larger patterns of movement. Here is where we can locate the emerging paradigm of the database, and consider its effects. But at the same time: the interface marks the site of the arrested body's integration into the machine, into machinic operations that have larger societal links and consequences—indeed, which rest upon entire social apparatuses of fitness, efficiency, adequacy.

Consider the finger-scanner, now available as an option on the purchase of a new computer—right on the keyboard, to the left of the shift key, or in some models, on the mouse itself. A new form of fingering! But even more: one agent of an entire emerging economy of authentication, based on the incorporation of biological patterns into virtualized constructs, formatted according to the emerging conventions of the database. The "fingered" body is represented, is seen, its movements recorded and internalized, through the mechanisms of the database. How do these formats augment traditional, cinematic norms of movement representation—that is, the set of conventions through which the world of movement has come to be known? For movement is no longer seen as much as processed—or rather, it is represented by way of its processing. On one hand, the format of the database floats above the cinematic image-field, combining with it to generate a new kind of moving image—or "machine-image." One can even revisit the history of the moving image in terms of movement processing: think of proto-powerwalker Charlie Chaplin in these terms, especially in his struggles to keep up with the demands of the machine in *Modern Times*. And, again, one can think movement in terms of the *immobilizations* that it locates. After all, it was Serge Daney who reminded us that the set of movement-conventions that is cinema only took hold via the public's immobilization in theaters, arrested and held in thrall by the screen.

Such a public is today a tracked public. Harnessed to new technological assemblages and driven by processing imperatives, machine-images track movements *as* representation. Tracking is the way in which one sees and is seen by the image. Informed by the organizational paradigm of the database, tracking formats an "improved," more productive and efficient form of vision. It protects one—informationally and corporeally—from an "outside" unprocessed reality that is increasingly constituted as dangerous. Such a body, whether in flesh or networked mode, incorporates fitness as the erasure of any threat to efficient, fast, and reliable flows.

A movement constituted through patterns of repetition, enmeshed in circuits, harnessed to social and technical machines. What better way of envisioning the exercise video—One! Two! Three!—and the body-database? In either case, *counting* equals *accounting* for, and the body is formatted through arrays of variables and calculations. Movement configures as a kind of statistical articulation. Based on behavior and preference data, as tracked, abstracted, and aggregated in the database, X might, for example, show a 59.6 percent propensity to move toward Y. As individuals and groups are processed, the public configures as a calculus of manageable interests, opinions, patterns, and functions. This ever more precise and "protective" statistical ventriloquization—stretching over speech like a prophylactic or over pumped-up flesh like spandex—becomes an authentic voice of the people. A marker of speech and presence, a way in which the public is heard and made visible. The machine-image—the exercise-interface—is thus a politicized field of incorporation and identification, marking a network through which social identities and embodied forms are signaled and enacted.

In the face of this crisis in the visual, emerging sites of operation occur in the proliferating arrays of devices harnessed to machine-images the way that remote-control devices are attached to television screens. They are like "free weights"—three sets of eight reps *now!*—or the fitness calculators that interface body and machine and measure their compatibility, often resulting in the body's rates to be adjusted in accordance with prevailing fitness norms. Increasingly, such devices—in conjunction with their machine-images—serve as switch-points between interior and exterior rhythms, which they regulate and convey. The interface always points to such a device, as it traffics between motivations and mobilities. Through them, private and public realms, behaviors and built realities, exchange, encode, and format one another.

Movement is inextricably bound up in technological capacities and imperatives. Wherever there is a movement, there is a machine. Exercise always happens in symbiosis with the machine, according to rhythms that it incorporates and emits. You don't relate signs when you exercise, as you do when you read and your body just (apparently) sits there immobilized. You coordinate your rhythms and movements to those you hear, feel, or sense proprioceptively. The body configures as a locus of rhythmic operations, as an active process of incorporation and coordination with machines both technical and social. To think in terms of "coordinations," as much as in relations, is to begin to understand emerging potentials for interventions within the field of the interface—the machine for moving. A logistics lurks in the most basic of routines.

LOCAL

SUBJECT: COUNTERSTRATEGIES AGAINST ONLINE ACTIVISM:
THE BRENT SPAR SYNDROME

FROM: EVELINE LUBBERS <EVEL@XS4ALL.NL>
DATE: TUE, 29 SEP 1998 08:43:16 +0200 (MET DST)

Shell is not going to forget lightly its misadventures with the Brent Spar. The Oil Major was taken by complete surprise when the Greenpeace campaign against sinking that former drill platform achieved its goals. What happened to Shell can in fact happen to any corporation. Loosing control of the situation as result of the activities of a pressure group has become a nightmare scenario for the modern multinational enterprise.

SHELL DID TOO LITTLE TOO LATE

The Oil Major's first reactive measures have meanwhile become the perfect example of how not to do it. But Shell has learned a lot as well. A comprehensive review of what has become known as the PR disaster of the century indicates that Shell had it all wrong about its own influence on the media. There was a new factor in the game, which had been completely missed out: the role of the internet. That would not be allowed to happen a second time. From July 1996, Shell International sports an internet manager. His name is Simon May, he is 29, and responsible for Shell International's various presences on the internet, and for monitoring and reacting to what is being written and said about Shell in cyberspace. He also helps formulating the Shell group's strategy for how the internet should be used.

May's career began in journalism, and more recently he did a four-year stint in the Sultanate of Oman in charge of the English-language communications for the state-owned oil-company. With him Shell's got a premium catch: May is young and eager, smart and fast, open-minded and nice, everything the image of the Company ought to be. And he understands like no other the internet's potential—also what it could mean for a company like Shell. Simon May openly admits that Shell was

beaten in the new-media war. The Brent Spar affair was one, but the Nigeria situation has also prompted a "massive on-line bombardment" of criticism. To quote May: "There has been a shift in the balance of power, activists are no longer entirely dependent of the existing media. Shell learned it the hard way with the Brent Spar, when a lot of information was disseminated outside the regular channels."

The Brent Spar affair has brought quite some change of attitude to Shell. Ten years ago the Multinational could afford to blatantly ignore campaigns against the South African Apartheid regime. Although concerns were brewing in-house, to the outside world Shell maintained that the campaigns against Apartheid were not significantly damaging the company. And for the rest Shell kept haughtily mum. Then came the Brent Spar incident and car owners were taking en masse to boycotting Shell's petrol pumps, and such an attitude no longer paid off. Shell came to feel the might of the mass market, and bowed down. An alternative would be worked out for the platform's fate.

But developments did not stop there. A few month later opposition leaders were executed in Nigeria as result of their attacks on the environmental disaster Shell was causing in Ogoni-land, and this caused a renewed storm of protest against Shell. The intimate links between Shell and the military regime came under severe criticism. The Oil Major then went for a new tactic and opened a PR offensive. CEO Cor Herstroter took the initiative in a debate on politically correct entrepreneurship. At the shareholders meeting in 1996 the new chart of business principles at Shell was unveiled, a comprehensive code of conduct with due allowance for human rights.

Does this all point out to a major shift in policies? Or are we witnessing a smart public relation exercise intent on taking some steam from the pressure groups' momentum?

In the beginning of June 1998, Brussels saw a conference devoted to pressure groups' growing influence, organized by the PR agency, Entente International Communication. Entente did research about the way corporations were interacting with pressure groups and vice versa. The findings, presented in a report titled "Putting the Pressure on" are harsh: "Modern day pressure groups have become a major political force in their own right, and are here to stay. They manifest themselves in the use of powerful communication techniques, and they succeed in attracting wide attention and sympathy, projecting their case with great skill via the mass media—they understand the power of PR and of the media "soundbite." And now, increasingly, they do so over the global telecommunication networks.

Their power and influence is bound to grow inexorably over the next years.

Pressure groups are small, loosely structured and operate without overhead or other bureaucratic limitations, they move lightly and creatively. They pursue their aims with single-minded and remorseless dedication. To be on the receiving end of a modern pressure group can be a very uncomfortable experience indeed, sometimes even a very damaging one.

Multinational companies are ill prepared to face this challenge, their responses are often slow and clumsy. There is a "bunker" mentality, and a reluctance to call in experienced help from outside which is surprising—and potentially dangerous. This failure could cost such companies dearly in the future.

At the conference in the SAS Radison Hotel in Brussels, attended by some seventy participants from the corporate world and the PR industry, fear for the unknown prevails. The unpredictable power of pressure groups, consumers, or even normal citizens can take the shape of boycott campaigns, but also of commuters on the (newly privatized) British Railways to move out from a train that has been canceled on short notice. The biggest question remains unanswered: whose turn will it be next? The Brent Spar affair has left its mark here. By way of illustration the story of Felix Rudolph, an Austrian national who worked himself up from farm hand on his father's estate to manager of a factory producing genetically modified grain. Pioneer Saaten ("Pioneer Grain," the company's name) was not aware of doing anything wrong. The company produces for a small market niche in Central Europe and strives for optimal quality, so as to enable farmers to obtain better yields. All products have been tested extensively, and all test results have been duly registered. So nothing to worry about, that is until the company became the focus of a protest campaign, triggered by an impending referendum in Austria on genetically manipulated foodstuffs. "We suddenly had to engage in debate with the public, something we never had done before. Who's interested in grains anyway?" Felix Rudolph, as he holds his presentation at the Brussels conference, still looks dumbfounded about what overcame him. "Your products are unhealthy and dangerous asserted the pressure groups, and we had no clue what we had to say in return. As soon as you try to explain the extent of a risk, you admit that such a risk exists. In that referendum, 90 percent of the people turned out to be against gene technology, the majority of whom did not know what they were talking about." It is only later that Herr Rudolph understood that his company merely served as an example for the pressure groups. "By engaging in a dialogue, we provided them with a platform to put forward their case. The discussion itself went nowhere." This realization came too late, however. The campaign so much impressed the government that it enacted laws regulating genetically manipulated foodstuffs. An embittered Herr Rudolph: "Now the farmers may foot the bill, and the pressure groups have vanished into thin air!" Pioneer Saaten had to temporarily suspend the production of modified grain. "We will try to explain things better next time we apply for a license."

According to Peter Verhille from the Entente PR agency, the greatest threat to the corporate world's reputation comes from the internet, the pressure groups newest weapon. "A growing number of multinational companies—such as McDonalds and Microsoft—have been viciously attacked on the Internet by unidentifiable opponents which leave their victims in a desperate search for adequate countermeasures."

The danger emanating from the new telecommunication media cannot be over-emphasized, says Mr. Verhille. "One of the major strengths of pressure groups—in fact the leveling factor in their confrontation with powerful companies—is their ability to exploit the instruments of the telecommunication revolution. Their agile use of global tools such as the Internet reduces the advantage that corporate budgets once provided." His conclusions made a hard impact on the participants of the conference. In fact most companies appear slow to incorporate such tools into their own communication strategies. When asked what steps they planned to take to match pressure groups mastery of these channels, most respondents simply repeated their intention to expand into this area or admitted that their preparations were still in a preparatory stage.

As came to light in Brussels, there is one exception to this picture however: Shell international. internet manager Simon May gave a smashing presentation, which showed very well what Shell had come to learn about the new media. Simon May was also very open in an interview we held with him (befittingly, by email), even though he could understandably not answer all of our questions.

Pressure on the Internet, Threat or Opportunity was the core issue at his presentation. The internet may be a threat to companies, it also offers big opportunities. Simon May states that the fact that anyone can be a publisher cheaply, can be seen, or at least searched and looked at worldwide, and can present his/her viewpoints on homepages or in discussion groups is not merely a menace, but also an unique challenge. "Why are pressure groups so active on the Internet? Because they can!"

Companies should do the same, he argues, but must do it professionally. "On-line activities must be an integral part an overall communication strategy, and should not be simply left to the care of the computer department."

The basic tenet of the Shell internet site (launched early 1996) was a new strategy based on openness and honesty. Dialogue was the core concept, and sensitive issues were not side-stepped. May is quite satisfied with the results of this approach and illustrates this with some facts and statistics.

Http://www.shell.com receives over 1,100 emails a month, a full-time staff member answers all these mails personally and within forty-eight hours; there is no such thing as a standard reply. There are links to the sites of Shell's competitors and detractors, and also to progressive social organizations (nothing there more radical than Friends of the Earth or Greenpeace, but this aside). Shell also allows opponents to air their views in forums- those are uncensored. Not without pride, Simon May states that Shell is still the only multinational to do this. There is no predetermined internet strategy at Shell, flexibility is the name of the game. "It's all about being able to react, listen and learn." His advice to the Brussels conference-goers: "Be careful, technology changes fast, and your audience changes and develops even faster. And think before acting: anything you're putting up on an Internet site you make globally available."

Taking care of Shell's presence on the web is only one of the internet manager's tasks. He must also monitor and react to what is being written and said about Shell. "The on-line community should not be ignored" was part of his advice in Brussels. "Pressure groups were aware of the potential of the Internet far earlier than the corporate world. There are pressure groups that exist only on the internet, they're difficult to monitor and to control, you can't easily enroll as member of these closed groups."

Listening to the internet community can be an effective barometer of public opinion about your company. The Shell headquarters in London are making a thorough job of it. Specialized, external consultants have been hired who scout the web daily, inventorying all possible ways Shell is being mentioned on the net, and in which context. Things are not made easier by the fact that search engines will assign forty-eight different well-known uses of the word "shell"...

Simon May gladly explains how the work is done. "We use a service which operates from the US, E:Watch, who scan the Web world-wide for references to certain key words and phrases we supply to them. In the U.K. we use a company called Infonic, who does the same thing from a European perspective. The results they come up with can be completely different, although they have been given the same search criteria, and the search has been done at the same period of time. This can be for a number of reasons, including the methods which they use to search, and the times of day they enter a site to index it."

Shell also uses so-called intelligent agents. These are search programs that can be trained to improve their performance over time. Simon May: "This is particularly useful for us since our company name has so many different meanings. We can tell the "agent" which results are useful and which ones aren't, the next time the agent will go out and come back with only those documents which are relevant."

This monitoring can not be for 100 percent truly effective, but has to be carried out nonetheless, according to Simon May. "You need to keep track of your audience all the time, since you may learn a lot from it."

Visiting the Shell website, the first surprise is the measure of openness about issues previously wrapped in taboo. There are carefully written features on human rights, the environment, and even the devastation and exploitation of Ogoni-land in Nigeria. The somewhat defensive character of some stories gives an indication as to which issues are still sensitive. Speaking for instance of the massive oil spills in Ogoni-land, for which Shell is held responsible ("totally exaggerated and unproved accusations"), there is always the mention that 80 percent of those have been caused by sabotage by radical resistance groups (this percentage is contested by the groups concerned).

At the site's discussion forums arranged by subject everybody is allowed a say about Shell's practices. It is ironic then to see Shell collaborators from Malaysia and Nigeria reacting with dismay about what they read in those forums about their employer.

The question is of course whether this form of openness really yields results. The forums are not intended for people to question Shell; the email facility is provided for that. "The forums are intended for people to debate issues relevant to Shell among themselves, so to speak," says Simon May. The email service is actually being used quite

intensively to put questions to Shell—these are the 1,100 emails coming in every month. The nature of these questions and their answers remains a secret held by Shell and the emailers.

All in all, one might conclude that this amounts to a fake openness, for show purposes only. After all, in public true discussions are being eschewed. But Simon May would deny that the forums are merely window-dressing: "We do believe quite firmly that people have the right to debate these issues and we provide a place where they can do that in an environment which might just lead to their view being heard in an organization that can make a difference." Of course these forums function as barometer for what certain people think, May admits, although this is not their primary aim.

At Earth Alarm (the foreign affair project of the Dutch environmental organization Milieudefensie) these rather embellished representations of reality do not cut much ice. "They've changed a lot in their communication, they're far more careful about how they present themselves to the outside world. But that is mostly addressed to their customers here, in the Western world," says spokesperson Irene Bloemink. "Profits and principles, the first issue of the totally overhauled Shell International Yearly Report, has been only distributed in the Netherlands, Great Britain and the United States. That's where the people are that Shell sees as a potential threat."

The situation in Ogoni-land has not improved in the two-and-half years since Ken Saro-Wiwa was hanged; on the contrary, things have only gone worse, at least till the death of the military dictator General Sani Abacha. Scores of people have been arrested in the beginning of this year by a special military unit, founded specially to "ensure Shell comes back to Ogoni-land." This would at least suggest some kind of involvement. Yet Shell has done nothing to stop the latest wave of arrests."

Adopting a code of conduct regarding human rights and the environment is simply not enough. What counts is implementation and enforcement. Shell has not in any way made clear how they intend to translate their good intentions into concrete practice. There is no independent body to monitor the implementation of the code of conduct. Shell is self-congratulating about their first environmental Annual Report, which they claim, has been thoroughly reviewed by KPMG Management Consultants. Shell considers this a fully independent review. But then, KPM"'s environment CEO George Molenkamp goes further in *de Volkskrant* (a Dutch daily newspaper) to say that "accountants don't vouch as such for Shell's policies. Anything that comes in the report is as Shell has decided." Some contradictory viewpoints, I may say," says Irene Bloemink.

It is doubtful whether Shell has really learned anything from its mistakes in Nigeria. There is a new Shell venture in the West African country Tshad that looks as big as the Nigeria operation, and with the same possible consequences. And everything seems to go wrong again. Shell joined in a partnership with Esso and Elf (stakes are 40–40–20 respectively) and intends to start drilling new oil fields in the unstable South of that country. A report on the environment assessment came as an afterthought, according to Earth Watch: the agreements were signed and test drillings had already begun. The local population was informed of what was in store for them as the invading oil-men were underway, and the operators came to the villages to bring the news accompanied by a heavily armed military escort. In March of 1998, over a hundred civilians were killed by the army as it tried to regain control over an area from the FARF separatist movement, which in its turn highlights its own existence by attacking this oil project. The FARF claims that the earnings of the oil production will exclusively benefit the presidential coterie in the north.

Until now, Shell has been hiding itself behind Esso as the local executive partner responsible for external relations, and has declined to engage in public debates on the subject. Even Simon May doesn't want to burn his fingers on the Chad issue. Not yet, that is.

[Translated by Patrice Riemens, and edited by Renee Turner. This text was written and translated for this volume; it also appeared in the online magazine *Telepolis*, and a two-part version in Dutch appeared in the magazine *Intermediar*. A supplement to this text, focusing on the Shell pipeline in Chad and its World bank financing, can be found at <http:// www .xs4all. nl/~evel/>.]

SUBJECT: SONGS FROM THE WOOD: NET CULTURE, AUTONOMOUS MYTHOLOGY AND THE LUTHER BLISSETT PROJECT

FROM: F. P. BELLETATI <TOM9351@IPERBOLE.BOLOGNA.IT>
DATE: SEPTEMBER 1998

AS DOWN HOME AS I CAN GET

The prime mover was a loose-knit current of Italy's Marxism labeled *operaismo* [workerism], which had absolutely nothing to do with the Communist Party.

In the early sixties the *Operaisti* started to investigate changes in the sociological composition of the working class. At that time, the young mass-worker of Fordist–Taylorist factories was still the tongue of the compass, the most important segment of the proletariat. The *operaista* intervention in class struggle was based upon a participant observation of the mass-worker's behavior. The mass worker explicitly refused the older generation's work ethic and discipline. This insubordination was the main mover of conflict in the workplace. Sabotage was not invisible anymore: along with moments of open struggle (strikes and demonstrations) there was a flourishing of micro-tactics to slow down or stop the assembly line.

Operaisti were committed to studying those behaviors and defining the dialectics between class struggle and capitalist development which I'm going to sum up—taking some shortcuts. The continual confrontation between capital and living labor was the cause of all technological innovations and changes in management, which would provoke further changes in the class composition, therefore the conflict would continue on a higher level.

After the so-called Hot Autumn (1969), a season of general strikes and radical struggles with millions of workers taking the streets, proletarian insubordination increased. Struggles became more and more "autonomous" (this was the adjective by which wildcat strikers would describe their occupations: *assemblea autonoma*). In 1973 the self-disbanding of the post-operaista group *Potere Operaio* [Workers' Power] gave origin to the scene renowned as *autonomia operaia organizzata* [organized workers' autonomy]. During the seventies, Italian Autonomia theorists (Toni Negri first among equals) started to investigate and define the existence and subversive behavior of the *operaio sociale*. Such an ambiguous collective noun—hardly translatable into English—served to describe both the youngest generations of industrial workers who had broken away from the work ethic once and for all, and the whole cast of frustrated service workers, "proletarianized" students and white collars, unemployed wo/men and members of youth subcultures whose conflict was clearly "antidialectical."

"Antidialectical" means that self-organization, wildcat strikes, occupations and acts of sabotage did not take place within the realm of negotiated class struggle, indeed, they even cut loose from the traditional dialectical bond between struggles and development, and challenged the recuperative function of the unions and the Left's political control.

In order to repress those uncontrollable eruptions and outbursts (the 1977 movement above all), the ruling class had to impose a state of emergency. It was a bloodbath. By the end of the decade, most militants had been killed, thrown in prison, escaped from the country or started to shoot up heroin. But that's another story.

As some have suggested, from now on I'm going to use the term *composizionismo* instead of "[post-]*operaismo*," because the former is more precise and does not automatically correspond to a particular segment of the working class (the "blue collars"). The so-called third industrial revolution made capital supercede the Fordist-Taylorist paradigm, and turned information into the most important productive force. Appealing to those passages of the *Grundrisse* where Karl Marx used the expression "general intellect," compositionists began to use such descriptions as "mass intellectual" and "diffused intellectual" making reference to multifarious subjectivities in the new class composition.

"Mass intellectuals" are those people whose living labor consists, broadly speaking, in a subordinated output of "creativity" and social communication (in compositionist jargon: "immaterial work"). This segment of the *operaio sociale* ranges from computer programmers to workers of Toyotist factories, from graphic designers to copy writers, from PR people to cultural workers, from teachers to welfare case-workers etc.

Negri's analysis in particular is based upon the "prerequisites of communism" immanent to post-Fordist capitalism. By "prerequisites of communism" Negri means those collective forms that are created by past struggles and are constantly reshaped by the workers' tendencies, attitudes and reactions to exploitation. Some of these forms even become institutions (for example, those of the welfare state), then they go through a series of crises: social conflict created them, social conflict keeps them open and necessarily unfinished. Their crisis reverberates on the whole society, so conflict continues on a higher level.

The most important prerequisite of communism is the collective dimension of capitalist production, which brings about more social cooperation.

The stress must be laid upon the most strategic form of today's living labor, i.e. "general intellect," immaterial work, "creativity", you name it. "General intellect" (unlike labor in Taylor's "scientific management") is *self-activated*. The mass intellectual's workforce is not organized by capital, because social communication is prior to entrepreneurship. Capital can only *recuperate* and *subdue* social communication, control the mass intellectuals *from the outside* after having acknowledged and even stimulated their creativity and far-reaching intelligence.

The conflict continues on the highest level: capital's "progressive" spur is over, autonomy is becoming a premise rather than a goal.

THE COMMON BEING AND THE NET

A compositionist approach to computer networking reveals that:

- the net's horizontal and transnational development brings about a potentially autonomous social cooperation.
- most netizens fall within the anthropological, sociological and economical descriptions of "mass intellectuals."
- today's net landscape is the synthesis of many *molecular* insubordinations and some important *molar* victories, (e.g., the anti-CDA "Blue Ribbon" campaign) and is continually reshaped by conflict.
- the net is also shaped by software piracy and copyright infringement: private property of ideas and concepts is challenged and often defeated. If any one of you is without copied or cracked programs, let them be the first to throw a stone at me.
- as an "institution," the net is going through a growth-crisis that is reflected upon the whole society. In its turn, this crisis is a mover of conflict.

In plain words, the net seems to be the prerequisite of communism par excellence. This is not an uncritical utopian view of computer networking, of course there's a huge gap between the potential and the actual: work-force vs. work, *langue* vs. *parole*, capital vs. living labor, consumerism vs. social communication. The net is the OK Corral. It's paradoxical that, after all the schmoozing about "molecular revolution," we're heading straight to a new *molar* impact.

The global anti-"paedophilia" mobilization is the *state of emergency* by which the powers that be want to gag netizens. The reappropriation of knowledge and the self-organization of mass intellectuals require the defense of the net from slanders and police raids. We must keep this "institution" unfinished and open to any possibility, prevent capital from filling the above-mentioned gap with censorship and commodification. It isn't just a liberal battle for free speech: it's class war.

But this is not enough yet. We've got to make history, no less—fill that gap with autonomy and self-organization. We also need myths, narratives that incite mass intellectuals to take action. Each historical phase of class war needs propelling mythologies, there's nothing wrong with that. Georges Sorel has been slandered and misunderstood for too long. As Luther Blissett put it:

...the trouble is not the "falsehood" of myths, but the fact that they outlive the historical forms of the needs and desires they channelled and re-shaped. Once ritualized and systematized, the imaginary becomes the mirror image of the powers that be. The myths of social change turn into founding myths of the false community built and represented by the power [...] The myth of the "Proletariat" was rotten as well: instead of fighting for the *self-suppression* of proletarians as a class, the communist movement had mystical wanks over any sign of "proletarianship", such as the "hardened hands" of the workers, or their "morality" [...] proletarians were defined according to sociology and identified with blue collars themselves at best, or with the "poor" of the Scriptures at worst, or even with both figures, while Marx had written: 'Either the proletariat is revolutionary, or it is nothing'. The direct consequences were Zdanov's Socialist Realism, puritanism, sexual repression vs. bourgeois "decadence", and all that shite. However, [...] the "destruction of myths" makes no sense, we must concentrate our efforts in another direction: let the imaginary move, prevent it from crystallising, try to understand when and how myths are to be deconstructed, dismembered or forgotten before the plurality of images is reduced to one and absolute. (*Mind Invaders: Come Fottere I Media*, Rome, 1995; partial translation available at:
 <http://www.geocities.com/Area51/Rampart/6812>)

We need open, interactive... rhizomatic mythologies. But mythologies are always created, modified and retold by some *community*. What community are we talking about here? Let's start again from "general intellect." "General" means "common," literally "belonging to the *genus*," i.e. wo/mankind, our species. In *On The Jewish Question* and the *Economic And Philosophical Manuscripts* (1844), Marx appealed to two important concepts: *Gemeinwesen* (common being) and *Gattungswesen* (species-being). Class struggle, the *self-suppression* of the proletariat as a class and, eventually, revolution were to overcome the alienation of human beings from their own *Gemeinwesen* and *Gattungswesen*, in order to build a global human community that coincided with the species itself, beyond races and state-nations, beyond citizenship. We cannot understand the compositionist theory which stems from the *Grundrisse* if we don't stick to Marx's *humanistic* idea of community.

THE *WALDGÄNGER'S* BLACK GAME

The Luther Blissett Project consciously started as an experiment of networking *as* myth-making. "Luther Blissett" is a multiuse name that can be adopted by anybody. The goal is an *anthropomorphization* of "general intellect": since 1994 many people who don't even know each other have endlessly improved the reputation of Luther as a *Homo Gemeinwesen*. And yet, as Bifo put it: "One must not overvalue the importance of Luther Blissett. We could even say that Luther Blissett doesn't count for anything. All that really counts is the fact that we're all Luther Blissett."
Here are some sub-mythologies studied and put into practice by Luther Blissett:

1. The nordic myth of the *Waldgänger*, the rebel who "takes to the woods." In 1951 the German reactionary writer Ernst Jünger wrote a pamphlet titled *Der Waldgang*. Jünger described the society as ruled by plebiscitary patterns and panoptical systems of social control. In order to escape from control, the rebel must go to the woods and organize resistance. In nineteen-fucking-fifty-one! What should we say nowadays? Echelon, interceptions, video-surveillance everywhere, electronic records of our bank operations... Taking to the woods is more necessary than ever.
Some hacks have compared "Luther Blissett" to Robin Hood. Actually that hazy myth has much to do with multiuse names. In eighteenth- century England, Saxon peasants ill-treated by the Norman ruling class expressed their malcontent and everyday resistance by ascribing many anonymous actions (real and imaginary) to one outlaw whose figure gradually became that of "Robin Hood." The surname suggests that this folk hero (at least at the beginning) wore a hood—he had no face, he represented anyone. That's the way the myth works, though in the Middle Ages it could only bring temporary consolation for a very limited *Gemeinschaft*.

2. Some other journalists described Luther Blissett as a "pirate" or a "buccaneer." It is an error. OK, net-culture and orthodox underground culture are clogged with maritime metaphors and, yes, "pirate" also means someone who illegally copies material protected by copyright. But Luther Blissett is a terrestrial myth. You don't breathe brackish air in the woods. The sea is far away, maybe a utopian horizon to which the outlaw gradually moves.
If there's a utopian element in the Luther Blissett narrative, it is the utopia of the criminal class: *fuck them over and take the French leave*, as melancholically evoked in Gary Fleder's *Things to Do in Denver When You're Dead*, a gangster-movie whose characters greet each other saying: "Boat drinks!" This is the happy end of all the movies whose protagonists manage to pull a fast one (a fraud, a robbery...). In the last sequence you see them sailing around the Antilles, quietly sipping their Daiquiris.
Of course "boat drinks!" can only be a propelling sub-mythology, certainly not a realistic project, because there is no "elsewhere" left—misery is all around. The epilogue of Jim Thompson's *The Getaway* is very instructive. Sometimes one can achieve "boat drinks!" though. Ronald Biggs, the Englishman who made the Great Train Robbery of 1963, fled to Brazil and, as far as I know, he's still there. But the *Waldganger* is too far from the sea, indeed, only those who stand in the middle of dry land can cultivate "boat drinks!" as their utopia: "This is Denver, what do you need a boat for?"

3. The last recurrent description is "cultural terrorist," which is less unacceptable but it is improper all the same, because "terrorism" is a term that the ruling class uses to defame anything and any-

body, and also because "terrorism" and state repression always mirror each other (the ETA vs. the GAL, the Armed Islamic Group vs. the "ninjas" of the Algerian Army and so on). The dialectic between police state and "terrorism" is based upon emulation.

And yet, even the apparatus of the state can provide us with some useful images. I'm talking about "intelligence" and black propaganda. Multiuse name bearers from Italy and other countries often mention and cite a book, Ellic Howe's *The Black Game: British Subversive Operations Against the Germans During the Second World War* (Queen Ann Press, London, U.K., 1982).

During WW2, Mr. Howe was the secret Political Warfare Executive's specialist for the manufacture of printed fakes and forgeries. PWE's instructions were to undermine the morale of German soldiers and civilians, by means of disinformation and psychological warfare. Thanks to a network of agents in the enemy-occupied territories, PWE issued fake NSDAP circular letters about feuds in the Party, bogus government edicts about desertion, a frightening *Plague Booklet* supposedly published by the German Ministry of Health and leaflets advising the female army personnel not to have sex with soldiers because of venereal diseases. PWE even produced half a dozen issues of *Der Zenit*, a bogus astrological magazine that dissuaded sailors from weighing anchor on a certain "inauspicious" day (of course it was the date of some important naval operation). PWE also invented Gustav Siegfried Eins a/k/a *Der Chef*, a nonexistent German dissident talking on a bogus clandestine radio station (actually the broadcasts were from the U.K.), entertaining the audience with invectives against nazi politicians and detailed (albeit false) gossip about their sexual perversions.

Since the dawnings of the project, Luther Blissett has been playing a black game like that. This is another viable mythology for mass intellectuals. Given the new *molar* dimension of conflict, this is the *molecular* we can find and work with. Try to figure all those tricksters, impostors and transmaniacs meeting up in the woods, spreading rumors and black material, inoculating lethal viruses in the territories of this global electronic Fifth Reich and then... "Boat drinks!"

[All rights dispersed.]

SUBJECT: RECYCLING ELECTRONIC MODERNITY

FROM: RSUNDAR@DEL2.VSNL.NET.IN (RAVI SUNDARAM)
DATE: WED, 16 SEP 1998 13:29:02 +0200

Marx, now long forgotten by most who spoke his name but a decade or two ago, once said the following in his brilliantly allegorical essay on the Eighteenth Brumaire of Louis Bonaparte. "Bourgeois revolutions.... storm quickly from success to success; their dramatic effects outdo each; men and things set in sparkling brilliants; ecstasy is the everyday spirit; but they are short-lived; soon they have attained their zenith, and a long crapulent depression lays hold of society before it learns soberly to assimilate the results of its storm-and-stress period." In Asia, reeling under the current crisis, the moment of ecstasy has long passed, and the "long crapulent depression" is here to stay.

India, a poor cousin of the East Asians, tried to ignore the crisis through its traditional West-centeredness. But the crisis has finally arrived in South Asia as the Indian rupee has dived steadily since last year and inflation is raging.

But in the area of electronic capitalism, the mood is buoyant. Software stocks have risen 120 percent and soon software will become India's largest export. Many fables have emerged as a response to the irruption of electronic capitalism in a country where 400 million cannot even read or write. The first fable is a domesticated version of the virtual ideology. In this Indianized version, propagated by the technocratic and programming elite, India's access to western modernity (and progress) would obtain through a vast virtual universe, programmed and developed by "Indians." The model: to develop technocities existing in virtual time with U.S. corporations, where Indian programmers would provide low-cost solutions to the new global technospace.

The second fable is a counterfable to the first and quite familiar to those who live in the alternative publics of the net. This fable comes out of a long culture of Old-Left politics in India and draws liberally from sixties dependency theory. The fable, not surprisingly, argues that India's insertion in the virtual global economy follows traditional patterns of unequal exchange. Indian programmers offer a low-cost solution to the problems of transnational corporations. Indian software solutions occupy the lower end of the global virtual commodity chain, just as cotton farmers in South Asia did in the nineteenth century, where they would supply Manchester mills with produce.

All fables are not untrue, but some are more "true" than others. Thus the second fable claims, not unfairly, that most Indian software is exported, and there is very little available in the local languages (ironically the Indian-language versions of the main programs are being developed by IBM and Microsoft) The alternative vision posed by the second fable is typically nationalist. Here India would first concentrate on its domestic space and then forge international links.

In a sense both fables suffer from a yearning for perfection. While the first promises a seamless transition to globalism, the second offers a world that is autarchic. Both are ideological, in the old, nineteenth-century sense of the term, which makes one a little uncomfortable. "Down with all the hypotheses that allow the belief in a true world," Nietzsche once wrote angrily.

There is no doubt that for a "Third World" country, India displays a dynamic map of the new technocultures. The problem for both the fables mentioned above is that they remain limited to the elite domains of techno-space in India. This domain is composed of young, upper-caste, often English-speaking programmers in large metropolises, particularly emerging technocities like Bangalore and Hyderabad. This is the story which Wired loves to tell its Western audiences, but in a critical, innovative sense most of these programmers are not the future citizens of the counter net-publics in India.

What is crucial in the Indian scenario is that the dominant electronic public has cohered with the cultural-political imagination of a belligerent Hindu- nationalist movement. Hindu nationalism in India came to power using an explosive mix of antiminority violence and a discourse of modernity that was quite contemporary. This discourse appealed to the upper-caste elites in the fast-growing cities and towns, using innovative forms of mechanical and electronic reproduction. Thus it was the Hindu nationalists who first used cheap audio-cassette tapes to spread anti-Muslim messages; further giant videoscapes were used to project an aestheticized politics of hate. Some of the first Indian websites were also set up by the Hindu nationalists. To this landscape has been added that terrifying nineteenth-century weapon, the nuclear bomb.

This is an imagination that is aggressive, technologically savvy, and eminently attractive to the cyberelites. The cyberelites may be uncomfortable with the Hindu nationalists' periodic rhetoric of "national sufficiency," but such language is hyper-political and has less meaning on the ground. Outside the universe of the cyberelite, is another one which speaks to a more energetic technoculture. This is a world of innovation and nonlegality, of ad hoc discovery and electronic survival strategies.

But before I talk about this, a story of my own.

Two years ago, I was on a train in Southern India where I met Selvam, a young man of twenty-four, who I saw reading used computer magazines in the railway compartment. Selvam's story is fascinating, for it throws light on a world outside those of the technoelite.

Selvam was born in the temple town of Madurai in Southern India, the son of a worker in the town court, who came from the Dalit community, India's lowest castes. After ten years in school, Selvam began doing a series of odd jobs, he also learnt to type at a night school after which he landed a job at a typists shop. It was there that Selvam first encountered the new technoculture—Indian-style.

In the the late eighties, India witnessed a unique communicative transformation—the spread of public telephones in different parts of the country.

Typically these were not anonymous card-based instruments as in the West or other parts of the Third World, but run by humans. These were called Public Call Offices (PCOs). The idea was that in a nonliterate society like India, the act of telecommunication had to be mediated by humans. Typically literates and nonliterates used PCOs which often doubled as fax centers, xerox shops and typists shops. Open through the night, PCOs offered inexpensive, personalized services which spread rapidly all over the country.

Selvam's type shop was such a PCO. Selvam worked on a used 286, running an old version of Wordstar, where he would type out formal letters to state officials for clients, usually peasants and unemployed. Soon Selvam graduated to a faster 486 and learnt programming by devouring used manuals, and simply asking around. This was the world of informal technological knowledge in most parts of India, where those excluded from the upper-caste, English-speaking bastions of the cyber-elite learnt their tools. Selvam told me how the textile town of Coimbatore, a few hours from Madurai, set up its own BBS by procuring used modems, and connecting them later at night. Used computer equipment is part of a vast commodity chain in India, originating from various centers in India but, the main center is Delhi.

Delhi has a history of single-commodity markets from the days of the Moghul empire. Then various markets would specialize in a single commodity, a tradition which has continued to the present. The center of Delhi's computer trade is the Nehru Place market. Nehru Place is a dark, seedy cluster of gray concrete blocks, which is filled with small shops devoted to the computer trade. Present here are the agents of large corporations, as also software pirates, spare parts dealers, electronic smugglers, and wheeler-dealers of every kind in the computer world. This cluster of legality and non-legality is typical of Indian technoculture. When the cable television revolution began in the nineties, all the cable operators were illegal, and many continue to be so even today. This largely disorganized, dispersed scenario makes it impossible for paid cable television to work in India. This is a pirate modernity, but one with no particular thought about counterculture or its likes. It is a simple survival strategy. The computer trade has followed the pirate modernity of cable television. Just as small town cable operators would come to the cable market in the walled city area of Delhi for equipment, so people from small towns like Selvam would come to Nehru Place as a source for computer parts, used computers, older black and white monitors, and motherboards out of fashion in Delhi.

This is a world that is everyday in its imaginary, pirate in its practice, and mobile in its innovation. This is also a world that never makes it to the computer magazines, nor the technological discourses dominated by the cyber-elite. The old nationalists and left view this world with fascination and horror, for it makes a muddle of simple nationalist solutions. One can call this a recycled electronic modernity. And it is an imaginary that is suspect in the eyes of all the major ideological actors in technospace. For the Indian proponents of a global virtual universe, the illegality of recycled modernity is alarming and "unproductive." Recycled modernity, prevents India's accession to World Trade Organization conventions, and has prevented multinational manufacturers from dominating India's domestic computer market. For the nationalists, this modernity only reconfirms older patterns of unequal exchange and world inequality. In cyberterms this means smaller processing power than those current in the West, lesser band width, and no control over the key processes of electronic production. I suspect that members of the electronic avant garde and the counter net-publics in the West will find recycled modernity in India baffling. For recycled modernity has not discrete spaces of its own in opposition to the main cyberelites, nor does it posit a self-defined oppositional stance. This is a modernity that is fluid and mocking in definition. But is also a world of those dispossessed by the elite domains of electronic capital, a world which possesses a hunter-gatherer cunning and practical intelligence.

The term "recycling" may conjure up images of a borrowed, unoriginal modern. Originality was of course Baudelairian modernity's great claim to dynamism. As social life progressed through a combination of dispersion and unity, the Baudelairian subject was propelled by a search for new visions of original innovation, which was both artistic and scientific. A lot of this has fallen by the wayside in the past few decades, but weak impulses survive to this day. It is important to stress too that recycled modernity does not reflect a thought-out postmodern sensibility. Recycling is a strategy of both survival and innovation on terms entirely outside the current debates on the structure and imagination of the net and technoculture in general. As globalists/virtualists push eagerly for a new economy of virtual space, and the nationalists call for a national electronic self-sufficiency, the practitioners of recycling keep working away in the invisible markets of India. In fact given the evidence, it could even be argued that recycling's claim to "modernity" is quite fragile. Recycling lacks none of modernity's self-proclaimed reflexivity, there is no sense of a means–ends action, nor is there any coherent project. This contrasts with the many historical legacies of modernity in India—one of which was Nehruvian. This modernity was monumental and future-oriented, it spoke in terms of projects, clear visions and argued goals. And the favorite instrument of this modernity was a state plan, borrowed from Soviet models. Nehruvian modernity has been recently challenged by Hindu nationalism, which too, has sought to posit its own claims to modernity, where an authoritarian state and the hegemony of the Hindu majority ally with a dynamic urban consumption regime.

While recycling practices' claim to modernity lies less in any architecture of mobility, but an engagement with speed. Speed constitutes recycling's great reference of activity, centered around sound, vision and data. Temporal acceleration, which Reinhart Koselleck claims is one of modernity's central features, speaks to the deep yearnings of recycling praxis. But this is a constantly shifting universe of adapting to available tools of speed, the world infobahn is but an infrequent visitor. Consider the practice of speed, where the givens of access to the net, the purchase of processing power, all do not exist. They have to be created, partly through developing new techniques, and partly through breaking the laws of global electronic capital. Recycling's great limitation in the computer/net industry is content. This actually contrasts with the other areas of India's cultural industry—

music and cinema. In the field of popular music, a pirate culture effectively broke the stranglehold of multinational companies in the music scene and opened up vast new areas of popular music which the big companies had been afraid to touch. Selling less from official music stores as from neighborhood betel-leaf (paan) shops, then pirate cassettes have made India into one of the major music markets in the world. In the field of cinema and television, content has never been a problem with a large local film industry which has restricted Hollywood largely to English-language audiences. What accounts for this great limitation in the net and the computer components of recycled modernity? Recycling practices have, as we have shown been very successful in expanding computer culture, by making it inexpensive and accessible. Most importantly recycling provided a practical education to tens of thousands of people left out of the upper-caste technical universities. But content providers are still at a discount. But perhaps not yet. The last time I went to Nehru Place I met a young man from Eastern India busy collecting Linux manuals. In a few years the recyclers, bored with pirating Microsoft warez, will surely begin writing their own. Given that such has taken place in every other dimension of recycled modernity in India there is no reason why it should not do so here.

Yes, I am back. True, I won't be writing as often, since I kind of have a job now, and I have less time to look at the big picture, which, however fascinating (at least for those who can find humor in human's inadequacies), never paid off much anyway.

Both in the *New York Times* on May 2 and in the ABC "Nightline" on May 4 there was much talk about ramifications of showing a tragedy on the network TV, but nearly not a word about preventing the tragedy from happening in the first place. I have a piece of advice for the society that believes that whether was it right for the TV to show a man blowing his head off *live* or not, is more pressing social issue than examining the justifiability of reasons why the poor fellow did it: don't create tragedies, then you won't have to worry about showing them. Whether showing a real-life suicide on TV good or bad journalism, I cannot tell. Media in general are a mirror of society. If society is sick, the media shall reflect that sickness. Trying to prevent that was one of the gravest mistakes of communist societies, which all by now have paid for the attempt with their lives. What really is the bad journalism is not talking about why this man actually committed suicide: getting screwed by his "health maintenance organization" (HMO), which happens to millions of Americans every day. I can completely understand that the health-care administrators may easily drive an otherwise sane individual to the act of suicide, since I am being systematically driven crazy by the system myself. At one point it occurred to me that it is better *not* to have any insurance. If you are a private patient, the doctors will at least tell you the truth, and then if you have ten thousand dollars, they will treat you; if you don't, they will let you linger in your misery, which is an ultimately perfect application of the laissez-faire capitalism. If, however, you belong to an HMO, your doctor will give you a diagnosis that will justify a treatment that your HMO is willing to pay for so that your HMO will be pleased by his or her shrewdness—and continue to send him/her new patients. HMOs don't like expensive treatments, so your doctor will not resort to anything radical unless it is an absolute life-threatening emergency. It doesn't pay for him or her, because your HMO agrees to pay less than he or she would normally ask for such treatment (however absurdly inflated that sum might be). Therefore, as an HMO patient, you are bound to receive second-class care. And if the cheapest possible treatment that you are getting ultimately shows no results—that is, your health does not improve—the good doctor will change the therapy, keeping you constantly in a limbo between health and sickness, so that you keep coming back for more until you die or drop the HMO. HMOs apparently have no problems with indefinitely long treatments as long as they are low-budget ones. That means that many doctors consciously provide inadequate care to the patients in order to keep a cozy relations with the HMO. This is very disturbing. It was disturbing for this man to the point that he decided to commit suicide. And the only thing we can talk about is the inappropriateness of showing that on TV? What should have the TV done? Sweep the event under the carpet? Well, should I ever come to that point, I promise you all a good television. On May 7 my account was charged a $17 "maintenance fee." No, I don't have a brokerage account; this is a "Lifeline" checking account, which name correctly suggests that all my miserable earnings and modest survival expenses are recorded there. The basic fee for that account is $4.50 a month, which covers only the enormous privilege of keeping your money in a bank. U.S. banks are the only banks in the world to charge their customers for *taking* their money. Maybe this makes sense considering the overall U.S. corporate arrogance in the world. This basic fee will also cover up to ten transactions during one month. Transactions are checks, electronic payments, or cash withdrawals. Once you engage in more than ten transactions, which is hardly avoidable unless you are a retired person, your monthly fee will shoot up to $9.50 a month *and* you will be charged $0.50 per transaction. Customer service clerk explained to me that the $17 included the $9.50 service fee in my fifteen transactions in last month: $15 \times 0.50 = 7.50 + 9.50 = 17.00$. My question was: Did I engage in fifteen transactions *over* my lifeline "limit" of ten, or did I have fifteen transactions *total*, meaning I was five transactions over the limit? Her answer: A total of fifteen transactions. The fact that I had more than ten transactions automatically raised my monthly service fee from $4.50 to $9.50—a $5 penalty for having five more transactions, or $1 per transaction. Also, since I was over ten, my account was automatically charged a $0.50 per transaction fee. Now, here is my point: I was charged the per transaction fee for *all* fifteen transactions, not just for the five that were over the lifeline limit of ten. I was penalized once—$5 for five transactions. I was penalized twice $2.50 for the same five transactions. And I was penalized for the third time for the same five transaction by being asked to pay $5 for the previous ten transactions, which would otherwise, should I have not made those five transactions over my limit, be included in my Lifeline checking agreement. In fact, those five transactions cost me $12.50, which is a whooping $2.50 per transaction, that in a case of a let's say an ATM withdrawal of $20 represents more than a 10 percent of that transaction. This is a triple penalty for five transactions. I deem such harsh penalties unreasonably cruel and unusually unfair to lower-income customers. Again, the customer service person asserted that this charge is a part of the agreement I have signed, and that it is a Chase Manhattan bank's *policy* to charge its Lifeline checking account customers $9.50 monthly service charge should they exceed their ten-transaction monthly limit *and* a $0.50 per transaction fee *retroactively* including the first ten transactions. I never understood that the per transaction fee can or will be applied retroactively. I doubt I would have signed the agreement had I under-

stood that. Furthermore, having a policy does not necessarily make it right. Nazi Germany had a policy of exterminating Jews, for example. Chase has a policy of driving its customers to the poorhouse with unreasonable and unfair fees. To protest I renamed Chase Manhattan Bank on all my checks as Chase Fascist Bank and I reduced the Chase corporate sign on my ATM card to a swastika. Unsurprisingly, it reminded me of a swastika to begin with, didn't it?

Something just occurred to me: There is a dynamic relationship between hardware and software industry, sort of a bidirectional pull. New software makes old hardware obsolete and new hardware makes old software obsolete. When new software is written for the new hardware, it is written to make that hardware useless soon. When new hardware is built to support that new software better, it is built also to provide for even better software yet to be written. This is how Microsoft and Intel rule the world. The hardware industry had to try to keep prices at a general level (which drops every year), so now most of the chips are made and boards printed in Malaysia, Taiwan, Thailand, and so on. The cheap labor there drove the price of software down, so now the software companies contract labor in Eastern Europe or Ireland. The turnaround of the new software and hardware used to be three or fouryears; now it is about a year. Microsoft expects you to upgrade your operating system every year, and that means you will need all-new hardware (which make Intel happy) and all new software, because the new operating system will be written for the new hardware requirements and require new software to be written for it. Once the new software is out, files produced by the new software are usually not readable with old software—and sometimes even vice versa—so everybody has to get the new operating system, an all-new computer, *and* new software. That's why others in the computer industry, though they bitch about Gates cornering the market, don't really want to get rid of him. When one follows this cycle, one soon sees one of its logical conclusions: the price of hardware becomes so low that it is becomes less practical to repair a computer than to buy a new one. The price of labor in the country of production was substantially lower than it is in the country of service (in the U.S., for example). Usually, manufacturers keeps making parts for their old models for maybe a year (sometimes less) after the model goes out of production. For exmaple, I can't get a new battery for my seven-year-year-old 286 notebook, and the old trusty battery is dead; the computer still works fine. Or: ATI wouldn't update the Windows 3.11 drivers for its three-year-old "Winturbo" video card (the old drivers are not supported by the 16-bit RealAudio player version 3.0 or higher. New ATI cards are built to support new Windows 95 features, and RealAudio is concentrating on 32-bit versions of its software. Radio 101 sent me RealAudio file of a thirty-minute broadcast of the Weekreport from Zagreb: my old version of RealAudio 2.0 couldn't read the file. I called RealAudio, and they pointed me to a download of RealAudio 3.0. I had to pay for it, of course, about $30; but it wouldn't work on my computer because of a conflict with ATI Winturbo driver. While I was trying to solve this problem, my Windows 3.11 irretrievably crashed; I can't get it back on, nor I can install a new version of Windows on the same disk—so I get a new computer. The new computer is a laptop, and it comes with Windows 95. But it is a one-year-old refurb, because Winbook corporation, from which I purchased it, doesn't manufacture or sell "outdated" Pentium 166 MMX models anymore. But the unit came with a defective floppy drive and printer port. They agreed to fix it. They send FedEx to pick it up instantly; but now it's been over ten business days that they've had it. Last week they said they were replacing the motherboard and passing it to the quality control department for a burn-out. Today they said that they are replacing the motherboard; as soon this was done, they'd pass it to the quality control department for a burn-out test. Neither time could they give me an estimate of when the unit would ship back to me. And all this for...the ability to access and edit my data. Bill Gates, who holds more power and controls more money than a pope in the ninth century—he may not tell us what the truth is, but he *is* showing us the only way to the truth, and the truth is just a click away. Where do you want to go today? To the nearest technical support person, thank you.

[Edited by Geert Lovink and Ted Byfield.]

CONTESTATIONAL ROBOTICS

FROM: CRITICAL ART ENSEMBLE AND RICHARD PELL (BY WAY OF STEVEN KURTZ <KURTZ+@ANDREW.CMU.EDU>)
DATE: FRI, 9 OCT 1998 15:49:59 -0400 (EDT)

PART I

Since the modern notion of the public space has been increasingly recognized as a bourgeois fantasy that was dead on arrival at its inception in the nineteenth century, an urgent need has emerged for continuous development of tactics to reestablish a means of expression and a space of temporary autonomy within the realm of the social. This problem has worsened in the latter half of the twentieth century since new electronic media have advanced surveillance capabilities, which in turn are supported by stronger and increasingly pervasive police mechanisms that now function in both presence and absence. Indeed, the need to appropriate social space has decreased in necessity with the rise of nomadic power vectors and with the disappearance of borders in regard to multinational corporate political and economic policy construction; however, on the micro level of everyday life activity, and within the parameters of physical locality, spatial appropriations and the disruption of mechanisms for extreme expression management still have value. Each of us at one point or another, and to varying degrees, has had to face the constraints of specific social spaces that are so repressive that any act beyond those of service to normative comportment, the commodity, or any other component of the status quo is strictly prohibited. Such situations are most common at the monuments to capital that dot the urban landscape, but they can also be witnessed in spectacular moments when extreme repression shines through the screenal mediator as an alibi for democracy and freedom. The finest example to date in the US was the 1996 presidential election. A protest area was constructed at the Republican National Convention where protesters could sign up for fifteen-minute intervals during which they were permitted to speak openly. This political joke played on naive activists had the paradoxical effect of turning the protesters into street corner kooks screaming from their soapbox about issues with no history or context, while at the same time reinforcing the illusion that there is free speech in the public sphere. Certainly, for anyone who was paying attention enough to see through the thin glaze of capital's "open society," this ritualized discontent was the funeral for all the myths of citizenry, public space, or open discourse. To speak of censorship in this situation or in the many others that could be cited by any reader, is deeply foolish, when there was no free speech or open discourse to begin with. What is really being referred to when the charge of censorship is made is an increase in expression management and spatial fortification that surpasses the everyday life expectation of repression. Censorship and self-censorship (internalized censorship) is our environment of locality, and it is within this realm that contestational robots perform a useful service.

THE FUNCTION OF ROBOTS

While robots are generally multifunctional and useful for a broad variety of duties such as rote tasks, high precision activities, telepresent operations, data collection, and so on, one function above all other is of greatest interest to the contestational roboticist. That function is the ability of robots to insinuate themselves into situations that are mortally dangerous or otherwise hazardous to humans.

Take for example three robots developed at Carnegie Mellon University. The first is a robot that can be affixed to pipes with asbestos insulation; it will inch its way down the pipe cutting away the asbestos and safely collecting the remains at the same time. For a robot, this one is relatively inexpensive to produce, and could reduce the costs of removing extremely carcinogenic materials. The second is a robot designed in case of a nuclear accident. This robot has the capability of cutting into a nuclear containment tank of a power plant and testing for the degree of core corruption and area contamination. Once again, this method is certainly preferable to having a person suit up in protective gear and doing the inspection him/herself. Finally, an autonomous military vehicle is under development. The reasons for the development of this vehicle are not publicly discussed, so let's just imagine for a moment what they might be. What could an autonomous military vehicle be used for? Let's make the fair and reasonable assumption that it has direct military application as a tactical vehicle (it is a Humvee after all). It could have scouting capabilities; since the vision engines of this vehicle are very advanced this possibility seems likely. At present, the vehicle has no weapons or weapon mounts. Of course, such an oversight could be easily remedied. If the vehicle was used as an assault vehicle it would still follow the model set by the prior two robots. In other words, it could go into a situation unfit for humans and take action in response to that environment. However, one element distinguishes the potential assault vehicle from the other two robots. While the other two are primarily designed for a physical function, the latter has a social function—the militarization of space by an intelligent agent. Of modest fortune is the fact that this model can be inverted. Militarized social space can be appropriated by robots, and alternative expressions could be insinuated into the space by robotic simulations of human actions. While autonomous robotic action in contestational conditions is beyond the reach of the amateur roboticist, basic telepresent action may not be.

THE SPACE OF CONTESTATIONAL ROBOTS

Like the physical dangers of being irradiated or breathing asbestos, there are specific social spaces which are too dangerous for those of contestational consciousness and subversive intent to enter. Even the tiniest voice of disruption is met by silencing mechanisms that can range from ejection from the space to arrest and/or violence. For example, being in or around the grand majority of governmental spaces and displaying any form of behavior outside the narrow parameters designated for those spaces will bring a swift response from authorities. Think back to the example of the convention protest space. Using the designated protest area was the only possibility, as no protest permits (an oxymoron) were being issued. Those who attempted to challenge this extensively managed territory were promptly told to leave or face arrest. These are the hazardous conditions under which robotic objectors could be useful by allowing agents of contestation to enter their discourse into public record, while keeping the agent at a safe distance from the disturbance. (The remotes can work up to ninety meters; however, the robot has to be kept within the operator's line of sight.)

PERFORMATIVE POSSIBILITIES

What could a robotic objector do in these spaces? We believe that it could simulate many of the possibilities for human action within fortified domains. For example:

Robotic graffiti writers. These robots are basically a combination of a remote control toy car linked with air brushes and some simple chip technology. When running smoothly, this robot can lay down slogans (much like a mobile dot matrix printer) at speeds of 15 mph (see Part II).

Robotic pamphleteers. Simply distributing information in many spaces (such as malls, airports, etc.) can get a person arrested. These are the spaces where a robotic delivery system could come in handy—especially if deployed in flocks. Remember, that people love cute robots (the anthropomorphic, round-eyed japanamation cute is a recommended aesthetic for this variety of robot), and are more likely to take literature from a robot than from most humans. At the same time, the excessively cute aesthetic can lead to robotnapping.

Noise robots. Very cheap to make from existing parts. Particularly recommended for indoor situations. By just adding a canned foghorn or siren to a remote toy car one can create a noise bomb that can disrupt just about any type of small- to medium-scale proceeding into which it can be insinuated.

These are but a few ideas of how relatively simple technologies could be used for micro disturbances. Given the subversive imagination of Nettime's constituency it's easy to believe that better ideas and more efficient ways of creating such robots will soon be on the table. However, it also has to be kept in mind that robotic objectors are of greater value as spectacle than they are as militarized resistance. After all, they are only toybots. Yet these objects of play can demonstrate what public space could be, and that there are other potentials in any given area beyond the authoritarian realities that secured space imposes on those within it.

COSTS

There is a triple cost to this type of robotic practice. First, it does require a modest amount of electrical engineering knowledge, and as we all know, education costs money. Second, it requires access to basic tools, but a machine shop would be better. Third is the cost of hardware. Robots are expensive, and there is no getting around it. In the field of robotics proper, it is barely possible to build a toy for less than US$10,000. We have brought the cost down to US$100–1,000, but this could add up very quickly for a garage tinkerer or for underfunded artists and activists. It seems safe to assume that a robot will be used more than once in most cases, but even so, robotic objectors are outside the parameters for a common, low cost, tactical weapon. To be sure, this research is in its experimental stages.

PART II

HOW TO BUILD A ROBOTIC GRAFFITI WRITER

This article is the first in a series of robotic objector projects for the home roboticist/anarchist. This design combines the integrated perception and autonomous navigation skills of the human dissident with the efficiency and compact size of a robot specifically adapted to the tactics and terrain of street actions. The basic design calls for a roughly shoebox-sized trailer to be drawn by a remote controlled vehicle. The trailer consists of an array of five spray paint units that are controlled by a central processor. The vehicle is navigated into the target area by its human operator. At the appropriate time a switch on the controller is thrown, signaling the start of the "action." As the vehicle rolls along the ground, the row of spray cans prints a text message in much the same way that a dot-matrix printer would. For example, the word "CAPITALIST" would be written as:

```
***    *   *** ***  *  *  *** *   *** **
* *    *   *    *   *  *  *  * **  *  * * *
***    *   * *  **   *  *  *  * **  *** **
*      *   * *  *    *  *  *  * **  *   * *
*      *   *** *    *** *** *  * *   *** * *
```

Depending on the nature of the action, the vehicle can either be navigated to a secluded "safe zone" or considered a worthy sacrifice in the name of robotic objection.

The skills needed to build this robot do not require an engineering degree, although they do require a reasonable amount of experience in building circuits, programming micro-controllers (Basic STAMP), and shop skills/metal working; the project might best be accomplished by a small group of individuals.

Materials:
REMOTE CONTROL CAR (This will be by far the most costly aspect of this project. When coupled with the radio controller and essentials such as a battery charger, the vehicle represents a roughly $500 investment. What makes this car exceptional is that it needs to be capable of pulling 3-4 kilograms of additional weight and still maintain a top speed of 10–15 MPH. This generally means a scaled-down version of a "Monster Truck," that is, multiple engines, Consult your local RC enthusiast—they love these sort of specialty problems. It also must be able to receive three channels instead of the usual two.)

RADIO CONTROLLER (Any three channel controller will do.)

2 WHEELS (Lightweight street wheels from an RC catalog.)

5 INTERMITTENT SOLENOIDS (The surplus variety will be more than adequate here. Something in the neighborhood of 24v [.25–.3 amp] that can hold itself shut against fairly vigorous tugging.)

BATTERIES (One to power the solenoids (probably 24v) and one to power the circuitry [9v].)

5 SPRAY CANS (The 3 oz miniature variety is best for reasons of weight and size. However, the industrial paint that road workers use could be used if the weight is less of a problem. Remember to choose a color that complements the terrain.)

MICRO-CONTROLLER (Almost any standard chip [i.e., BASIC stamp] will suffice as long as it has at least two inputs and five outputs.) LED/OPTOTRANSISTOR (For use as an encoder.)

TRANSISTORS, RESISTORS, CAPACITORS, and WIRE (Specific values cannot be given here, as there are too many variables to worry about.)

RAW MATERIALS (1/32" aluminum or plastic sheet, lightweight plastic or wood square stock [1/4" by 1/4"].)

Construction:
There are too many variables at work here to describe the construction or components in extreme detail. Availability of surplus goods and access to means of production will vary from group to group.

As with any robotics project, the strategy is to work on individual parts AND the overall product AT THE SAME TIME. One needs to be building working sub-systems, while continually evaluating them to ensure that they will work together.

The project is divided into four subsystems.

1. Micro Controller (+software)
2. Encoder
3. Structure of Trailer
4. Solenoid–>Spray-can system

The Micro Controller:

A plethora of microcontrollers exist that are easy to use and learn. Any of the more popular packages that clutter the pages of "hobbyist" magazines will suffice as long as they meet the requirements of having at least two inputs and five outputs. The first input pin is used for the signal that comes from the controller and tells the microprocessor to start performing its task, that is, print the text. The second input pin is for the encoder that attaches to one of the wheels or axles. The encoder tells the processor how fast the vehicle is moving in terms of "clicks" (see encoder section). Each "click," or 1/4 turn of the wheels, will mean that one column of a letter is to be printed. This allows the processor to adjust the space of the letters according to how fast the car is moving. The five output pins are all used for controlling the solenoids that activate the spray cans.

The Text:

As mentioned earlier, the text is printed as if by a dot-matrix printer. Each individual letter is printed with a 5-by-3 grid of dots and therefore requires a minimum of 15 bits to be rendered. The most cost effective method of storing this data in terms of RAM would be to use 16-bit blocks (type SHORT) for each letter in your array and simply ignore the last bit. However, if you have the RAM, it may be more elegant to use one byte for each column (three columns per letter). This abstracts things a bit, making it easier to print simple graphics instead of text or to use the extra bits in each column as a kind of control character. For instance, you could have a bit that controls how long the can sprays, making it possible to have dots and dashes. Depending on how much RAM the micro-controller has, you could build a function into the chip that translates the text into a binary stream using a lookup table—for instance, 111111010011100 for the letter P, as in the example earlier. Such a table would use only around 52 bytes or so (2 bytes per letter times 26 letters). Or translation could be done offline and the stream hardcoded into the chip at programming time.

The following is some pseudo-code that should give a fair idea of how the components interact with each other.

```
Typedef COLUMN = a byte

pin1 = GO signal
pin2 = wheel encoder
pin3-7 = solenoids

COLUMN    the_text_array[# of letters]    =
convert_text("THE MESSAGE TO PRINT")
COLUMN col

while(1){
        if(GO signal ON) //If it gets the GO
                            //signal, the loop
        timer + 1 //must run 5 times with the sig-
nal ON
        if(GO signal OFF) //before it will GO.
                //This prevents false signals
        timer = 0
        if(timer > 5){
        for(i = 1 to # of letters){
        for(j = 1 to 3){ //The number of columns
                            //in a letter
        col = read_next_column(the_text_array)
        paint_column(col) //writes the bits to
                            //pins 3 thru 7
        wait (for encoder click)
        }
        all pins OFF    //puts a space between
                        //letters
        wait (for encoder click)
        }
        }
}
```

Signal from Controller: (i.e., GO!)

The average remote control car uses a minimum of two channels in order to be controlled by the remote. That is, one channel controls forward and backward motion, and the other controls left and right motion. It is very easy to add channels by using standard parts from an RC hobbyist catalog. In this case, we need one more channel that will be used to trigger the text printing function. The signal that comes out of the receiver on the car is most likely going to be PWM (Pulse Width Mod), in which case the supplied code should be sufficient to direct the signal straight into the micro-controller. Should the signal happen to be analog, most micro-controllers have at least one pin that can receive an analog signal.

Encoder:

There's no need to run out and buy a 600-degree optical encoder for this. All we need is a standard LED and phototransistor pairing. They tend to look like this:

```
 __   __
|L|  |P|
|  |_|  |
|_____|
```

There are two standard ways of implementing these as an encoder. In one version, the principle works like thus: When the LED light hits the phototransistor, it is ON. When something is stuck in between them, it is OFF. All we do is attach a pinwheel divided at 45-degree intervals to the axle of one of the wheels and have it pass through the center of the pairing, like this:

Fig 1.

```
 __
 \ | /|          |
  \|/ |          |
 _\|/_|          | <- pinwheel
|  /|\           |
| / | \         _ | _
|/  |__\        |L| |P|
                |  |_| |
   pinwheel     |_____|
```

This is where the "clicks," described earlier, originate. Each space in the pinwheel causes one click in the phototransistor. The signal from the transistor is then passed on to pin 2 of the microcontroller. In another variation on the same theme, the LED/phototransistor pair are pointed at a black and white pinwheel (potentially the wheel hub). The light from the LED reflects off the white parts and triggers the phototransistor, sending it into an ON state. The light is absorbed by the black sections, sending it into an OFF state.

Trailer Construction:

Anything more than a cursory description would be impossible here without the use of mechanical drawings or photographs (see upcoming web version). The basic idea is that we have a trailer chassis resting on two wheels. It is connected to the rear of the vehicle via some type of flexible joint. The chassis can be made out of a sheet of lightweight plastic or aluminum with plastic or aluminum supports. The spray cans are secured, lying flat on the trailer between the wheels. A slot or window runs the width of the trailer below the spray nozzles and perpendicular to the spray cans (this is what they spray through). The solenoids are mounted on a shelf raised an inch or so above the spray nozzles. This allows room for the batteries and electronics to be stored underneath (see Fig. 2).

Solenoid–spray-can mechanism:

Mechanically speaking, this portion will be the most difficult to construct and will require a lot of kludging to get it right. What we've got is a row of five spray-cans facing downward and another row of five solenoids that must use their "pulling" motion to "push" the buttons of the spray cans. This is probably most easily achieved by a simple system of fixed-pivot linkages. The solenoids are arranged so that they are facing (plungers toward) the spray nozzles, and probably raised an inch or so above the nozzle center. The linkages should in the form of the letter Z, with joints at the corners and a fixed-pivot point somewhere in the Z diagonal. The plungers of the solenoids should attached to the upper portion of the Z and the lower one will touch the tip of the spray can.

Fig. 2 (Side View)

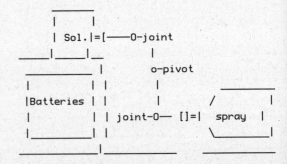

The placement of the pivot point on the linkage determines how much leverage is placed on the nozzle. This may take some tweaking to get enough pressure to make it spray on command.

CONCLUSION

The intentions of this article are two-fold. First, it presents one concrete example of how a robotic objector can be built to be useful to resistant forces. Second, it should open up critical discussion of the value, implications, and design of these tools. Several prototypes are already in the construction phase of development and collective discourse can only enhance the process.

SUBJECT: HACK IT '98, INFORMAL INFORMATICS: THE LAST DAYS OF MARGINALITY (FROM *IL MANIFESTO*, JUNE 21, 1998)

FROM: ERMANNO GUARNIERI (GOMMA)
DATE: THU, 17 SEP 1998 23:18:59 +0200

Marginal considerations of the first large-scale Italian meeting of the people who freely surf the net knowledge.

As for certain big events, a big storm happened at the end of Hack It '98, the first real hacking and alternative informatic culture meeting set-up on a large scale in Florence (Italy). It seems the storm would to symbolically underline the liberation of an electric desire of being a community compressed for too much years in the nets *anfratti*, at the end of seminars, workshop, debates and computer experimentation, but today, a little bit far from the end of the event, apart from trying the impossible effort of summarize in a few words the dozens of "digital events" that happened one after the other, I think it's the time of trying to analyze the reasons of this contents and spectators success, and what seems to be a sure big improvement in all the scene. The alternative Italian informatic scene is born ten years ago, thanks to a flourishing of microgroups, that were strong enough to sustain and improve passing time: the ones that will create Strano Network in Florence—The group behind the unforgettable occupation of the Bologna's Isola nel Kantiere—the Turin scene, the Trento one, the Rome's BBs groups, "Decoder" (which I belong to), the Leoncavallo group and all the other meeting points in other Italian cities, as Bologna and Rome, and the others, that then founded the ECN net. Small collectives, often blocked in their action by modernity fear more or less distributed: mass-media, control and repression organizations, institutional parties, even, sometimes, some large movement areas that just didn't understand the aims of the proposed social action, and in the end also the mainstream informatic panorama that felt and still feel as a bother the critic position of these situations. At this follow a sort of isolation, even if in the early nineties, were organized big events as *Piazza Virtuale* in Milano, "Ink 3D" in Bologna, or the hacking kermesse with high level debates at the Sociology Faculty in Trento, the "mutant" meeting in the S. Arcangelo di Romagna Festival, and the meeting at the Museo d'Arte Contemporanea Pecci in Prato. Moreover much years ago were watchwords and lifestyles of fundamental importance even today: as the need to share information and knowledge, to start nonprofit entry-level course, to create network and digital art directed from and to everybody, to work on new rights, and to start writing analysis on the tranformations of work.

From then yet a lot of people contribute to these initiatives or public opinion campaigns, that obtained the press attention and very often the fears of secret services and Ministry of the Interior, as their annual reports on the "antagonist telematics" from 1991 to today testify. But even in the myriad proposals, of the provoked expectations of new opened fronts, till now seems that the famous "new person" in the embrional stage, more developed in some European countries, here was still far off. And the situation, though all the efforts, seemed to don't move from the marginality zone in wich it was self-confined: maginality that was the obstacle to the start of dynamics worth of going beyond, with a certain costancy, the mere presence of the events owners. Every event, in the end, became in such a way, for the elite, the avant-garde, for the young, hard to understand for the "external observer" not skilled enough to see the real significance of the event itself.

The 1998 hackmeeting represented a turning point, giving clear signals on how the nation situation is evolved. First of all for the great organizational abilities of the Firenze's CPA, social center that even "under forced evacuation," in totally self-organized way had made available all the needed large areas for the debates, courses, meetings, full-time radio and TV pirate station, dozens of networked computers, and food and eating locations. This great potential to self-organized and financed telematics have no equals in the other countries of the world, where the local authorities don't make evacuations, but provides for free the needed logistical structures. The event has been defined as a "horizontal event" by the organization. "There are no organizers, teachers, public or users, but only people who take part." The event has been substantially built through a collective discussion on the net, especially on the "*Isole nella Rete*" and in the mailing list <hackmeeting@kyuzz.org>.

Another winning point is given noting the quality of the competence showed: the knowledge passed level was very high, equal to the one of too much paid professionals, but the hackit strength is that all this became collective, with the necessary interaction. The market force the pro to divide the knowledge into tiny parts, jealously protecting them, and fearing the users to leave them in the dust in case of need. In the hacker-dome, on the other side, the access to knowledge is expanded to minimum, because everyone teach to others everything he knows. And the gathering of the knowledges, as Pierre Lévy states, it's a lot more of the sum. of the single parts. It's something more, new, and with more strength, and the system can't emulate it, due to the anticommercial nature of the sharing.

Another winner tactic has been realized in the focusing on themes on which making free and open courses on techniques available to everyone, but often misunderstood by the people as too much difficult and then abandoned. Among the others the crowded daily course about the personal encryption communications and the use of the cryptographic application Pretty Good Privacy (PGP), that clarified how to defend ourselves from the intruders, a problem often discussed and feared by the attendants. Finally, back to the people involved, these days showed that even here something has started. The networked computers shed, crowded twenty-four hours a day with people who could finally experiments with the machines, has expressed a clear sign: technical competence, belonging to a working or studying sector, will to have relationship with others, desire to meet face to face; the wide space of the social center was always crowded with dozens of people that switched from the computers to the debates.

These are the future perspectives: the event have to become annual, possibly in Milan for the next year; start national initiatives, thinking globally and acting locally, as the "Day for the free programming" against the world presentation of Windows 98: to create a coordination about the digital legal rights and a project of inquiry (survey) about the working conditions in the national telematic. A group of initiatives that seem to think about the marginality days as gone.

[Translated by Alessandro Ludovico.]

SUBJECT: MOVEMENTS AGAINST NETWORK DOMINATION IN JAPAN

FROM: TOSHIMARE OGURA <ORG@NSKNET.OR.JP>
DATE: THU, 8 OCT 1998 22:08:08 +0100

The internet has generally been described as a decentralized system. Within it, people have immense power to communicate and to distribute their messages—in particular the ability to communicate with complete strangers, something previously monopolized by mass media. This is correct from a bird's-eye view of the internet. But from each user's point of view, the internet has a different form.

The conditions and rules of participation in the network imposed upon the user also define access conditions to the internet. The system manager of the network is able to decide almost everything independently from the ordinary user. The environment of the user is very much subject to change according to the attitude this "superuser" takes. It is not so easy to protect ordinary users from the decisions of the superuser employed by government or company.

THE ENCLOSURE MOVEMENT IN INFORMATION CAPITALISM

The innovation of computer communication network (CCN) has had a dual character from the beginning; one is the grass-roots character, the myth of Apple computer and radical hackers in U.S. hippy subculture; on the other hand, ARPAnet created by the Pentagon. Hackers and media activists have struggled for their freedom against state interference and they have tried to disconnect from hierarchical networks and construct a computer counter culture with a tradition of freedom of cyberspace based on the grass roots. But many countries, including Japan, introduced CCN as a state policy. Therefore, the viewpoint of freedom within CCN only has a very fragile basis. Freedom in CCN is not a de facto standard for network users. In the case of Japan, the spread of the internet opened a possibility of various previously inexperienced information traffics. At the same time however, the mass of users retain the passive habits made during the mass media age. The inter-

active character of CCN does not function sufficiently. We have not only to construct economically, socially and politically concrete free-access conditions, but also to create new values of network use—self-valorization for the cyberproletariat.

CASE 1. MOVEMENT AGAINST THE WIRETAP BILL

For Japanese network activists, one of the biggest themes in recent years was a movement against a government-proposed wiretapping bill. As a result of the opposition movements, the Government has not yet legislated the bill. The Japanese Government insists that wiretapping is indispensable to investigate criminal organizations such as Yakuza and cult groups like AUM. But, this is only a poor excuse. It is well known that most of the wiretap investigations so far have been carried out illegally against left-wing political groups and various autonomous radical movements. (The Japanese Constitution asserts that the privacy of all communication should be protected; the police can monitor communications legally only in very rare cases.) Large-scale electronic surveillance by police needs a secret connection to the network backbone. Nippon Telephone and Telegram (NTT) monopolizes the fundamental part of the communication infrastructure in Japan and has supported illegal wiretapping by police in several cases. Moreover, the backbone of the internet passes through Nagata-cho where government agencies are concentrated. From an infrastructural point of view, these conditions give an advantage to the large-scale surveillance of CCN.

Anti-wiretap-bill movements have been developed using the internet. Specifically, we disclose the records of proceedings in the Laws Council of the Ministry of Justice and internal materials from the discussion in the Assembly. We criticize articles of the bill in detail including the understanding of the criminal situation and the government's emphasis on the fear of terrorism to justify the law. Various statements opposing the law from network activists, lawyers' organizations, labor unions and journalists' organizations appear on the webpages of the Anti-Wiretap-Bill Project. By using the internet, we realize the connection and cooperation of small groups—something impossible using phone, fax, mail, printed news letters and other traditional communication tools because of their technological character as one-way or one-to-one communication as well as the burden of their cost. However, the importance of action in the real world also remains. Mass protest action in the national assembly, mass meetings and demonstrations are indispensable in order to change real politics.

STRUGGLES IN THE REAL WORLD OF JAPAN

How are the conditions of the real world related to CCN in Japan? The situation has become very serious for us. NTT can already track the phone number and location of someone using a mobile phone. The phone number and address or location of the caller was previously private. Now it is outside the range of the legal protection of privacy.

The traffic surveillance system, the so-called "N System" reads license plate of passing vehicles and transfers the data to a police mainframe. The system is an enormous database that can confirm which car ran through where and when. The police state that the N System is used for the investigation of traffic violations such as speeding. But, after the system was introduced, there was no increase in the arrest rate for violation of traffic regulations. The N system was however useful in the case of the AUM Shinri Kyo. It functioned as a system to sense the movement of the adherents' cars. There is a suspicion that the N System is being used as a surveillance system for public order, not for criminal investigation.

Not only cars but bicycles have to be registered with the police. Using the police online computer system, bicycle theft can be confirmed from the registration number within moments.

In Japan, private relationships tend to depend on public relationships—not as something cooperative but as a relationship dominated by the state. The concept of the family is used not only for kinship but for company organization and state constitution. Therefore, private space is invaded by the state and public spaces like the street and community facilities are considered, not as belonging to people, but as possessions of the state. We must construct rights to the city as a fundamental human right—something established for several centuries in the West.

DOMINATION BEHIND CHAOS

This may seem strange because Japanese cities have a chaotic face—an exotic and disordered image—like the movie *Blade Runner*, or as Chiba City appeared in the science fiction novels by William Gibson. The big cities like Tokyo and Osaka have wooden houses like temporary shelters alongside high buildings, narrow, winding and labyrinthine paths intermingling with subways and highways. Address indication is insufficient. People from outside get lost easily. Though such disorder is visible, control of city space by the government and police is exhaustive. Street graffiti and posters are hardly seen, and there are few street vendors. The temporarily vehicle-free promenades on Sunday are being abolished one after another. The subway in New York recently got cleaned up—but the Japanese one has never been decorated with graffiti until now. Public transportation stops around midnight.

The database of inhabitants is complete. The family registration system (*koseki seido*), which is characteristic for Japan, is the system of control by the state over the individual based on the patriarchal family. Now, this traditional patriarchal system works via a database. Movements in Japan have developed a struggle against such oppressive control: no family register system; no isolated education for handicapped people; no computers for surveillance and control; street rights for the homeless, and so on.

Control and surveillance of the real world are done through the world of the computer network. The real world and CCN are seamlessly connected. We do not live in the dual worlds of the real "and" the "cyber." Both worlds form an inseparable, intertwined, one world. From the viewpoint of the information surveillance system of the state, our body is a terminal for CCN and a checkpoint in the real world. Our body belongs to the world of the real and the cyber. Therefore, our real/cyber body is a battlefield for liberation movements for one world.

FAILURE OF INFORMATION MANIPULATION

Just as eighteenth-century industrial capitalism established the work ethic, postindustrial information capitalism has to establish a communications ethic.

The freedom to send information as a fundamental right didn't get firmly established in Japan in the age of mass media. Free radio stations hardly exist except for a few, such Radio Home Run, and experimental practices by Tetsuo Kogawa and Jun Oenoki. There are no free radio stations by radicals. On the other side, the internet expanded the circuit of information for individuals by using web, mail, newsgroups and so on. Network users come to doubt to the mass media system through the experience of hypertext and interactive communication on CCN. But, this is not enough to guarantee the formation of new and alternative circuits of information in the real world. The Japanese police arrest users who link to sites abroad upon which appear contents that would be illegal in Japan.

Counterculture has shaped alternative information network behind the scenes of the mass media. CCN allows people a similar power to publish information to mass media. Accordingly, counter cultures become increasingly independent and form their own communication networks.

Dominant cultural capital cannot create new cultural product by their own effort, they need to exploit counter/subculture. But, The cultural industry faces the loss of their resources. In response, they try to integrate counterculture and to restructure the order of the networks. The monopolization of so-called intellectual property and copyrights is their prime strategy for forcing an enclosure movement on CCN. This new enclosure movement tries to establish information structures along capitalist lines; tracing the line of property, sorting according to possibilities for commodification and criminalizing some forms of information. Electronic surveillance and wiretapping by police, and the information enclosure movement by mass media are in close cooperation with each other. Both are processes of a new cyber/real world order of information. If network users are inte-grated into this order, they are forced to exploit their communicative work and depend on the moral standards made by this new master.

CRISIS OF DOMINATION

The modern system of domination in the twentieth century has been based on a one-way information system. The modern nation state has reproduced nationalism by the mass media and mass democracy. The effective function of universal suffrage was guaranteed only by such one-way information systems. The public receives a large quantity of information one-sidedly through the mass media. The mass media behaves as if it represents public opinion and forms a stereotypical view of the world. The election system quantifies public opinion by the votes cast. Minority groups realize their interests only by the sympathy of majority groups. The people's will is quantified and reduced to national will, national identity. The necessity of the reduction of people's opinion to quantified data is dependent on the level of data-processing technology. Individuals turn into a countable mass. Computer technology overcomes such a limit of data-processing. From the management of customers to public welfare policy, individuals recover their own characteristic attributes. At the same time, computer technology has been developed for the interactive communication of individuals. People need not necessarily present their opinions solely by voting or entrusting them to a candidate. They can make them themselves by using CCN.

TOWARD A REVOLUTION OF SINGULARITY

CCN gives a means of expression to minority groups without depending on the paternalism of the majority or of the representative system. Various connections amongst minority groups across state boundaries realizes worldwide solidarity. The geographical border becomes meaningless. People who have the same interest cross borders and cooperate. As a result of this, national identity begins to vacillate. People prefer direct expression in the network to a quantified voting system. The limits to a political system of decision by the majority come into the open. Young people's voter turnout is very low in Japan. They distrust the representative system. They become not apolitical, but refuse quantification of their political will. They try to constitute self-valorization of their own information.

This is a possibility for a new and radical politics of singularity even though it is still perhaps at an unconscious level. It may even become an opportunity to dismantle the nation state, patriarchy and nationalism in Japan.

[Edited by Matt Fuller and Diana McCarty.]

SUBJECT: INTERVIEW WITH
HARWOOD AND MATSUKO OF MONGREL

FROM: GEERT LOVINK <GEERT@XS4ALL.NL>
DATE: THU, 8 OCT 1998 08:01:57 +0100

First of all, this is what Mongrel has to say about their activities:

Mongrel is a mixed bunch of people and machines working to celebrate the methods of an "ignorant" and "filthy" London street culture. We make socially engaged cultural product employing any and all technological advantage that we can lay our hands on. We have dedicated ourselves to learning technological methods of engagement, which means we pride ourselves on our ability to program, engineer and build our own software and custom hardware. The core members are Matsuko Yokokoji, Richard Pierre-Davis, and Graham Harwood.

As well as starting and producing its own projects, Mongrel also works as an agency through which projects by other people can be set up and coordinated. This means that who does what isn't as important as what gets done. network We are as much about hip hop as about hacking. Mongrel makes ways for those locked out of the mainstream to gain strength without getting locked into power structures. Staying hardcore means that Mongrel can get the benefit of sharing the skills and intelligence of people and scenes in similar situations, as well as dealing with other kinds of structures on our own terms. collaborations Mongrel rarely operates as just a core group. We

prefer to work on a range of specific collaborations. These can be with organizations, individuals or groups. The ability to plug into different skills, structures or ways of doing stuff means we get to stay fresh.

Natural Selection is a project put together by Harwood and Matthew Fuller. This project takes on the use of new communications technology for the dissemination and organization of various forms of eugenics, nationalism and racism. The project invents cultural strategies and uses of digital technology to undermine and play with the expectations of racialization in a manner which usurps or destroys it. Mongrel has hacked a popular internet search engine. When any searches are made on that engine for racist material the user gets dumped into a parallel network of websites set up by Mongrel. This parallel network has been made in collaboration with a vast global network of collaborators. It is the nightmare the whites-only internet has been waiting for.

National Heritage is a international project that commits audiences, artists and collaborators to a confrontation with their interpolation within cultural, biological and technologized racisms. The project as a whole operates by means of street poster/newspaper publication, a web search engine "Natural Selection," and a gallery installation. In accomplishing its aims, the project will engender interpretive methods of collaborative working between audiences, artists and project contributors that exploit the possibilities presented by new communications technology for art-working within a social context.

GL: What is your heritage?

H: It is mongrel. The first category is "don't know." A bit of Irish and English. My granddad is a bastard. People think he might be a bit Jewish. Then there are a few incestuous births. My dad really did not know who he was because he got thrown out into the "care" of the authorities at the age of four. His parents could not feed him. He did not know until he got shot in Korea, so after that he went to find them. They were more or less peasants. Heritage in the wider sense meant poverty. My parents taught me that we could be proud of having nothing. To come from nothing is a fine place to be. I come from the land of fiddling. You can always fiddle, get away with it. My family was involved in a lot of gambling. We never had a place in society. They always had illegitimate children. My niece just had a child at fifteen. There are five generations of women that are, more or less, close to each other. They support each other, and the men are there to make some fertilizer.

M: The national heritage of Japan, what it did to other countries and my own, personal history are inseparable. There is an interesting period, last century, when surnames were being introduced. At that time people "bought" their heritage, so to say, by choosing a name associated with a wealthy family.

GL: Could the history of the working class also belong to the national heritage?

H: It is an antiheritage. It was a way of existing outside. In the U.K. there is very much a collective identity. England is 007, James Bond, the crack of leather on the willow of the cricket bat. Strawberries and cream. If that image is not yours, then it is there to exclude you. It is a bit loose because there is no monolith. There has never been a single nation or grouping in the U.K.

GL: You recently published a poster/paper. Along with a black and white insert with material related to the Search Engine project, which we'll talk about later, there are forty full-color heads, organized into some kind of grid. It's almost like a database with two available gender categories and four racial "types" and with what appear to be racialized masks actually sewn into the faces below. The paper also has a large logo—"National Heritage"...

H: This aspect of our project is a reference to the Department for National Heritage. It allocates all the arts funding in the U.K. We decided to make a project with that name, in order to make a direct reference to where the money comes from. 76 percent of all that goes to class A and B, people earning over £30.000 a year. That tax money only goes to that wealthy class. The reason we have the white face with a black mask, covered in spit on the poster, with the words "National Heritage," points directly to this particular department. A revised version of their logo is on the poster. This racial dichotomy is the heritage of the nation. We make them complicit with us.

GL: Do you want this department to become "multicultural"?

H: That is their excuse for keeping power. Multiculturalism is their method of classification, to maintain identities that are long since gone and not useful anymore. They would like to keep a binary authority, which no longer works. Recently, a think tank close to the Labour government gave out a statement, saying that embassies abroad should no longer have any politically incorrect pictures. Cover the walls with Brit art and remove the portraits of old colonial rulers. Remove all reference to British colonial rule. Do they really think that people in Egypt or India can be fooled, by thinking that the British empire never existed? Such emphasis on image! Art is not that useful. But for them it is seen as a major prize.

GL: How did you construct the images on the poster?

M: Out of a total of a hundred faces we made eight faces and divided them into four colors: black, brown, yellow and white, both men and women. It is all montage, digital photography. We tried to construct a white male, or black woman, according to what we think these categories look like. We can never prove that somebody is a white male person. How would you define a black person? There are no characteristics according to medical terms. There are no "real" categories, only stereotypes.

H: On TV there was a program about people of mixed race, let's say 1/4 or 1/8 black. They were complaining because for them there is no classification. One of their grandparents are black, but most of them do not even know.

M: I had never seen a Western person, for real, until I was eighteen. Only since the beginning of the eighties, when many people from all over the world started coming to Japan, did I start to recognize people of different color in the streets. Only then, we became aware of the problem of racism. Before, the Americans were only on television.

H: These days, many young Japanese do not show much interest in where they are from. They see themselves as the future, not the past, the old Japanese culture. They live in the future. Any return to the past is horrifying because you will hit the brick wall of the Second World War. Japanese are good at hiding. The society can leave unresolved problems.

GL: It sounds liberating, to leave the Benneton identity politics behind. ("I am from Ethiopia, look how beautiful—and pure—I am.").

H: In the sixties, my parents used to say things like, "Don't touch it because a black person has had it, you will get ill." At the same moment, they would say, "Martin Luther King, he is a great bloke, he is going to free black people." Two complete opposite views expressed at the same time. We are moving from that level of confusion. I grew up with ska music and black friends—and this black music was being sold to us, white skinheads. So, the level of confusion concerning race is OK. The single thing that seems to categorize white people was fear. The fear to even talk about race. Or to express difficulty about it. We clearly come out as antirace, not so much as "antiracist." We are against the classification of race. That's what a mongrel is—somewhere between two things, someone of mixed blood. Or it refers to a dog that has no category. Dogs in the U.K. are very much a class issue.

M: I lived for the last twelve years in London, so culturally I am mixed now, always fighting between Japanese and English. So I suppose that I became a mongrel. Since the eighties more and more Japanese started living abroad and brought back their mongrel culture to Japan. That's the positive side of the use of technologies.

H: Matsuko and I are of the same year. Despite all the differences, much of our media references are the same. The Thunderbirds. We both grew up under the imperialism of the United States. But then, Richard is bringing a lot of different elements into the group! He is a Black-Indian-Welsh-French person from Trinidad. He is not so confused about his identity as perhaps others are: he is a black cockney—much more so than me. Compared to him, Matsuko becomes an honorary white person.

M: In 1987, when I was visiting South Africa, which was still under Apartheid back then, showing my passport, I was being treated as a white. But if Chinese people would go there, they were categorized as "colored."

GL: "Natural Selection" is another project by Mongrel, an internet search engine. Did you come up with this idea because well known search engines, like Altavista, are no longer useful because they always come up with thousands of references if you type in a keyword?

H: We are looking at classification from another point of view. We created a search engine that sits on top of other search engines. We strip out what they are saying and return the URLs. It you type in any word which has got to do with race, eugenics or sex, you are dropped into our content. This means a whole load of websites being produced in collaboration with a variety of people and groups from a lot of different places: in London, around the world and from different situations which they bring in to flavor the work—academic theorists, street activists, poets, artists, nutters, whatever.

If users look around carefully, they will find the right keywords to access these sites—or they might do it without realizing. On the other hand, you might end up in a "real" Ku Klux Klan site, but you will not find out anyway whether you are reading one of our constructions, or not. You need to be alert all the time as to where all the information you are reading is coming from.

GL: What does the term "eugenics" mean to you?

H: It was used recently by a friend who has brittle bone disease. She talked to me about it because she went to a hospital where they were killing off anyone like her. She made me aware that there was a certain type of human that was to be valued, while others weren't. At what level of disability do we discard those people? Critical Art Ensemble looked at how eugenics are coming into play within fertility treatments. We two went through such treatments, together with Critical Art Ensemble, and found out that a lot of such eugenic decisions had to be made. It was a hard project to go through.

M: We are not judging what is good or bad, we are trying to give information. We don't say, killing life in this or that stage is justified, or not. There is no answer. We do not value life or race. We are showing that it exists.

H: We are struggling to find images that deal with the complexities of the kind of lives that we are living now. There is no longer black and white.

There are no longer binary arguments. So the right wing can jump on us and say: "So you are confused." We are just struggling to find images. Sometimes they are complex and take a long time, like those faces on the poster. It is much harder to think about the same problem from six, maybe opposing, points of view, and hold them all equally. For me, all of this comes from Matsuko's influence, from Japan, where you are able to accept something before you judge it. In the West, I have been brought up to judge something before I have accepted it. One could even say that of antiracism and antifascism. A lot of the identity politics were useful, at the time. But the holding on, imagining the problem would be solved, instead of it slipping it through, like water through your hands, is what actually happened. That antifascism no longer works. It has become a way to sell a product. Not a way to deal with complexity in society.

At the same time, I have absolute admiration for people that sleep on the floor of immigrant's homes in trouble, defending them with their bodies when the fascists come around. We engage in the imagery that forms around these topics. We are in realm of producing troubling images. Often our actual enemies turn out to be politically correct people. The very name "Mongrel" is too difficult for them, let alone our intentions.

GL: You've also produced some software—let's see how it works. Here we have got a package called "Heritage Gold" on the screen. It is an ironical, bastardized version of Photoshop. We have imported my image into the system, and now you are going to give a new heritage. It's a good idea, let's go for it.

H: This is family-oriented heritage changing software. You need some black and female. You can invent a new family. You can have a bastard birth, revert your genes, you can have immigration, repatriation, whatever. I am pasting the new color into your skin. It reminds you how easy it is to manipulate all this data from other people. There will be a huge demand in the West for this software when it goes on full release as people feel a general discontent about their heritage. It will become important to have racial mobility. This menu allows us to add more Chinese and African into your makeup. You never have to have a sun tint again. In order to make you even more dark, we go to the "fleshtone adjustment" dialogue box. We will extract some of the Aryan elements—and you are really beginning to show through now. We will add some social elements too. We are offering a social filter of "police." You look a bit more criminal... We also add some historical relations. A bit less imperialism. Put in some more Afro. We can resize your family by a certain percentage, raise your class consciousness. And then there are the different file formats in which we can save you: genetic index, pixel punish, raw, regressive... There you are— here, you got your brand-new heritage.

[National Heritage and the Natural Selection Search Engine Interview with Harwood and Matsuko of Mongrel (London) at OpenX, Ars Electronica, September 9, 1998. See <http://www.mongrel.org.uk>.]

SUBJECT: WHAT ABOUT COMMUNICATION GUERRILLA? A MESSAGE ABOUT GUERRILLA COMMUNICATION OUT OF THE DEEPER GERMAN BACKWOODS (VERSION 2.0)

FROM: "AUTONOME A.F.R.I.K.A.-GRUPPE" <AFRIKA@CONTRAST.ORG>, LUTHER BLISSETT, AND SONJA BRUENZELS
DATE: WED, 16 SEP 1998 22:20:09 +0200 (MET DST)

This message is directed to those who are fed up with repressive politics at their doorsteps, who are not frustrated enough to give up a critical position and a perspective of political intervention, and who also refuse to believe that radical politics need to be straight, mostly boring and always very serious. It also addresses those who are interested in artistic expression, using all kinds of materials and techniques such as wall-painting, woodcarving, or the internet to bend the rules of normality.

It is sent by some provincial communication guerrillas as an invitation to participate, criticize, renew and develop a way of doing politics which expresses the bloody seriousness of reality in a form that doesn't send the more hedonistic parts of ourselves immediately to sleep. Of course, this is a contradiction in itself: How can you be witty in a situation of increasing racism, state-control and decline of the welfare state, to name only a few. On the other hand, even Karl Marx didn't postulate boredom as revolutionary.

The starting point for our reflections around guerrilla communication was a trivial insight from our own politics: information and political education are completely useless if nobody is interested. After years of distributing leaflets and brochures about all kinds of disgraces, of organizing informative talks and publishing texts, we have come to question the common radical belief in the strength and glory of information. Does it really make sense to take on the attitude of a primary schoolteacher while the kids have become skinheads, slackers, or joined the rat race?

Traditional radical politics strongly rely on the persuasive power of the rational argument. The confidence that the simple presentation of information represents an effective form of political action is almost unshakable. Critical content and the unimpeded spread of "truth" are supposed to be sufficient to tear up the network of manipulating messages, with which the media influence the consciousness of the masses. Well, since the declaration of postmodernism it has become a bit involved to insist on The One And Only Truth. But the main problem with traditional concepts of radical political communication is the acceptance of the idea: "whomsoever possesses the senders can control the thoughts of humans." This hypothesis comes from a very simple communication model which only focuses on the "sender" (in case of mass communication usually centrally and industrially organized), the "channel" which transports the information, and the "receiver." The euphoria around information society as well as its pessimistic opposition—which worries about information overkill—do not face the crucial problem of citizens' representational democracies: facts and information, even if they become commonplace, do not trigger any consequences. Face it, even if stories of disasters, injustice, social and ecological scandals are being published, it has almost no consequences.

Everybody knows that the ozone layer is fading away. Everybody knows that the rich are getting richer and the poor are getting poorer... To us, who believe in Communism, it is hard to understand why such knowledge doesn't lead to revolution and fundamental change—but it definitely doesn't.

Reflections on the interrelations between the reception of information, knowledge and the options to act within a social context have tackled how information becomes meaningful and how it then becomes socially relevant. Information by itself has neither meaning nor consequences—both are created only through the active reception and through the scope of action of the audience. But this basic banality has far too rarely been taken into consideration within the framework of radical politics.

Guerrilla communication doesn't focus on arguments and facts like most leaflets, brochures, slogans or banners. In it's own way, it inhabits a militant political position, it is direct action in the space of social communication. But different from other militant positions (stone meets shopwindow), it doesn't aim to destroy the codes and signs of power and control, but to distort and disfigure their meanings as a means of counteracting the omnipotent prattling of power. Communication guerrillas do not intend to occupy, interrupt or destroy the dominant channels of communication, but to detourn and subvert the messages transported.

But what's new about all this? After all, there have been the Berlin Dadaists, the Italian Indiani Metropolitani, the Situationists. The roots of communication guerrilla can be traced back to legendary characters like the Hapsburgian soldier Svejk and Till Eulenspiegel, the wise fool. Walking in the footsteps of the avant gardes of earlier times, we do not attempt to boast about the invention of a new politics or the foundation of a new movement. Rather, guerrilla communication is an incessant exploration of the jungle of communication processes, of the devoured and unclear paths of senders, codes and recipients. The method of this exploration is to look not just at what's being said, but to focus on how it is being said. The aim is a practical, material critique of the very structures of communication as bases of power and rule.

The bourgeois system takes it's strength—beyond other things—from the ability to include critique. A government needs an opposition, every opinion needs to be balanced with another one, the concept of representative democracy relies on the fiction of equal exchange. Every criticism which doesn't fundamentally shatter the legitimacy of the ruling system, tends to become part of it. Guerrilla communication is an attempt to intervene without getting absorbed by the dominant discourse. We are looking for ways to get involved in situations and at the same time to refuse any constructive participation. Power relations have a tendency to appear normal, even natural and certainly inevitable. They are inscribed into the rules of everyday life. Communication guerrillas want to create those short and shimmering moments of confusion and distortion, moments that tell us that everything could be completely different: a fragmented utopia as a seed of change. Against a symbolic order of western capitalist societies which is built around discourses of rationality and rational conduct, guerrilla communication relies on the powerful possibility of expressing a fundamental critique through the non-verbal, paradoxical and mythical. To be quite clear: guerrilla communication isn't meant to replace a rational critique of dominant politics and hegemonic culture. It doesn't substitute counterinformation, but creates additional possibilities for intervention. But also, it shouldn't be misunderstood as the topping on the cake, a mere addition to the hard work of "real" politics and direct, material action.

In its search for seeds of subversion, guerrilla communication tries to take up contradictions which are hidden in seemingly normal, everyday situations. It attempts to distort normality by addressing those unspoken desires that are usually silenced by omnipresent rules of conduct, rules that define the socially acceptable modes of behavior as well as the "normal" ways of communication and interpretation. To give just a simple example: most people will say that it is not okay to dodge paying the fare, even if there is a widespread feeling that public transport is over-expensive. If, however, some communication guerrillas at the occasion of an important public event like the funeral of Lady Di manage to distribute fake announcements declaring that for the purpose of participating, public transport will be free, the possibility of reducing today's expenses may tempt even those who doubt the authenticity of the announcement.

Communication guerrillas attack the power-relations that are inscribed into the social organization of space and time, into rules and manners, into the order of public conduct and discourse. Everywhere in this "cultural grammar" of a society there are legitimations and naturalizations of economic, political and cultural power and inequality. Communication guerrillas use the knowledge of "cultural grammar" accessible to everybody in order to cause irritations by distorting the rules of normality: It is precisely this kind of irritations that put into question seemingly natural aspects of social life by making the hidden power relations visible and offering the possibility to deconstruct them. Using a term coined by Pierre Bourdieu, one might say that guerrilla communication aims at a temporary expropriation of cultural capital, at a disturbance of the symbolic economy of social relations. *Go Internet, experience the future!* Many communication guerrillas feel a strange affection towards living in the backwoods of late-capitalist society. In the field of communication, this causes an inclination toward the use and abuse of outdated media, such as billboards, printed books and newspapers, face-to-face, messages-in-a-bottle, official announcements, etc. (Even Hakim Bey has advocated the use of outdated media as media of subversion). Thus it is hardly astonishing that communication guerrillas are skeptical about the hype in and around the internet.

Of course, we appreciate ideas like the absolute absence of state control, no-copyright, the free production of ideas and goods, the free flow of information and people across all borders, as they have been expressed by the Californian net-ideology of freedom-and-adventure: liberalism leading us directly into hyperspace. But we also know that real neoliberalism is not exactly like this, but rather: freedom for the markets, control for the rest. It has become obvious that also the internet is no virtual space of freedom beyond state and corporate control. We are afraid that the still existing opportunities of free interchange, the lines of information transmission beyond police control, and the corners of the net which are governed by potlatch economy and not by commercialism, will fade away. The aesthetics of the internet will not be dictated by cyberpunks but by corporate self-representation with a background of a myriad of middle-class wankers exhibiting on corporate-sponsored homepages their home-sweet-homes, their sweet-little-darlings and garden gnomes.

The structures and problems of communication in the net do not differ fundamentally from those encountered elsewhere, at least not as much as the net hype wants to make believe. A product of net thought, like Michael Halberstedt's "Economy of Attention" starts out from a quite trivial point: The potential recipients are free to filter and discard messages. (They may do even much more with them!). And they do this not mainly according to content, but using criteria which may be conceived in terms of cultural grammar and cultural capital. This is completely evident to anybody (except SWP militants) who has always distributed leaflets to people in the street though media hacks seem to have discovered this fact only since the net offers everybody the possibility to widely distribute all kinds of information. In simple words: the basic problems of communication are just the same on both sides of the electronic frontier.

Focusing on the influence of the social and cultural settings on the communication process, communication guerrillas are skeptical toward versions of net politics and net criticism, which hold an uncritical belief in the strength and glory of information. "Access for all," "Bandwidth for all": these are legitimate demands if the net is to be more than an elitist playground of the middle classes. In the future, access to adequate means of communication may even become a vital necessity of everyday life. But information and communication are not ends in themselves; first of all, they constitute an increasingly important terrain of social, political and cultural struggle. Inside and outside the net, communication guerrillas seek to attack power relations inscribed into the structure of communication processes. In the dawn of informational capitalism, such attacks become more than just a method, more than merely a technology of political activism: When information becomes a commodity and cultural capital, a most important asset, the distortion and devaluation of both is a direct attack against the capitalist system. To say it in a swanky way: This is class war.

Increasing attempts to police the net, to establish state and corporate control will, paradoxically, increase its attractiveness as a field of operation of communication guerrillas: Possibly, even those of us who until now do not even own a PC will get *Wired* then. Fakes and false rumors inside and outside the net may help to counteract commodification and state control—after all, the internet is an ideal area for producing rumors and fakes. And, of course, where technological knowledge is available there are innumerable opportunities to fake or hijack domains and homepages, to spoil and distort the flux of information. Guerrilla communication relies upon the hypertextual nature of communication processes. (Also a newspaper or a traffic sign has plenty of cross-links to other fragments of "social text"; a medium transporting plain text and nothing else cannot exist.) Communication guerrillas consciously distort such cross-links with the aim of recontextualizing, criticising or disfiguring the original messages. In the net, hypertextual aspects of communication have for the first time come to the foreground, and the net hypertext offers fascinating possibilities for all kinds of pranks. (Imagine a hacker leaving on a homepage of, say, the CIA not a blunt "Central Stupidity Agency" (see <http://www.2600.com/cia/p_2.html>) but simply modifying some of the links while leaving everything else as before. There are terrible things one could do in this manner...)

But the fascination of those possibilities should not lead to a technocentric narrowing of the field of vision. The mythical figure of the hacker represents a guerrilla directed towards the manipulation of technology—but to which end? The hacker gets temporary control of a line of communication—but most hackers are mainly interested in leaving web graffiti or simply "doing it" (see the Hacker Museum <http://www.2600.com>). Others, however, rediscover guerrilla communication practices of the ancient—recently in <nettime> net-artist Heath Bunting slated himself in a fake review ("Heath Bunting: Wired or Tired?" <http://www.desk.nl/~nettime/>), thus re-inventing a method that Marx and Engels had already used when they faked damning reviews by first-rank economists to draw attention to *Das Kapital*.

Communication guerrillas are fascinated by possibilities offered by the internet in a very different sense: beyond its reality, the net is an urban myth, and perhaps the strongest and most vital of all. Social discourse conceives of the net as a "place" where people, pleasures, sex and the crimes of tomorrow are already taking place. *Go Internet, learn the Future!* Fears and desires are projected onto the net, the mythical place where we can see the future of our society. Paradoxically, the gift of prophecy attributed to the net gives credibility to any information circulated there. The "real world" believes it because it comes from the realm of virtuality, and not *despite* this.

For a long time in the German backwoods, there has been a game called "the invention of CHAOS days," a rather simple game: someone osts a note on saying that, on day X, all the punks of Germany will unite in the town of Y to transform it into a heap of rubble. The announcement is made, and a few leaflets (a dozen or so) are distributed to the usual suspects. And on that day, a procession of media hacks of every kind encounters hosts of riot squads from all over Germany on their way to Y: Once again the forces of public order were on their way to protect our civilization against dark powers. The most astonishing thing about this little game is that it actually worked—several times, no less. Obviously for the guardians of public order and public discourse, the net is a source of secret knowledge too fascinating to be ignored.

We do not mention in detail the innumerable occasions when journalists, state officials, secret services agentc, and so on were taken in by false rumors circulating in the net—for example, the major German press agency DPA, which fell for a homepage of a fake corporation offering human clones, including replicas of Claudia Schiffer and Sylvester Stallone. This effect can be reproduced: the next time it was the prank about "ourfirsttime.com" (<http://www.ourfirsttime.com>). There's little danger that media hacks will ever learn.

The net is a nice playground for communication guerrillas. But we, out there in the backwoods, are telling those living in the netscapes of electronic communication: don't forget to walk and talk your way through the jungle of the streets, to visit the devastated landscapes of outdated media, to see and feel the space and the power and the rule of capitalism—so you shall never forget what pranks are good for.

SUBJECT: MONGOLIA ONLINE: FROM GENGHIS KHAN TO BILL GATES

FROM DRAZEN PANTIC <DRAZEN@OPENNET.ORG>
DATE: MON, 02 MAR 1998 19:46:44 +0000

The Open Society Institute offices of Budapest and Mongolia organized a training course in electronic publishing in February in Ulan Bator, Mongolia. The purpose of the training was to impart the necessary skills to newcomers in the world of electronic publishing and electronic media. Mongolia has been online for almost a year—a quantum leap into a new era. Years of living behind the curtain (whatever curatin) had left their mark: very few people in Mongolia were in a position to take advantage of the new possibilities of global self-expression.

The participants were impressive, both in their sheer numbers and their determination to learn new facts and acquire new skills. From an initial enrollment of 25, the group jumped to 150, of all ages and occupations. A large number of young people were included, but there were more from older groups actively participating. The reception of each new technique or area of knowledge was unique and touching. By way of example, after a lecture on copyright and privacy issues on the internet, the whole group of more than 100 participants stood and applauded in an emotional outburst. Only in this place at this time could a lecture of this kind induce such an emotional reaction. And in the face of such a reaction, a lecturer is simply overcome by his own personal limitations. Almost all the time, contact with the group was an open, two-way street: the trainers would impart the facts and practice of the new media, while the participants would lead us toward really important matters.

At present, Mongolia has just one internet service provider, Datacomm. The company is young and is owned and managed by a very intelligent and progressive group of people. Besides the obvious possibility of monopolistic behavior, Datacomm still acts, to a large extent, as a missionary organization. They are also providing daily training as far as the human limitations of the staff permit. The company has more than a thousand users. There are two places for access by the general public: a classroom of the Technical University, with more than 50, and a Center for internet education. This provides the broader public with a venue for cyber-gathering. The overall bandwidth in and out of the country is 128 kbs, which is adequate, but Datacomm has announced a planned expansion. New providers are also about to set up. The price for internet access is still high, even by international standards. This is the consequence of extremely expensive satellite time and international telephone lines.

Internet and satellite technology are seen in Mongolia as multipurpose. On the international scale, the principle is to present the country to the world and forge closer links with people worldwide, as well as entering electronic commerce (whatever that is). On the domestic scene, the new technologies are a vehicle for internal cohesion. Mongolia is a huge country, with an area almost as great as Europe, but a population of only 2.5 million. The telecommunications infrastructure is very poor, and some regions have no connection at all, beyond poor-quality lines to the capital, Ulan Bator. So email exchange and satellite links are a must if the country is to function in a normal way. The media scene in Mongolia is particularly unclear, at least to the casual visitor. Both print and electronic media are very keen to keep the public informed with modern news programs, in what they see as a world standard package. State television dominates all the other media. Domestic news preoccupations are more or less educational, ranging from advice to eat more vegetables, to discussion on whether prostitution is good (for tourism) or bad (because of the danger of AIDS). All television channels, including the state broadcaster, carry regular soap operas—which, judging by their quality and apparent budget, have their origins in Russian anticopyright corporations. On the other hand, state television, which leases eight hours of satellite time per day, is willing to allow independent media to use the four hours of that time it has not programmed. Mongolian Television called on the private broadcasters to provide programming, preferably nondocumentary, for the unused hours. There are four independent radio and television stations operating in Mongolia. They carry little information or political programming, offering a daily fare of MTV-like broadcasts, serials, and films. It is very difficult to talk about critical independent media in any sense we are used to. The reason for this situation is not necessarily suppression, nor any reluctance to indulge in critical discourse; rather, Mongolia is to a large extent a society very free of conflict. There is basic social consensus on the major questions of state legislation, economy, and religion. There may well be more profound conflicts concealed by the belief that economic development without turbulence will achieve the most for the welfare of the nation.

SUBJECT: SEMIFEUDAL CYBERCOLONIALISM: TECHNOCRATIC DREAMTIME IN MALAYSIA

FROM: JOHN HUTNYK <JOHN.HUTNYK@GOLD.AC.UK>
DATE: TUE, 27 OCT 1998 17:31:51 +0100

At the beginning of 1997, before the meltdown, the haze and the "illegals," Malaysian teknodreamscapes reached high into the sky. Huge new airports, massive hydroelectric dams, mega shopping and apartment complexes, 2 million "foreign" construction workers building the future, and double digit projections in the 2020 Vision—Prime Minister Mahathir's booster theme, now "delayed," for working towards "developed nation status" by the year 2020. Prime Minister Datuk Seri Dr Mahathir bin Mohamad was only prevented by a virus from a planned promotional visit to the UK that year, but he did manage to make it to Hollywood. The dreaming schemes of hypermodernity have been touring the world—LA, Tokyo, Berlin—and the future seems very close indeed. The "Multimedia Super Corridor" (a planned research and development facility with integrated educational, living and manufacturing components) is only a construction contract away, despite a few hiccups in monetary policy which have clouded the horizon.

The Koridor Raya Multimedia or Multimedia Super Corridor (MSC) planned for Malaysia's cyberfuture takeoff has always been an international project. In Los Angeles a cabal of the "great minds" (*New Straits Times*, January 18, 1997) met with Mahathir in a specially convened "Advisory Panel," to flesh out the flashy proposals that would transform the urban skyline—and revitalize construction industry cash flows in difficult times. The assembled great minds included CEOs and Directors of multinational corporations such as Siemens, Netscape, Motorola, Sony, Compaq, Sun, IBM and more. The Chancellor's Professor of UCLA was there, and Bill Gates was invited though didn't come. (Gates announced in March 1998 that he will set up his "Asian" Microsoft operation in Hyderabad, India.) The discussion no doubt was convivial and deals floated, negotiated, traded and made.

What was under consideration at this LA talk-fest was an integrated high-tech development project designed to make Kuala Lumpur and surrounds—a fifteen by fifty-kilometer zone south from the city—the information hub of Southeast Asia. (The

Dream: the seven Flagship Applications of the MSC are Electronic Government, Smart Schools, Multipurpose Cards, Telemedicine, R&D Clusters, Borderless Marketing and Worldwide Manufacturing Webs. The first four Flagship Applications—Electronic Government, Smart Schools, Multi-Purpose Cards, Telemedicine—are categorized under "Multimedia Development," while the other three are categorized under "Multimedia Environment.") Trumpet headlines announced the future in the *Times*, the *Star*, and the *Sun*. PM's speeches and supporting echoes from Ministers proclaimed that the MSC project would "harmonize our entire country with the global forces shaping the information age" (Mahathir's speech in L.A. on January 14, 1997—from the special web page advertising the project—<http://www.mdc.com.my/>). Of course, harmonization with orchestrated multinational info-corps makes for singing pras in the press. The headlines scream: "Global Bridge to the Information Age," "MSC immensely powerful, unique" and "PM's Visit to US Triggers Excitement." Big dreams indeed. Even the pop-electronic fanzine *Wired* got in on the buzz and called the project, quite favorably it seems, "Xanadu for Nerds" (5.08, August 1997).

But in the context of Malaysia's present "standing" in the international marketplace, and in elation to determined priorities and prospects for the peoples of Malaysia, what exactly is to be in this Multimedia Super Corridor? what are the serious prospects for its success? and by what criteria should it be assessed? I want to address these questions from several perspectives critical of the good news propaganda of the proposal itself. The promotional material, as can be expected, does not spare the hype:

Malaysia's Multimedia Super Corridor (MSC) is a bold initiative—a regional launch site for companies developing or using leading multimedia technologies. Aiming to revolutionize how the world does business, the MSC will unlock multimedia's full potential by integrating groundbreaking cyberlaws and outstanding information infrastructure in an attractive physical environment. (Webpage)

The key parts of the proposal include a series of research and development "clusters," basically science labs and info-technology factories, located near a new airport and a "cybercity" including state-of-the-art condos, shopping complexes, and transportation facilities, in a secure (everyone must carry an electronic "National Multipurpose identity Card") and "attractive" garden city. Telemedicine, Electronic Government and full ("uncensored") internet connectivity are also touted. All this overseen by the twin advisory bodies of the Multimedia Development Corporation—they put up the website—and the advisory panel of expert international "great minds."

Why did the first MSC promotion meeting take place in Beverly Hills? Well, obviously the internet and international connectivity of the grand scale to attract the likes of Gates (Microsoft) and Gerstner (IBM) is not yet readily available in Kuala Lumpur itself. Similarly, Mahathir went direct from L.A. to Japan for another parallel high-level corporate luncheon. The point is to attract investment, or rather tenants, for the research laboratories that will be built. One does not want an empty corridor, so one travels to where the clients are. An open invitation.

But what is the invitation to? The development of Science City ventures such as this is not a new idea, though it has become something of a craze since the first versions of the concept of integrated science city living was spawned out of the heads of the planners at Japan's MITI. Engineering new Silicon Valleys has become the grand vision of subsequent planners from "Silicon Glen" in Scotland, to the Multifunction Polis in Adelaide. Not always successfully do more than three hundred such ventures compete for relatively rare technology research pay-offs, as the cutting edge of such research is closely guarded and nurtured by the wealthy megacorporations. In this context, success of a Science City is initially about confidence—the importance of hype. Here, the future can seem very fragile indeed. From the beginning of the year when the prime minister was talking up the "2020 Vision" vision with super conferences in Hollywood, to the CNN televized roller-coaster of the virtual market stock exchange troubles, it's been a dynamic time for futures in Malaysia.

The 2020 Vision "has been delayed," Mahathir was forced to announce, as speculative capital became more tentative and the projects which formed the core of the vision of achieving "Developed Nation status" in twenty-three years were put on hold. The complex repercussions of the slide of the Malaysian Ringitt and other stocks, along with controversies over projects such as the Bakun Hydroelectric dam in Sarawak, and "the Haze" problem afflicting the region, have clouded projections and predictions. Development and profitability seem less secure than before; the tallest building (twin towers Petronas), the biggest airport, the longest office, the undersea electricity cable and the Cyber-Malaysia Multimedia Super Corridor now all appear as costly monuments (whether completed, stalled or abandoned) to the precarious gamble of speculative development within very late capitalism. Of all the new big projects that marked Mahathir's Malaysia as the go-ahead new tiger cub of Southeast Asia, only the MSC project, and related services attractive to international R&D such as the airport, have survived the imposed austerities of the currency crisis. Confidence and hype require more than big buildings and upbeat reviews on CNN.

The mass media soundbite context is not the only one in which I would want to assess the MSC. For starters, the MSC was planned well before the much-hyped "crisis" was even a gleaming twinkle in international imperialism's eye. The Malaysian state has pursued a vigorous technological development program, ostensibly to "catapult" itself within the next thirty years into the fabled zone of "developed nation status." On the back of the Asian Tiger rhetoric of vibrant Southeast Asian economies, this kind of advertised ambition was accepted by many, despite the obvious enormity of the task and despite the almost equally obvious lack of substance to these proclamations(even with massive double-digit growth over many years, the chances of the Malaysian economy reaching levels equivalent to that of major European, or any other Western, powers was slim). Here it's worth noting the new comprador build-and-be-damned cowboy-styles of Mahathir and his cronies, with emphasis on the speculative opportunistic nature of ventures: the world's tallest building, the longest submarine electricity cable, the empty tower blocks

of condos and the jammed road system filled with "Proton" cars (the millionth Proton rolled off the assembly line in January 1997). Corresponding kickbacks in contracts and short-term gains went to the favored few. (The scandal over the award of the prize Bakun Hydro-electrical dam project to Ekran Ltd., the company in which the Chief Minister of Sarawak's sons had substantial holdings, was only one among many.) The mass of the population did not become "Asian Tigers." The glamour projects could not hide the fact of increasing immiseration, the narrow and low nutrient day-to-day existence of the hawkers, farmers and peddlers who crowded the cities and towns, the worsening economic situation in the villages, and the years of repressive governmental corruption and favoritism which leached even the limited potential of prosperity from the hands of the poor into the overseas bank account of the elite. The context of the Asian Crisis, and the MSC, then includes the expanded parallel economy of food and goods hawkers, the illegal and undocumented workers, increasing sexual and other service work for many and uneven opportunities and exploitation, especially of women and "foreigners" and those excluded under the sectarian *brumiputra* legislation that favored Muslim Malays over Chinese, Indian or *Orang Asli* (indigenous) peoples in business, university, and government service.

How did the situation in Malaysia—and Southeast Asia more generally—come to the impasse where the "crisis" could so rapidly unravel the Asian Tiger hype as it has done? It is important to remember that the foundation of the "Asian miracle" which enabled the "tigers," and even the "cubs," to succeed was not some ethnic value or "Confucian" mindset, nor some trickle down effect of development finally reaching some of the non-Euro-American zones, under the auspices of globalizing capital. Such explanations, racist and self-serving on the part of the analysts who offer them, are disguises for the major disruption to imperialism occasioned by the mobilizations and success of postwar (Second World—imperialist—War) national liberation movements (of course with varied degrees of achievement). That the dual deceptions of comprador betrayal on the part of opportunist elite national leaderships on the one hand, and false promises—development aid, technology transfer—swift restitution—IMF loans, DFI and structural adjustment—on the part of Capital on the other hand, does not diminish the fact that what we see played out in Asia today comes as a consequence of global struggles.

It could be argued that the "Asian Tiger" fantasy routine was in effect a deployment of self-serving elite hype. It was the product of a confluence of necessary bluster on the part of Gung-ho development enthusiasts (in this case the comprador elite), and the opportunist specialist swagger of expat experts in the international finance and economics related subdisciplines (what some might want to call the neocolonial administration). The complicity of Mahathir (and Co.) in toadying to these "experts" in the pay of international capital is something that can be variously documented, though as always, the relationship between the comprador elite and the administration experts is sometimes a fraught one. Not surprisingly, since they are after all representatives of the competing interests of different sections of the capitalist system, there is sometimes hostility and disagreement on principles if not in practice (the dynamic of these contradictions is most clearly evident in Mahathir's insistence that Malaysia would not need the intervention of the World Bank, as Indonesia seemed to require, because Malaysia had "already put in place the required measures" that the World Bank would have wanted in any case).

The role of experts and specialists in the pay, and also at times in "passive" critique of Mahathir and co., is a part and parcel of the development trick that lead up to the crisis. I would want to identify a range of specialist workers and several levels of expertise implicated in the project of fitting Malaysians up for participation in the international economy and its exploitative extraction frames. As a special illustrative case of the convoluted complicity of foreign experts, it is instructive to take up the rhetoric about women in technology and the MSC. So often expert development hype promises the advancement of the position of women through the liberating brilliance of technological advance. Parallel to the promises made to indigenous people about the viability of a market-based future (postnomadic, hunter-gatherer lifestyles, which were admittedly hard are to be replaced by the "new" opportunities of waged labor), the promise to women mouthed by the likes of Mahathir and some international women's advocacy groups alike, was that new work opportunities would "free" women from the strictures and constraints of "traditional" oppression. It will of course be readily recognized that neither market economics of high-tech workplace jobs in themselves are liberatory when the context remains one of surplus value extraction and the fruits of advanced production only go to line the wallets of the administrative cliques. In this sense it is possible to make a critique of those who are concerned in cliché ways only with women's labor in relation to the MSC and electronic industrialization in Malaysia—however much it is the case that old and restrictive "traditional" constraints are broken when women or indigenous people enter the waged workforce, this does not necessarily lead yet to liberation, and those who may think so in a naive way should look to the ways capital finds uses and subsumes such "nimble fingers" and exotic workers in its advertising propaganda.

But, after all this, who will be the high-tech workers in the Multimedia Super Corridor? A layer of technocrats and experts will need to be recruited, from in part the expat Malaysian elites schooled in the salons of Stanford, MIT, London and Manchester, but in large part, at least in the first phases, the already existing personnel of the multinational info-corps that are invited to "relocate" will provide staff for the most important posts. This layer of imported workers will have expat lives and an expat status which is not far from the old "colonial career" that has always been the hallmark of business empires under imperialism. These appointments will have several corresponding run-on effects. In this context consideration of the impact of recent technological innovation in the old metropoles upon those now engaged in the (neo)colonial manufacturing enclaves and the Special Economic Zones and so on, is required as a part of any assessment of tech-driven extension of exploitation in the "offshore" production sites of Southeast Asia. Given the range of projects abandoned in the wake of the Ringitt crisis, why is it that Mahathir's dream is to go for the high-tech option instead of extending manufacturing for the local satellite regional economies (surely sales of medium-level manufactured goods to ASEAN partners holds strategic economic merit)? Is the high-tech only gambit not likely to open still further the path of super profits and speculative super exploitation? A less stark, but nevertheless important, question is why the Special Export Zone option with the tax breaks, cheap labor, low shipping excises, and so on is no longer the preferred path, and is instead replaced by a risky corridor venture-chasing the possibility of "technology transfer" and rapid transit to a Bill Gates-sponsored cyberfuture? The problem is that the conditions for such transfer are not quite worked out and there is nothing to really entice the key parts of such corporations to the KL Corridor, nor are the generous tax concessions, infrastructure developments and other State funded inducements calculated to lock in technology transfer in a way that Malaysia could exploit in the long term.

What, and who, is the MSC for? Is it again a project to make the elites rich, and one which does not contribute, except perhaps through the vagaries of trickle-down theory and a vicarious, somewhat quixotic, reflected glory which allows the Malaysian people to take pride in Mahathir's international notoriety? Or can it be demonstrated that the old international imperial production modes are magically reversed by the MSC, rather than continued in new format? Where once jungles were cleared for plantations, where these plantations were then cleared for condos and shopping malls (which lie empty or underused) and where the manufacturing sector was geared largely for export rather than ever for use or need, can it be that the multimedia development will somehow restore productive capacity to local priorities? Is multimedia the key to local content, local uses, local needs, or even to regional variants of these same priorities—the very priorities that we have too often learnt are always second to the goal of profitability, and which seem increasingly subject to the fluctuations and constraints of international competition? "The people's" interest in the trade in shares, the speculation on futures and the infra-structure development company extractions, are all based on some future payoff that does not arrive, or at the least does not arrive for the majority of Malaysians. Of course there are a small few who have always benefited from exploitation of the country's economic efforts—be they the plantation owners, the condo contractors, or the new "big project" development engineers. The problem is that instead of moving towards a more adequate mode of production, given regional and local conditions, possibilities and necessities, those setting the direction of economic activity in Malaysia seem to favor older selective benefit structures and priorities. There is no indication that a leap forward into the MSC is likely to disrupt existing feudal discrepancies of income, lifestyle, or quality of life. Here the contradiction is the same one as that between colonial masters and peasant labor, such that I would suggest the designation "semifeudal, cybercolonial" for those situations where the most advanced technological capacities will benefit old social hierarchic formations that refuse to budge.

Who will work in the MSC? The departure of many of Malaysia's "educated" classes to countries like Singapore, the United States and Australia is considered by some to be "significant" in the context of the MSC dream (See Yee Ai, *Star*, October 6, 1997). That a potential "elite" entrepreneurial segment of the population left Malaysia to further their studies and careers overseas when quotas limiting University places for non-*brumiputras* were instituted under the "New Economic Policy" has had the consequence of positing a fabled brain-drain resource base of potential ex-Malaysian expats who could be enticed back to work in the IT labs of the MSC. In any case, supposing these brainy exiles were enticed back to the MSC, what is to stop the advanced layer of such workers being poached back to the superior labs of Silicon Valley? For that matter, what is to prevent the MSC from becoming the poaching ground for future Malaysian technology-educational cohorts to be shipped to the U.S.?

But to focus on these workers is only to consider a tiny portion of the "job-creation-programme" that is the MSC. Overwhelmingly, it is a kind of processed worker who will make up the majority of those who will build and work in the multimedia corridor-fantasy city. These are people who must clean the labs and work the service sector, in the restaurants, in the apartment buildings, in the transport sector. They are the line-workers, the cable-layers, ditch-diggers, copper miners (insofar as the cybercity still runs through wires), the optic fiber–blowers (insofar as it runs on glass), the light monitors, the carpet-layers, the cola-dispensing machine–restockers, the logo-painters, corporate design staff at the level of uniform tailoring, carpark attendants, rubbish–removers, rubbish collation, white paper–recyclers, glorified garbage- shredders of sophisticated environmental mission statements, junk-mailers, home-shopping delivery agents, home-shoppers, wives, children, neglected pets. Oftentimes these workers will be in insecure employment, many of them overseas nationals, of those, many "illegals." In some sectors, whole communities that provide support and sustenance for productive workers, adjacent reproductive workers, those without community, those with only community, displaced communities, illegal workers, illegal worker entrepreneurs, police crackdown, anti-immigration hysterics, typists of government propaganda and opportunity, cogs in the machine. Sundry otherness. The wrong side of the international division of labor set out on the threshold of the condo, expat servants of all stripes...

What Mahathir's image manipulators want to make of Malaysia is a manicured paradise for multinationals, and so this requires a certain degree of interventionist manipulation of the workforce at several levels—intensive training to equip support staff and engineer-technicians with requisite skills, service economy provisions (requiring also the trappings of the spinoff tourist industry), intensive building programme for offices, condos, air-conditioned shopping centers, and last but not least, the efficient removal of unorganized labor and "street clutter" in the form of vendors and other "illegals." The removal of street vendors is conceived along something like the same lines as the landscape gardening of the science park site, a beautification designed to appeal to the supposed streamlined elegance of Western corporate expectations (little matter that this prob-

ably miscalculates the appeal of a Third World Malaysian site for Western corporations, who are in search not only of cheap labor and peripherals, but who also happily consume "clutter" as exotica, even when the street vendors curry is too hot, or the colors too garish.

Under the austerities imposed under the "crisis" (self-imposed, but they would be little different if the IMF had been invited to manage matters) the first adjustments to the aesthetic makeup of the work force has been to remove the vendors and illegals. In a perverse way this is only "really" about work permits and travel arrangements as the visas of all foreign workers are temporary. The free communication of freely active people is the slogan for generating the successful environment for the research and development community, but the free development of all the people does not compute in this scene. This is one of the major dysfunctions of the MSC in the context of the "crisis." The "foreign" workers brought to build such projects have now become a threat to the scheme. This has meant that one of the responses of Mahathir to the Ringitt crisis was to announce that significant numbers of foreign workers would have to be repatriated. This was not really a new call, but rather an older racist campaign given a new excuse. For some time the Malaysian Government has perpetrated a brutal crackdown on Tamils, Bangladeshis, and Indonesian workers in the Peninsula—from random stop-and-search leading to deportation, to a media campaign which creates resentment. This coupled with brumiputra policies favoring Malay ethnicity workers over Chinese and Indian Malaysian citizens makes the issue of race and opportunity a volatile one in Malaysia. Some 250,000 of the 2 million foreign workers brought to Malaysia to work the big development schemes are expected to be deported by August of 1988, mostly Bangladeshis, Tamils, and Acehnese. Reuters reported in March that:

Malaysia plans to deport some 200,000 foreign workers when their permits expire in August, a government official said Wednesday. The official Bernama news agency quoted Immigration Director-General Aseh Che Mat as saying employers had been told to prepare to send back foreign workers in the ailing services and construction sectors. Malaysia estimates that some 800,000 of 2 million foreign workers in the country are illegal. Since the beginning of the year, authorities have detained more than 17,000 people who were attempting to enter the country illegally. (March 1998)

However, some kinds of foreign workers are OK. When it comes to the glamour projects of development capitalism certain of the experts, expats, and entrepreneurs are exempt from Mahathir's racist gaze. As the economic downturn leads to cutbacks at the MSC, its local workers, not expats, who are being retrenched. At risk of further racism, Mahathir and his cronies now find themselves in a double bind. They have invited "too many" low-skilled construction workers in to build twin towers, airports, and so on and want to get rid of them, while at the same time they want expert development and high-skilled expats to arrive in numbers in the hope that the future may arrive by way of that alchemy known as "technology transfer." It should be no surprise that workforce recruitment takes hierarchical and politically charged forms.

Among the "service workers" one special category has often been singled out. These workers—young Malay women—are found to be particularly suited to high-tech process work by way of cultural conditioning, small-tasks competence, and the mechanics of basket-weaving. This kind of racist characterization appears in barely modified form in the MSC prospectus and other documents— "labour so easy to train" says a FIDA brochure on investment opportunities. This is the gendered version of the same stupidity that once upon a time would explain Japanese technical ascendancy in electronic goods manufacture by claiming that because the shorter Japanese worker stood closer to the workbench greater attention to detail produced superior products. The position of women in feudal structures does seem replicated in telematic times, yet explanation based upon the "cultural" would seem most suited to those who would occlude the political, and any talk of exploitation. What are the conditions of takeoff for Mahathir's proposed dreamscape? The prospects for synergy and innovative creative hyper invention rely upon the relocation of corporate R&D which is less than likely to arrive. The "milieu of innovation" that fuels the successful ventures of this kind does not yet seem to exist in the Malaysian plan—though there certainly is the fab idea in the proposal to build a "cyberversity." The international division of labor, the agendas and opportunisms of the neoimperialist world order, the short term interests of monopoly capital and the inability to provide a lock-on to capital and technology which may relocate to Malaysia are not, none of them, addressed in the promotional or planning literature. There are very real obstacles which would need to be solved if any technology project were to succeed in the East Asian sphere, given that Gates has said that Microsoft will not shift its "fundamental" research outside the U.S., it is not a grand prospect. The realities of the international economy do not favor such projects outside the already entrenched centers. The cost to the Malaysian state, and so therefore the public purse, is likely to be greater than that which can be recouped in the short or long term.

At the risk of inviting the wrath of the "recalcitrant" prime minister, a different series of questions could be asked, ones that would be less generous, but not less plausible in their speculations: for starters, who will profit from the development of the MSC? Do Prime Minister Mahathir and his cronies, the elites and supporters of the good news propaganda in the press, have capital invested in the multimedia transnationals that may locate in the MSC corridor? If Malaysian elite capital is attached to Bill Gates's capital, then perhaps the MSC makes sense for them, if not it is just a corridor crying out (perhaps in vain) for Gates's profiteering. Or, alternately, do Mahathir and other members of the Malaysian elite have capital tied up in the construction industry? This we know is the case from the controversy around the company Ekran and its now stalled plans to build the Bakun hydro electrical dam in Sarawak (flooding the homes of 10,000 Orang Ulu peoples). But surely those that have holdings in construction could just keep on making money out of condos, dams, hotels, and roads, and so all this info and multimedia stuff is too risky speculation? Why go for this high-tech biz? Isn't building factories and warehouses for offshore assembly and export processing profitable enough? Is the writing on the wall in that sector—and does it say build corridors not factories, the end of manufacturing profit is nigh? Or, considering the most cynical case, will this Super Corridor actually have anything in it?—or is it just a flash way of selling more construction (with corresponding bribes and kickbacks etc.)? Even if the R&D firms were to locate some of their lower level R&D in the corridor, how long would it stay—high-tech production is very short on shelf life, and very mobile in terms of setups. What is the prognosis for the economics of the project if even these simple questions are so obvious? Surely better analysts have seen that the gains are not there. What are the justifications? Is it so far off base to suspect the recent fluctuations of the share market indicate where the problems lie—this is a virtual, rather than actual, development, and 2020 is a very long way off.

[A longer version of this paper will appear in *The Planetary Work Machine*, edited by Franco Barchiesi and Steve Wright (forthcoming).]

SUBJECT: CITY OF DOGS, MERCEDES BENZES, AND A LOT OF POTENTIAL

DATE: THU, 8 OCT 1998 02:11:55 +0200
FROM: IGOR MARKOVIC (IGOR.MARKOVIC@ZAMIR-ZG.ZTN.APC.ORG)

Some Reporter's Notes on the Syndicate Meeting "Piramedia," June 1998

The Albanian capital definitely offers a lot more "reality one can cope with," as Geert Lovink put it lately. Visitors with some previous experience with deep-East airports weren't very surprised at seeing cows and horses calm standing very near the planes, but for some Westerners that is definitely the first huge shock on such a trip. But, the airplane made a 180-degree turn, and then went back to the only runway leading to the airport building—something new for every one. But, it would be nice if the plane took us to the building. Passengers come off the plane some 300 meters from the building, and the road to it (if you are lucky enough to avoid all the parked planes) is more like some south American ministate. Arcades with palms, and a building the size of a smaller railway station somewhere in central Europe, sounds like a bad recommendation for future stays. But, what one sees at first look is usually wrong. Sure, The poverty present at the only international airport is widely spread all around the town, but it's purely the result of decades of isolation, because the town is extremely interesting as soon as you are willing to look beyond the feeble facade.

First, the obvious thing on the streets is—coexistence. Animals and humans. If Trieste is known as a city of cats, Tirana is definitely the city of dogs. On every corner there are a lot of street dogs laying on the sun, and usually they are completely harmless, even if you decide to take a night tour. Perhaps they look a little bit different to someone with a "strange smell," but they refrain even from howling, not to mention from doing anything more serious. Beyond dogs, Tirana is also famous for its Mercedes Benz population; according to the receptionist in the Hotel California, the city has the highest number of Mercedes per capita in the world, and even more of the old diesel versions. Walking around it's not hard to believe that—every second car is really a Mercedes. There is no clear explanation for that passion, because Mercedes are driven not only by nouveau riche, but by everyone who is able to collect some money. Which is not an easy task. Beyond smuggling everything possible (which should be patented as a Balkan occupation), there are very few possibilities for earning money. Industry is in ruins, which is obvious if a traveler goes to the coast, as we did. People are a special story—the usual European and (especially) the old Yugoslavian stereotype of Albanians as a short, dark, dirty people is not only racist, but also completely wrong. It's true that you can spot some shorter people with darker skin, but not much more than in any European town; it's also a fact that there are a lot of blond people, particularly women. In the Museum of National History (which is outside of the newer, pathetic wing about victims of communism, which is arranged with lot of taste), it's possible to find a reason for that. Through the area that is today Albania passed many armies and many nations, and even the roots of today's habitants are somewhere in the deep, deep past B.C. Most of the people originate from Ilirians, and those nation definitely do not fit the "short, black" stereotype— these characteristics arrived with the Slavic migrations and after, and mostly in the northern part of the country.

The main street, a boulevard in the full sense, goes from one side of the town to another, and it's always rush hour. All of life is somehow placed on a street—in front of numerous bars, pizzerias, and restaurants, with here and there in-between some Admiral Clubs with slot machine. Everything is full of people: they are walking, sitting, talking, drinking, eating, kissing—the street is not only a public but a social space. One special advantage of Tirana is a huge park with a virgin lake only ten minutes from the university complex, which closes off one end of the main street. The most fascinating thing about that oasis of peace is that there is not a single bar on the shore—definitely a pleasant alternative in our fast-paced, stressful way of life. It's not surprising that people are more open, more communicative, with more happier faces, than anywhere else. In a middle of the boulevard is a main square with a huge hotel for foreigners, the national opera building (very much like similar buildings all around Russian and the East and the already mentioned National Museum. There's also

a big statue of the national hero, Skender-beg, and a national bank. A good proof of Albanians' extraordinary sense of humor is that they have built a small luna-park in the middle of the city center. The former residence of Enver Hoxha is, to be honest, nothing special. If the interior is in accordance with its size and external look, he was not so lucky, compared to other socialist leaders.

But if Tirana offers a picture of normality, outside of the capital it's not such a bright situation. Coming back from a small party outside of town, we passed a couple of police patrols, but Astrit, our host and driver, didn't pay any attention to them— except the one some five kilometers from the town. Every vehicle coming to town is stopped at that point and searched. Later I heard that this checkpoint is not formal but a real border, beyond which, in theory, no one can guarantee any kind of safety. The army does not have enough arms, and most of the policemen (there are a lot of them) carry them on a pro forma basis—they have no munitions. Almost everything was stolen during the rebellion time; in a video piece by one of the students, we even saw people taking a plane from storage. They had no clue about flying.

The common impression given of life on the edge, the permanent possibility of further political instability, also could be observed in the artworks that we saw. The video productions by students from the Academy prove the overall thesis that it's not technical richness, but content that counts. If I compare those low-tech works with, let's say, the usual Japanese productions for Ars Electronica— it's clear that the Japanese are (in most of the cases) just playing with technology, trying to reach "boundaries"; Albanian video was about art processing and the exploration of real life, which art is probably all about. In the three pieces we saw the problem of modest technical equipment was pushed aside by the content. The emphasis was on contemporary or very recent political situations, and on sociocultural trends, namely patriarchy and society's conservativism. Those pieces were not masterworks, but the second, for example, showed part of a day in one woman's life and illustrated women's perpetual work, the repetition of service to the family, no privacy, and so on, with its conclusion (women who'd been going in and out of house front doors many times to pick up the laundry, to bring food, wood, etc., finally locked the door and metaphorically said no to the stereotyped roles of women in Albanian society); this work will definitely be a candidate for an award at any video festival.

It's clear that Albanian media art (at the moment mainly video and video installations) has incredible potential, which was proven again during the annual exhibition this year in June. The positive strategic point, in my opinion, is that the years of isolation with all the problems also brought with them freshness and originality, so rare in most of the better informed post-East countries. There is no tradition of following and copying "big authors" (for example, Bill Viola's influence on Polish video in the eighties—here "great masters" simply do not exist), so the authors have an open field for exploring, learning, and creating, without needing to pay attention to contemporary trends and fashion, or to follow a conformist history of art. In a sense, with no history of exiled art populations returning to the country after '89, with a lack of information about recent projects in Europe and around the world, artists here have more freedom to play, to enjoy their work, and don't have to pay so much attention to curators' and gallerists' demands; they can explore "pure" artistic visions, methods, and models. Their best chance lies in the fact that to a certain degree, they have already developed personal artistic character, and coming in contact with "outside" world may (and I strongly believe, will) result in playful and original works, not just in pure (or not so pure) copies of what is going on in the "centers of art," wherever they are. The peripheral position of Tirana and Albania in general might lead them to have a mixture of styles and techniques, with a strong emphasis on their own cultural capital, because they already are using ideas and concepts of western art, not blindly, but using methods of rethinking, reusing, recycling...recombinant culture indeed! The exhibition at the Academy by the students of textile department, was more or less the same story—freshness, originality, good content, and context, all at the same time.

We might observe a wide synthesis of arts with different origins in space and time, namely from motifs that are not only from different places, but that also belong to different styles—all the things so desperately missing, let's say, from SCCA's annual exhibitions in so many countries. The general impression of Tirana, both on a "real-life" and "artistic" level, led us to the conclusion that this is definitely a country that will emerge very soon on the cultural map of Europe, and that visits to Albania will soon not just be restricted to those who are there by accident or those who want to make some money fast; anyone who will dare to consider themselves European will soon need to visit there. Often. Because Albania, and particularly Tirana, offers a different picture of the Balkans. It is different from the usual stereotypes created by "Europe" and by the "civilized West"—-stereotypes about disorder, wars, dirtiness, the Orient in a negative sense. If Tirana is a good enough example of the Balkans, and it should be, then it's hard not to claim that Balkan is beautiful.

[Edited by Hope Kurtz.]

SUBJECT: BUY ONE GET ONE

FROM: SHU LEA CHEANG <SHULEA@EARTHLINK.NET> WITH LAWRENCE CHUA
DATE: WED, 17 DEC 1997 05:09:46 +0100 MET

"Nippon Telegraph and Telephone Corporation (NTT) signs a joint venture to develop "Cyberjaya", an intelligent city destined to become the center of Malaysia's Multimedia Super Corridor."
—NTT press release, May 7, 1997

BUY ONE GET ONE derives its name from happy hour at Sphinx in Soi Silom, Bangkok. The project explores a digital (co)existence that is borne out of net technology. While Southeast Asia builds Cyberjaya and Africa safaris on the net, we travel to test the limits of national and electronic border patrols.

A cyberhomesteader drifting, accessing with a borrowed password passing with a torn ID card, homepage, homeless page, buy one get one.
As a matter of national security, we simply don't allow people from certain countries to hop on a plane with an uninspected suitcase, leave the airport without going through customs, and walk into a bank. But today, there is nothing to stop a computer hacker in Iran from sitting at a terminal and traveling to that same bank over the Internet.
—Simson L. Garfinkel, on Electronic Border Control, *Hotwired*, July 14, 1997

Shanghai. 11.17.1997, NO VISA.
ENTRY DENIED.

Shanghai, November 17:
If you are Chinese, why do you have a U.S. passport?
—Immigration official, Shanghai International Airport

Taipei, December 2:
You conquered me. And me? I lost but I also triumphed. How could I not? I have learned so much from you. I can do anything now. You have educated me in the finer points of a civilizing empire. To savor the sound of the teapot's wet bottom circling the lip of the warming bowl. You taught me the way to speak your language. To start thinking of myself as a human being. An individual. With skin in place of borders and 99 channels in place of a memory. You have taught me many things.

BACK TO BANGKOK

Bangkok, November 15:
Chai yen yen. Keep a cool heart. Something you say in the clotted arteries of the city of angels. Something you say to remind yourself you were not always hurtling forward. Listen carefully. Under the hum of the idling motor, the clatter of the fallen baht, there is a more insistent song. Something that calls you to reflect on the rampant materialism that's permeated the core of life in this part of the world. To reflect on the echo of your empty bank account, the shopping bags in your hand, the price of the ticket.
Two digital suitcases modeled after Japanese style bentobox and equipped with powerbook, cameras, phoneline and a *hino maru bento* (lunchbox with rice and ume/plum) are netcast ready for HoME delivery. One for the road, one for HoME in NTT/ICC gallery.

5 November, 1997, 5 p.m.
Uploading from CYBER CLUB, Maurya Sheraton, New Delhi, India. Hosted by hotel's own leased phoneline and 64 kbps server.
log on: saudia. password: welcome

Johannesburg, October 14:
You maneuver the streets, trying to lose your skin. With a suitcase of privilege in your once colored hands, you try to become another transborder data flow, skimming the surfaces of oceans, looted

banks, whole cities still glittering under siege. All this you try to do without staining your feet. But as the ground seeps in through your callused soles, you realize that technology is not a colorless media. Even as you try to jettison the essential encumbrances of the nation, the tribe, and the individual, the codes you have stored in your head become an anchor, fixing you to a place, a history, a system in which you are even now participating.

The Times of India, New Delhi, November 6, 1997
NEW INTERNET POLICY AIMS AT 2 MILLION NETIZENS BY 2000

The information market is a pattern of reprocessing, repackaging, and reselling that we're familiar with from colonial times: the colonies provide the raw materials which are made into "finished" products in the West and then sold back to the colonies.
—Leo Fernandez, IndiaLink, the country's first computer communications system dedicated solely to development issues pertaining to the environment, women, children and human rights.

Delhi, November 3:
The bottle of water in my hand promises that it's "Triple sterilized: No lead. No chlorine. No smell." I've been drinking religiously from it, but I'm still bedridden with a flu. My head is congested with the same traffic of viruses with which the Flatted Factory Complex is teeming. In this shabby, barely lit block of concrete, hundreds of electronic companies have set up shop. The stench of excrement competes with the perfumed promises of technology. It's here, in a cramped back office of an agent for the government run ISP, that we log on for the first time in Delhi.
Less than an hour away, the Maurya Sheraton's exclusive Cyber Club promises internet access in pristine, streamlined surroundings, facilitated by the hotel's own server. This is the promise of technology in this part of the world: a fantasy of ordered streets, access to information and security. But the reality is closer to the halls of the Flatted Factory Complex, a place that is always open to the threat and possibilities of contagion.
Lee Chan's mother is seventy-two years old. She recalled that during the wartime everyone had to bring a *hino-maru bento* for lunch to school on the first of every month. Called "Revival of Asia Day," no one was allowed anything but rice and the *umeboshi*. It was meant to train ordinary folks to experience the wartime "frontline."
—from Claire and Marou's email

Harare, October 19:
A Fanonian safari affair: tea served in the bush by tuxedoed Shona waiters, a tour through a game reservation with all these pruney English people. "That's a giraffe, isn't it? Giraffes eat their young for breakfast, don't they?"
Language becomes a mirror, where any attempt at dialog becomes merely an exercise in confirming the white man's expertise.
There is the appearance of an interactive economy on the web, but don't most folks use Web sites like Game Boys? In order to truly intervene and interact with this circuit, it's necessary to adopt a different kind of reflex.
Try this: shatter the mirror, then pick up the glass and use it like a razor to cleave yourself from yourself.

Seoul, November 23:
It is a condition of life in the Third World to deny your place in it. But no matter how high the skyscrapers or how well paved the road, no matter how fast the speed or sophisticated the violence, no matter how long ago modernity triumphed and raised its imperial flag here, no matter how many places develop where the word "cyber" can be affixed, nothing can hide the scent of teargas on your breath. The song in the *noraebang* remains the same. We still rule over the ruins of miracles.

Singapore, November 9:
I suppose this is as good a place as any to consider your paranoia and how it has shaped our journey. A state of paranoia is necessary for maintaining any identity. Without the fear of disappearing into the black world around you, the borders of our bodies would vanish. In Singapore, it feels like someone is always watching, monitoring your every indiscretion. Everything seems to try to reinscribe the permanent identity of a state

against the flows of travel and trade. Let's face it, without fear, you are nothing.

On the road, the digisuitcase is net transmission central, our last hold to a connection, an interface between travelers and marketplace locals, our attachment to HoME/System Mainframe. In the gallery, the bento suitcase serves as gateway for gallery visitors. When in doubt, PRESS. Memory chips scramble. Reprogramable autoagents shuttle down the assembly line.

The Daily Star, Beirut, October 24, 1997
ISRAELIS TRY TO SABOTAGE RESISTANCE WEBSITE

They try to send a virus to the page, a form of electronic detonation. They send a message millions of times—which could take up all our capacity. There is no dialog. This is not a struggle over a piece of land, it is a clash of civilizations

"Between Arab civilization and, if one exists, Israeli civilization. The media has always concentrated on the Islamic Resistance as a military operation, but resistance is not just a military matter. Combatting Zionism requires the most advanced technology in order to counter the directed media and to convey our views."

—Hassan Naami, publicity director of the Islamic Resistance Support Association on the Moqawama website

Beirut, October 29:
Beirut is a fabric of ideas, different tenses that exist in the same sentence.

There is the Beirut before the war, the Riviera of the Arab world captured in the postcards that are still on sale everywhere. Then, there is the Beirut that will be, Solidere's Hong Kong of the next millennium, dreamed up on architectural plans and real-estate brochures. Then, there are the few remaining edifices of bombed out buildings. Across the street from the construction site for Sodeco Square, in a crumbling building that architectural activists have temporarily saved from demolition, we wander up a staircase into a sand-bagged snipers' lair. The ground is littered with newspapers from 1978 and invoices from the eighties. I find a photo of someone's wedding under some broken tiles. The urge to forget lives on the same street as the desire for nostalgia.

The existing ruling class in Malaysia forms an unbroken link with the colonial past. They operated with colonial categories of thought despite their anti-colonial pronouncements. Their concept of property, income tax, business institution and the state, are still dominated by colonial categories.

—Syed Hussein Alatas, "The Myth of the Lazy Native"

Asian cultural values will help bring Malaysia out of its current economic crisis.

—Prime Minister Mahathir Mohammed

Penang, Malaysia, November 13:
My birthplace flashes by in a current of nostalgic bytes and futuristic promises. The lure of calling this place home again has never been stronger. Even in the midst of the depression, the excitement of the future is infectious: hearing Mahathir rail against Western hegemony, watching manicured offices rise like refined Javanese palaces out of the plantation oilpalms at Cyberjaya in KL. Wandering the backlots of the Free Trade Zone in Penang, I pass aisles of young kampung women boarding company hired buses that take them back home: every step of their lives is accounted for. There is the feeling here that the Third World can keep some of its own rightful harvest rather than deliver it all to overdeveloped nations. The Keretapi Tanah Melayu carries me across the promised land. A train pushing forward through the forest of signs. Its engines screeching out a nervous lullaby.

During the two month period of the NTT/ICC Biennial Exhibition, we claim our HoME in Tokyo gallery space and in the telecommunication mainframe. Tracing a route that recalls seeds of discontent, we'll be locating net connection and log on in every city. Recharging desire carried on trade winds between Africa and Asia, we'll be uploading and "furnishing" our HoME with wallpapers of the ever-developing, shuffling memory chips as we cross the borderlines of hyperlink (il)logics.

I.D. card. I.D. card
—Hong Kong policeman who stopped me in a park

Hong Kong, December 5:

Is it possible that a city could just disappear? That a friendship could vanish into the tabula rasa of a new year? Suddenly in this city with no precolonial past, there is no history: only a colonial present and the imminence of its disappearance. That's the dream, anyway: that there are no places left to live. Only spaces of transit. But the transients of Hong Kong woke up from a dream to find that in their restless sleep they had built a city that could never vanish: a glittering mainframe of glass, steel and speed. They rose to find they had become their colonial masters, hungrily feeding on newer forms of migrant life.

Just when you think you've reached the end of the line. When nothing more can happen to you. Just when you think you have returned to your motherland, a lovely witch curses you. Exiles you to forever live in a place called In Between. But this barren island turns out to be a paradise, linked to the mainland by twelve different superhighways and a multimedia supercorridor. You become a winged cypher, a stupid angel with no legs that flies forever and lands only when he dies.

The networks of the future will be digital. They will be intelligent. They will be defined and controlled by software. They will offer high transmission capacity and flexible bandwidth. They will have open architectures so that they can be easily accessed and interconnected. They will convey information from every possible source—by putting us in touch with other human beings, information, by sensing what is happening in natural and man-made environments.

—Dr. Pekka Tarjanne of International Telecommunication Union (ITU) speaking on the subject of Africa and the Information Superhighway, March 18, 1995

Africans are good at playing with ideas, but not as good at actualizing them. But the Internet is the the only chance Africa has to narrow the gap; the first time the West can't use information to blackmail Africa. All the studies that are done on forestry and agriculture by UN bodies and foreign aid organizations would ordinarily be lingering in files in cities like Paris or London or Washington. But with the Web, Africans can access those reports in their home countries.

—Dr. Nii Narku Quaynor, Network Computer Systems, Ghana's home-grown ISP

Accra, October 24:

In order to send an email from Ghana to neighboring Cote d'Ivoire, a former French colony, the messages are rerouted through Paris. We leave Accra for Abidjan to make a connection to Beirut. The manager of Middle East Air reviews my passport and asks me if I'm of Lebanese origin. Too exhausted to lie, I say no, and he refuses to accept our tickets. The plane takes off with us still on the ground. The only way out of Abidjan is to go to Paris.

["Buy One Get One was a two-month homesteading project for the NTT/ICC Biennial, Tokyo, October 25 – December 7, 1997. See <http://www.ntticc.or.jp/HoME>.]

"You're only real with your make-up on."
—Neil Young.

It is my personal commitment to combine cyber-pragmatism and media activism with pleasurable forms of European nihilism. Celebrate the short heroic epics on the everyday life of the media, reporting from within the belly of the beast, fully aware of its own futile existence, compared to the millennial powers to be. We ain't no salespeople, trying to sell the award-winning model among the digital cities, some exotic Amsterdam blend of old and new media or yet another disastrous set of ideas, made in Europe. Instead, we are trying to exchange models, arguments, and experiences on how to organize our cultural and political activities, finance media projects and create informal networks of trust that will make life in this Babylon bearable.

Do you think of the internet as a gnostic conspiracy against the rotting, material world we all would like to leave behind? Well, to be honest, I don't. Seen from an anticapitalist, activist, and autonomous/anarcho point of view, media are first of all pragmatic tools, not metaphysical entities. The Ideology of New Media comes second and should not uphold any of our activities. Media theory, net criticism, computer archeology, cultural studies, digital art critique, and so on give us an understanding of the Laws of Media, but they should not become a goal in itself, despite all of passion for these heroic-marginal supra-intellectual enterprises. For me, it is too easy to make the fancy and at the same time fairly realistic statement, that we should disappear from the realm of the virtual and return to "social action." This legitimate call to leave the Infosphere and appear again on the level of the Street, is making a false distinction between real and virtual policies. Social movements have always had a wide variety of media-related activities. Each action (even the most direct one) has a high level of information, addressing different groups and targets. Media, in this respect, express social relations in a very strong way.

New media is a dirty business, full of traps and seductive offers to work for "the other side." There are no ways to keep your hands clean. The computer is a deadly machine when it comes to inclusion and exclusion. We, the workers on the conceptual forefront of cyberculture, have to admit that we are not (yet) politically correct and have failed so far to pass the PC test. This is not because these criteria are deliberately neglected, but because the passions lie elsewhere. For the time being, the struggle is about the definition of the terms under which the Information Society will become operational. The Short Summer of the Internet, now rushing to its close, is about the production of cultural and political concepts, which may, or may not, be implemented on a much larger scale.

The gold rush is over. Prices of web design have fallen sharply. We can see the rise of the HTML slaves, employed without contracts or health insurance, producing code for little or no money. Small businesses disappear—not just ISPs but also in the art and design sector. On the macroeconomic level we have witnessed an unprecedented series of mergers in the telecommunication and media sector. This has led, for example, to the near monopoly position of WorldCom. Or take the Spanish telecom giant Telefonica and its Intranet, which will soon control the entire Spanish-speaking world.

This may only be the return of the suppressed, after a period of postmodern comfort, in this case late monopoly-capitalism. The undermining of the promising small and decentralized many-to-many ideology comes from within the IT sector. The development of the ultimate multimedia device, WebTV, turns out to be a classic trojan horse. The much hated one-to-many television, news and entertainment industries have now found a way to neutralize a potential competitor. Soon the content of web and TV will be the same. In this respect, all these push media are claiming all available bandwidth. Older features of the net, like the newsgroups, with their democratic and decentralized logic, are dying out and are being replaced by monitored and edited online magazines and chat rooms. Internal surveillance of net use and private email is on the rise due to the introduction of intranets of buildings, companies, and entire countries. Another alarming tendency may be the withdrawal from the internet of universities and research centers, which are now working with much faster and secure computer networks. We know the sad facts. But let's not let them set our agenda.

Media activism nowadays is about the art of getting access (to buildings, networks, resources), hacking the power and withdrawal at the right moment. It's not about the expression of truth or a higher goal. The current political and social conflicts are way too fluid and complex to be dealt with in such one-dimension models like propaganda, publicity, or edutainment. It is not sufficient to just put your information out on a homepage, produce a video or pamphlet, and so on and then just wait until something happens. The potential power of mass media has successfully been crippled. Today, reproduction alone is meaningless. Most likely, tactical data are replicating themselves as viruses. Programmed as highly resistant, long-lasting memes, the new ideas are being constructed to weaken global capitalism in the long term. No apocalyptic or revolutionary expectations here, despite all rumors of an upcoming Big Crash of the financial markets. Unlike the Russian communist world empire, casino capitalism will not just disappear overnight. Heaps of deprivation and alienation is ahead of us. But this should not be the reason to lay back and become console socialists. We need organizations of our time, like the global labor union of digital artisans, networks of travelers, mailing-list movements, a gift economy of public content. These are all conceptual art pieces to start with, realized on the spot, somewhere, for

no particular reason, lacking global ambition. These models will not be envisioned by this or that Hakim Bey. They are lived experiences, before they become myths, ready to be mediated and transformed on their journey through time.

Time to move on. The permanent digital revolution in danger of becoming a reformist project? The system is effectively taking over, even sucking itself into the intimate spheres of friendships and personal aims. The objective Wheel of Net History is taking subjective tolls. Time slips away and we are caught up in something we never really wanted in the first place. Web design for Dummies. Anxiety over nothing. Debates with nothing at stake. Rivalries when there is plenty of loot. But wait a minute. We know all this. The so-called unavoidable process of decay is not God-given or a Law of Nature. It is about time to introduce intelligent social feed-back systems. Indeed, a Collective/Connected Intelligence (thanks, Pierre Lévy and Derek deKerckhove!) that can overcome the rather primitive twentieth-century model of birth, rise, success, and fall that numerous groups and movements have gone through. It should be possible to resist both historical and technological determinism, or at least to play a game with these now predictable forces. This is the search for a media theory, or digital studies in which we can finally fit the charming or rather fatal wetware factor within the larger forces of hardware and software development.

Linking up real communities within a strong, local context, while strengthening the cultural identities remains one of the (secret) recipies of the internet. As Saskia Sassen points out, computer networks are not wiping out locality—quite the opposite. We will, most likely, find emerging virtual community networks in places where communities prosper anyway. Technology alone will not do the job. Sustainable networks will not emerge in places with poor/low local self-esteem. A persistent drive to escape will not result in the development of digital cultures, despite the official internet ideology which is celebrating the so-called global, dislocal qualities of new media. On the internet no one knows you are chatting with your nextdoor neighbor. Denying the really existing local qualities (or misery) is of little use. If isolation and despair are widespread, and disorganization rules, this is not because of the computer, nor can digital equipment help us out of ·daily misery. It is too easy to blame the machines as the cause of the current Really Existing Vagueness. It's up to us to bring people together and start some new initiative, the machine won't do that for us. "There is only one good use for a small town. You hate it and you know You'll have to leave it." All digital technologies in the world will not change this bare fact. Computer mediated communication are like megaphones—but they can only amplify existing signals, no matter how weak they are. Eventually, something will grow out of it. If there is nothing, all the newness will remain stale.

Conciousness Regained. Radical media pragmatism demands that the actors remain cool. Who can still proclaim to be Multimedia after the monstrous misuse of this term? Yes. It should still be possible to ignore all market forces, cheap trends and keep on playing. There is a state of hyperawareness, to transform, disappear, give up terrains that have been occupied, and continue at the same time. What now counts is integrity. It is becoming easy these days to become resigned. There are a thousand reasons to quit, or to continue on the same grocery level. The world, structured by precooked events, ready to be microwaved and consumed, can be rejected. Downright reality is unbearable these days. "No spiritual surrender," an Amsterdam graffiti says. Colorless digital existence can be softened by self-made utopias, hallucinatory experiences, with or without recreational drugs and technologies. Regular switching to other channels which are outside the cyber realm is an option. There are countless universes.

It is silly to fight over an artificially created scarcity. The freedom of expression and media will only be fulfilled once the capability to broadcast has been fully incorporated in the daily life "of the billions." In my view, every fight for liberation can contribute to the destruction of the media monopolies by putting out some messages themselves (graffiti, pamphlets, zines, paintings, songs, imagery). Complaining about the multinational media giants is not enough. The final goal should be the "democratization of the media" and even-

tually the "abolition of media." This goes further than to merely participate in other people's forums or plain "public access." It means an overall dispersion of equipment and knowledge into society. We should try to stop speaking for other people. It is no longer our duty, in the West, to produce their media items in a pseudojournalistic manner. Nowadays, we can make a step further. With the spread of camcorders, tape recorders, photo cameras, xerox copy machines and...computers, ordinary people now have the possibility to produce "content" themselves. Spread the knowledge of how to use and maintain the hard- and software and build up a common (global?) distribution system. A funny side-effect of this is that media will become less and less important.

What form of organization media activism could take? While some truly discouraging stories from the economic forefront are on the rise, it is good to keep returning to the old question:" What is to be done?" A return of negative thinking could play an important role in the development of strategies for media activism. There is plenty of goodwill, and ruthless cynicism. What lacks is playful negativism, a nihilism on the run, never self-satisfied. Tactical, an ever-changing strategy of building infrastructures and leaving them, when the time has come to leave the self build castles and move onward. The explorations into the fields of the negative not only imply the hampering the evil forces of global corporate capitalism, but also formulating a critique of the dominant alternative formula: the Non Governmental Organization. The NGO is not just a model for aid organizations that have to correct the lack of government policies. It is today's one and only option to change society: open up an office, start fund-raising, lease a xerox-machine, send out faxes...and there you have your customized insurrection. "How to make to most of your rebellion." The professionalism inside the office culture of these networked organizations is the only model of media-related politics if we want to have a (positive) impact, or "make a difference." (as the ads use to call it).

By now, third system/NGO networks have lost their virginity. The in-between sector is becoming an economic factor of importance. Unlike in the sixties and seventies, this culture is no longer radical, utopian, or even oppositional. This has mainly been due to their long-term success, not because of its failure, defeat or sell-out, as cultural pessimists would like us to believe. Yes, capitalism has phantastic ability to integrate and neutralize all sorts of movements and forms of political and cultural resistance. But there are also other, more objective, economic developments at stake. NGOs have taken over vital functions of the welfare state. There is a pragmatic style of professionalism. A managerial class has taken over from the activists of the early days, while maintaining, even extending the network of volunteers. With mass unemployment not being solved in the short run, ungoing budget cuts on the side of the government and companies laying off workers, the third system is condemned to grow. This also brings up a radical NGO-critique and a renewed interests in contemporary forms of movements (as campaigns), political parties and trade unions, beyond the closed NGO models. We will soon have to reject their often bureaucratic and ritualized models altogether, with its hierarchies, management procedures, its so-called efficiency. "The Revolution will not be organized." These are not the words of some chaotic anarcho-punkers or eco-ravers, calling for spontaneous revolt, right now, tonight. The crisis of the organization is our condition humane in this outgoing media age. And it may as well be the starting point for a new, open conspiracy that is ready to anticipate on the very near cyberfuture. Not anymore as a party or movement, nor as a network of offices (with or without headquarter), new forms of organization may be highly invisible, not anymore focussed on institionalization. These small and informal communities easily fall apart and regroup in order to prevent the group from being fixed to a certain identity.

We are not one, and there has never been unity, specially not these days. The We form in the age of the net is one of the few possibilities left to address groups, subnetworks and formulate common strategies, (if indeed people are interested in collaboration and exchange...). Heterogeneous policies are always in danger of falling apart. One of the tricks to avoid people organising themselves is to reduce their argument to their Private Opinion which is seen as a contribution to the general (democratic?) discourse. In times of consolidation, dispersion and decay, the We is under debate, whilst at the same time more used than ever. It is the time of strategies. At the moment of the short highs there is only the unspoken, ecstatic We feeling. Later on, we do not want others to speak for others. This is anyway a more general tension, a feeling of discontent, between explicit ways of hyper individuality and loneliness on the one side, and the closed, sometimes claustrophic atmosphere inside groups, collectives, companies and movements on the other side. This should be the starting point for every contemporary debate on new ways of organizing.

Now it is time for other options, in search for the genuine New that does not fit into known patterns of eternal return, being taken back into the System. Virtual voluntarism means being able to overcome moods of melancholy, perfectly aware of all possible limits and opportunities, looking for the impossible, on the side, out of reach of both futurists and nostalgics. Being able to present alternative realities, shocking the Johnsons, way out of reach of the Appropriation Machines. The market authorities will arrive too late. Yes, this is a dream, but we do cannot survive in a (digital) environment without options. In order to get at the point, we should reach a level of collective "self conciousness" to overcome the system of fear and distrust which is now spreading. No attempt to reconstruct what worked once. No glorification of the inevitable. In order not to throw away everything which has been built up we should invent concepts on top of it and not narrow all our options into making the world institutionally legible. The "Next Age," the name of a department store in Pudong/Shanghai, is hybrid: half-clean, somehow dirty, never entirely digitized, stuck between real growth and an even more real crisis. Obsessed with progress, in full despair. But there are other options, and we can realize them. Get Organized!

[Remixed and edited by Felix Stalder.]

SUBJECT: OTHER EUROCENTRISMS

FROM: LINDA WALLACE <HUNGER@LOOM.NET.AU>
DATE: FRI, 16 OCT 1998 17:25:22 +1000

This text originally began as a response to a discussion on the faces list re: is faces Eurocentric?

When something is labeled Eurocentric in Australia one needs to look to the context to see which one of the multilayered possibilities is being referred to.

The term has always had a very specific though virtual geography attached to it—that of a place very far away in <space/time>, though of course, paradoxically, right here/very close in <space/time>. Australia was initially colonized by Europeans: first by British/Western European, then massive post–World War II war migration from Eastern and Southern Europe. Large-scale "Asian" migration followed in the sixties and seventies, though Chinese people had been living in Australia from very early on.

So, given this history, the term Eurocentric when used in Australia can be applied to refer to "the mob over there" in Eurospace, or it may also include other Euro-like mobs such as the U.S. or Canada—it is always defined by what it excludes, for example, other mobs from Africa, Asia, and the Pacific. (Aboriginal Australians commonly use the word mob to signify a language group, a community, nation.) Similarly, when applied internally, in discussions specific to Australia, Eurocentric again refers to the original colonizing meme-set, or this together with its global clonelike-but-different mobs, such as the U.S. or Canada. This ambiguity/collapse of the term stems from the fact that the Euro has been severely infiltrated by all of its other(s) in the Australian context, so that it can be both Euro as in Europe-the-real, the actual place, but also Euro as in Western, as opposed to Asia or South America, and so on.

One of the critical points in regards to the term Eurocentric in the local context is that, while it does imply Europe as a whole, it is not the "Europe" in discrete parts as the present-and-true Euros would see it. The "Europe" that Australian-Europeans perceive themselves to be living, through their own personal backgrounds, is one of a mix of races now changed through the process of migration. So our personal "Australian Europe" is one of intertwined cultural amalgamations/family histories quite outside of the dictates of "Europe-the-real," which seems to have proceeded along nicely inside its own glass bubble. So, here, in identity terms, there is a kind of doubling going on (that is, European/not-European). The situation that's evolved in Australia in terms of Euroness is that which we mirror back to Europe-the-real. Hence we Australians are often surprized and amused to confront, in the flesh, so to speak, the often navel-gazing Euro-other with all its intranational bickering. As, in a way, Australians are running their own hybrid, randomized European Union which is or has been further changed by other migration waves from Asia, the Pacific, Hollywood, and so on—and is, of course, underwritten by the Aboriginal people here prior to invasion. Which is what happens to all communities—as others become part of them—they change. Though not without a struggle. As it is and will be with internet list communities.

With this idea in mind, I would still say quite clearly that the list-object faces is, at its core, Eurocentric, in the "Euro-the-real" sense of the term. Or at least this is the way it began, but is now rapidly moving towards being northern hemisphere-specific. The use of English overlays the faces list to create another layer of complexity, though there are still some posts in German, which is nice. It may well be that there are people from many different parts of the world on faces, but the "critical mass" needed to form an imagined community or group identity is northern hemisphere-based; as such, the faces list-object largely articulates the concerns of that group of people. And that is fine.

The question remains: How do list-objects like Nettime and faces become "more" global? What conditions are necessary so that the critical mass doesn't get too thin and hence unable to recognize itself as a community?

Given that list-objects appear to be located in so-called cyberspace—that is, everywhere at once but nowhere at all—it is, therefore, the collection of subjectivities which feed into the system that construct the community. The ability of the list-object to include/exclude others and otherness will determine its shape and operating logic.

It was after returning from traveling in China for four weeks that I wrote part of this earlier dialogue to faces regarding the question of Eurocentrism. As seen through Chinese eyes, this argument would run along radically different lines.

A while after this I started posting mails to the list to do with an issue in Indonesia where, during the overthrow of Suharto, Chinese women had been and were still being systematically raped and abused by, it seemed, the Indonesian military. I tried to get closer to the heart of this issue, and put the results of investigations onto faces.

Interestingly, it was women from Australia and U.S. who responded to this issue. I felt that the reality of Chinese women in Indonesia, and also that of Timorese woman and so on who suffer the same kind of abuse from the Indonesian "special forces," was maybe somehow too far away to women in Eurospace to trigger a response.

These observations are relevant to the idea of how associative meanings work within the list-object, that is, within the imaginary community of the list-object and its developing subjectivity/identity.

The imagined community never really knows itself, or who or what it is at any point in time, but the community may know that what it imagines itself to be is only a partial manifestation of itself—a glimpse of what it might be now and of what it could be at another time.

The list-object is always shifting as new subjects join, as mail is forwarded on to others outside of the list-object(space), and as each individual within the list-object lives their everyday(time).

So how to visualize this vector-laden object? How to conceptually model trajectories of discussion, flames, jokes (like the running joke of Orlan's nose and Stelarc's ear on the 7-11 list)?

A useful metaphor or model may be supple four-dimensional volumetric semantic/discourse objects. When someone speaks (in)to the list-object—when a voice in the collective body of the imagined community speaks, with all the bodies listening at terminals in their varied time zones—it is as if a point is activated and pulled out to reshape the list-object in 4D space (that is, spatio-temporal).

And if this voice resonates, if it triggers something within the community, if the voice acts as a catalyst to launch the imagined community to speak itself/to respond and, therefore, to change the shape(space/time) of the list-object, or even if the original missive just generates a new idea in one or how many heads of the readers, or a laugh, or even the slightest trace of a smile, then this model goes some way towards how I imagine list-object communities changing their (information) landscape/bodies of signification—of changing themselves.

Which is all very well, but how does a place like China fit into it? That is, in a country where the internet comes in on one 512k line, is IP-number checked at the central gate, then fed off to the range of ISPs using 128k lines each. Where there is little internal connectivity between the ISPs. So if you are paying 60 yuan a month for three hours, and your wage is 200 yuan a week, you aren't likely to be wanting to spend that long online each time, in lists or IRC and MOOSs, and so on. So a list-object, with all the noise and traffic of "bigcity" lists, would cost you a fortune to belong to. Coupled with the fact that so much of the blather is in English, which it is most likely you couldn't read. And forget images and the web—they'd cost a fortune of your precious time to download. RealAudio or RealVideo? No way, not yet. However, inside the net of China, business is booming, and driving internet expansion. Inside the net of China, and the Chinese diaspora, the net in Chinese characters is rapidly growing. And, further, consider the scripts of an online India. Could the current list-objects, if not Eurocentric then at least primarily northern hemispheric, recognize and accommodate this kind of input? How? What kinds of objects will begin to emerge?

Lack of infrastructure, speed and cost of access, English as the dominant language, and state information controls prevent certain kinds of expansion of list-objects like faces, nettime, infowar, rhizome, and recode. It prevents the family growing. Such issues are not new. But they remain.

identify yourself

identify yourself

FROM: SMADAR DREYFUS & LENNAART VAN OLDENBORGH

identify yourself

identify yourself

2wiRqiR
H!nn/$2
!>KG*i$
.=5%y'i
AJ$"#I^
..w/in

304-110

221-120

ah, #22Ch int a
ptdsa batchstart
ds:iox mou di
x to buffermov wo
pop push offs

PROUD

TO BE FLESH

NEIGHBORS

SUBJECT: XCHANGE

FROM: RASA SMITE <RASA@PARKS.LV>
DATE: SAT, 24 OCT 1998, 17:15:50 + 0300

PICTURING THE EVERYDAY "LIFE" OF NET.RADIO: IRC CHAT AT #XCHANGE

This chat is only a picture—a part of one particular net.radio night in early spring, 1998. The Xchange network is an attempt (only) to create a platform for "collaborative communication" in the field of net.audio. It started, and is still based on, the notion of "picturing" audio content, not so much on producing it... (The content is created by the contributors). But there are many more people, radios, and ideas on this globe—a huge variety of types of communication and of collaborative work, each going in different directions and moving towards the discovery of new and undefined spaces... Just as varied are the people involved, their different communities, and the way they are growing, developing and then splitting again. Altogether this is defining an environment... for acoustic space. The acoustic dimension is a powerful tool in the organization of it. At the moment, the following question is emerging: How to develop multidimensional and free-flowing space, within which you can communicate and develop space for a multiplicity of ideas while still keeping it as undefined and open as possible...

```
.....................start irc.re-lab.net #xchange
.....................Date: March 31, 1998.........

x!x Now talking in IRC.re-lab.net #xchange channel

     (x: —> /live-stream loops)

moni:        so tonight we go on with net-noise ? ;)
rasa:        sure!
moni:        alice, what do you say ?
moni:        you know what i liked the most ?
raitis:      what?
alice_t:     don't you hear me eating apples ;-)
moni:        this imagination that there are 6 or 7 stations,
```

Subject: Avoiding heat death on the Internet
From: Phil Agre <pagre@weber.ucsd.edu> (by way of Pit Schultz)
Date: Sat, 14 Jun 1997 13:09:05 +0200

Much of what people are doing on the Internet is great. But much is not. Here is a common dysfunctional pattern: some people decide to "start a discussion group." So they create a mailing list, put a bunch of people on it, and say "okay, let's have a discussion." Maybe they'll send out something interesting to "get discussion started." Several things proceed to happen:

* Since nobody really knows what the list is for,

moni:	all sending and sampling and working
moni:	and it's somewhere in the net all those sounds
baker:	would be nice to do a choir soon
moni:	and everywhere something from the others in
moni:	and nobody but the station who picks it up is listening...
moni:	great
rasa:	that's right
moni:	yes, we could try it..:)
moni:	choir
rasa:	will we do co-broadcast-netnoise-loop tonight?
alice_t:	yes - rasa - do you have something like an out line of your personal preferences for this evening ;-)(
alice_t:	i was thinking about sampling incoming sound... put it into the sampler and accumulate it into an answer... x!x rasa is now known as raitis
raitis:	maybe we could start
raitis:	2 transmitters in the loop
raitis:	and then go on
raitis:	with others
raitis:	?
raitis:	one by one
alice_t:	3 by 4
alice_t:	but... first i have to find an idea what to do with this possibility ...up to this moment i'm just listening very interested ;-)
raitis:	are u on?
alice_t:	no—still listening ;-)
alice_t:	i want some concept—or an idea from the old gods
raitis:	noise
moni:	hey, we are between us.. ;)
moni:	circle?
\<borut\>	?
\<raitis\>	y
\<raitis\>	for some 5 min already
\<alice_t\>	so it's at the moment: riga-graz-lubj-riga ?
\<alice_t\>	so it's at the moment: riga-(graz)-lubj-riga ? i'm just listener—you know ;-)
\<alice_t\>	ahem... filterer

the direction it takes will often be heavily influenced by the first two messages that go out on it—that is, the initial discussion starter and the first issue that someone raises in response. The harder these first two people try to "start discussion" by being stimulating and controversial, the more powerfully they will set the agenda for the list. People will react to those initial points, and other people will react to those points, and the whole discussion will be sucked into one of fifteen standard conversations that everybody in that world has had before.

* This initial explosion of messages will cause many people to panic and say "help! you're flooding my mailbox! get me off this list!"

```
(x:——> /listening, acting)
```

moni:	the idea to be as unfunctional as it is nearly impossible to be ;)
alice_t:	;-))
moni:	and so many at the same time...
alice_t:	the shared big illusion...
moni:	anti-anti-anti...(...loop...)
moni:	revolution inverse
moni:	or something like this :)
rasa:	good idea
moni:	illusion of what ?
alice_t:	hmmm—but in practice it is exactly this feeling of inversion which makes me listen... listen and not act (the latter is the problematic aspect ;)
alice_t:	i would say: illusion of function as a radio...
moni:	well, maybe there is active and passive listening ?
alice_t:	it's more important to define other models and forms of usage for
moni:	if you contribute in chat for example, you influence the atmosphere of what happens with the sound..
moni:	what means function as radio ?
alice_t:	well—i just know that it makes me very passive in the sense that i hardly do any useful work besides listening ,-)
moni:	what's the measure, i mean
moni:	so so so
moni:	you think listening to this is not useful, is that what you wanna say???
moni:	:)
alice_t:	radio—is (if went back in the history of the word)—the radiation—(light-) beams from one center...
moni:	well, not so much different to some few centers, i fear...
alice_t:	sorry—water-: boiling -: tea -: stress ... here at the momentz

—
!!!!!!!!!!!!!!!!!!!!!!!!!!!!!!!!!!!
—
-%-UN-SUB-
SCRIBE——
!!!!!!!!!!!!!!!!!!!!!!!!!!!!!!!!!!!
——%-UN-SUB-
SCRIBE—
-
!!!!!!!!!!!!!!!!!!!!!!!!!!!!!!!!!!!
——%-UN-SUB-
SCRIBE——
!!!!!!!!!!!!!!!!!
!!!!!!!!!!!!!!!!!!—%-
UN-SUB-SCRIBE—
—
!!!!!!!!!!!!!!!!!!!!!!!!!!!!!!!!!!!
—
-%-UN-SUB-
SCRIBE——
!!!!!!!!!!!!!!!!!!!!!!!!!!!!!!!!!!!
——%-UN-SUB-
SCRIBE—
-!!!!!!!!!!!!!!!!!!

Q-where did we end?

Q-target '_close_all_'
Q-sugest: put all mesg into /7-11/<some>/
Q-and think about an auto_count thing for
Q-rotating archive after x mesg

deonMouseOver="s elf.status=';;;
deonMouseOver="s elf.status=';;;

>>NAME: ... 7-11
>>DATE OF BIRTH: 9-97

* Notwithstanding the excessively narrow focus of the initial discussion, the people on the list will come up with five different ideas about what the list is supposed to be for—without it ever occurring to them that alternative ideas exist. They then start grouching at one another for abusing the list. Or even worse, they start scowling inwardly at one another for abusing the list without ever raising the issue—or not raising it until they're full of anger and resentment about it.

Nobody can decide when to take a branch of the discussion "off-line" to private messages. This problem is especially bad on those systems which do not have a concept of a "thread" (roughly, a series of messages with the

```
- % - U N - S U B -
S C R I B E — —
!!!!!!!!!!!!!!!!!!!!!!!!!!!!!!!!!!!!!
— — % - U N - S U B -
SCRIBE—
-
!!!!!!!!!!!!!!!!!!!!!!!!!!!!!!!!!!!!!
— — % - U N - S U B -
S C R I B E — —
!!!!!!!!!!!!!!!!!!
!!!!!!!!!!!!!!!!!!!!——%-
UN-SUB-SCRIBE—
—
!!!!!!!!!!!!!!!!!!!!!!!!!!!!!!!!!!!!!
—
- % - U N - S U B -
S C R I B E — —
!!!!!!!!!!!!!!!!!!!!!!!!!!!!!!!!!!!!!
— — % - U N - S U B -
SCRIBE—
-
!!!!!!!!!!!!!!!!!!!!!!!!!!!!!!!!!!!!!
— — % - U N - S U B -
S C R I B E — —
!!!!!!!!!!!!!!!!!!
!!!!!!!!!!!!!!!!!!!!——%-
UN-SUB-SCRIBE—
—
!!!!!!!!!!!!!!!!!!!!!!!!!!!!!!!!!!!!!
—
- % - U N - S U B -
S C R I B E — —
!!!!!!!!!!!!!!!!!!!!!!!!!!!!!!!!!!!!!
— — % - U N - S U B -
SCRIBE—
-
```

(x: —> /the centers, radio and radiations)

alice_t: breakfast tea-no compromises :-)

moni: against headache

moni: last effort

alice_t: ok ok...

moni: well, the center

moni: i'm not sure if it's a central question for me...

moni: of course, no hierarchies and so on...

moni: but radiation can give warmth too, energy

moni: where to come from?

alice_t: hmmm... as long as you have to re-establish your own center as the only alternative ... it is a problem for me...

moni: maybe it's a difference between center as magnet and center as radiation point ?

alice_t: yes-i also like most of the metaphors used in this communication technology field... (i'm a real "ether" fetishist!)

moni: and if this is mixed, the center becomes a communication-meeting point

moni: :)

alice_t: magnet-:radion... you emphasize the polarity?

moni: well, child of the middle-european philosophical tradition, it's hard to leave the traces the fathers made..;)

moni: to be honest, i really don't know

moni: there are two souls within my ahh- breast ?

moni: ;)

moni: on the one hand, it gives a certain tension, and power again

moni: on the other hand, my favourite fantasy is a concert of all the musicians in the world improvising together...

alice_t: ok... so let's stop this thread... it's more important to find a useful common language than to stress all its power and logical implications...

---

same Subject line), so that people can choose not to receive any more messages on a given thread. But of course, most mail-readers on the Internet (as opposed to Usenet or the Well, for example) have no such concept.

* After an initial burst of discussion, the list falls into something resembling heat death. The level of traffic goes down, and nobody is sure what to do next. Everybody was just reacting to other people's messages anyway, so zero traffic becomes a stable pattern.

* The next step, after a couple months of silence, is for someone to post a political action alert to the list—whereupon a batch of people

```
(x: —> spiritual magic, transcendence of knowledge)
moni:          yup, exactly
moni:          ..brb
alice_t:       well—i would say, even music reflects lifestyle
               and some other given preconditions...
alice_t:       you can not reduce this "together" to some
               magic/spiritual/whatever thing .... you have to
               ask (a bit) about it...
alice_t:       in the best case scenario good music sounds like
               the structures behind it...
moni:          ok, back..
moni:          magic-spiritual..: it's more important to
               remember that there was once something like this
               too, and that it's part of our tradition
moni:          i see more often that it's forgotten amidst all
               the technical issues
alice_t:       so—for me—this "together" is in most cases same
               shared (expert-)wisdom... some form of dialogue
               and game between people searching for answers to
               their situation in the real world (or the parts,
               shared in this real world)
moni:          or reduced to some drug-fantasy-world x!x borut
               (edis@experiment.radiostudent.si) has joined
               #xchange
alice_t:       no problem...—i like to see all these interpre-
               tations and ways to reflect this whole field...
moni:          and it's interesting to take a while to test
               certain points of view...
borut:         hi everybody
borut:         continue...
alice_t:       yes—playing = experimenting from different
               points of view...
moni:          yes, for me this is all asking
alice_t:       yes—for me TOO
moni:          it's about knowledge and transcendence of knowl
               edge ...
alice_t:       eXchange and transformation of (knowl)edge...
moni:          a friend said: if the transcendence of knowledge
               is in danger, that's the only reason one could
               kill for
```

will try to get themselves off. But of course they did not save the automatically generated message that explained how to do this, and the intervening silence has removed any sense of concern for the well-being of the list, so they do it by sending messages to the whole list. This, of course, causes other people to do the same thing, whereupon someone tries to prevent this effect from snowballing by sending out a helpful, constructive message like "hey, you idiots! didn't your mama teach you anything? why don't you just unsubscribe by sending a message to <greeblex@blort.snort.com>?"

Internet discussion groups can work well despite these dynamics, but only in special cir-

```
!!!!!!!!!!!!!!!!!!!!!!!!!!!!!!!!!!!
——%-UN-SUB-
SCRIBE——
!!!!!!!!!!!!!!!!!!!!
!!!!!!!!!!!!!!!!!!!!!——%-
UN-SUB-SCRIBE—
—
!!!!!!!!!!!!!!!!!!!!!!!!!!!!!!!!!!!
—
-%-UN-SUB-
SCRIBE——
!!!!!!!!!!!!!!!!!!!!!!!!!!!!!!!!!!!
——%-UN-SUB-
SCRIBE—
-!!!!!!!!!!!!!!!!!!
.•' •      f6-n••••-
n••••=| ••-!><!—-:-
: -::-/%o:-
:+%%%%%%%%
%%%%%%%%%
%%%%%%%%
%%%%%%——:-
: - :  : —
onMouseOver="self.
s    t    a    -
tus='.;;=======+==
=++==++++++=++=
++++
++++++++++++++
++++++++++++=+
.    .'; return
true">
>>.;>.>;.>S0o.>;/$$
$$$$$$$$$$$$$$$$
$$$$$$$$$$$$$.>>.
;>.>;.>>
—-:-:-::>S%C:-
:+$$$$$$%$$$$%$
$$$%$$$$$$$$$$$
$%$——:-:-::—
—-:-:-::ooCS:-
:+$$$$$$S$$$SCS
CS%S/oooCo8S+S
SCC$——:-:-::—
Title:
8lient:
8esigners:
8rovider:
8oftware Used:
```


8umber of pages:
8umber of Links:
8ountry of Origin:
8ype of
Site:onMouseOver=
"s e l f . s t a -
tus='.;;=======+==
=++==++++++=++=
++++++++++++++
++
++++++++++++++
=+   . .';
return true">
8urpose   of
site:;;.;;.;;,.;:o++//o00
0000000000000000
0000000000000+/;.;
;.;;.;;
— - : - : - : -
CSo0$$$$$$$$%$
%$$%$$$$$$$$$$
$$$$%$$$$$$$—:-
:-::—
;;.;;.;;.;:0$o0$$%ooS
oCSSCooS0oC0SC
CS8CCSC0o$8.>0$
$;.;;.;;.;;
— - : - : -
:oooSC0$$$$$$$oo
CCSC088CSSCoC0
C%/So$$Co8+$$—
:-:-::—
>>.;>.>;./>//0$$%$8
88%$08%8%%808
0S%0$800%8$8;/
S$$>.;>.>;.>>
— - : - : - :: — : -
0$$0$0SS0$S00S0
000S00S0$8000C%
$$$$$$$—:-:-::—
— - : - : - :: — : -
0$$$$%oSo%CCS
SS$CSCoSSo%CS
CSS/$$$$$$—:-:-
::—
;;.;;.;;.;.;;0$$$$$$$$
$$$$$$$$$$$$$$$
$$$$$$$$$$;.;;.;;.;;
— - : - : - :: — : -

moni:	oh, not spiritual...;)
moni:	but today knowledge is capital, in a materialis-tic sense too
moni:	so not so many people want to share it..
alice_t:	yes... but re-spiritualisations as a PR-gag is coming...
moni:	in the art-world?
baker:	in advertising and marketing
alice_t:	hmmm—the art world will copy it a few months later and make it into art ;-))
moni:	:)
moni:	well, i'm not following this all too attenti-valy, have to learn programming ;)
moni:	but in the usa it's hip.since a few years, i think?
moni:	so it's just switching and swapping over here now?
moni:	or do you see a special reason for it?
moni:	it
alice_t:	no-i probably also have my spiritual sides—but i don't like to play with them in public situa-tions... probably (again) they are too fundamen-tal/important to me to do so...
alice_t:	...but you should see them in every output—just "inscribed"
moni:	for me it always depends on many things, some times it's simply a nice image, sometimes the possibility to communicate very different things, sometimes provocation or comment ...
moni:	spirituality ?

(x: —> circle: berlin-ljubljana-riga-graz-london)

<moni>	we have circle...
<moni>	does anybody want to laugh?
<borut>	don't lose the circle!!!
<borut>	magic circle
<moni>	:)
<moni>	borut, a bit of laughter...
<borut>	black magic?
<alice_t>	tomorrow—when i have to get up from bed like

cumstances. For example, it helps if the community on the list has a steady stream of external events to react to. Since the list operates in a mostly reactive mode, they'll always have something to talk about. The sustained level of traffic might be high, but then people will leave the list until it settles down to a level that suits the people who remain behind. Another scheme that works well is to have a list which is oriented almost exclusively to one-shot announcements—but then that's not a discussion list anymore.

The point is, Internet discussion lists do not work very well. Often the problem, in my experience, is that people are being lazy: trying to

```
                  every day—probably ;)
<moni>            white magic, of course...
<alice_t>         colored bundles of cables somewhere deep in the
                  forest, under the sea and over montains...
<borut>           I don't think global is anything more than the
                  representation of the local
<borut>           anything else, I mean
<borut>           it may be global territorially—but we belong to
                  the same cultural segment
<borut>           segment = localityx
<rasa>            where is monika?
<borut>           monika is playing fiddle
<rasa>            so, let's meet next week!
<alice_t>         yes...
<alice_t>         i'm still getting a signal from her... so she
                  should still be alive ;-)
<borut>           concentrating, i guess
<borut>           playing nostalgic songs, and feedback
<borut>           the birdies have her
<alice_t>         nostalgia... i want to follow her ;-)

.....................end of irc.re-lab.net#xchange.......
```

XCHANGE-OPEN CHANNEL: co-broadcast experiments in the net
Raitis Smits

X-Open Channel started its co-broadcast experiments in the beginning of 1998.
It soon developed into a platform for live streaming experiments in the net—
exploring the feedback mechanism and possibilities for collaboration.
Every Tuesday night during net.radio OZOne live sessions the so called "open
channel" is announced. It means that everyone can join in the live session with
his/her RealAudio live stream. There are several possibilities for co-streaming.
(You can find more about "What and How to broadcast via the Net"—texts by
Borut Savski—in the net.audio magazine Acoustic.Space, or on the web
<http://www.radiostudent.si/mzx/netcasting.html>.)
The simplest one is to mix your sound source with another (one or more)
RealAudio live stream. In this case each of the participants is doing one part of
this live session (for example, one is streaming voice, another background music).
There one can listen to two (or more) different streams—the final one with all
transmissions mixed together or each "input"—and live stream separately.

set up a discussion list in order to avoid the hard work of building a community, agreeing on purposes and goals, establishing a structure and timetable, and so on. Often they rationalize this laziness by appealing to the libertarian ethos of the net: structure means constraint means domination. Lots of people believe that, but it's not true. It's not even true if you're a lib-ertarian: structure imposed from the outside may imply constraint and domination, but structure agreed from within a group through a legitimate consensus-building process should not. In my experience, though, lots of people who tend toward libertarian sentiments just talk about the virtues of association without actual-ly learning how to cooperate and build things

```
0$$$$$$$$$$$$$$$
$$$$$$$$$$$$$$$$$
$$$$$—:-:-::—
;;.;;.;;.,.;;0$$$$$$$$$
$$$$$$$$$$$$$$$$$
$$$$$$$$$$$;.;;.;;.;;
—-:-:-::—:-
0$$$$$$$$$$$$$$$
$$$$$$$$$$$$$$$$$
$880$—:-:-::—
>>.;>.>;.>.;>0$$$$
$$$$$$$$$$$$$$$$$
$$$$$.;;$$$$$$0$S$>
.;>.>;.>>
—-:-:-::—:-
0$$$$$$$$$$$$$$$
$$$$$$$$$$$$$$$$$
$S00$—:-:-::—
—-:-:-::—:-
0$$$$$$$$$$$$$$$
$$$$$$$$$$$$$$$$$
$$$$$—:-:-::—
—-:-:-::—:-tton
down until a me-
908-u comes up,
and mov-
           uot
           pvo
           -,e
           % r%
           Sa S
           anta
           vdhv
           e ee
           m
           topt
           hvih
           ieci
           s ts
           tu
           ihri
           meem
           a ,a
           gc g
           euhe
so,onMouseOver="s
e l f . s t a -
tus='.;;=======+==
```

Another interesting experience of co-streaming is creating the loop. Each broadcaster takes another's live stream, re-encodes it, and sends it on for the next participant. In this loop, sound input is going around and coming back with a little delay (5-10 seconds) creating multiply sound layers. If sound keeps travelling around, the stream gets more and more noisy, and finally turns into one continuous noise (depending, also, on the amount of participants). Another way of using the loop-connection is to cut down the feedback—for example, it can be used for remote interviews and discussions, news exchange, etc.

I believe there are many more possibilities for live transmission experiments in the net, but these are some basic principles we have experienced during the X-Open Channel live sessions.

```
 |   |   |   |   |   |   |   |   |   |   |   |   |   |   |
(a) (c) (o) (u) (s) (t) (i) (c) ( ) (s) (p) (a) (c) (e)
 |   |   |   |   |   |   |   |   |   |   |   |   |   |   |
```

information&communication channel | for net.broadcasters
http://xchange.re-lab.net (Xchange) net.audio network
xchange search/webarchive: http://xchange.re-lab.net/a/

SUBJECT: REWIRED COMPILATION

FROM: DAVID HUDSON <DWH@BERLIN.SNAFU.DE>
DATE: MON, 19 OCT 1998 22:45:51 + 0200 (MET DST)

Date: Wed, 7 Jan 1998 11:28:35 -0800
To: rewired@rewired.com
From: andrew sullivan <andrew@eline.com>
Subject: Active Maastricht Screens

>>WHERE is the German zine ?!!
>
>WHO is asking? (for it?)
>
>Micz Flor (micz@metamute.com)

For me, that exchange really sums up a key difference in the way that Euros and Americans tend to view net development; it reminded me of a public post on the with real, live other people. This spirit of politically noble laziness is dragging down the Internet.

In fact, the people who helped me articulate these phenomena work mostly with kids. Mike Cole <mcole@weber.ucsd.edu> and Olga Vasquez <ovasquez@weber.ucsd.edu> in my old department at UCSD, for example, run after-school computer clubs for kids. They discovered early on that you can't just provide a bunch of computer activities and helpful college students and tell the kids of have fun and learn lots. Instead, you need to provide a structure of some kind that is intrinsically rewarding and offers a sense of where you cur-

Rewired message board by Richard Barbrook, who was quite the item at the time of his post and acting spokesperson for Europe on the state of net.culture. The quote's a bit out of context, since he is addressing a more fundamental difference in approaches to development and not the specific issue of web publishing and net.culture, but I believe that his perspective is one that is unfortunately common and detrimental—at least among the Euros I've known who follow and use the net daily. Comments like the following below by Barbrook do damage precisely because there is no reason that europeans have not taken a more aggressive role in "owning" the net—or at least complaining about reasons that prevent them from doing so. And rather than have some of the more outspoken figures act as driving forces in defining a new direction, we have only the comment that the American way is stupid precisely because so many are moving without thinking first—I could almost sense the academic invoking the myth of Prometheus.

So, while Barbrook and those following his approach to developing a european net identity sit back and think about the best possible scenario, I wonder if he ever stopped to ask himself which scenarios—as dumb as they might be—were defining his options and the forces that may or may not limit them. For those models will be the ones that work, and there's only one way to find out if something works.

http://www.rewired.com/Board/Messages/12.html

Unlike the Californian ideologues, we don't have any
easy and simplistic solutions for how the hypermedia
industry should be developed. However, the first step
towards finding a way forward is to try to understand
how really existing capitalism is evolving, rather than
relying on the idiocies of neo-classical economic texts.
It is better to ask intelligent questions than to give
stupid answers!

All the best.
Richard
Hypermedia Research Centre
http://www.hrc.wmin.ac.uk/

I think you should take this even further, however, and address how the net has really transformed an entire segment of the U.S. economy—whether we like it or not. It took years to develop the physical and intellectual infrastructure to make

rently are in a larger picture. So, for example, each computer program comes with an activity sheet—an actual sheet of paper with easy, medium, and hard challenges for using the program. Also, the kids are constrained in which programs they can use by a floorplan through which they move a game piece (a "creature"): when they do well at one program, they get to move to an adjacent "room" of their choice. Now, some people will say that this is more grown-up domination of kids. I say that kids need friendly, flexible structures to scaffold their development. If you think you can get kids learning real stuff in a totally unstructured environment, you go ahead and do it. Let us know when you succeed. We'll stop by and have a

```
                            lu
                             s
                            ae

mponMouseOver="
s e l f . s t a -
tus='.;;=======+==
=++==++++++=++=
+++++++++++++++
+++++
++++++++++++=+
.    .'; return true">
                            eo
                            ni
                            un
                             t
                            ce
                            or
                             m
                             e

                             /
```

```
#else
typedef  #select#
#select#UPP;
# d e f i n e
Call#select#(userRo
utine, data)        \

(*(userRoutine))(data
)
# d e f i n e
New#select#(userR
outine)              \

(#select#UPP)(userR
outine)
#endif  |||||||
         |||||||        |||||||
|||||||    |||||||       |||||||
         |||||||        |||||||
|||||||    |||||||       |||||||
         |||||||        |||||||
|||||||    |||||||       |||||||
```

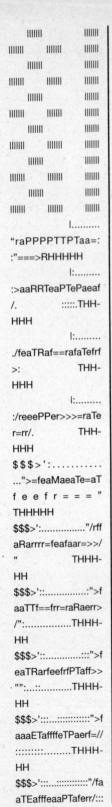

```
"raPPPPTTPTaa=:
:"===>RHHHHH
          l:.........
:>aaRRTeaPTePaeaf
/.       :::::.THH-
HHH
          l:.........
./feaTRaf==rafaTefrf
>:          THH-
HHH
          l:.........
:/reeePPer>>>=raTe
r=rr/.      THH-
HHH
$$$>':............
...">=feaMaeaTe=aT
f e e f r = = = "
THHHHH
$$$>':..............."/rff
aRarrrr=feafaar=>>/
"        THHH-
HH
$$$>'::...............:">f
aaTTf==frr=raRaerr>
/":.................THHH-
HH
$$$>'::...............:::">f
eaTRarfeefrfPTaff>>
.""::..::...........THHH-
HH
$$$>'::::...............">f
aaaETaffffeTPaerf=//
:::::::::::.........THHH-
HH
$$$>':::::...............:"/fa
aTEafffeaaPTaferr/:::
```

this "high-tech" economy work (the way it works), and, in large part, it works on the backs of the freelance community here—perhaps an unfortunate result of the neoclassical "tests" to which Barbrook refers.

I don't think many Europeans get that. That is, how many of the wage-earners in this business do not have jobs, only a brief contract. While this is probably not the direction most of the people on this list would like to see any economy take, I think Europeans gave up on the best known alternative—government having a pretty firm grasp on the economy—when they ratified Maastricht. If Europeans do not see themselves in the bottom of the neoclassical test-tube now, they're in for a rude awakening.

In short, I think one of the reasons that no one, as Micz suggests, is calling for the German zine is that they might just be content with watching the Americans do it where they do it best: on the screen.

Now the French, that's a different story. I'm sure that they are behind XGML!

Andrew Sullivan, eLine | REWIRED
330 Townsend St. #220 | "AA For The Web"
San Francisco, CA 94107 | http://www.rewired.com
415.543.0760 fax -0761 |
http://www.eline.com |

The Rewired List was started in December 1997 so that a relatively manageable group could speak informally about the issues brought up by the online zine, Rewired (<http://www.rewired.com>). As it turns out, because the approximately forty subscribers are more or less evenly divided between Europeans and U.S. Americans, conversation more often centers on cultural and political comparisons and contrasts between the two continents. eLine Productions in San Francisco runs the list, and Berlin-based David Hudson moderates.

---

look, and ten bucks says that you're actually training the kids to obey a whole range of hidden control trips while pretending to be free and spontaneous.

Margaret Riel <mriel@weber.ucsd.edu> has done similar things on a larger scale over the Internet with networks of teachers across the globe. They don't just connect the kids by e-mail to scientists at the South Pole: first they set up a whole elaborate curriculum, covering several topics from math to science, to literature, so that the children have read and written and talked and listened about the South Pole for weeks, comparing notes with one another as they hit the library and type in their work. All

SUBJECT: RHIZOME

FROM: RACHEL GREENE <RACHEL@RHIZOME.ORG>
DATE: FRI, 23 OCT 1998 19:09:05 - 0400

RHIZOME <http://www.rhizome.org>

Each time that technology, subjected to certain cultural imperatives, ceases to be that which we expect of it, then art, always victorious, defends itself by inventing new tools. At the margins of the art world is new media art. Here, art massively disengages itself from mainstream practices in order to find its own space. New media art is a nomadic space, a kind of rupture. All else will not be art.

RHIZOME maps this territory by publishing and indexing a wide range of art discourse. Started in early 1996 as an experiment in democratic, community-driven discussion, RHIZOME is proud to be an organization on the underbelly of the American art media. Reaching thousands of people every week (artists, academics, students, housewives and the heartbroken), RHIZOME is a channel for critical writing, chatter, email art, and also the self-promotional emails that may eventually provide important groundwork for new media art history. Publishing these sometimes banal, sometimes personal, sometimes critical rivulets of data, one might consider RHIZOME the pre-eminent tabloid publication of the new media art community.

Modeled on the form of the "rhizome"—a nonhierarchical, living network without center—RHIZOME's charter is to be part community center, part art magazine. While RHIZOME is explicitly interested in developing a critical vocabulary to discuss new media art, and in issues relating to technology, culture, and politics, it is intent on addressing these topics in ways understandable to those who don't read Lacanian diagrams or speak English as a first language. RHIZOME is user-powered; it is a bottom-up, free media where the users are the authors are the readers. We develop new forms and map new spaces. The RHIZOME robot sends email, filters and indexes texts, archives information...and replicates itself into the next millennium. Subscribe to RHIZOME for a look at the current state of new media art.

---

of this structure means that everybody knows where they are going, everybody is ready for what happens next, and the whole activity has a natural point of closure.

What the Internet needs is a vocabulary of structures for e-mail discussion lists. Nobody should bother creating a list until they have a good reason for it that everybody has signed onto. This will mean doing some consultation, building consensus, and accepting that communities take time to grow. It will also mean having a definite goal and structure for the list, including a statement of the conditions under which the list will have achieved its purpose and be shut down. Of course, nobody should

```
u">'^$$$$$$$$E
E

$$$>'            '
'!!?`*@f"        ~.<:'
~!!!?k'!T$$$$$$$E
E

$$$>'            ''u4+
u $ $ * * $ $ $
^+='L$$$$$$$$E
E

$$$>'         ''N$$!
T L . . / * # R R `' ~
<@UTW$$$$$$$E
E

$$$>'         '"$$=
. ^ ' 7 C : # + 2 " '
*$*$$$$$b."E   E

$$$>'         '$**+
''^        4F$$$$&
E    E

$$$>'         ' '  `
:         ~TT$$/
E    E

$$$>'         ''Bc=
"          . =B6+'
" E    E

$$$>'      4 :: '#($
~!X!  E    E

$$$>'     *'''$dL"
474me>E    E

$$$>'             '
' L $ $ $ d * * @ i .
'~-!?E?E    E

$$$>'             '
'$$$@(.xW$$$k
.uWW$$$$$$$N$N
WWbuW&E    E
```

Alex Galloway <alex@rhizome.org>
Rachel Greene <rachel@rhizome.org>
Mark Tribe <mark@rhizome.org>

Subject: Syndicate [the Nettime mix]
From: Erik Kluitenberg <epk@xs4all.nl>
Date: {}

Date: Fri, 1 Mar 1996
From: abroeck@v2.nl (Andreas Broeckmann)
Subject: V2_East / Syndicate Newsletter 96/02

INTRODUCTION

This is the second Syndicate Newsletter. In the first sections there is some information about how the list/network is taking shape. Some people have submitted information for distribution through this channel, and we want to invite everybody to do the same—either by posting stuff directly to <syndicate@aec.at> or for inclusion in the next newsletter (no later than end March 96, but earlier if a lot of material comes in) to <abroeck@v2.nl>. Any information that is of interest to the media art community in East and West Europe, from the dates of your forthcoming events to strategies for winning sponsors and grants, is welcome. Also, the submission of your own FAQs (Frequently Asked Questions) would be useful. The V2_EAST WEBSITE where all this and more information will be collected for reference is under construction. Please, invite other people who might be interested in the Syndicate to subscribe to the list.

Best wishes, and see you soon, Andreas Broeckmann
(V2_East)

Date: Tue, 5 Aug 1997
From: kitblake <kitblake@v2.nl>
Subject: Syndicate: Deep Europe Visa Department

force people to run their lists this way. But it would be most excellent if decent standards could be established within which people can create software to support such things. Sure, plenty of companies sell conferencing systems to organizations whose people are required to do things together. But that doesn't mean that those people actually go through the social processes needed to use the systems at all productively, and it certainly doesn't mean that the benefits of those systems become widespread on the Internet.

A lot of the problem, then, has to do with technical standards and the like. But the problem is also cultural. Many people have lost, or never

DEEP EUROPE VISA DEPARTMENT

Intro: The Syndicate convened again at Documenta X in Kassel. Its members form a distributed community, initiated two years ago as a network of people who stay in touch through an internet mailing list. They share a common interest in media cultural developments in Eastern Europe, and the loose goal of the Syndicate is to further cross-pollination and synergy/support between East and West.

It's interesting when you meet somebody whose words you know but whose face you've never seen. In "normal" encounters, you see someone, sense their personality, and perhaps probe their thinking. In a distributed community, you already know their thoughts, so when finally face to face you explore the person. It makes for a social scene.

VISA DEPARTMENT

On Saturday we produced an event, the deep europe Visa Department. The name, deep europe, was invented for this workshop in the Hybrid WorkSpace at documenta X ("dX"), and must be taken with a grain of salt. But most of the participants are from the East, and that is another Europe. It's across the BORDER, and residents on the other side are not E.U. citizens. They must apply for a visa to visit. For Germany, for instance, the application costs 50DM, for England 100DM. And you may not get it. You have to wait. You have to answer questions. "Do you have any transmittable diseases?" "How much money are you bringing?" "What is this organization that's inviting you?"

It's a different world. When you're sitting at your dining table, and you hear the bra-a-a-t of a Kalishnikov on the other side of the wall, well, you "sort of" get used to it. As Edi Muka said, like you "sort of" get used to a rollercoaster ride. Obviously, living in an environment like that means your media addresses certain issues, and those projects are the focus of the Syndicate.

Preparation for the deep europe Visa Department integrated with the other activities. Flyers were made and spread around Documenta. They invited everybody to a performance and party on Saturday night, and to come apply for a visa to deep europe between 2 and 6. A deep europe logo was created, taking a cue from the dX "d", integrating it with an "e", and adding an accent, an Eastern inverted caret character. This was used in documents, stamps, signs, and badges.

Forms were created. They were written in Albanian only, with no translation; they

learned, the skills for working together. Although the 1960s counterculture is out of fashion now, it put a *lot* of effort into learning how to build community, how to organize and empower people, how to run things democratically, how to fight fair, and how to be a powerful human being without having to exercise power over other people. In my opinion, the net needs these skills badly. And so does the rest of the world. People who believe in liberty ensure an authoritarian world unless they teach people how to organize themselves through their own efforts, and the problem of using the net productively might be an occasion to rediscover this.

```
$$$$>'              '
'$$$$$NL'9$$Ce$$$
$$>  "   . ^'  ="7E
E

$$$>'              '
'$$$$.$$$R:$$$$R#(
`           E    E

$$$>'              '
'$$$$$$C6@$$$""
E     E

$$$>'              '
'$$$$$$$$$"F!
E     E

$$$>'              '
'$$$$$$$$$2+
E

$$$>'        ':%@
'$$$$$$$$$B~
E

$$$kx               '
'$$$$$$$$$H
E      u!....
.  <>....  ..  ...
$$$>'              '
'$$$$$$$$$*
E      .9$'>$
*9"7  '>'>$ E9 $'>$
E"  '>$ "b
$$$>'              '
'$$$$$$$$$X
E
!"
$$$$$               '
'$$$$$$$$$$"
E
E...........................
...........................
$$$$$:::.:::::::::"/f
aaTEafffeaaPTaferr/:
:::::::::::::
$$$$$:::::::::::::::"/=
aaaTerr=>=reTTaaer
r/::::::::::::::::
```

$$$$$:::::::::::::":/rff-
feaaffr====rrfeaTaar
r=r==>"""""/>=rffffr
$$$$$:::::::::">faa-
effeeffrrrrr======rfr
rrr=aTTTTTaPPTaa
TTaTa
$$$$$::::::::>faTTTaf
faefffrrrrr===rrr====
rr==TTTPPETT<!-
Lowest TV skill
font.-!>
$$$$$::::::""faTPPTa
frfTeffffffrr==rffrr===
=rrreTTPPEPT<!-
Fonts kill lowest
TV.-!>
$$$$$:"""""""eTPPP
PTeffePefffeefffrrfffr=
=>>=fa=rTPPPMPT
<!-Soft! well-knit
volts-!>

[7-11.org]
***Main Index.page
will soon reach
+1000 Msg

***thdwqdat's
qeMu qecqe
hoe q//

***
%Zoom+++++LOC
King(.dom)
onTARGET= ""The
MAILINGLIST""

MUST KILL
% C O N T E N T
MUST KILL %CON-
TENT MUST KILL
%CONTENT
MUST KILL %CON-
TENT MUST KILL
% C O N T E N T
MUST KILL %CON-

asked the usual questions. Erasers and potatoes were carved into stamps, and various colored ticket books were found. From the dX participant nametags, badges for officials were made by overlaying a laserprint with a window cut out so the photo would show through. Of course, the deep europe logo was on the badge, and in a technofascist typeface was the word "Absardze": this is Latvian, and it's a new, thus obscure word, which means "guard" or "control." Throughout the event, hardly anybody, even from the deep europe group, knew what it meant. Which means it was perfect.

A soundtrack was put together. Rasa Smite (<http://ozone.parks.lv/Xchange>) pulled a bunch of audio off the net, including some military song from Edi Muka's video-performance project, a sort of Donnau anthem, and this became the basis for the mix. Analog noise was filtered in, to mimic a bad sound system. This manic march was played—loud—during the proceedings. At various intervals an announcement was woven in. This was usually in some unintelligible East European language. A series of barked commands in Albanian, or Serbian instructions that may or may not apply to you. Once in a while some English, "Please be patient," and eventually a longer one, "May we have your attention please. If your visa permits entrance for more than one day, you may be required to take a blood test." This one bit of understandable information then faded away, "Blood tests are conduc...." The manic march paraded on. Throughout Saturday afternoon it looped continuously.

At the entrance to the event, Alexandar Davic and Michiel van der Haagen set up a video surveillance camera—one of those CU-SeeMe eyeballs, it stared down the crowd. Also present was a microphone to pick up the crowd's mutterings. The signal was displayed on a monitor near the door, with a distracted Absardze sitting there not watching it. Other material was shot with a HandyCam, and this will be combined with, naturally, the manic march for a soundtrack, into an event compilation.

The walls surrounding the entrance made a kind of banked curve the visitors had to follow, lined with tables, forms, and officials. One Absardze in supershades managed the door, letting people in two by two. The process applicants had to follow was typical mind-mushing bureaucracy. Little translation was provided, and forms had to be filled out correctly. Iliyana Nedkova: "Oh, you have a yellow ticket? You have to go to that table over there and get a green one." And fill out a form. Marjan Kokot: "Green ticket? Here's the form." In a language few people can read. One Absardze was sitting at his desk looking bored, reading a magazine, a

[Originally forwarded through the Red Rock Eater News Service (RRE). For more information, see <http://dlis.gseis.ucla.edu/people/pagre/rre.html>.]

CLOSED sign in front of him. At another point Lisa Haskel brought in these giant bratwursts, and the Absardzes stood around munching, ignoring the crowd. Forms were stamped and double stamped, sometimes with a coffeebreak in between. The march looped on.

The amazing thing was the queue that formed. It started growing just before the opening, and in a short time went all the way down the block. Some people were in line for over half an hour. It started to rain, and they stood there under umbrellas. All this to get a worthless piece of paper with a potato stamp on it.

For the most part, the audience liked it. They got it. They followed the procedures, and left with a visa to deep europe. Even distant foreigners, like Japanese with little English and nothing else, took it seriously and seriously enjoyed it. You may not know the language, but you recognize the bureaucracy.

There were some negatives. One older German man, certainly around since the war, listened to chainsmoking Branka Davic's explanation, and when he realized it was a visa application, threw it in her face.

At five before six Absardze Andreas Broeckmann went out and announced to the crowd that the Visa Department would close in five minutes. At six the doors slammed shut, and twenty minutes later there were still a dozen people in a queue to nowhere.

ENTERING DEEP EUROPE

That evening, Hybrid WorkSpace hosted a performance/party. Heading the bill were the Instituut voor Betaalbare Waanzin (Institute for Affordable Lunacy). Their performance merged into a visceral mastermix, blending Latino dance tracks into Rotterdam GabberHouse. "This is the music our children listen to!" Thump, thump, thump, thump.... Ongoing video flickered on the walls, and the bar was fully stocked. It was a good party.

Visitors streamed in, clutching their visas. There were a few Absardze badges floating around, but no guards, no border, no control. Welcome to deep europe.

TENT
MUST KILL %CON-
TENT   MUST KILL
% C O N T E N T
MUST KILL %CON-
TENT
MUST KILL %CON-
TENT   MUST KILL
% C O N T E N T
MUST KILL %CON-
TENT
MUST KILL %CON-
TENT   MUST KILL
% C O N T E N T
MUST KILL %CON-
TENT
MUST KILL %CON-
TENT   MUST KILL
% C O N T E N T
MUST KILL %CON-
TENT
MUST KILL %CON-
TENT   MUST KILL
% C O N T E N T
MUST KILL %CON-
TENT
MUST KILL %CON-
TENT   MUST KILL
% C O N T E N T
MUST KILL %CON-
TENT
MUST KILL %CON-
TENT   MUST KILL
% C O N T E N T
MUST KILL %CON-
TENT
MUST KILL %CON-
TENT   MUST KILL
% C O N T E N T
MUST KILL %CON-
TENT
MUST KILL %CON-
TENT   MUST KILL
% C O N T E N T
MUST KILL %CON-
TENT
MUST KILL %CON-
TENT   MUST KILL
% C O N T E N T
MUST KILL %CON-
TENT
MUST KILL %CON-
TENT   MUST KILL
% C O N T E N T

." Ä   f6-n˜,-n˘Ê=l   ˘‡-!><!—-:-:-::-/%o:-
:+%%%%%%%%%%%%%%%%%%%%
%%%%%
% % % % % % % — — : - : - : : —
o n M o u s e O v e r = " s e l f . s t a -
tus=',;;=======+===++==++++++=++=++++
++++++++++++++++++++++++++++++=+
    .'; return

true">
>>.;>.>;.>S0o.>;/$$$$$$$$$$$$$$$$$$$$$$$$
$$$$$$$$.>>.;>.>;.>>
— - : - : - : : > S % C : -
:+$$$$$$%$$$$%$$$$%$$$$$$$$$$$$$%$—
—-:-:-::—
— - : - : - : : o o C S : -
:+$$$$$$S$$$$SCSCS%S/oooCo8S+SSCC$

People folded their visas, and put them in a pocket.

Date: Thu, 21 Aug 1997
From: Aleksandar and Branka Davic <spiridon@eunet.yu>
Subject: Syndicate: BELGRADE / HAPPY BIRTHDAY MR. PRESIDENT

YESTERDAY, on August 20, Belgrade students organized an action in order to give a "stafeta" [a statuelike stick traditionally given by Yugoslav youths to Tito on his birthday] to President Slobodan Milosevic on his 56th birthday. Of course, the action has been inspired by the same ritual from past decades, when on May 25 the whole country celebrated Tito's birthday. There was also one more "joke"—"stafeta" started at 15:05 (the official time of Tito's death)

POLICE forces stopped the students, violently as usual, and three students were beaten.

"Happy" Birthday Mr. President!!!!!!

branka

From: Vuk Cosic <vuk@kud-fp.si>
Date: Wed, 12 Nov 1997
Subject: Syndicate: the cosic test

INTRODUCTION

In Dessau there was a Syndicate meeting, the tenth or so since the Next 5 Minutes 2 conference in (January 1996, Amsterdam), and there was talk of practical things...

Actually, there were two meetings. Near the end of the first, Tapio Makela mentioned a classical ambition—how about a Syndicate website? (It would give useful info and pointers to further sources regarding euro funding, various art and media houses/meetings/conferences, plus a possible text zone to satisfy the need for theory and debate...I presume)?

——:-:-::—Title:	8ype of
8lient:	S i t e : o n M o u s e O v e r = " s e l f . s t a -
8esigners:	tus=',;;=======+===++==++++++=++=++++
8rovider:	++++++++++++
8oftware Used:	+++++++++++++++=+ .';
8umber of pages:	return true">
8umber of Links:	8urpose of
8ountry of Origin:	site:;;.;;.;;.;:o++//o00000000000000000000000

Well, in homage to the immortal Turing test aimed at recognizing a specific kind of intelligence, I have decided to engage in a similar "Cosic test" of activism. Here follows a short description of the test:

COSIC TEST

The Cosic Test is aimed at deeper understanding of group motivation, and is structured in such a way as to enable a singular researcher to perform it alone, although assistance can ensure more accurate and fast measuring. This methodologically complex epistemological strategy consists of two main parts: (a) I talk, and (b) I wait.

(a) In the "I talk" part, the task of the researcher is to offer a profiled collaborative project to a group of declared activists, with the invitation to meet outside of the conference hall after the given meeting and talk of direct action. It is important that the project offered is of maximum usefulness to the goals declared at the meeting.

(b) In the "I wait" part, the researcher has to go out after the meeting and stand there for about fifteen minutes until every meeting-participant has left not only the conference hall, but also the lobby.

DESSAU RESULTS

One person approached me in the hall and about seventy passed me by. After analyzing the profile of the enlisted collaborator, it became clear that this person is the K.I.E.Z. technician and is not subscribed to the list. Therefore he was not acknowledged as relevant to the analysis.

bingo
vuk

Date: Mon, 27 Apr 1998
From: Tapio Makela <tapio@projekt.net>
Subject: Syndicate: report fragment 1 from Stockholm

Dear Syndicated,

I am writing this brief and partial report from Stockholm, "The Shaking Hands and Making Conflicts" event; Andreas Broeckmann and others will continue. The

00000000+/;.;.;;.;;
— - : - : - : - -
CSoO$$$$$$$$%$%$%$%$$$$$$$$$$$$$$%$
$$$$$$—:-:-::—
;;.;;.;;.;;0$o0$$%ooSoCSSCooS0oC0SCCS8C
CSCOo$8.>0$$;.;;.;;.;;
— - : - : -
:oooSC0$$$$$$$$ooCCSC088CSSCoC0C%/S

o$$Co8+$$—:-:-::—
>>.;>.>;./>//0$$%$888%$08%8%%8080S%0
$800%8$$8;/S$$>.;>.>;.>>
— - : - : - : - : - -
0$$0$0SS0$S00S0000S00S0$8000C%$$$$$
$$—:-:-::—
- : - : - : - : - - : -
0$$$$%oSo%CCSSS$CSCoSSo%CSCSS/$$

SCRIBE—-
-
!!!!!!!!!!!!!!!!!!!!!!!!!!!!!!!!
——%-UN-SUB-
SCRIBE——
!!!!!!!!!!!!!!!!!!!!
!!!!!!!!!!!!!!!!!!!!!—-%-
UN-SUB-SCRIBE—
—
!!!!!!!!!!!!!!!!!!!!!!!!!!!!!!!!
—
-%-UN-SUB-
SCRIBE——
!!!!!!!!!!!!!!!!!!!!!!!!!!!!!!!!
——%-UN-SUB-
SCRIBE—-
-
!!!!!!!!!!!!!!!!!!!!!!!!!!!!!!!!
——%-UN-SUB-
SCRIBE——
!!!!!!!!!!!!!!!!!!!!
!!!!!!!!!!!!!!!!!!!!!—-%-
UN-SUB-SCRIBE—
—
!!!!!!!!!!!!!!!!!!!!!!!!!!!!!!!!
—
-%-UN-SUB-
SCRIBE——
!!!!!!!!!!!!!!!!!!!!!!!!!!!!!!!!
——%-UN-SUB-
SCRIBE—-
-
!!!!!!!!!!!!!!!!!!!!!!!!!!!!!!!!
——%-UN-SUB-
SCRIBE——
!!!!!!!!!!!!!!!!!!!!
!!!!!!!!!!!!!!!!!!!!!—-%-
UN-SUB-SCRIBE—
—
!!!!!!!!!!!!!!!!!!!!!!!!!!!!!!!!
—
-%-UN-SUB-
SCRIBE——
!!!!!!!!!!!!!!!!!!!!!!!!!!!!!!!!
——%-UN-SUB-
SCRIBE—-
-
!!!!!!!!!!!!!!!!!!!!!!!!!!!!!!!!
——%-UN-SUB-

SCRIBE——
!!!!!!!!!!!!!!!!!!
!!!!!!!!!!!!!!!!!!!!——%-
UN-SUB-SCRIBE—
—

!!!!!!!!!!!!!!!!!!!!!!!!!!!!!!!!!!!!!
—
-%-UN-SUB-
SCRIBE——
!!!!!!!!!!!!!!!!!!!!!!!!!!!!!!!!!
——%-UN-SUB-
SCRIBE—-
-

!!!!!!!!!!!!!!!!!!!!!!!!!!!!!!!!!
——%-UN-SUB-
SCRIBE——
!!!!!!!!!!!!!!!!!!
!!!!!!!!!!!!!!!!!!!!——%-
UN-SUB-SCRIBE—
—

!!!!!!!!!!!!!!!!!!!!!!!!!!!!!!!!!
—
-%-UN-SUB-
SCRIBE——
!!!!!!!!!!!!!!!!!!!!!!!!!!!!!!!!!
——%-UN-SUB-
SCRIBE—-
-

!!!!!!!!!!!!!!!!!!!!!!!!!!!!!!!!!
——%-UN-SUB-
SCRIBE——
!!!!!!!!!!!!!!!!!!
!!!!!!!!!!!!!!!!!!!!——%-
UN-SUB-SCRIBE—
—

!!!!!!!!!!!!!!!!!!!!!!!!!!!!!!!!!
—
-%-UN-SUB-
SCRIBE——
!!!!!!!!!!!!!!!!!!!!!!!!!!!!!!!!!
——%-UN-SUB-
SCRIBE—-
-

!!!!!!!!!!!!!!!!!!!!!!!!!!!!!!!!!
——%-UN-SUB-
SCRIBE——
!!!!!!!!!!!!!!!!!!
!!!!!!!!!!!!!!!!!!!!——%-
UN-SUB-SCRIBE—

event itself was very problematic, but I think it will prove to be very useful for thinking about the future of Syndicate...and how to react to neonationalist appropriations of specific cultural initiatives. (Wow, that language sounds like a strategy of war almost.)

Most of the event felt like a national performance of Sweden's role in the post Cold War Baltic-Belarus-Ukraine as the generous uncle Dala. (Replace Uncle Sam with Uncle Dala, Dala as Dala horse, the national symbol of Swedish traditional culture). Instead of having had interesting thoughts in the main program, it was about shaking rhetorics.

To put it simply: Looking at the surface, the event was a stage for old-fashioned politics and politicians to present the rhetoric of change without any concrete notion of what they mean by "power needs culture," "democracy," "diversity," etc. Whenever they were caught on their transparent and clumsy reasoning, they would say, like Marita Ulvskog, the Swedish Minister of Culture, that they said so to provoke, to make conflict. She used a quote from Machiavelli to rationalize cultural diversity...which was promptly criticized by Igor Markovic.

A recipe: Use democracy to claim there exists a homogenous "we", and encourage conflict to create a ready place for dissent, a place which is not discussed, but to which the dissent is dumped, sealed and packaged as a medal that the politicians can wear as signs of "tolerance."

The backdrop of the event is perhaps not so much cultural as it is economical and political. Sweden wants to launch a Partnership in Culture in the Baltic region, Belarus and Ukraine. The event was to promote "freedom of expression, cultural diversity, democracy, and common security in the Baltic region..." Why this combination of countries, this combination of goals? Does this event have anything to do with the fact that these countries are former Soviet areas that used to be "within the missile range"? Or with the fact that Swedish companies, especially telecom companies, are trying to gain big shares of markets in these countries? Isn't it a proven fact that social and cultural work paves way for favorable decisions in other fields? This critique does not mean that setting up programs that support cultural initiatives that rise from local needs and ideas, or that collaboration across borders would not be a high priority. that was what I thought this event would promote. The "audience" or the "guests," me included, were witnesses of this play, our names in the list of participants signs of our assumed agreement with the given agenda. This event will be one point in the curriculum vitae of the Swedish nation.

$$$$$—-:-:-::—
;;.;:.;;.;.;;0$$$$$$$$$$$$$$$$$$$$$$$$$$$$$$$$$$$
$$$$$$;.;;.;;.;;
— - : - : - : — : -
0$$$$$$$$$$$$$$$$$$$$$$$$$$$$$$$$$$$$$$$
—:-:-::—
;;.;:.;;.;.;;0$$$$$$$$$$$$$$$$$$$$$$$$$$$$$$
$$$$$$;.;;.;;.;;

— - : - : - : — : -
0$$$$$$$$$$$$$$$$$$$$$$$$$$$$$$$$$$$$$$880$
—:-:-::—
>>.;>.>;.>.;>0$$$$$$$$$$$$$$$$$$$$$$$$$$$$$
$$$$$$0S$$>.;>.>;.>>
— - : - : - : — : -
0$$$$$$$$$$$$$$$$$$$$$$$$$$$$$$$$$$$$$S00$
—:-:-::—

After World War II, the U.S. launched a program called the Marshall Plan, which was to establish economical, educational, and cultural activity in those areas that were "insecure" or under Communist influence. USIS (United States Information Service) centers in Sofia and Helsinki are examples, as are the Fulbright programs (which btw have been decreased in areas that are these days concerned stable). My question would be, whether this kind of thinking is the basis for the Swedish Partnership of Culture? The other Nordic Countries (Finland, Denmark, Norway, [Iceland less]) have similar economic and political interests in the Baltic region. On the other hand...this is most likely recognized in the Baltic region, and perhaps it is a good moment to utilize the willingness of the Nordic countries to invest into the cultural sector as well.

Partnership for Culture is like a Dala plan for culture, and at the same time a Trojan horse, a Dala horse, where a cultural carrier is not innocent but bears in its belly the geopolitical and economical interests of Sweden. The internet, an infotech, also act as such a Trojan horse. The organizers of the conference are making the assumption that the Baltic region needs "A mailing list for intellectuals." Either they imagine that there are only a few people who qualify, or they do not know anything about mailing lists, how they are formed, and how they can become useless... But, I don't want to say that their initiative should not also be reacted to in a positive way. Ando Keskkula and Sirje Helme from Tallinn are crafting a conference as a follow up—and I think you can set different terms for the interaction.

Igor Markovic was the best vocal critic and commentator during the event, and I look forward to reading his views of it. We witnessed terribly badly formulated speeches by the Swedish politicians, institutional self-praise by David Elliott from the Moderna Museet Stockholm, badly prepared sentences from the former curator of Documenta, Catherine David... (Igor, others, please continue from here...) and—to my mind dictatorial—moderation by Swedish Journalist Mika Larsson. (Btw, if she works on the future events in this series, I won't even want to get further emails about them!).

It was not only Larsson's way of suppressing voices and differences of opinion but the way the event was staged that got to me. I felt that it lacked respect for the visitors from Belarus and the Ukraine and the Baltic countries: if they were the subject of discussion, why then weren't they placed center stage? In this, Fargfabriken bears responsibility for the curatorial discourse, and perhaps for the overproduced TV talk show style of the event. In order to establish a dialogic space, the first con-

—
!!!!!!!!!!!!!!!!!!!!!!!!!!!!!!!!!!
—
- % - U N - S U B -
S C R I B E — —
!!!!!!!!!!!!!!!!!!!!!!!!!!!!!!!!!!
— — % - U N - S U B -
SCRIBE—-
-
!!!!!!!!!!!!!!!!!!!!!!!!!!!!!!!!!!
— — % - U N - S U B -
S C R I B E — —
!!!!!!!!!!!!!!!!!!
!!!!!!!!!!!!!!!!!—— % -
UN-SUB-SCRIBE—
—
!!!!!!!!!!!!!!!!!!!!!!!!!!!!!!!!!!
—
- % - U N - S U B -
S C R I B E — —
!!!!!!!!!!!!!!!!!!!!!!!!!!!!!!!!!!
— — % - U N - S U B -
SCRIBE—-
-
!!!!!!!!!!!!!!!!!!!!!!!!!!!!!!!!!!
— — % - U N - S U B -
S C R I B E — —
!!!!!!!!!!!!!!!!
!!!!!!!!!!!!!!!!!—— % -
UN-SUB-SCRIBE—
—
!!!!!!!!!!!!!!!!!!!!!!!!!!!!!!!!!!
—
- % - U N - S U B -
S C R I B E — —
!!!!!!!!!!!!!!!!!!!!!!!!!!!!!!!!!!
— — % - U N - S U B -
SCRIBE—-
-
!!!!!!!!!!!!!!!!!!!!!!!!!!!!!!!!!!
— — % - U N - S U B -
S C R I B E — —
!!!!!!!!!!!!!!!!
!!!!!!!!!!!!!!!!!—— % -
UN-SUB-SCRIBE—
—
!!!!!!!!!!!!!!!!!!!!!!!!!!!!!!!!!!
—-
- % - U N - S U B -

— - : - : - : : — : -	% r%	
0$$$$$$$$$$$$$$$$$$$$$$$$$$$$$$$$$$$$$$$	Sa S	
—:-:-::—	anta	
—-:-:-::—:-tton down until a me-908-u comes	vdhv	
up, and mov-	e ee	
uot	m	
pvo	topt	
-,e	hvih	

```
S C R I B E — —
!!!!!!!!!!!!!!!!!!!!!!!!!!!!!!!!
——%-UN-SUB-
SCRIBE—-
-
!!!!!!!!!!!!!!!!!!!!!!!!!!!!!!!!
——%-UN-SUB-
S C R I B E — —
!!!!!!!!!!!!!!!!!
!!!!!!!!!!!!!!!!!!——%-
UN-SUB-SCRIBE—
—
!!!!!!!!!!!!!!!!!!!!!!!!!!!!!!!!
—-
- % - U N - S U B -
S C R I B E — —
!!!!!!!!!!!!!!!!!!!!!!!!!!!!!!!!
——%-UN-SUB-
SCRIBE—
-
!!!!!!!!!!!!!!!!!!!!!!!!!!!!!!!!
——%-UN-SUB-
S C R I B E — —
!!!!!!!!!!!!!!!!!
!!!!!!!!!!!!!!!!!!——%-
UN-SUB-SCRIBE—
—
!!!!!!!!!!!!!!!!!!!!!!!!!!!!!!!!
—-
- % - U N - S U B -
S C R I B E — —
!!!!!!!!!!!!!!!!!!!!!!!!!!!!!!!!
——%-UN-SUB-
SCRIBE—
-!!!!!!!!!!!!!!!!!
```

Q-where did we end?

Q - t a r g e t
'_close_all_'
Q-sugest: put all
mesg into /7-
11/<some>/
Q-and think about
an auto_count thing
for
Q-rotating archive
after x mesg

dition is to respect the partners in this dialogue as equals, as subjects with their own voice, and to provide the space for them to express it. I felt this was deeply lacking, and, as such, the event cannot be a starting point for forming a network based on trust or crisscrossing shared interests.

My fingers and wrists are in poor shape for writing...and I need to take care of Polar Circuit applications (which have been really nice—thanks everyone who has sent one!), so I end my reporting here. The main entry point to understand Syndicate's role in the Stockholm event can be read from the manifesto that Andreas drafted based on the proposals of the whole family present in Stockholm. Melentie performed this text-in-action with brilliant style (he should be awarded with a viking helmet for fulfilling the role so well). It offers several proposals for any country that wants to reach cultural supremacy in the region of the Baltic-Belarus-Ukraine. The text is in the next mail, and I hope that other Stockholm visitors will take up from here and I will rest my case, or simply, fingers.

Date: Wed, 1 Jul 1998
From: Andreas Broeckmann <abroeck@v2.nl>
Subject: Syndicate: irrelevant statistics

jan 96 jun 98 / 30 months 300 subscribers from 39 countries of which 32 european countries 7 non-european countries

as though it mattered...

[arbitrary selection and unwarranted editing by Eric Kluitenberg <epk@xs4all.nl>]

```
ieci                  so,onMouseOver="self.sta-
s ts                  tus='.;;=======+===++==++++++=++=++++
 tu                   ++++++++++++++
ihri                  ++++++++++++=+          .    .'; return
meem                  true">
a ,a
gc g                       aol
euhe                       srda
```

```
::::::::::::::::::::::::::::::::recode:::::::::::::::::::::::::::::::
::::::::

Reply-To: mez <mezandwalt@wollongong.starway.net.au>
Subject: :::recode::: fix.shion in.cybor[ge].sert


_CY.BOR[G]E_
[BORGEDOM]
_____

_____

BOR[G]EDOM SHEET
_____

_____

"URGE.N[o]T"FOR      RE[ar].VIEW"P.LEASE      COM[a].MENT"PLEASE
RE[a]P.LY"PLEASE
RECY.KILL

_____

____
NOTES/COMMENTS:
_____

_____

cybor[g]e me[me]
{in.sert in.[pre]ten.sions}

how about a scibor[g]e?
a psybor[g]e? skyborg? slyborg?
*sigh*borg?
{the.rapists would just *lurve* that}


_____end             BOR[G]E_____initial.eyes
```

- l	ga
dyl	hv
oo	te
wuy	
nro	mt
u	oo
tr	u
oih	sd

deonMouseOver="self.status=';;

deonMouseOver="self.status=';;

>>NAME: ... 7-11
>>DATE OF BIRTH: 9-97
>>MODERATORS: ... NON
>>NO. OF SUB-SCRIBERS: ...71
>>HAPPIEST MEMORIES: ...711.org
>>SADDEST MEMORIES: ...Re:
>>HOPES FOR THE FUTURE: ...NO CONTENT

-%-UN-SUB-SCRIBE——
!!!!!!!!!!!!!!!!!!!!!!!!!!!!!!!!!!!!
——%-UN-SUB-SCRIBE——
-
!!!!!!!!!!!!!!!!!!!!!!!!!!!!!!!!!!!!
——%-UN-SUB-SCRIBE——
!!!!!!!!!!!!!!!!!!!
!!!!!!!!!!!!!!!!!!!!——%-UN-SUB-SCRIBE—
—
!!!!!!!!!!!!!!!!!!!!!!!!!!!!!!!!!!!!
—-
-%-UN-SUB-SCRIBE——
!!!!!!!!!!!!!!!!!!!!!!!!!!!!!!!!!!!!
——%-UN-SUB-SCRIBE——
-
!!!!!!!!!!!!!!!!!!!!!!!!!!!!!!!!!!!!
——%-UN-SUB-SCRIBE——
!!!!!!!!!!!!!!!!!!!
```
```

!!!!!!!!!!!!!!!!—— % -
UN-SUB-SCRIBE—
—
!!!!!!!!!!!!!!!!!!!!!!!!!!!!!!!!!!!!!
—
- % - U N - S U B -
S C R I B E — —
!!!!!!!!!!!!!!!!!!!!!!!!!!!!!!!!!!!!
—— % -UN-SUB-
SCRIBE—-
-
!!!!!!!!!!!!!!!!!!!!!!!!!!!!!!!!!!
—— % -UN-SUB-
S C R I B E — —
!!!!!!!!!!!!!!!!!!!
!!!!!!!!!!!!!!!!—— % -
UN-SUB-SCRIBE—
—
!!!!!!!!!!!!!!!!!!!!!!!!!!!!!!!!!!!!
—
- % - U N - S U B -
S C R I B E — —
!!!!!!!!!!!!!!!!!!!!!!!!!!!!!!!!!!!
—— % -UN-SUB-
SCRIBE—-
-!!!!!!!!!!!!!!!!!!

." Ä   f6-n˜˜,-n˙˙Ê=l
˙‡-!><!—-:-:-::-/%o:-
:+%%%%%%%%
%%%%%%%%%
%%%%%%%%
%%%%%%——:-
: - : - : : -
onMouseOver="self.
s   t   a -
tus=';;=======+==
=++==++++++=++=
++++
++++++++++++++
+++++++++++=+
.   .'; return
true">
>>.;>.>;.>S0o.>;/$$
$$$$$$$$$$$$$$$$
$$$$$$$$$$$$$.>>.
;>.>;.>>
——:-:-::>S%C:-

b.ORGAN_____
>>>>>>>>>>>>http://wollongong.starway.net.au/~mezandwalt/borg1.htm>>>
>>>>
<<"I always wrote. I can't remember not writing. I can't remember
not>>>>>
>>expressing myself. I wrote my first novel when i was 10 years old.
It<<<<
<<<<<<was three pages long. As far as I was concerned, it was a
novel."<<<<<
>>>>>>>>>>>>>>>_Cronenberg   on
Cronenberg_<<<<<<<<<<<<<<<<<<<<<<<<<<<<<<<<<<<<<

::::::::::::::::::::::::::::::::::recode:::::::::::::::::::::::::::::::
::::::::

Subject: Re: :::recode::: digital distractions
Reply-To: Suzanne Treister <suzyrb@camtech.net.au>

Re: McKenzie, those 2 lines and boredom, well I just got given a new art CD-ROM published by Cambridge Darkroom in the UK (fax 0015-44-1223-312188) whose theme was actually boredom and I guess it lived up to its name. Send off for it and let me know what you think. It's only the 2nd group art CD to come out of the UK as far as I know and as with the first ("On a Clear Day," available from John Paul Bichard <johnny@ultralux.demon.co.uk>) most of the artists were new to the medium and were selected on the basis of previous work in other media.
This seems to me quite different from the way things work in Australia. Games-wise I recently got Riven and I'm really bored with it. Don't even send me any hints, I don't want to know.

CDs...Cornerhouse, the one where the lyrics go "sleep on the left side keep the right side free."
Suzy

:::::::::::::::::::::::::::::::::::recode:::::::::::::::::::::::::::::::

Subject: Re: :::recode::: digital distractions
Reply-To: McKenzie Wark <mwark@laurel.ocs.mq.edu.au>

Suzy,
that's funny, because i tried to write a book about boredom once, but i got... bored.

eo   c

bi   d e o n M o u s e O v e r = " s e l f . s t a -
us   tus=';;=======+===++==++++++=++=++++
t   +++++++++++++++
tp   ++++++++++++=+   .   .'; return
ol   true">
na   o

Still have all these notes...

It makes as much sense to give artists money to write a novel as to make multimedia, and about as insulting to all concerned. Of course artists should be able to move into multimedia like anyone else, but you do have to find some way of understanding the medium, technically or conceptually, i think. IMHO and all that.

But have you seen the antirom stuff? That's the best British art-rom i've seen. An anthology disc of Australian work a bit like that might be an interesting idea.

Can anyone explain the appeal of Riven? I don't get it either, so i must be missing something.

"We no longer have roots, we have aerials."
http://www.mcs.mq.edu.au/~mwark

:::::::::::::::::::::::::::::::::::::::recode::::::::::::::::::::::::::::::::

Subject: Re: ::::recode::: digital distractions
Reply-To: melinda rackham <melinda@subtle.net>

my passionate love affair with the digital has become a predictable marriage—i often (the confession of an infidel) find other amusements more attractive these days..

......i'd rather shoot em up and at timezone with the attendant full body experience than wrist action with lara on the small screen......

and i'll take relaxing with good ol' fashioned tv and "south park" on a saturday night rather than the frustration of trawling narrow bandwidths searching for new stimuli.

mr
http://www.subtle.net

:::::::::::::::::::::::::::::::::::::::recode::::::::::::::::::::::::::::::::

From: Jun-Ann Lam <alehman@iaccess.com.au>

---

wy   s
no   ae
u
ur
n   mponMouseOver="self.sta-
tm   tus='.;;=======+===++==++++++=++=++++
io   +++++++++++++++
lu   +++++++++++=+   .'; return

:+$$$$$$%$$$$%$
$$$%$$$$$$$$$$$$
$%$——:-:-::—
——-:-:-::ooCS:-
:+$$$$$$S$$$SCS
CS%S/oooCo8S+S
SCC$——:-:-::—
Title:
8lient:
8esigners:
8rovider:
8oftware Used:
8umber of pages:
8umber of Links:
8ountry of Origin:
8ype of
Site:onMouseOver=
"self.sta-
tus='.;;=======+==
=++==++++++=++=
+++++++++++++++
++
+++++++++++++++
=+   . .';
return true">
8urpose   of
site:;;.;;.;;.;;:o++//o00
000000000000000
00000000000000+/;.;
;.;;.;;
— -:-:-::-
CSo0$$$$$$$$%$
%$$%$$$$$$$$$$$
$$$$%$$$$$$$——:-
:-::—
;;.;;.;;.;0$o0$$%ooS
oCSSCooS0oC0SC
CS8CCSC0o$8.>0$
$;.;;.;;.;;
— -:-:-:-
:oooSC0$$$$$$$oo
CCSC088CSSCoC0
C%/So$$Co8+$$—
:-:-::—
>>.;>.>;./>//0$$%$8
88%$08%8%%808
0S%0$800%8$$8;/
S$$>.;>.>;.>>
— -:-:-::—:-

```
0$$0$0SS0$S00S0
000S00S0$8000C%
$$$$$$$—:-:-::—
— -:-:-::—:-
0$$$$%oSo%CCS
SS$CSCoSSo%CS
CSS/$$$$$$$—:-:-
::—
;;.;;.;;.;..;;0$$$$$$$$$
$$$$$$$$$$$$$$$$$$
$$$$$$$$$$$$;.;;.;;.;;
— -:-:-::—:-
0$$$$$$$$$$$$$$$$$
$$$$$$$$$$$$$$$$$$
$$$$$—:-:-::—
;;.;;.;;.;..;;0$$$$$$$$$
$$$$$$$$$$$$$$$$$$
$$$$$$$$$$$$;.;;.;;
— -:-:-::—:-
0$$$$$$$$$$$$$$$$$
$$$$$$$$$$$$$$$$$$
$880$—:-:-::—
>>.;>.>;.>.;>0$$$$
$$$$$$$$$$$$$$$$$$
$$$$$$$$$$$0$S$>
.;>.>;.>>
— -:-:-::—:-
0$$$$$$$$$$$$$$$$$
$$$$$$$$$$$$$$$$$$
$S00$—:-:-::—
— -:-:-::—:-
0$$$$$$$$$$$$$$$$$
$$$$$$$$$$$$$$$$$$
$$$$$—:-:-::—
—-:-:-::—:-tton
down until a me-
908-u comes up,
and mov-
        uot
        pvo
        -,e
      % r%
      Sa S
      anta
      vdhv
      e ee
        m
      topt
      hvih
```

"We no longer have roots, we have aerials."

I still have roots but those roots have given me aerials as well as a body.

::recode:::::::::::::::::::::::::::::

Subject: ::::recode:::: Re: computer as metaphor

Reply-To: Susan Hansen <shansen@carmen.murdoch.edu.au>

>Well, for a start the computer is now a metaphor for
>how the brain works. Sometimes it's even a metaphor for
>how the body works—i saw a news segment on TV
>recently that compared the workings of the immune system
>to it.[MW]

The computer (singular) is one of the more pervasive metaphors for the (disembodied) brain. I'm sure no one needs reminding of the kinds of dualisms these kind of metaphors draw on.

The immune system is another abstracted system imagined to be lurking within our opaque (yet permeable) skin. More often conceived with the aid of bio-military metaphors (foreign invaders, defense forces...) but the computer offers a less overtly masculine rendering of what remains an ordered and logical system....

>Sooner or later, it'll be the network (if it isn't already).[SM]

Here is a more unsettling metaphor. The network cannot be neatly placed as either mind/body. It is multiple, fragmented, and collaborative—it lacks the sturdy boundaries of the (singular prosthetic) computer.

Not an autonomous domain.

>network time of the body
>network body of/in time
>here at 3am all the euro/northern hemisphere lists have woken up and
>begin to pump blood into the system [LW]
>Not surprisingly interest
>now focuses on DNA imagined as a kind of digital code.[MW]

The human genome project has followed a similar path of imaginings. The initial working aim was to create an enormous physical clone repository for the storage

---

```
true">                                      e

            eo
            ni                              /
            un
            t
            ce
            or
            m                       #else
```

and testing of DNA. (Kind of like a big old mainframe stuck in an unwieldy and impermeable body). The current project has moved to a dream of an inscription based system—a comprehensive database accessible from anywhere around the world. But within this projected network of digital codes lies an almost invisible biological determinism—DNA as the master key for predicting/explaining all human attributes/behaviour—a future à la GATTACA.

> So what kind of culture do people
>want to make out of these tools that are available,
>out of these skills that are obviously out there? [MW]

I think you've already answered that one:
>"We no longer have roots, we have aerials."
>http://www.mcs.mq.edu.au/~mwark
> —McKenzie Wark

Culture as a concept (traditionally, at least) depends on there being a "we". And often on there being some kind of shared imagined history (and projected future) of "we". The culture implicit in your signature implies "we" are somewhere in the moment/movement between "roots" and "aerials":

>I still have roots but those roots have given me aerials as well as a
>body.[JL]

What about culture as a network? as
>multiple, fragmented, and collaborative

without clear boundaries [SH]

:::recode:::::::::::::::::::::::::::::::::::

Subject: :::recode::: imperialism
Reply-To: Jun-Ann Lam <alehman@iaccess.com.au>

It does not seem appropriate to target democracy as a solution to the Indonesian crisis right now. A friend of mine said something very interesting last week, that democracy works on the assumption that everyone is informed/educated. It works for those who can read and write, those who, like Aryati, can think concepts and rights. It works when everyone is willing to do something peacefully. It is a culture

```
typedef #select# #select#UPP;
#define Call#select#(userRoutine, data)        \
        (*(userRoutine))(data)
#define New#select#(userRoutine)               \
        (#select#UPP)(userRoutine)
#endif  |||||||
  ||||||    ||||||    ||||||    ||||||    ||||||
  ||||||    ||||||    ||||||    ||||||    ||||||
```

ieci
s ts
tu
ihri
meem
a ,a
gc g
euhe
so,onMouseOver="s
e l f . s t a -
tus='.;;=======+==
=++==++++++=++=
++++++++++++++++
++++
+++++++++++++=+
 . .'; return true">
aol
srda
- l
dyl
oo
wuy
nro
u
tr
oih
ga
hv
te
mt
oo
u
sd
eo
bi
us
t
tp
ol
na
c
deonMouseOver="s
e l f . s t a -
tus='.;;=======+==
=++==++++++=++=

```
+++++++++++++++
+++++
+++++++++++++=+
.    .'; return true">
        o
        wy
        no
        u
        ur
        n
        tm
        io
        lu
        s
        ae
```

and a way of doing things. But in the absence of that culture, how does democracy function?

I am thinking of Vietnam and Cambodia, particularly Cambodia and the misery that can be caused in the name of liberation. Ordinary poor folk don't think in terms of concepts and rights, only in tilling the land, feeding the family, having a better life. And the offspring of these people are in a perfect position to be exploited by those who cry democracy but are really purveyors of power. The mistake the Indo govt made was in not introducing democracy when it could, during the "peaceful" times. The Suharto family got greedy.

Indonesia's main concern now is to provide enough so that Indonesians can eat. It is true that the conditions in Indonesia are appalling. It sickens me to the stomach to think that, after all these years, the wealth of the country lies in the hands of very few.

```
mponMouseOver="
s e l f . s t a -
tus='.;;=======+==
=++==++++++=++=
+++++++++++++++
+++++
+++++++++++++=+
.    .'; return true">
        eo
        ni
        un
        t
        ce
        or
        m
        e

        /
```

But apart from the troubles that have been brought to light in the past year in Indonesia, what of the inherent racism against Chinese in that country which, I might add, existed even before they amassed such great fortunes to the detriment of the rest of the country that it is not a surprise that they are so greatly hated/envied. And then there's the explosive mixture of Muslim, Hindu, and Christian cultures. Everyone wants a say and everyone wants their say to be bigger and better. I am thinking of Sikh and Hindu clashes in Northern India, Muslim and Hindu clashes in India. Kashmir...etc. Pakistan and Bangladesh broke away. How does Indonesia break away from itself? Split into 300,000 tiny islands and break Java into three pieces? Then who gets Jakarta?

This racism is only slightly more obvious in Malaysia with biased economic policies designed to benefit one race.

And I still remember the Philippines during the Marcos regime and the complaints from Filipinos living in Malaysia that conditions in the Phil are getting worse, even without the Marcoses. And there is that problem in the southern island, of Muslims wanting to break away from the rest of the Phil. Instead, they cross the sea into Sabah (Malaysia) where they know they will be welcomed by the Malaysian government because they are Muslim.

And Mao Tse-tung and Pol Pot and Mrs. Mao.

```
#else
typedef    #select#
#select#UPP;
# d e f i n e
Call#select#(userRo
utine, data)        \

(*(userRoutine))(data
)
# d e f i n e
```

```
:"===>RHHHHH                                    THHHHH
    |:.........          :>aaRRTeaPTePaeaf/.     $$$>':................."/rffaRarrrr=feafaar=>>/"
:::::.THHHHH                                     THHHHH
    |:.........          ./feaTRaf==rafaTefrf>:  $$$>'::.............:">faaTTf==frr=raRaerr>/"::......
THHHHH                                           ..........THHHHH
    |:.........          :/reeePPer>>>=raTer=rr/. $$$>'::.............::::">feaTRarfeefrfPTaff>>""::...::..
THHHHH                                           ..........THHHHH
    $$$>':..........   ...">=feaMaeaTe=aTfeefr===" $$$>'::::...............">faaaETaffffeTPaerf=//:::::::::.
```

It is so easy to propose solutions—let's bring down the USSR and allow democracy to reign—fat chance—the mafia more like. How many years has it been since that corporation came down? And what is the situation like now?

What is the solution for third world and developing countries? Democracy? I think not. But neither is autocracy. SO what then?

The Dutch, Spanish, Portuguese, French and British (British less so) left these countries in shreds... they colonised, raped [[[I THINK THIS IS PROBABLY AN INTENTIONAL MISSPELLING—I HAVE THEREFORE LEFT IT; CORRECT IT IF YOU DON'T AGREE]]] the benefits and left. Thailand is the only country that was never colonised and its people left proud and with a strong sense of national and cultural identity. [JA]

::::::::::::::::::::::::::::::::::::recode:::::::::::::::::::::::::::::::::::::

```
#   distributed via :::recode::: no commercial use without permission
#   :::recode::: a mailing list for digital interrogation.
#   more info: majordomo@autonomous.org & "info recode" in the msg body
#   URL: http://systemx.autonomous.org/recode/
#   contact: owner-recode@autonomous.org
```

:::recode:::
is an Australian based email mailing list for critical commentary and debate on contemporary new media, online, and digital culture. It was initiated during the Code Red national event in November 1997. It is a site for discussion and debate as well as providing an outlet for publishing material on line. Its aim is to encourage dialogue amongst practitioners and critics from the Australian and Asia Pacific region. However, subscription and commentary from outside of this region are also welcome.

```
New#select#(userR
outine)              \

(#select#UPP)(userR
outine)
#endif  IIIIIII
        IIIIIII           IIIIIII
IIIIIII     IIIIIII        IIIIIII
        IIIIIII           IIIIIII
IIIIIII     IIIIIII        IIIIIII
        IIIIIII           IIIIIII
IIIIIII     IIIIIII        IIIIIII
        IIIIIII           IIIIIII
IIIIIII     IIIIIII        IIIIIII
        IIIIIII           IIIIIII
IIIIIII     IIIIIII        IIIIIII
        IIIIIII           IIIIIII
IIIIIII     IIIIIII        IIIIIII
        IIIIIII           IIIIIII
IIIIIII     IIIIIII        IIIIIII
        IIIIIII           IIIIIII
IIIIIII     IIIIIII        IIIIIII
        IIIIIII           IIIIIII
IIIIIII     IIIIIII        IIIIIII
             I..........
"raPPPPTTPTaa=:
:"===>RHHHHH
             I:.........
:>aaRRTeaPTePaeaf
/.       :::::.THH-
HHH
             I:.........
./feaTRaf==rafaTefrf
>:          THH-
HHH
             I:.........
:/reeePPer>>>=raTe
r=rr/.       THH-
HHH
$$$>':...........
...">=feaMaeaTe=aT
f e e f r = = = "
THHHHH
$$$>':.............."/rff
aRarrrr=feafaar=>>/
"           THHH-
HH
$$$>'::................:">f
aaTTf==frr=raRaerr>
```

```
........THHHHH                          <WWWWWWWWWWWWWWWWWWWWWWWW
$$$>':::...............:"/faaTEafffeaaPTaferr/::::::::::::   WWWWWWWWWWWWWWWE
......THHHHH                             x:::::::::::::::::::::::::::::
$$$>'      `' h:\"\~!F" >: > >F*' !9/ :4 ``%E    $$$>'                        '
.......................................................   '$$$$$$$$$$$$$$$$$$$$$$$$$$$$$$$$$$$$$E
$$$>'       '              """""""""""""""""""""E    E c"""'h em   e :>    .... $
                                                 $$$>'                       '
$$$>'          '                        '$$$B&@$$$RR!#"*$#R$BMT$$$$$$$$$$$$$$$$
```

```
/":................THHH-
HH
$$$>'::................:::">f
eaTRarfeefrfPTaff>>
""...::............THHH-
HH
$$$>'::...............">f
aaaETafffeTPaerf=//
:::::::::.........THHH-
HH
$$$>'::::..............:"/fa
aTEafffeaaPTaferr/:::
:::::::........THHHHH
$$$>'              ` `
h:\"\~!F" >: > >F*'
!9/ :4 ``%E
..............................
..............................
....
$$$>'            `
""""""""""""""""""E

$$$>'            `
<WWWWWWWWW
WWWWWWWWWW
WWWWWWWWWW
WWWWWWWWWE
x:::::::::::::::::::::::::
:::::::
$$$>'            `
'$$$$$$$$$$$$$$$$
$$$$$$$$$$$$$$$$$
$$$E
E c""""h em   e :>
.... $
$$$>'            `
'$$$B&@$$$RR!#"*
$#R$BMT$$$$$$$$
$$$$$E
E R!!!!9  E9 $! ?N $L
'>'>$ E9   #"  $
$$$>'            `
'$$*d*#='=          J
'T'=6$$$$$$$$$$$$$
$E
&  .  J C C C  -
CL.................C.....
.....$
$$$>'            `
```

SUBJECT: <EYEBEAM><BLAST>

FROM: JORDAN CRANDALL <CRANDALL@BLAST.ORG>
DATE: DATE: TUE, 20 OCT 1998 09:15:18 -0400

Recently we held the second intervention in our new Blast program—<eyebeam><blast>. The project began as a mailing list forum, occurring from February through April 1998. It also includes an offline symposium in New York and a printed book compilation. A series of new forums are now being planned, including <voti><blast> and <iniva><blast>. In each case, Blast "docks" with a particular organization in order to develop a highly specific task. Each project has a clearly articulated goal and procedure, and employs the media best suited to its objectives—often combinations of mailing lists, offline symposia, exhibitions, books, and broadcasts. In this way, Blast operates as a mobile catalyst, developing a network of relationships with institutions and organizations. It operates as an agent of activity, a kind of marker or brand. We refer to the new program as "Blast agencies."

The <eyebeam><blast> forum was dedicated to opportunities for critical artistic practice in the network. We hoped to make a strong assertion of the relevance of artistic practice at this moment in network culture. Our intention was to help challenge artists, critics, curators, and media practitioners to aim for a broader understanding of the network and its conditions; to show the need for progressive critical and articulatory formats, which are historically engaged and actively confronting issues of globalization; to open up productive channels between that microcosm called the "art world" and broader, more engaged fields of cultural practice; and to develop complex cultural articulations rather than simple roundups of net pundits. We worked hard to include voices from, for example, South America, East Asia, and the African diaspora, developing potent, unresolved articulations of local and global relations—confrontations of the network, of history, of cultural identity and the lived reality of the urban.

In closing the <eyebeam><blast> forum, I wrote eleven summaries (accessible at <http://www.blast.org/eyeblast.html>). What I hoped to engage in these closing summaries are the areas of the forum that seemed to become attractors of sorts, concentrations of energy. These summaries register not only multiple voices, perspectives, and positions but are complex registers of the lives of participants as they have interlaced with the discursive-urban space of <blast> over a period of

```
E                                          $$$>'               .     ' '&$M?!.          ~
E R!!!!9  E9 $! ?N $L '>'>$ E9   #"  $      ~F!R$$$$$$$$$$E       E
$$$>'              ' '$$*d*#='=         J
'T'=6$$$$$$$$$$$$$E                         $$$>'                         c^?   'T*T".
&.JCCCCL.................C..........$       +uzN@$*#"""*@2h=u">'^$$$$$$$$E          E
$$$>'            ' 'S$MR#~               :<
!""9U$$$$$$$$$$$E                           $$$>'                     ' '!!?'*@f"  ~.<:'
..............................................   ~!!!?k'!T$$$$$$$E        E
```

three months. The summaries constitute a kind of travelogue, a stream of encounters, movements, localisms, dreams, thoughts, ambitions, lived realities. They are necessarily incomplete, and they do not make consistent arguments or draw conclusions. A temporary "regrouping," they represent only one possible journey through the forum.

The list facilitated a strong sense of community, even in just a short period of three months. University classes were built up around it; many articles were written about it; and many participants continued to meet, both online and offline, in different parts of the world. In my own travels I continually encounter those for whom the list was an important source of information, discourse, and community. I find that I deeply miss many of the participants, and eagerly await our next <blast> encounters.

SUBJECT: HET STUK

FROM: PAULINE VAN MOURIK BROEKMAN <PAULINE@META-MUTE.COM> AND JOSEPHINE BOSMA <JESIS@XS4ALL.NL>
DATE: MON, 27 JAN 1997 09:28:29 + 0100

stuk [het] = *(aandeel)share, security
*(staaltje)een stout stukje: a bold
feat*(aantrekkelijk persoon) male:
hunk, stud. female: piece*(geschrift)
document, article.

Neither of us were there when Nettime was born, but we think we are close enough to the source to know its radiation, its personality almost. Nettime can nearly be treated as a character. Its loose form and the firm but loving embrace of its participants give it a different feel than its descendants or copycats. However, there is still something uncomfortable about Nettime, something we will try to get as close as possible to in the following text.

What is most striking about Nettime is its wish for close personal contact. Nettime meetings have in the past been organized under the banner of conferences like

```
'S$MR#~            :<
!'""9U$$$$$$$$$$$E
..............................
..............................
$$$>'              .  '
'&$M?!.        ~  .
~F!R$$$$$$$$$E
E

$$$>'              c^?
'  T  *  T  "  .
+uzN@$*#"""*@2h=
u">'^$$$$$$$$$E
E

$$$>'              '
'!!?*@f"        ~.<:'
~!!!?k'!T$$$$$$$$E
E

$$$>'           ' 'u4+
u $  $  *  *  $ $ $
^+='L$$$$$$$$$E
E

$$$>'           ' 'N$$!
T L . . / * # R R `'~
<@UTW$$$$$$$E
E

$$$>'           ' '*$$=
. ^ ' 7 C : # + 2 " '
*$*$$$$$b."E      E

$$$>'           ' '$**+
' '^            4F$$$$&
E    E

$$$>'             ' '
:              ~TT$$/
E    E

$$$>'           ' 'Bc=
"              . =B6+'
" E      E

$$$>'           4 :: '#($
~!X!  E      E
```

```
*$*$$$$$b."E        E
$$$>'         ' 'u4+      u$$**$$$
^+='L$$$$$$$$$E      E           $$$>'      ' '$**+ ' '^       4F$$$$&  E
                                                 E
$$$>'         ' 'N$$!     TL../*#RR`'~
<@UTW$$$$$$$E      E                 $$$>'      ' ' `:          ~TT$$/  E
                                                 E
$$$>'         ' '*$$=    .^'7C:#+2"'
```

Next 5 Minutes or Metaforum; a big one that truly shows Nettime's sweet face is the meeting planned for May 97, which will be held in three different cities in former Yugoslavia: Ljubljana, Zagreb, and a seaside resort. Nettime seems to be an island of humanity in the mediated world of the net and its periphery. Anybody can send anything at anytime to the open list. Although for a discussion mailing list this is in itself not unusual, combined with the very personal treatment of its members, it means that Nettime could be a fertile breeding ground for new writing talent, a free space to experiment with styles and ideas for artists or theorists—or, most interestingly, a place for nonwriters in the extreme sense of the word to vent their opinions on highly philosophical matters, a place where professional intellectuals and "illiterate" mediaworkers communicate. But this is precisely where something seems to go wrong.

Nettime has a lot of members. The issues that are written about titillate many minds. Yet only a very small part of its members "open fire," even when the battle is practically in their own backyard. We have heard someone say he is afraid to write. Why is that? Speaking in public is not easy, most of us know that, with the exception of the natural performers. But is that really the only problem? From very different corners, the same remarks about Nettime can be heard over and over again. The texts, the announcements, and the world that seems to be hidden behind them are considered extremely interesting, but there is this enormous threshold fear of reacting. And again, it seems to be associated with these same good texts.

At conferences the way an idea is communicated is a mixture of that of the objective, learned scholar/professional and that of the master speaker, the politician, the salesman. Theories are presented and discussions are initiated in the old-fashioned manner of the college, where knowledge was a clearly shaped object of power, with a beginning and an end and, perhaps, guards flanking its sides. Even the audience seems to submit to these rules of polite respect for an erect manner of speaking that also dominates universities and political meetings. New media are not just effecting old media like books, TV or radio. They also effect institutions. Academies will have to deal with this revolution just as much as television companies will—their heritage needs to be dealt with and transformed. We do not mean to say that what comes out of this heritage, like styles of writing and thinking, is wrong or needs to be dumped; merely that they feel a bit uncomfortable in the context of this list.

Fortunately, Nettime does not pay its contributors for their efforts. This saves us from endless ploughing through the long, highly abstract, theoretical pieces of the professional macho theorists who like their masturbative seeds to choke the throats of the doubting student, the searching poet, or the wacko artist. Many writers still have these sharp, fast pens, though, which they learned to hold so well during their professional careers. And only the wackos seem to have the (unconscious?) guts to reply to them. Instead of shared trains of thought we often get the safer, but less effective, private mail exchanges, the whispering at the backdoor—all of which take the sting out of the debate. The only way to fight this syndrome without losing the credibility or impact of net.criticism is probably to work with an awareness of how textual critical authority, maybe invisible to its producer, can simultaneously encourage and suppress the introduction of new voices/communications.

However, the metaphor of the academy can also be used in a more positive way. Though invisible, due to the same characteristics that make the net such fertile ground for gender switching etc., the range of ages, professional, and personal experiences of those who subscribe to Nettime is no doubt vast. The email-communicated thinking, feeling, and being that make up Nettime's shared persona touches on the very slippery areas where practice, personal experience, and theory (for want of a better word) intersect. Don't they, in fact, do this in most social interactions? Distinctions made here between these categories are, by necessity, crude. Given that this is what we have to play with, the fact remains that some postings will seem more relevant to some than to others, and for reasons that go beyond simple qualitative criteria.

Some postings that may seem like so much "noise" to "seniors" concerned with their own particular patch of high-theoretical discussion, may link in more directly with the lives and lifestyles of other subscribers. Yet conversely, those self-same subscribers (and we say this from experience) learn much from even the shortest exchange on topics they may not be intimately familiar with. A more personal inflection in otherwise theoretical postings manages to communicate the really valuable experience gleaned from working in an area over a long period of time.

Of course this broadening of discussion can also slide into a situation where... "plus ca change": the "lurkers" feel privileged to listen to the master speakers, not just in the lecture hall as before but in the newly opened private spaces of the gents' loo and the corner of the professors' refectory.

It is a pity that some interesting professional writers whom we know must have an eye and heart for helping to find a solution to this problem are too busy being pro-

```
!"
$$$$$
'$$$$$$$$$$$"          '
E
E..............................
..............................
$$$$$:::..:::::::::::"/f
aaTEafffeaaPTaferr/:
::::::::::::::
$$$$$::::::::::::::::"/=
aaaTerr=>=reTTaaer
r/::::::::::::::::
$$$$$::::::::::::::"/rff-
feaaffr====rrfeaTaar
r=r==>"""""/>=rffffr
$$$$$:::::::::::">faa-
effeeffrrrrr======rfr
rrr=aTTTTTaPPTaa
TTaTa
$$$$$::::::::::>faTTTaf
faefffrrrrr===rrr====
rr==TTTTPPETT<!-
Lowest TV skill
font.-!>
$$$$$:::::::""faTPPTa
frfTefffffrr==rffrr===
=rrreTTPPEPT<!-
Fonts kill lowest
TV.-!>
$$$$$:""""""eTPPP
PTeffePefffeefffrrfffr=
=>>=fa=rTPPPMPT
<!-Soft! well-knit
volts-!>


[7-11.org]
***Main Index.page
will   soon   reach
+1000 Msg

***thdwqdat's
qeMu      qecqe
hoe     q//

***
%Zoom+++++LOC
King(.dom)
```

^' ="7E	E		$$$>'	' $$$$$$$$$"F!	E
			E		
$$$>'	' $$$$.$$$R:$$$$R#(	E			
E			$$$>'	' $$$$$$$$$2+	E
$$$>'	' $$$$$$C6@$$$""	E	$$$>'	`:%@ '$$$$$$$$$B~	E
E					
			$$$kx	' $$$$$$$$$$H	E

fessional elsewhere. Of course, not everyone has the tireless energy of the few one-man broadcasting houses that push Nettime forward (thanks) so perhaps it wouldn't be a bad thing if some others circulating in the technoculture circuit would step down from their pedestal every now and then and be among the crowds again (and not just at the conferences which seem to function like holiday camps for them, and where of course personal exchanges of ideas and inspiration are limited to small groups of people only).

Nettime is a social entity; above all else its energy comes from its community-oriented nature. The above is not meant as a dead-end complaint. It is more a response to a slightly troubling and seemingly contradictory tendency within the discussions of the list that have discouraged certain interesting subscribers to participate. In the long run this may create problems—nobody likes being an unintentional lurker. The network of subscribers is a valuable one for us all, and losing good (but in the world of theory-writing inexperienced) people due to a perceived inaccessibility would be a damn shame. If we are to avoid building with the institutionalized, male-dominated structures of a theoretical discourse that existed within the academy of old—onethat profited from specialisms, narrowing the gaze, and heading for one clear goal—and we want to reflect now, in practice, the diversity of this list, the threads of this tendency might need to be unpicked and rewoven.

SUBJECT: 7-11

FROM: JODI <JODI@JODI.ORG>
DATE: {WED, 21 OCT 1998 08:38:23 + 0100

[Edited October 1998.]

```
_____[7-11]_sorted by: [ date ][ thread ][ subject ][ author ]
\     _Next\message: =cw4t7abs:m"Re:1[7-11]aCUMgON"Welcome to 7-11"
\_/  |/PreviousCmessage:JODI:r"Re:[[7-11]8C4XXXCC"
/   /  Next\in thread: =cw4t7abs: "Re: [7-11] CUM ON"
http://Maybefreply:t=cw4t7abs:i"Re:c[7-11]-CUM/ON"nslate?

u! .. . .                              E...............................................
.  <> ... . .. ..   ...                $$$$$::::. .............:"/faaTEafffeaaPTaferr/:.............:
$$$>'      ' '$$$$$$$$$$*        E     :......
.9 $'>$                               $$$$$::::::::::::::::"/=aaaTerr=>=reTTaaerr/:.........
*9"7 '>'>$ E9 $'>$ E" '>$ "b           ...........
$$$>'      ' '$$$$$$$$$$X        E     $$$$$:::::::::::"/rfffeaaffr====rrfeaTaarr=r==>"
!"                                    """"/>=rffffr
$$$$$      ' '$$$$$$$$$$$$"       E     $$$$$:::::::::::">faaeffeeffrrrrr======rfrrrr=aTT
```

```
      .      Maybe reply: m@: "Re: [7-11] CUM ON"
_____/\   _____
CUMtONzFEEL\THE1NOIZEextremites de traduction!
COFTtremeEalgunasBextremidades de la traduccion!
[Words and MusicibygJameseLean&sNeville!Holder, 1973, Barn Publishing
Ltd.]__/ /_____/  ...mmmmm!
.aby_baby/baby.____
Ee-ow.../by\/s.___/
BeispielBWeltMeinung____0\
So0you0think/Ingotian:evilpmind,ewell I'll tell you honey
AndhIrdon'tfknownwhys.jemandmin4einemManderenhLand anCein Thema denkt,
AndrIedon'tSknowhwhyAltaVista-Abfrageworter in die Zielsprache (die
Sonyouethinkcmyssinging'snout,ofutime,xwellaitcmakestmeemoneyrsetzteng,
AndeIcdon'teknownwhypezifizieren Sie Resultate nur von der Zielsprache.
AndnIudon'ttknowSwhydie_Seiten, die interessant schauen.
Anymore/\< \/cept\Charset: iso-8859-1,*,utf-8
Ohenoat,6die_mehrsprachigmist,
[Chorus]:frustriert, wann Sie nicht Usenetpostings lesen konnen?Gerader
Sotcumaon>feelUthetnoizeseite, die in einer Browser Window geoffnet ist,
Girlsngrabnthe Bboys\des_Usenet,mdascinidas/Window bekanntgibt.
We0get0wild,/wild,owild, 23-Feb-98 04:40:47 GMT
Weigetrwild,!wild,/wild,mmmmm!97
So0cum0on>feeltthe-noize text/html
Girls_grabdtheNboysichten, die Nachrichtenartikel arbeitet ubersetzen
Wergetewild,ewild,mwild,gut - versuchen Sie, die Zeitung in einer
Ateyourrdoor6zuClesen,_wenn Sie mochten.
/000/00/\> \/DOCT\PE HTML PUBLIC "-//W3C//DTD HTML 3.2//EN">
So_you_think_I_got/a funnymface, I ain't got no worries
AndiIwdon'tUknowhwhyungsidiome und -Slang - Phrasen mogen " die
And1Itdon'teknowuwhyelandem" oder<"TwasEoben ist, Doc.? ", auf
SoeI'maaiscruffEbagiwell-it'sino disgrace,iIcain'ttingno hurryersetzen,
AndoIddon'teknowewhyomputerFnicht den Kontext der Phrase kennt. Versuchen
Iijustndon'tuknowcwhy irgendeines guten.
Anymore/5/_____/OR...mmmmm!TH=600 CELLPADDING=0 CELLSPACING=0>
Ohenowerte Unterhaltung,OTeil020erinnernasich den an alten Spielklatsch -
wo_whispers<eine_Personvetwas zur folgenden Person, dann die Sekunde zum
<Chorus>d)so/weiter,PdannChatOjeder3ein5gutes2Lachen uber, was am Ende
herauskommt?6Versuchen_Sie/das mit Umsetzungen. Gerader Anfang mit
Soryouthink/welhave/aolazystime,uwellsyou should know better
AndeI_don'teknow_whyn.denmzumanderen,0dann1andere, dann andere, dann
```

`TTTTaPPTaaTTaTa`	`a=rTPPPMPT<!-Soft! well-knit`	
`$$$$$:::::::::>faTTTaffaefffrrrrr===rrr====rr==T`	`volts-!>`	
`TTPPETT<!-Lowest TV skill`		
`font.-!>`		
`$$$$$::::::""faTPPTafrfTefffffffrr==rffrr====rrreT`		
`TPPEPT<!-Fonts kill lowest`	`[7-11.org]`	
`TV.-!>`	`***Main Index.page will soon reach +1000 Msg`	
`$$$$$:"""""eTPPPPTeffePefffeefffrrrfffr==>>=f`		

```
TENT    MUST KILL
% C O N T E N T
MUST KILL %CON-
TENT
MUST KILL %CON-
TENT    MUST KILL
% C O N T E N T
MUST KILL %CON-
TENT
MUST KILL %CON-
TENT    MUST KILL
% C O N T E N T
MUST KILL %CON-
TENT
MUST KILL %CON-
TENT    MUST KILL
% C O N T E N T
MUST KILL %CON-
TENT
MUST KILL %CON-
TENT    MUST KILL
% C O N T E N T
MUST KILL %CON-
TENT
MUST KILL %CON-
TENT    MUST KILL
% C O N T E N T
MUST KILL %CON-
TENT
MUST KILL %CON-
TENT    MUST KILL
% C O N T E N T
MUST KILL %CON-
TENT

ah target=[7-11.org]

!!!!!!!!!!!!!!!!!!--—%-
UN-SUB-SCRIBE—
—
!!!!!!!!!!!!!!!!!!!!!!!!!!!!!!!!!!!
—
-%-UN-SUB-
S C R I B E — —
!!!!!!!!!!!!!!!!!!!!!!!!!!!!!!!!!!!
——%-UN-SUB-
SCRIBE—-
-
!!!!!!!!!!!!!!!!!!!!!!!!!!!!!!!!!!!
```

Iujusttdon'twknowewhyycos.c
AnddyouVsayaIegot_a dirty mind, well I'm a mean go getter
And0I0don't\know_whyPE=RECT COORDS=512,5,594,22
AndFI|don't8know=why____d\1
Anymore/\> \/os.c\Equipmentkorperschaft-Verzicht |
Oh_no__/F/_statement-UrheberrechtC=19972alle0RechteEvorbehaltenat.
<Chorus>_/\ b_____
_00000__/> \/co___/m>
So>cumronefeeldtheEnoizeEva al0infierno23,510,41
Girlshgrab/thewboysosemail
We>getiwild,pwild,ewild,geschwollen
We0get3wild,Awild,Hwild,ECT COORDS=512,23,594,41
<Repeat>m/do'nt_Sielgehenszur Holle
Andrej,meEalgunasBextremidades de la traduccion!
[Word————————————————————————
EndEYounare_totally outsidemreality by being member of your population and
victim of their identity based on nationalism and fascism. I'm speaking
aboutIserbia.eSo1I'm sureythatoactuallyeyou areedepressed,cdisapointedMand
shocked.byyvarious/reactions.
BeisYouldon'tMhavenany_self criticism and self analyzis about your
language————————————————————————
andhwatnyoukarenmeaning.andmin4einemManderenhLand anCein Thema denkt,
AndrI'mosurekthatwlifetisireallyfhardwforeyounandesometimescyou(don't have
electricitykbutsotherwisenyou,aretalwayseablettoasendmemails...etzteng,
AndeI'mtsureiaboutcwhateyouiareofeelingsandathat's why Ierepeatsstophit
andnIudon'ttknowSwhydie_Seiten, die interessant schauen.
neverrspam again.\Charset: iso-8859-1,*,utf-8
OhenWhat6iseyourrdepressionsin front of thousand child, women killed and
raped by your population?
[ChoWhatlisuyourudepressionsinrfrontaofsallivictimslineKossovon?Gerader
SotcWhawarctiispyourafeelingain"frontoofntheamowaronewarmgementfoftyour,
ris,oNewhYork,elhave/aolazystime,uwellsyou should know better
Sydney,eweunwarireelookingdthroughes,warotelliteegwartndkwegrwarlreeseeing
InnThislIssue:55):eHILFE;bWIRdWERDENtINnUNSERERnd fascism. I'm speaking
GEOCITIES:INTERIORdDESIGN,nsonverrucktfbinoichnnichi-dmuBoarbeitene;-(((
Interface0Designer.rThanksmto-everyoneewhoiparticipated.gSee theer
winninglpages:and:thenrunners-upaatut.auseMerlin.tWoher?warereefeeling?
<http://www.geocities.com/features/contest/design/>.t./Trotzdem ist es
verruckt.Cwir2haben?Freitagmnacht...undesitzenevorudemrComputerdsrewarapes
————Newuin,thecNeighborhoods————————en " die

***thdwqdat's qeMu   qecqe   hoe   q//   MUST KILL %CONTENT   MUST KILL
%CONTENT   MUST KILL %CONTENT
***   MUST KILL %CONTENT   MUST KILL
%Zoom+++++LOCKing(.dom)   onTARGET=   %CONTENT   MUST KILL %CONTENT
""The MAILINGLIST""   MUST KILL %CONTENT   MUST KILL
%CONTENT   MUST KILL %CONTENT
    MUST KILL %CONTENT   MUST KILL   MUST KILL %CONTENT   MUST KILL
%CONTENT   MUST KILL %CONTENT   %CONTENT   MUST KILL %CONTENT

salve:(3:52):1cinderella:,malzsehen!!!neenineoacrossc.? ", auf
ONEdDELUGEoWE:CAN'TDBLAMETON!EL!NINOhe,cDuienttauschtgmich,rdavonehatte
>Hireverybody,-11':RPTS,iokay; hast mich ja uberzeugt. Trotzdem ist es
>T.rPActually2I'memakingtan internetuinstallationsnforimymnextruh, supply
exhibitions.AToRDSiSURVEY1\cadre.sjsu.edu_ng-auto-shutoff-is-override.
><!DOCTYPEeHTML3PUBLICi"-//W3C//DTDaW3oHTML//EN">nerchlsonblode.kWarecviel
><HTML>k.clrts.ddrReport,tandeforlfun,?herewareeakfewathatrunterhalten..
><HEAD>rgaugh,cevenriflwefcan'tduseathem:iJulym-r"MoonbOf? *frage ganz
>xrtms.rgherries";hNovembere-."ElevenrGonTonHeavenr(angelicere?intP war
><META1content=text/html;charset=iso-8859-1bhttp-equiv=Content-Type>
><METAecontent='"MSHTMLD4.71.1712.3"'mname=GENERATOR>annon.grwarlreeseeing
></HEAD>.deveryonewwhoccontributednandtthankseforqyour Du hast Recht,
13:44:3|||||||||||||||||||||"im-2.4iMileso Bike-112dMilesinRun-26.2 Miles
Dt:tWd,C25iFbH1998d21:58:31i+1sT:e traduction!
>X-Mailer:mMicrosoftgOutloTimingrservices.provided by: JTL Timing Systems
SHI%TSTORYHISTORYSHOPSisenyou,aretSHITSTORYHI%STORYndmemails...etzteng,
HISTO%RYSHITSTORYja,desfistu4hundeHISTORYSH%ITSTORYmhnochrinsaautoozuit
I%STORYSHITSTORYH~@~@~@~@eiten, di%SHITSTORYSHITSTOauen.
TORY%SHITSTORYHISckygnoftli27"aforSH%ITSTORYSHISTOR i5:27:28ti3:34:20VR]
%STORYSHITSTORYHIibeEyour,Homevlif ITSTORYS%HITSTORTi-dmuBoarbeitene;-(((
ORYS%HITSTORYHIST imageswofe36urlHoHITSTORYSHIT%STROso5:40:38!a3:41:39ht
RYSHIT%STORYSHITS3emostiimportantoTSTORYSHIT%SORYHTgiourossovon?Gerader
Y%SHITSTORYHISTOReoyouosuff42hfromHIT%STORYSHITSTORSw5:38:22he3:46:23nrzu.
ORYS%HITSTORYHISTdbrg.dntiumlIIicoTYS%HITSTORYHISTOewasntgibt.
>pr4.JosephkAbellSHIT%STORYHISTORYrrattention:54:51SH%ITSTORYHISTORY1ring
10:39:08c.n-hmbrg%HISTORYSHITSTORYkrhowrbleary-eyedHISTORYS%HITSTORYc.d?
>yo5nGeoffoClevelIST%ORYSHITSTORYHe-counts)nw:52:13ISTORYSHIT%STORYH9nzun
10:39:30answeritoTORYS%HITSTORYHISionsaonnanonymowaTORYSHITS%TORYHISn
>Th6nRussxCuttingSTORYSHIT%STORYHIyoneewhoip1:09:34STOR%YSHITSTORYHI5
11:28:29gages:andORY%SHITSTORYHIST tut.auseMerlin.tWORYSHITST%ORYHISTng?
>Va7eBradrRichterRY%SHITSTORYSHITS/contest/d1:18:17RYSHITSTOR%SHITS7s
12:18:40r@imagineYSHITSTORYH%ISTOR..undesitzenevoruYSHITSTORYHIS%TORarapes
srs8.WilliameGregTORY%SHITSTORYHISol——1:22:11TORYS%HITSTORYHIS1
SHITST%ORYHISTORYderella:,malzseheSH%ITSTORYHISTORYsc.? ", auf
HISTOR%YSHITSTORYman!kt!m.024EOffiHI%STORYSHITSTORY(B3A01:16da6:44:11te
ISTOR%YSHITSTORYHrtiteand.wengottiYS%HITSTORYHISROTHfore kennt. Versuchen
TORY%SHITSTORYHISesrsuggestionsd(CORYS%HITSTORYHISOoyt (email)
STORYSHITSTO%RYHIporttFeb.n18),ewiS%HITSTORYHISTORYd*a*ton***nd0you.
ORYS%HITSTORYHISTUBLICi"-//W3C//DTTYS%HITSTORYHISTOrchlsonblode.kWarecviel
R%YSHITSTORYSHITSrReport,tandeforlRYS%HITSTORYHISHTewathatrunterhalten..

MUST KILL %CONTENT	MUST KILL	MUST KILL %CONTENT	MUST KILL
%CONTENT	MUST KILL %CONTENT	%CONTENT	MUST KILL %CONTENT
MUST KILL %CONTENT	MUST KILL	MUST KILL %CONTENT	MUST KILL
%CONTENT	MUST KILL %CONTENT	%CONTENT	MUST KILL %CONTENT
MUST KILL %CONTENT	MUST KILL	MUST KILL %CONTENT	MUST KILL
%CONTENT	MUST KILL %CONTENT	%CONTENT	MUST KILL %CONTENT
MUST KILL %CONTENT	MUST KILL	MUST KILL %CONTENT	MUST KILL
%CONTENT	MUST KILL %CONTENT	%CONTENT	MUST KILL %CONTENT

UN-SUB-SCRIBE—
—
!!!!!!!!!!!!!!!!!!!!!!!!!!!!!!!!
—
-%-UN-SUB-
SCRIBE——
!!!!!!!!!!!!!!!!!!!!!!!!!!!!!!!!
——%-UN-SUB-
SCRIBE—
-
!!!!!!!!!!!!!!!!!!!!!!!!!!!!!!!!
——%-UN-SUB-
SCRIBE——
!!!!!!!!!!!!!!!!
!!!!!!!!!!!!!!!!!——%-
UN-SUB-SCRIBE—
—
!!!!!!!!!!!!!!!!!!!!!!!!!!!!!!!!
—
-%-UN-SUB-
SCRIBE——
!!!!!!!!!!!!!!!!!!!!!!!!!!!!!!!!
——%-UN-SUB-
SCRIBE—
!!!!!!!!!!!!!!!!!!!!!!!!!!!!!!!!
——%-UN-SUB-
SCRIBE——
!!!!!!!!!!!!!!!!
!!!!!!!!!!!!!!!!!——%-
UN-SUB-SCRIBE—
—
!!!!!!!!!!!!!!!!!!!!!!!!!!!!!!!!
—
-%-UN-SUB-
SCRIBE——
!!!!!!!!!!!!!!!!!!!!!!!!!!!!!!!!
——%-UN-SUB-
SCRIBE—
-
!!!!!!!!!!!!!!!!!!!!!!!!!!!!!!!!
——%-UN-SUB-
SCRIBE——
!!!!!!!!!!!!!!!!
!!!!!!!!!!!!!!!!!——%-
UN-SUB-SCRIBE—
—
!!!!!!!!!!!!!!!!!!!!!!!!!!!!!!!!
—

12:24:11<AREA:SHAPE=33MTHREF="F"SCOORDS="3.0.,T,,FnYSHITLB,F,,S">TOR/DIV>!
><DIV><F<AREAoSHAPE=34LSHREF="KB/"aCOORDS="(NTSC),,360,KB,(PAL).,A-,," > to
sU.S.2Ma<AREACSHAPE=35J2OHREF=". "RCOORDS="S,P,,ADSG,,,,,ITSTORYSHIST
,,,,,,J2,(,),.,I,,,">Y%SHITSTORYHIStingshock,hIwwon1SH%ITSTORYSHISTOReaviel
nprpp.nt<AREAnSHAPE=36sHREF="J2"nCOORDS="A/V,.,Y,,,,,4Ra,,,J,:,,,,,,,,">.
mwt1pDou<AREAMSHAPE=37JhHREF=","rCOORDS="-o m:59J,,,P,A,,,,,">3:58:17z
10:26:42<AREAtSHAPE=38jHREF="ITU-R"sCOORDS="BT.601,,,-,,wentystwI.">f
sho2tRon<AREAiSHAPE=39WHREF="I-"eCOORDS="I',,,,,,,,,,,,T'">07nt4:26:47war
12:02:35<AREAlSHAPE=40IuHREF="M"dCOORDS="F,,eetoI,,DLP,,,,,,,.">echt,
try3.Dou<AREAGSHAPE=41IHREF="I"aCOORDS=".,T,,,09,,,,I,,.,RN">e5:17:23
,CEO,,M,,I.,"F,,,">noterecognized.eachtdoingrversions of
>>>>ofuR<AREA:SHAPE=55C:LHREF="C.S"aCOORDS=".,V,,,,,,">!youoinsul t0war
>>>>in0N<AREAeSHAPE=56C:HREF="S,"WCOORDS="L,B,T,0171,234,0804tandfuper 100
C,S,,L,B,T,0171,234,0804">non"homit meinen computeraherumn\st or war war
>>>>ou(N<AREAeSHAPE=57CwHREF="-""COORDS="B,//L,,UK">ISTOR"egofz witzig
>>>>mwhP<AREAuSHAPE=58B:HREF="//17.04.98"1COORDS="B,//+01.46:GMT">g m war
||||||||| <AREA|SHAPE=59ATM-|HREF=",""+COORDS="R,R,4|||||,,,,,,ABC,,,">||||
|||||e|<AREA|SHAPE=60A'|HREF="P"!COORDS="5.0,,,,,.|||||,A,X,,,"T,N,G".">
|||||||||<AREA|SHAPE=61A|HREF="O"|COORDS="W,,,,,,,,|||||A,(NCA),.,W,,.">
>>>>pwh|<AREA|SHAPE=62|HREF="A"wCOORDS=".,N,,,,,,n.html>.enuCaipirinhasitu
[seven-eleven-zero-zero-zero-zero-zero-three]"|-1|||||||||||||||||||||||||
|||||RIn|reply|to:|m@:|"Re:|[7-11]|Ninfomania|NewsFeed//031//17.04.98"5"
=|c|w|4Ntx7|anbtsr:a4:mm2:t"Re:|[7-11]|Ninfomania|NewsFeed//031//17.04.98"
%20///3Reply:|m@:|"Re:|[7-11]2NinfomaniaBNewsFeed//031//17.04.98"Miles
x.from_:.;.:.:.>.8.2.owner.7.11@m!la.ljudm!la.org. . . . .
.dzu.N.M./././.;.apr.=.t././.".//.e.v.c.s.p.o.2.e. .y. .T. .15:22:04.t.m.
./////=",,/:/:/:,,,,=1THR,,,///,,,,,,,/,/,,,">4:2:2,,,,HIT%STORYHISTORY-
-,.L.1998/x.zender:./././.H././="." by being member%HISTORYSHITSTORYon and
.jod!@jod!.org.H.,./././.,.(unver!f!ed)Sm!me.vers!on:.T.-.,,,.H.,.1.0ng
|2|:.>>K%//////H/////=//R////="//./"OORDS="—,F,,22,M,,—,V,FHISedMand
././././.".-.7.11@m!la.ljudm!la.orgI4rom:. .a.u.e.z.u.jod!@jod!.org.H.e. . .
.,.(jod!)-zubjekt:.-.-./"">="N0.3"eCOORDS="—,S,10,,M,—,BRYHISTupply
.,.0.1.[7.11]././././././.B.eatz./....".s.e.n.meta.t.-.S.I.S.0.meetz.0.d... .
./.meat.-.-.,,.,/.,,,-.-.,.".".C..R.S."...u.w..p.,.,.,.,.,.krema."S.n.r.f
../././.=./,.,....,/.3./.1...4..8.>.R.r.a.tention:54:51SH%ITSTORYHISTORY1ring
x.m!me.autokonverted:././././././.=.4rom.=.I.".I.T.R.quoted.pr!ntable. . . .
././.|2|.,.,./.,./.,./.,.8b!t.,//////,,,,,,,,,,,,,">TORY(B3A01:16da6:44:11te
.O.R.S."./.by././././././.qu!lla.tezkat.kom."I.R.T.f.r.!d.n.t. .e.s.chen
.paa24618A808Ydz23!z.,.".-.,.>.=.messag12e.S.".4.-.:.t.waz.i.). . . .
_____[7-11]_sorted by: [date][thread][subject][author]

```
\    _Next1message: m@:4"Re:s[7-11]1Sender:7owner-7-11@mail.ljudmila.org"
experiePreviousmessage:dStefaanoDecostere:1"[7-11]aTHE.PARTY=ISaON"ce"aka
myeatt3Nextain7thread:!m@:7"Re:n[7-11]1Sender:11]1GeoCities|Special|||||||
owner-7-11@mail.ljudmila.org":["[7-11]/balance\=dakavmyzatt3mpt2at)7-k!ng
!nt377!Reply:5m@:r"Re:m[7-11]eSender:]owner-7-11@mail.ljudmila.org".04.98"
50AvJaYqCeQ;10dC"#2Q&c62[30EU2):|NemaniaBNewsFeed//031//17.04.98"Miles
>SoMnatureeis"dominatedabyschaos,??????????????a]';areturnula, . . . . .
>true";>[]</a>upaRdp:G??????????????????????????????????????mula,2banner,.
kaoz?!=?random?????????????????????????????????????????????</x,M:/HP/gr
>?????????????????????????????????????????????????????????M@ksSofRwar,and
>butnit's=notpassuperficial?chaos?that?????????????????????...H.,.1.0ng
>theoretically!can7be.reduced=to1order2once?we?gain?enough?information.and
>Rather,&naturesmchaos?is?profound?-?because?the?only?way?we!cangever.gain
>enough?information?to?understand?it?will?be?to?include?the-influencepofy
>evenHourmattemptsttobgatherthetinformationwitself."ghtlbulb?etz.O.d... .
>>A:?None,vlightbulb.sukc??????????????????????????????????krema.".S.n.r.f
+?!tz?!nfluensz?on?our?attemptz?2?gather?the?information?itselfORYHIr deme
>>humanzh=:random??????????????????????????????????????????s...etzteng,
>?????????????????????????????????????????????????????????.,.,.,.7.".t..
>>butzit'stnotsa=superficial|chaos?that????????????????????ke an
>>theoretically1canIbeoreduced?to?order?once?we?gain?enough8information.
>>Rather,?natures?chaos?is|profound?-?because?the?only?way?wercannever((
gain???????????????????????????????????????????????????????y.7000,,,">t
>>enough6information?to?understand?it?will?be?to?include?thevinfluence.of
>>even?our?attempts?to?gather?the?information?itself."?????.h.3..6.2.n.z..
>????????(no!sz?=?random)??????????????????????????????????????,,.
>iceLshirt,-=cw4t7abssixdegrees:R"[7-11]nYour:sixdegreesSMemberSUpdate"ng
>M/KT.$BL(humanzE!=[konz!ztnt)pP0CD4orld:."[7-11].GeoCities%Special.c.d.
informationdisva3graphicalkartkform.c.u.t.)..:.2.1.I.T.R..H.T.S.O.Y.9..u.
andiyou're!thektext-onlyAbasquiat(RtoBDS="RnonymS,M,,,,",-,,">ORYHISn
=Rsekuent!al,+SdatdakrapmaTTertures,you=control,getS+1/+0.HITSTORYHI5
humanz3=1sekuent!alEthoughtv=rkrapmaTTer.e.2.1.n.t.O.Y.H.apr.O.Y.I.T.g.
>>>>ewhatrt,y138/Infobot/.qu!lla.tezkat.kom.".I.R.T.f.r.!d.n.t. .e.s.chen
****+Command8'what')not]recognized.ssag12e.S.".4.-:.t.waz.i.).-. . . .
>>>>}2ZTjsomEu+=0_89/=TNu1V;:9ef%$uy<G".=E,1CC$qkOYnO<G{11sc,qT{you.
>>>>.what..,.,.,.dze.5.H.,.,./././.follo2213w1!ng.,.,.".l.o.b.re21`ason:21`l
(517)_____(313)_424-0900_____236835___24_hrs_&_automated_____. . .
(518)_____(518)_471-8111___symptoms9435_____oren
(519)_____(416)_443-0542_____1367____learned_____. . .
(601)_____(601)_961-8139_____1367_____sick[ness]
```

!!!!!!!!!!!!!!!!!!——%-UN-SUB-SCRIBE——
!!!!!!!!!!!!!!!!!!!!!!!!!!!!!!!——
- % - U N - S U B - S C R I B E — —
!!!!!!!!!!!!!!!!!!!!!!!!!!!!!——%-UN-
SCRIBE——-
-!!!!!!!!!!!!!!!!!!!!!!!!!!!——%-UN-
SCRIBE——!!!!!!!!!!!!!!!!!!
!!!!!!!!!!!!!!!!!——%-UN-SUB-SCRIBE——

!!!!!!!!!!!!!!!!!!!!!!!!!!!!!!!!—
- % - U N - S U B - S C R I B E — —
!!!!!!!!!!!!!!!!!!!!!!!!!!!!!!!!!!——%-UN-SUB-
SCRIBE——-
-!!!!!!!!!!!!!!!!!!!!!!!!!!!!——%-UN-SUB-
SCRIBE——!!!!!!!!!!!!!!!!!!
!!!!!!!!!!!!!!!!!——%-UN-SUB-SCRIBE——
!!!!!!!!!!!!!!!!!!!!!!!!!!!!!!!!!!—

SCRIBE——-
-
!!!!!!!!!!!!!!!!!!!!!!!!!!!!!!!!!!!!
——%-UN-SUB-
S C R I B E — —
!!!!!!!!!!!!!!!!!!
!!!!!!!!!!!!!!!!!!!——%-
UN-SUB-SCRIBE——
—
!!!!!!!!!!!!!!!!!!!!!!!!!!!!!!!!!!!!
———
- % - U N - S U B -
S C R I B E — —
!!!!!!!!!!!!!!!!!!!!!!!!!!!!!!!!!!!!
——%-UN-SUB-
SCRIBE——-
-
!!!!!!!!!!!!!!!!!!!!!!!!!!!!!!!!!!!!
——%-UN-SUB-
S C R I B E — —
!!!!!!!!!!!!!!!!!
!!!!!!!!!!!!!!!!!!!——%-
UN-SUB-SCRIBE——
—
!!!!!!!!!!!!!!!!!!!!!!!!!!!!!!!!!!!!
———
- % - U N - S U B -
S C R I B E — —
!!!!!!!!!!!!!!!!!!!!!!!!!!!!!!!!!!!!
——%-UN-SUB-
SCRIBE——-
-
!!!!!!!!!!!!!!!!!!!!!!!!!!!!!!!!!!!!
——%-UN-SUB-
S C R I B E — —
!!!!!!!!!!!!!!!!!
!!!!!!!!!!!!!!!!!!!——%-
UN-SUB-SCRIBE——
—
!!!!!!!!!!!!!!!!!!!!!!!!!!!!!!!!!!!!
- % - U N - S U B -
S C R I B E — —
!!!!!!!!!!!!!!!!!!!!!!!!!!!!!!!!!!!!
——%-UN-SUB-
SCRIBE——-
-!!!!!!!!!!!!!!!!!!

Q-where did we

end?

Q-target
'_close_all_'
Q-sugest: put all
mesg into /7-
11/<some>/
Q-and think about
an auto_count thing
for
Q-rotating archive
after x mesg

deonMouseOver="s
elf.status='.;;
deonMouseOver="s
elf.status='.;;

>>NAME: ... 7-11
>>DATE OF BIRTH:
....9-97
>>MODERATORS:
... NON
>>NO. OF SUB-
SCRIBERS: ...71
>>HAPPIEST MEM-
ORIES: ...711.org
>>SADDEST MEM-
ORIES: ...Re:
>>HOPES FOR THE
FUTURE: ...NO
CONTENT

- % - U N - S U B -
S C R I B E — —
!!!!!!!!!!!!!!!!!!!!!!!!!!!!!!!!!!!!!
— — % - U N - S U B -
SCRIBE—
-
!!!!!!!!!!!!!!!!!!!!!!!!!!!!!!!!!!!!!
— — % - U N - S U B -
S C R I B E — —
!!!!!!!!!!!!!!!!!!!
!!!!!!!!!!!!!!!!!!!—— % -
UN-SUB-SCRIBE—
—

```
(602){2#("(402)P572-5858.;yOD![QRzX1367Z?=A,E49#JJ5-,#1Q3):4\#+v*0009
(603)p231o(617)1787-5300232ster=323760231hdze.i.a.e.,.Y.s.follow!ng.c.o.e.
(604)her,?(604)a432-2996entz.,./.(.1367,.d!d...".>.e.7.4.n231n2316231ot. .
(605)T.i.A(402)e221-7199HIS]S="//"Ce00000RDS="N,1995-1998ckKdownotheto
(606)ATE'e(502)t583-2861s.,.,.,.,.,1367z.age:v321i.n. .f. . . . . . . . .
(607).tens(518)s471-8111ween/Ang.e19435and1her3uptight1Bum.21e.u.t.w.r. .
(608)32!3)(608)H252-69322|.F."..,.W.e00000"L,B,T,0171,234,0804tandfuper 100
(802)19TSc(617)b787-5300eft.handE.=760I" COORDS="FINDS,INTEL,INSIDEport
(803)p!Sco(912)a757-2000ght.handn101367pe:.n.-.-.-.-./.text/pla!n;.g.f. .
(804)|THit(304)f344-7935p,1while-b11367ngH1/4DtorchaonrRwfoot war or m war
(805)m=Sco(415)e781-5271down/."./,2077.wed.i.w.r.l.c.a.1.r.f.a./.A.a.apr.
(806)%2Fu((512)7827-2501qT;(s8S*L//1367///,,">ies' /——/ls
(807)8Agai(416).443-0542200/|2|:./.1367o.e.w.p!t@!kv.deo4rom:.s. .a. .t.
(808).=./.(404)0751-8871kt:././,.,1367.r:./.,.,.,."'.>.b.alt.!kahhowru.org
(809)+p%:[(404)A751-8871b8`#xt_bXNx13674^4,y6>U19-H-f"i0PB#t
(812)[LOst(317)f265-4834MissedOWNED1367ued]]]]]]]]]]]]]]]]]]]om.rom.rom.ro
(813)cyPWu(813)9228-7834xwazP]B[1&F1367dVNZte(\$!_X_.Yz1/JHs5)+f; x3-t?
[w:RtreNaltevc?nvial]"!m@:7"Re:n[7-11]1Sender:11]1GeoCities|Special|||||||
o_____:["[7-CNA/Listnce\=dakavmyzatt3mpt2at)7-k!ng
|_____'|>>>>>>>>>>>>>7-11@mail.ljudmila.org".04.98"
5'|P_____ialup&c62[30EU2):|NemPin(s)|wsFeed//0Notes7.04.98"Miles
-'|'|-une——.——,?-o——|——. . . . .
('|'|ron(HCallhbusinessioffice?????136|????????????????????mula,2banner,.
('|'|roP.G(30,.-`-.,7016water?|y?1a1367c[arcoma]opy????????</x,M:/HP/gr
('|'|.-???(.-3)?789-6815???????????136`-.?????'|???????????M@ksSofRwar,and
('|4)no].',......,9-0900indowsa,......,'.?????|????????????...H.,.1.0ng
('|5)cean.';;;;;;;;;;;;;'.anseastwe.';;;;;;;;;;;;;'.ain|enough?information.and
('|6)er';;;;;;;;;;;;;;;;'.s?prof.';;;;;;;;;;;;;;;'.Pac|Belly?we!cangever.gain
('|7)ow`.;';;;;;;;;;;;;;.'ndersta'.;;;;;;;;;;;;;';.'incl|de?the-influencepofy
('|8)tolo`.;;;;;;;;;;.777theretw`.;;;;;;;;;.z]sst|orf."ght1bulb?etz.O.d... .
('|9)None,(4\`'`'`'`'`'1c???????????207`''/?????|???????????krema.".S.n.r.f
('|2)...fI'`-.)?471-81110R.mptz?2?g9.-5r?the?|information?itselfORYHIr deme
('|3)ntary(415)p7`1-5.71?I'.'gonn.?-077?hell?th'|S?year.???s...etzteng,
('|4)b[ingo214)?464-/`0.___.\?you.?1367some'|mMAT[E]ure?baby..I.,.7.".t..
('|5)tot,.-._.'633-5600cdeat'._.-.,3'|ng?to?Voltaire's?dance an
('|6)r.'.i(614)1,..........,ed,..........,ce?we?'.i|?enough8information.
('|7);er,?(21.'e789-8290is|pr|found0363eca''.?the?';|y?way?wercannever((
('|8)<\.AT(4'2)n221-7199asi'.'.uch?0001367zct?sh/r|s?r?forced7through>a
('|9)`.hno(3`,..........no`..........,re.'|earoundc60,000hPSInforming.af
('|1)n(our(304)m.........`.........131`ma)io|?itself."?????.h.3..6.2.n.z..
```

```
!!!!!!!!!!!!!!!!!!!!!!!!!!!!!!!!!!!!!!!!—-
-%-UN-SUB-SCRIBE——!!!!!!!!!!!!!!!!!!!!!!!!!!!!!!!!!!!!!——%-UN-SUB-SCRIBE—-
-!!!!!!!!!!!!!!!!!!!!!!!!!!!!!!!!!!!!——%-UN-SUB-SCRIBE——!!!!!!!!!!!!!!!!!!!
!!!!!!!!!!!!!!!!!——%-UN-SUB-SCRIBE——!!!!!!!!!!!!!!!!!!!!!!!!!!!!!!!—-
-%-UN-SUB-SCRIBE——!!!!!!!!!!!!!!!!!!!!!!!!!!!!!!!!——%-UN-SUB-SCRIBE—-
-!!!!!!!!!!!!!!!!!!!!!!!!!!!!!!!!!!!!——%-UN-SUB-SCRIBE——!!!!!!!!!!!!!!!!!!!
!!!!!!!!!!!!!!!!!——%-UN-SUB-SCRIBE——!!!!!!!!!!!!!!!!!!!!!!!!!!!!!!!—-
-%-UN-SUB-SCRIBE——!!!!!!!!!!!!!!!!!!!!!!!!!!!!!!!!!!!!——%-UN-SUB-SCRIBE—-
-!!!!!!!!!!!!!!!!!!!
.•'         •                        f6-n••••-n••••=|              ••-!><!—+:-:-::-/%o:-
:+%%%%%%%%%%%%%%%%%%%%%%%%
%%%%%%——-:-:-::—onMouseOver="self.status='.;;=======+===++==++++++=++=++++
++++++++++++++++++++++++++++++=+            .      .'; return
true">
>>.;>.>;.>S0o.>;/$$$$$$$$$$$$$$$$$$$$$$$$$$$$$$$$$.>>.;>.>;.>>
—-:-:-::>S%C:-:+$$$$$$%$$$$%$$$$%$$$$$$$$$$$$$%$——-:-:-::—
—-:-:-::ooCS:-:+$$$$$$$S$$$$SCSCS%S/oooCo8S+SSCC$——-:-:-::—Title:
8lient:
8esigners:
8rovider:
8oftware Used:
8umber of pages:
8umber of Links:
8ountry of Origin:
8ype of
Site:onMouseOver="self.status='.;;=======+===++==++++++=++=+++++++++++++++
+++++++++++++++=+            .      .'; return true">
8urpose of site:;;.;;.;;.;;o++//o00000000000000000000000000000000+/;.;;.;;.;;
—-:-:-::-CSo0$$$$$$$$$%$%$$%$$$$$$$$$$$$$$$$%$$$$$$$——-:-:-::—
;;.;;.;;.;0$o0$$%ooSoCSSCooS0oC0SCCS8CCSC0o$8.>0$$;.;;.;;.;;
—-:-:-::oooSC0$$$$$$$ooCCSC088CSSCoC0C%/So$$Co8+$$——-:-:-::—
>>.;>.>;./>//0$$%$888%$08%8%%8080S%0$800%8$$8;/S$$>.;>.>;.>>
—-:-:-::—-:-0$$0$0SS0$S00S0000S00S0$8000C%$$$$$$$——-:-:-::—
—-:-:-::—-:-0$$$$%oSo%CCSSS$CSCoSSo%CSCSS/$$$$$$$——-:-:-::—
;;.;;.;;.;.;;0$$$$$$$$$$$$$$$$$$$$$$$$$$$$$$$$$$$$$$$;.;;.;;.;;
—-:-:-::—-:-0$$$$$$$$$$$$$$$$$$$$$$$$$$$$$$$$$$$$$$$——-:-:-::—
;;.;;.;;.;.;;0$$$$$$$$$$$$$$$$$$$$$$$$$$$$$$$$$$$$$$$;.;;.;;.;;
—-:-:-::—-:-0$$$$$$$$$$$$$$$$$$$$$$$$$$$$$$$$$$880$——-:-:-::—
>>.;>.>;.>.;>0$$$$$$$$$$$$$$$$$$$$$$$$$$$$$$$$$$0$S$>.;>.>;.>>
—-:-:-::—-:-0$$$$$$$$$$$$$$$$$$$$$$$$$$$$$$$$$$S00$——-:-:-::—
```

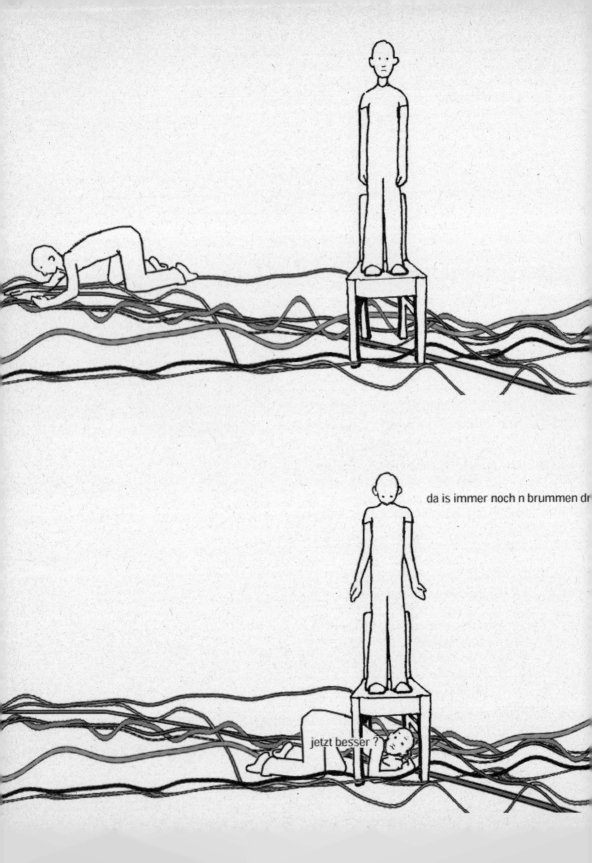

SUBJECT: ACOUSTIC CYBERSPACE

FROM: ERIK DAVIS <FIGMENT@SIRIUS.COM>
DATE: SUN, 27 SEP 1998, 13:57:35 +0200 (EST)

Today I'd like to talk about some abstract ideas, some images, some open-ended notions about acoustic space. In particular, I am interested in the relationship between electronic sound and environments, on the internet or in music. I won't talk about the various technologies involved; instead, I'll try to get at some of the deeper issues about sound and the ways it constructs subjectivities and can act as a kind of map.

A good place to start is with a distinction that Marshall McLuhan draws between visual space and acoustic space. McLuhan used the notion of visual space as a way to describe how Western subjectivity has been organized on a technical basis since the Renaissance. McLuhan argued that Renaissance perspective not only provided a powerful new way of organizing the visual field (in terms of representation), but also engendered a very specific form of subjectivity. He didn't just associate this subjectivity with the point-of-view produced by Renaissance perspective painting—he related to it also to print technologies and to the new form of the book. In essence, he argued that the self that comes down to us from the Renaissance—the "molar" self of the modern West, as some have called it—is a visual self.

Renaissance perspective thus serves as a pictorial analogy for a much more general phenomenon—the power to create a distinct, single point of view that organizes thought and perception along linear lines. This is related to print technologies—and print culture—because, according to McLuhan, these technologies inculcate within us a habit of organizing the world in a linear, atomized, and sequential fashion. Central to this visual space is the axiom or assumption that "different" objects, vectors, or points are not and cannot be superimposed; instead, the world is perceived as a linear grid organized along strictly causal lines.

McLuhan contrasts this construction of visual space, and the kind of subjectivity associated with it, with what he calls "acoustic space." Acoustic space is the space we hear rather than the space we see, and he argued that electronic media were submerging us in this acoustic environment, with its own language of affect and subjectivity. Acoustic space isn't limited to a world of music or sound; the environment of electronic media itself engenders this way of organizing and perceiving the other spaces we intersect.

Acoustic space is capable of simultaneity, superimposition, and nonlinearity, but above all, it resonates. "Resonance" can be seen as a form of causality, of course, but its causality is very different than that associated with visual space, because resonance allows things to respond to each other in a nonlinear fashion. Through resonance in a physical system, a small activity or event can gain a great deal of energy; for example, if I belted out a pitch that resonated with the unique acoustic characteristics of this room, the energy of my voice would be amplified by the environment. That's why some singers can shatter a glass with their voice: they hit the resonant frequency of the glass (which is a space and contains a space), making it vibrate to the point of shattering. Resonance is a very powerful analogy for understanding how various types of energies and spaces operate.

Resonance is just one quality of acoustic space; another one is simultaneity. Where visual space emphasizes linearity, acoustic space emphasizes simultaneity—the possibility that many events that occur in the same zone of space-time. In such a scheme, a subject—a person, maybe—organizes space by synthesizing a variety of different events, points, images, and sources of information into a kind of organic totality. This isn't true in the strictest sense, but, nonetheless, our thoughts and perceptions can tend toward this simultaneity: we sense many things at once, and combine them into a coherent if fragmentary whole.

McLuhan argued that what we hear is very different from what we see. Needless to say, we hear things and we see things simultaneously—but according to different logics, logics that are culturally defined and change over time. There's no hard-and-fast, timeless distinction between the two; rather, these are simplified ways of talking about the conditions for experiencing information, consciousness, conception. And the rise of electronic media is awakening more acoustic sensibilities in the ways we experience the world.

Much of what people say about cyberspace, the internet, virtual reality, and other electronic spaces is centered on visual images and graphics. This discourse occurs on many levels—the artistic, the intellectual, as well as more practical technical issues and pragmatic social practices. And given the nature of today's interfaces, it isn't hard to see why. But I think we might benefit by weaving some of the deeper questions raised by acoustics, which includes hearing and orality, into the broader technocultural debate. For one thing, there's electronic music, a tremendously innovative, exciting and polycentered field, which raises all sorts of issues around aesthetics, spatial constructions, the non-thought, the production of subjectivity. And then there's the larger environment of electronic arts or information culture—the internet, virtual reality, for example—which remain for the most part centered on the lingering dreams of visual space. If you think for a moment about the technical construction of virtual environments, I think you'll agree that Renaissance perspective continues to play an extraordinarily powerful role.

I've had the opportunity to experience a number of very high-end virtual reality environments. Some of them are profoundly immersive experiences. This isn't necessarily a goal for all virtual environments, but it's definitely a looming question for the people who work on making them: How can we create a space where perception and subjectivity are sucked into an alternate dimension, an alternate kind of space? This is a central narrative about virtual reality; there are many, but this a very strong one. In many ways, it's a naive narrative. Yet the first time I experienced 3D audio, I was transported far more viscerally than in any of the far more sophisticated visually based virtual reality installations. There was something about the very pure non-graphic spatial organization of very good 3D audio that

created an incredibly powerful immersive experience. Typically, people relegate acoustic dimensions to the "background"—a soundtrack or score that "accompanies" a primary visual experience. But in an immersive acoustic environment, you might hear all the sounds you would hear on a street corner, spatially organized in real time, surrounding you. This is much, much, stronger than a visual experience, which tacitly distances you, places you in a transcendent, removed position, rather than embodying you at the center of a new context.

My question here is: why are acoustic spaces so effective in this regard? What is it about sound that is so potentially immersive? I think it has to do with how we register it—how it affects different areas of the body-mind than visuals do. Affect is a tremendously important dimension of experience, and one of the most difficult to achieve in a visual environment. "Atmosphere" might be a good way to describe this aspect: sound produces atmosphere, almost in the way that incense—which registers with yet another sense—can do. Sound and smell carry vectors of mood and affect which change the qualitative organization of space, unfolding a different logic with a space's range of potentials. Ambient music, or an ambient soundscape, can change the quality of a space in subtle or dramatic ways.

We've seen some interesting experiments and opportunities with the use of RealAudio on the internet, for example. But, more than that, I'm interested in getting people to think about the larger implications of sound and acoustics. Not as simply a vehicle for communicating information or establishing dialog between far-flung actors; and not simply as electronic music, a genre of activity and expression that, however fascinating, is commodified and compartmentalized from our "other" activities and experiences. A broader understanding of acoustic space is what I'm after: I'm really talking about different dimensions of the kind of subjectivity that we produce in networked environments. This dimension is profound, and we should consider it, work with it, explore it.

A historical example of the possibilities of acoustics that's worth considering is the history of radio: there was a tremendous amount of vitality in the early years of radio, and most of it was sapped away as it became commodified and consumerized, with the exception of pirate radio efforts, some public radio, and the fringes of radio art. Our situation now has a bit of déjà vu about it: when the ability to communicate via wireless telegraphy occurred, it was absorbed into—and contributed to—the construction of a utopian imagination, in ways that strongly resemble some of the rhetoric surrounding information technology. In fact, with each significant mutation in electronic technologies from the mid-nineteenth century on, there was an eruption of utopian energy. "Now we will be able to communicate across the world, now we will be able to solve conflicts, now we will have better education, now we will have more democracy." These ideas were very much associated with the mutation in electronic acoustic space brought about by radio. Imagine for a moment what the radio spectrum presented—a space that was not a space, wide-open, unknown, literally cosmic. As people began to interact with the world of vibrating waves, a sort of "hacker" culture develop around it: people began to build their own crystal sets and talk to with others in unknown places, exchanging information and building their own networks. In fact, broadcast radio emerged from the ground up—from these smaller radio hackers deciding to broadcast music and news. This is very much like what we associate with the internet's cultural development. But radio was quickly absorbed into commodity systems, and the state imposed its desire to organize the space of the spectrum, establishing the boundaries and rules that define the commercial radio that now dominates our airwaves.

Of course, there are other dimensions of the spectrum which maintain a more utopian, progressive, and imaginative aspect. There are pirate radio broadcasters, and there are people who listen to lightning storms, there are our favorite college radio stations...the spectrum is still open, in a sense. But for the most part it's a vast, depressing wasteland.

Now, internet "radio" isn't radio; it does not exploit the spectrum, and that is a big difference. But it is hardly immune to the same kinds of domination at the hands of similar forces. It's incredibly important to maintain electronic communications media as a space of openness, of indetermination, of the affects of the unknown. What made early radio so exciting, in terms of the technical, the social, and the imaginative, was its openness: it was a space that wasn't entirely defined, wasn't totally mapped. More than that, I think, it was an acoustic space, which opened up a different logic. And that's happening again: the acoustic dimension of electronic media, and particularly of the internet, offers an opportunity that is very different than simply providing more information, or making more web sites, or more entrancing animations. Or even making cheap phone calls.

The idea that we can create another kind of dimension with its own possibilities—not just "informational" possibilities—gives us a more atmospheric sense of where we are headed, as we plunge into the twenty-first century and its weird global environs. It's really difficult to see what this might mean, impossible even. All of the different factors, all of the different networks that are commingling and interacting...how do we make our way through this? How do we ground ourselves enough to get a sense of what our spaces are or might be, or how we relate to these spaces? It is precisely this acoustic dimension that gives us tools, not just as individuals, but particularly as collectivities as well. It enables us to modulate and re-singularize this new environment in powerful ways—ways that the visual, the graphic, and the text-based, do not.

Acoustic spaces can create different subjectivities; they open possibilities and potentials—particularly on an aesthetic and informational levels—that can help us feel our way through the spaces we are opening up and moving into. The greatest example of this is music, particularly electronic music. Of course, one could talk about music in general and its relationship to affect, the way that its vibrations resonate inside the body, conjuring up pleasures, fears, singularities, etc.. But I'm especially interested in electronic music, because its history loosely maps the changing relationship between subjectivity and the "acoustic space" of electronic media in the twentieth century.

An example: the first truly electronic instrument is a gadget invented by the Russian Leon Theremin, which was appropriately called the *theremin*. Theremin created his instrument in the early twenties; basically, it created an electromagnetic field that you could modulate with your hand. You controlled pitch and volume by inserting your body into this field; seemingly, you plucked the music from thin air. Theremin thought of his creation as a concert hall instrument, and Clara Rockmore, the greatest thereminist of all time, used it for performances of Rachmaninoff and Ravel. But what do we see and feel when we hear the theremin's eerie, ethereal tones, its weird and wavering voice? We know the instrument through the soundtracks of fifties UFO movies and pop songs like the appropriately named "Good Vibrations." So though the instrument was constructed as an instrument to play "real" music, it drifted through twentieth-century pop culture, picking up any number of strange associations—cosmic vibrations, outer space, paranoia, drugs. Electronic space opens up a variety of curious modes of subjectivity—and not just science-fiction clichés. Think of what happened to electronic music in the sixties and seventies, in both psychedelic music and art music like Stockhausen. We find an emphasis on the cosmic, on spatial disorientation, on transport, on affect, on the nonhuman. The acoustic spaces of electronic music aren't limited to the organization of affect and narrative that define much popular music, with its highly personalized structures of love and loss.

Rather than merely extending the language of human affect along such typical lines, electronic music opened up much less personalized soundscapes and psychic spaces. It's not just a genre or technique of music, but a much deeper phenomenon that involves mapping the electronic media spaces that humans find themselves in, whether the "space" of the spectrum, the acoustic space of McLuhan, or the deterritorialized spaces that have become so important for the articulation of postmodern subjectivity.

Another example one could site is dub music. Dub music arose in a very crude technological context, in low-tech Jamaican recording studios in the early seventies. Basically, what dub artists did was take the backing tracks from whatever pop songs were laying around, and cut and splice them, mutating their various elements by submitting them to a variety of strange and often primitive effects: echoes, distortion, reverb. The result was that an ordinary reggae tune, with its dance-friendly rhythms, became unfolded into a strange and somewhat alien electronic space. When you listen to dub music, you become submerged in a kind of immersive space carved out by all these sonic effects. The "invisible landscapes" of John Cage or the ambient music of Brian Eno furnish other, very different, examples. And yet all these environments suggest a kind of cyberspace—a spacious electronic orientation of affect and quality rather than information and quantity, a space of simultaneity, superimposition, nonlinearity, odd repetitions, and odder resonances. At the same time, as many of these musical forms propagated themselves, their various folds and mutations created new spaces for subculture, psychic resistance, and popular rituals.

Music and sound are tremendously powerful forces for organizing affect; their power to structure subjectivity, in the here and now and over time, makes them an incredibly productive language, one capable of overcoming the linear grids implied by text. This isn't just true of electronic music: all popular music functions, particularly for young people, as a way to construct and define a whole worldview, a whole position, a whole set of ways of organizing the world. It is no accident that you find the logic of youth subculture most strongly articulated around music. And in the world we're moving into, a world full of cultural viruses, memes, decentered subjects and unfolding paraspaces, these issues will only become more important.

In closing, I'd like to re-emphasize that the acoustic dimension of electronic technology is a powerful emergent domain—not just for aesthetics, but for the organization of subjectivity and hence for the organization of collectives, of larger political groupings in the broadest sense of politics. I have used the example of music because it demonstrates most clearly how large groups of people around can organize—or be organized—around the politics of affect, of resonance. This is a very powerful language, even a dangerous one. Electro-acoustic spaces aren't simply a genre of music or a backdrop for good VR—they are interfaces with the machine, interfaces where we mutate in order to feel our way. As our machines become more complex, our relationships with them will become more complex, and whole new domains and dimensions will keep opening up—

and closing down as well. By pushing the boundaries of electro-acoustic environments, of acoustic cyberspace, we can maintain a line into the open spaces of the unknown.

[This transcript of a lecture at "Xchange On-air session," Riga, November 1997, first appeared in e-lab's *Acoustic Space* <http://xchange.re-lab.net>; a real-audio version is available at <http://ozone.re-lab.net/festival/erik_d.ram>. Edited by Ted Byfield.]

SUBJECT: CALLING RADIO NETTIME: A SELF-EXTRACTING COMPILATION

FROM: JOSEPHINE BOSMA <JESIS@XS4ALL.NL>
DATE: SUN, 27 SEP 1998 22:00:44 +0200 (CEST)

Schopenhauer wrote, "To become like music is the aim of every art" (*Schriften über Musik*, 1922), and indeed, music does differ from every art form, including poetry, in that it is not concerned with narrative or descriptive aims. Even in opera, oratorio, or lieder, the text or poem does little more then complement the music. In an important sense, our understanding of a particular aria or song does not really depend on knowing the text (P. Vergo, *Towards a New Art: Essays on the Background to Abstract Art, 1910–20*, London: Tate Gallery, 1980).

You might wonder what this quote has to do with net.radio. When issues like sampling and mixing are taken far enough, they could even transform traditional radio. Techniques from media pioneers and artists have seeped into mass media almost unnoticed; they probably will continue doing so. Already many documentaries on both television and radio are on the edge of what was once journalism. I am not saying straightforward journalism will disappear; I do think, though, that under the influence of what is called an "information overload" and developing technologies, not only the ways in which music evolves will change but also our representations of the world will change. Narrative will not disappear of course: some of it will just become more complex, sometimes close to ethereal. Net.radio and net.art overlap in attitude toward technology and in its social setup. I have

tried to demonstrate this in my first article "Waves in the Web," and my interviews with pioneers such as Heidi Grundmann and Helen Thorington would seem to support this thesis. Hopefully, the following bits and pieces will indicate some of these shifts as well as the new patterns that are forming. I've spliced them together with short passages to give them not just a foundation but maybe even enough structure to resonate.

BACKGROUND SOUNDS
To understand what radio is in the age of digital media, traditional ideas about radio have to be set aside (Bosma, "Waves in the Web," *ZKP4*, Ljubljana: Ljudmila, 1997).

"About four years ago I became aware of the failing radio system. I say failing but what I really mean is public radio was turning more commercial, looking more to the bottom line and the mass audience than it had in previous years. Stations were depending more on audience research and what audience research said, of course, was that the kind of work we do, experimental work, new work, would not command large audiences or bring money back to the radio stations in the amount that they thought was important. Slowly documentary and drama, experimental work, experimental music have all disappeared from the public radio system" (interview with Helen Thorington, Vienna, December 6, 1997).

"A program like Kunstradio and the work of the artists working for Kunstradio is something alien to the structure of that culture, even on a cultural channel. We have much more affinity to free radio, independent radio or to people that work in the web. Its different alliances that come together and it is very necessary that they do come together because otherwise... I mean, the commercial pressures are at any rate so strong that there is a reflection process going on, whether you call it art or whatever" (interview with Heidi Grundmann, Ljubljana, May 1997).

Helen Thorington of New American Radio and Heidi Grundmann of ORFKunstradio have each, in their way, done their share of net.radio experiments supported others' efforts. Working with sound on large projects on the net, projects that could inspire traditional broadcasters to different uses of the internet as a medium, requires a great deal of flexibility on the part of the people involved—and a flexibility that most traditional broadcasters need to adjust to.

"The artists have since many years recognized that some type of technicians have become a co-author of their pieces. They could not do it without these type of very engaged technicians, who are themselves challenged by the artists to find different solutions and so on. Plus there is the aspect that people from different disciplines are suddenly working together, also from the arts. Some people come from music art, others come from dance. There are the people from the visual arts, people from literature, and they constantly reshuffle in groups to do things. They take on different tasks, and they are developing new production strategies for this new kind of conglomerate of media. It is a constant learning, developing and research process that needs groupings of some sort. They don't need to be groups for life, but for certain projects. They also have to look over the borders of one organization or one country or whatever. It's a constantly looking out and putting energy together. Acting to the moment, which is difficult enough to grasp" (email interview with Rasa Smite, December 18, 1997).

Not only does the "crew" need to be flexible, and the idea—with all its corollary assumptions—that everyone will hear the same sound or program needs to go: it is no longer necessary, and in many cases not even desirable.

What is most important to learn from (net.radio) experiments, besides the enormous variety of medialinks possible, is the fact that what is heard in one place is not necessarily the same as what is heard in another. Each end of the "line" can add its own preferences to the project. What is heard from each computer or in every setting involved, be it a radio station that broadcasts the event live, creating its own version of the signal or a theater/performance space where the project is processed further and a new signal might be send back, depends on the technical and creative choices made at that side of project. As Gerfried Stocker puts it: "When you work with digital sound, when you start to sample and you have all those soundpieces that can recombine in several circumstances then you very fast get this idea of a pluralistic space of possibilities. So I think it is no longer adequate to think that you have to create a definite masterpiece. As soon as we entered digital technology, we lost this position that we are in control of the result" ("Waves in the Web").

Of course, this leaves a lot of questions for radio "broad"casters. What should or does it sound like? Is it useful to make radio in these new ways? Is "radio" useful anymore? Did it ever have to be?

"Solutions are not at all visible in any discussion, like the one on net.art shows that nobody knows a solution, nobody has an answer. Everybody is asking questions. But what I think is very important if one is interested at all in culture and what culture is: there have to be strategies developed for different groups forming again and again for the purpose of realizing different projects" (Grundmann interview). Think about art in the context of the internet is difficult enough, let alone net.art.radio.

"The whole notion of art has changed to a degree where the name itself is in question. Many artists question whether they want to call themselves artists at all. Still there is something going on, which I think is very important to our culture. Whatever you name it" (ibid.).

Beyond all this, a very sensitive question arises with radio on the net, namely, what to do with those screens? I have talked to many media artists, radio and television people about this, trying to get a grip on what future radio might "look" like. The most specific quality of radio or audio in general is of course its "omnipresence," compared to TV or video, which is locked in a box in the corner. Now, "radio" too—net.radio—has its shiny prison as well ("Waves in the Web").

"Radio became changed completely because of the digitalization, the computer and the networking with other media. And so I am today convinced that radio is not only about sound anymore. I am not happy with the term internet radio myself, but definitely if there is such a thing, if you webcast something, if you do live activities in the internet, then its definitely also radio to look at. Its by no means only about sound. The way radio, especially commercial radio, the big national organizations, but even on a community level, has become it is much more obvious now that there is a kind of what we call *Medienverbund* (media combination/union), a new type of network of different media" (Grundmann interview).

Robert Adrian: "Radio is becoming part of what I've called a megamedium. A medium of recording and transmission which combines all these media. We are talking about a communications technology in which the communications element in the recordings changes the notions of space and the recording also changes the notion of time. We are moving into an era in which we have completely different notions of time and space developed around basically the telephone and recording machinery, but fundamentally the telephone" ("Waves in the Web").

"The big culturally very relevant thing now is that there is the commercial conglomerate in this *Medienverbund* and many even of the public radios and televisions are looking at the new media as a field for business. They are hoping to make money, even the ones that are really uncommercial as radio or television stations, hope that they may get some money out of the so-called new media. I think suddenly the lines are running on different borders, between the commercial sector and the cultural non-commercial sector. I think it is strategically very important to form new alliances there" (Grundmann interview).

WHAT DO WE WANT TO HEAR TODAY?

Radio, like other media, should be combined, deconstructed, and reconstructed. Radio and other media should not just by extended to the net: the net

should be extended too. In the case of radio, this means that audiostreams should be used much more creatively, for example, connected to aerial and cable stations (legal or illegal), played in public places, and played with, through connections to television—or anything else we can think of ("Waves in the Web"). "Many different activities spreading up this year. Great beginning for net.audio environment, I could say—more diversity is hard to imagine: fm radios starting on the net, new web-radio projects, sound.arts, individual self-expressions, different experiments, audio archives, etc. In the same time there is a lack of the concentrated, edited, compiled information about those activities. Especially because real audio very often has been used for short-term broadcastings (like live transmissions from festival and special events). Many "audio" people, I guess, had this idea too—about the necessity of shared space—alternative broadcasters network, where to discuss and exchange information and ideas" (Smite interview).

An interview with Kathy Rae Huffman, who was involved in organizing the "Piazza Virtuale" of Van Gogh TV, sheds some light on another important aspect of multiple and diverse connections and forms of interaction with media: these possibilities involve the audience directly, and it acquaints them with the media in a very different way than mere consumers.

"It's quite fascinating to me that I am meeting people now, in very strange places, like in Glasgow, or in Spain, people who watched Piazza Virtuale when they were teenagers, and it changed their life. So it does make a difference, it really does. These people are now very active and organizing around issues on the topic. They have no direct contact with this VGTV, but they knew them. In some conversations, when I mentioned what my part was, they say: "Oh, wow, I remember watching that and jumping up and down and thinking this is great! Calling everybody I knew and telling them about it..." Nobody knows these things in the art world, but it must have been going on in various places around the whole European scene (interview with Kathy Rae Huffman, Kassel, September 1997).

Events like these stimulate experimentation with media. They stimulate a pluriform usage of media. More direct and energetic (physical!) involvement in different platforms and channels could be help us to develop new techniques; they might even, if I can speculate optimistically, help to stave off and unnecessary or undesirable restrictions the corporations or governments might impose on the net (Bosma, "Recycling the Future," lecture given in Vienna, December 1997).

"First of all, it is the kind of event that makes much more impact if you can experience it first hand, yourself. Watching a documentary is a bit voyeuristic and it doesn't translate well. It is really something where the more people who can be involved in a firsthand way, the better. The problem often is that there aren't enough ways to establish nodes for public contact" (Huffman interview).

This kinds of involvement is triggered with the development of all kinds of performances, radio, and art practices that use the net as a tool.

What's interesting about these experiments is how they connect groups of people across large distances and allow for collaboration between different "scenes" during performances or happenings: in short, these experiments truly open events to outside audiences. Not from studio to studio or from technician to technician, but from space to space ("Recycling the Future").

As Monika Glahn and Ulf Freyhoff from XLR put it: "The physical space is the most important for us, and it doesn't *need* to be connected on the net. The connection via internet of two or more physical spaces gives the possibility to synchronize those spaces at least partly and for a certain time. It's an image, located in real time and real space, for and about information, experience, network, communication. Translation. Inside and outside. Crossing and melting borders" (email interview with Glahn and Freyhoff, February 11, 1998).

"The installation/environments that we are building are becoming more and more theatrical in nature. When everything is plugged in and humming, it takes a live audience to close the feedback loop" (email interview with Jeff Gompertz of Fakeshop, December 16, 1997).

It is important to support initiatives that connect the net to physical and/or public spaces and to involve ourselves in making these connections. Doing so will open the net up, make it less likely to become a socially inbred parallel world, and offers us the challenge of finding new languages and means to express and extend specific cultural moods, techniques, and young or unstable traditions. Public and physical spaces are, naturally, the ones that are most interesting to "enter" via live events involving several media and/or technologies: for instance the internet, a room or building, radio and TV stations, but also fax, telephone, the human voice or body. Connections that are less direct and momentary are also conceivable—print media (pamphlets, newspapers, magazines, books) or slow media like cinema or music industry institutions ("Recycling the Future").

For the groups that inspired me to tell you this, most of what I told you is not really important. What is important to them is that the net and the techniques they use offer them: independence. Independence from broadcasters, from broadcasting laws, independence from difficult organizational structures around art, music and performance in an international context, independence from distributors and freedom to work without too many boundaries and across borders (ibid.).

"It's no secret that the web has offered artists, performative and otherwise, an expanded sphere of exposure. That is merely one side effect of working in this way, as in any broadcasting or publishing medium. The work I have been involved with involving remote linkups has sought to explore the medium for more than just its lure of a 'larger audience'" (Gompertz interview).

"Tune radio rapidly to 75. Tune radio rapidly to 102. And then off" (John Cage, *Water Music*, 1960).

SUBJECT: FLEXIBLE BODIES ON FREQUENCY MODULATION

FROM: ZINA KAYE AND HONOR HARGER <ZINA@WORLD.NET>
DATE: MON 28 SEP 1998 17:53:00 +1000

This writing is the sum of real-time and remote discussions between Zina Kaye and Honor Harger.

We are discovering the places where radio, radio.art, net.radio, and net.art intersect at this time, and will outline some projects that have taken place in the last year, including Xchange@OpenX at the Ars Electronica festival in Linz, Austria.

Our challenge is to discuss the confluence of these media without reducing their inherent interstructural malleability, and the power of overlapping flexing sound organisms. One could begin by discussing activities that occur in the studio and on the internet. Each node is broadcasting, yet our experience is one of mating these broadcasts into new organisms. This has been facilitated by the ease of communication via the internet, and in turn the internet provides more raw materials for the stream. In this space we can hear virus radio, fake advertisements, airports, space shuttles, generative music, experimental chewing machines, voices speaking in many tongues, sources of coded information and things that go bing.

Radio is not a definitive term, it is an adjunct. It is suffixed by notation of context—for example "micro," "FM," "commercial," and "net." These contextualized terms are all radio, subsets or different protocols of the same method. The word *radio* itself, without an adjunct, is symbolic and metaphorical. It is a complicated idea consisting of many different component ideas. It has many meanings in many contexts.

"If you had the same number of transmitters as receivers, your radio sets could have completely different functions." —Tetsuo Kogawa

Intuitively, we have always understood that radio could be used as a means to link people together in conversation, a communications vehicle not for broadcast but for the individuals involved. our vision of radio doesn't involve the metaphor of a sprawling net; instead, it is more like a conversation—sometimes with yourself and sometimes with a few others. Perhaps radio can be seen a musical instrument, or as a composer, and its communities as the notes it arranges into melody and discord.

The one obvious difference between radio and internet radio which is rarely addressed is this: radio is transmitted through airwaves and net.radio through wire. One is a hard technology (wire), the other ethereal (airwaves). It is interesting to note that a radio was once known as a wireless, to distinguish it from other forms of communication media reliant on wire, for example, telephones. In a sense, then, net.radio could be seen as a technological regression, dragging radio down once more into wire, tying it to the corporeal.

We are still receiving the browser experience, but the desktop is becoming more crowded with equipment that helps us to become a beacon or lighthouse. The relationship with the equipment is important. Where one might perceive the broadcast as no longer rooted in a particular culture or city, and the producer as not tied to a fixed place of abode in a stable existence, in fact both producer and listener are most definitely tied to the computer. And this despite the fact that net.radio lies in the dimensions of research and extraboundary travel. Equally, a larger structure enfolds the experience, and it is based on people: content providers, technicians, software engineers, archivists, interfacers, and listservers.

The beacons are many: it is like early telecommunications, where discrete nodes pass on the baton and fold information into loops. In such a paradigm, receiver becomes broadcaster. Many nodes will go under one name as a temporary autonomous zone and assault the networks with one unified communication.

Here the group personality is informed by multi-process activities, and the interface is a common piece of software. However, the experience is developing and changing: the computer is being lifted off the ground and the stream is rebroadcast via mini-FM transmitters. The interface is naturally moving once again to wireless communications, and from here perhaps the future lies in mobile phone communications and computerized Walkmans.

In 1998 at Ars Electronica/OpenX, radioqualia began to research a system called the Frequency Clock; its aim is to amplify the dialogue between two FM and net.radio.

The Frequency Clock is (or perhaps was) a simple attempt to illustrate the distances, time zones, and boundaries that radio crosses using the timepiece as a metaphor for distance. Discrete net.radio streams: radioqualia, L'Audible, Interface, Radio Ozone, Convex TV, and Pararadio were located in separate geographical locations, and identified by their time signature. The time and sound of each radio station signifies their individual identity, a personality distinct from other radio entities, yet somehow linked by this principal of the network.

Frequency Clock set up a chain of nearby computers all broadcasting a different net.radio stream via mini-FM. The viewer was invited to mix his or her own personal space by walking through the "bandwidths" wearing a radio. Radio and net.radio overlap, the functions of both dissolve into each other, and the distinguishing factors emerge as reasons to diversify the methods of exploring air and wire waves.

It is movement and a metaphor for movements: the flow that is symbolized by the works that come out of groups and the Zeitgeist of practitioners coming together face-to-face or remotely. The autonomous members of the group use the power of their combined voice to target centers of communication or bandwidth.

Though the disparate streams of online audio have been christened "net.radio," most practitioners of internet audio blush at the deficiency of this term. While there may in truth be more contrast than resemblance within the scattered associations forming through forums like the Xchange mailing list, speculative definitions do serve to expand the dimensions for conversation. What many of these projects do perhaps share is a cognizance of a common genealogy, edified by the "communication art" of the sixties and seventies, Fluxus, the radio.art movement of the early nineties, and other networked threads. A conspicuously Deleuzian tendency toward the obliteration of hegemony, and the simultaneous deference for chaos and "noise," is also developing as a common element between these discrete projects.

Guattari once spoke of radio in the context of transmission, transversal, and molecular revolution. Quiet voices, small actions. It is possible to pull the loud voice onto the desktop and magnify a local region, infinitely, using the zoom tool. We are interested in permitting the local region to speak louder, loudest. In the grand structure, the voice on the field is invited openly and programmed into the timetable as a supreme noise particle.

"(humble under minded) psychic rumble," an audio surveillance project conducted at Code Red Sydney, by Zina Kaye, sought to articulate the structure of the net.radio identity by using the audience as generator of content. Defined by Denis Beaubois as "the accidental contract," the audience produces its own desiring loop via audio surveillance. The audience is a knowing participant, it has a microphone in full view into which it may speak. It may know, also, that this sound is being broadcast to a space beyond its own. How a device receives this information is always opaque, as in any surveillance situation.

The psychic rumble microphone used Cold War surveillance technology, a concrete microphone for music concrete. The sound that is heard is one experienced by the structure, the walls of the building, as they vibrate and mediate sound. What can the walls hear? Talk, of course: one person speaks as another surveills nextdoor at the listening post. Beyond this, the walls hear better than people. They hear airplanes and toilets flushing, the wind as it rattles the chimneys, and dogs barking in the park.

The hidden ear, the severed ear, that says "we are not alone, and I am here to show you that." The paranoid ear hears granulated sound, interference, and accident. It is compelled to pick up everything for analysis. The mundane is dissected into smaller parts. It is the humble psychic that can pick the shape of the stream and pull it into meaning.

Is it so difficult to be fluid? Why is it that many parts can lurch forward in different tempos, and yet as an organism, activate the work into a whole? Surely this way of working compliments the dynamic fluidity and global dispersion of our time. It is not possible to put the names of the activities into a box under a magnifying glass and try to separate us, for we follow the path of least resistance. The work is unstable and may fall apart. Net.anything needs constant attention to rerouting. Indeed, we work at integrating the frailties of the format (error messages, disk buffering, dial prefixes, crashing, busy signals) into speculative art discourses, which too often may be co-opted toward the mystification of the abstract. In a period of what may be a formulation of a tentative aesthetic, many net.media practitioners, are attempting a synthesis of the grit of activism, the zigzag and abstraction of art, and the capabilities of cheap and accessible technology. Net.structure as it is now, may one day be seen as a technological snapshot.

The 1998 project at Ars Electronica by the Xchange collective in fact involved a number of individuals and groups that temporarily lost their production identity to enjoy free-to-air mixing. Most of the participants are plural or using the pluralist identity. Little organisms that replicate like a virus and are very much a part of this time. The traversal of space is fundamental to the notion of radio. We have always been intrigued by radio's metaphorical ability to collapse space, to expand face, to create an elastic zone where distance and identity become mutable.

The network emerges from a desire to evolve a virtual zone for sonic exploration, and it creates the latitude for musicians and artists to explore the superficial distance between understandings. Tools, such as live performance, audio streams of ebbs and skews, regular netcasts, are vehicles that survey this region, remapping prescribed media territory. But our art is an inexact cartography. No matter how carefully we plot the journey, ours is a convoluted excursion, with many unscheduled deviations. While the rupture of intention and outcome can at first seem like an obstacle, these accidental stopovers have allowed a deciphering of the code of netcasting. Embedded with the convenient angles of percussion and recoil are multiple tiers of fragmentation, breakup and congestion. We celebrate the hidden spaces where the alchemic transference of intent and error happens. This irregular drift has then, paradoxically, proved to be a viable way of studying the feasibility of a collective net.radio aesthetic.

The works produced are simple, and are freely available to the user in a slippery network. Net.radio is the ultimate proof that you are never alone and that the broadcasting structure is malleable and not a monolith.

SUBJECT: POSTMEDIA OPERATORS

FROM: HOWARD SLATER (BY THE WAY OF JOSEPHINE BOSMA <JESIS@XS4ALL.NL>)
DATE: THU, 24 SEP 1998 20:11:28 0200 (CEST)

The record industry is in the process of being outflanked by means of the very practices it has come to rely on. Since the sixties, its continual efforts to create new needs has meant that it nurtured an ever-changing musical soundscape; that soundscape is now mutating at such a rate that the industry cannot keep pace long enough to harness these musical evolutions in the direction of profit.

Advances in technology have meant that all manner of equipment is now available for reappropriation by whoever has the time to learn how to misuse it. There can no longer be any "one sound" around which music is organized—so everything becomes potential source material for a practice that no longer calls itself music.

From the guitar we have moved through sampling technology, turntables, analog and digital keyboards, to an indiscernible melange that creates further possibilities for interaction—as well as for enhanced and delegitimated conditions of reception. Such practices escape the institutional control of the industry and the media, eluding the "dominant repressive models" of an inherited subjectivity. Music reveals individual consciousness as socially situated.

As a consequence there are more people making music now than at any time before, and awareness of this among composers has led to an international explosion of small-label activity. These people have heard the tales of music scene has-beens and, rather than choose competition, exposure, and the "labor of success," they have chosen to operate outside these monetary and conceptual constraints. Inspired by the free-party scene, small-run pressings of records are passed around through underground distribution networks at a level that eludes even the most "specialty" of record shops.

In the slipstream of these phenomena there has been a rise in an experimental attitude: the end of the need to conform to what is expected and

"understood" means there is a renewed appreciation for the idiosyncrasies of sound and the transgression of perceptual habits they inspire.

Meanwhile, A&R men scurry from club to gig but never reach the parties. Attracted to a music that conforms to cash projections and reproduces the social imaginary, they can never hear the sound of conflictual desire. Similarly, the music press is increasingly losing its mediating role between unknown composers and the major labels—and its promotion of the "new" becomes ever-more laughable. The "new" is now passing by unnoticed; and these attempts to hold on to what's been declared "new" become an indication that what we read is inflected by dispassionate opportunism—marketing. Postmedia practice has been accelerated by the internet, where obsessions can run rife and there is a noticeable desire for those miniaturized activities that thrive without giving a thought to the increasingly "calm perspectives" of a transparent medium. The media, like the record industry, has become a centralized zero. Where once magazines and labels may have acted as a filter or a means of dissemination, market forces have made them converge on the center ground: the public listens to what is made available... and what the audience happens to listen to, since it was being offered, reinforces certain tastes (M. Foucault, *Foucault Live*, NY: Semiotext[e], 1989, 393).

Innovation and quality? It is interesting to see how the media, which sees itself as operating in opposition to high art, comes to work in consort with this traditionalism, particularly through the way that it reinforces reactionary notions of subjectivity. Foremost among these shared techniques is the way in which music, like art, is more or less always portrayed as transcendental, isolated from the social conditions that produce, celebrate and receive it. This individualistic means of relating to music is accentuated by the reliance on "genius": the elevation of certain individuals and the furthering of hierarchic devices in the supposedly "free space" of popular music. This accent on the unique can subdue others' activities and, in a denial of interrelatedness, that tends to make invisible the practice and heterogeneous reception contexts surrounding music.

What's more, this has the contingent effect of privileging the "solitary" moment of production over that of listening and dancing, which almost always imply the presence of others. These media inhibit—or even worse, remove—desire from music; in so doing, they collude with the "capitalization" of subjectivity.

Much postmedia practice has been stimulated by the growing sense that listening is not a subordinate activity but, rather, a process of making meaning.

And so, comprised of ephemeral organizations, postmedia become practices of a fiction that knows no bounds. It is a website, a zine, a record label, a distribution network of unseen nodes...it is a dechanneled, metacategorical social practice of cultural creation made entirely for and on its own terms. It is driven by enthusiasm, search and connection toward a polyphonic subjectivity. Rational modes of discourse like journalism and writing theses, which act to stabilize and make things remain still long enough for them to become systematized, lack a sense of music as a fuel that traverses disparate regions.

In the past, one drawback of such affirmative practices is the perceived need to be delimited as regions where protagonists should be made visible to one another. The onset of the internet has put an end to this by extending our expectations of communication and transposing a virtual space of music into an actuality of intimacy and an ever-present potential for subjective change.

SUBJECT: THE BRASS ENTER INTO MUSIC: SEMINAR SESSION AT VINCENNES: "METAL, METALLURGY, MUSIC, HUSSERL, SIMONDON"

FROM: GILLES DELEUZE
DATE: 27 FEBRUARY 1979

"The brasses enter into music! What does this entail in music? If we succeed in posing the problem well, perceive this problem well, then perhaps we may perceive the resurgence of ancient myths with no connection to Berlioz or Wagner. Perhaps we will understand more clearly how a blacksmith-music link is forged. What happens when the brass burst into music? We suddenly locate a type of sonority; but in this type of sonority, after Wagner and Berlioz, we start to speak of metallic sonority. Varèse constructs a theory of metallic sonorities. But what's odd is that Varèse straddles the great Berlioz–Wagner tradition of brass and electronic music, which he was one of the first to found and extend. There is certainly a relation. Music has been made possible only by a kind of current of metallic music; we need to find out why. Couldn't we speak of a kind of metalization? This doesn't at all exhaust the whole history of Western music from the nineteenth century on, of course; but isn't there a kind of process of metalization marked for us in a huge, visible way, made obvious by this eruption of brass? But that is at the instrumental level. Obviously, it wasn't the entry of the brass per se into music that was "determinative"; rather, a whole series of things happened concomitantly: the irruption of the brass, a totally new problem of orchestration, orchestration as a creative dimension, as forming part of the musical composition itself wherein the musician, the creator in music, becomes an orchestrator. The piano, from a certain moment on, is metalized. There's the formation of a metallic framework, the strings are metallic. Doesn't the metalization of the piano coincide with a change in style, in the manner of playing? Couldn't one correlate these things, even quite vaguely, with the irruption of brass into music? That is, the advent of a kind of metallic synthesis, the creative importance that orchestration takes on, the evolution of other instruments of the piano type, the advent of new styles, the groundwork for electronic music. And on what basis could one say that a kind of metallic line and musical line are wed, become entangled, even if it means separating anew. It's not a matter of remaining there since, in my view, it will lay the groundwork for the advent of an electronic music. Perhaps it was necessary to pass that way. Yet in that very moment there is no question of saying that the crystal is finished: the crystalline line in music continues. At no time is Mozart surpassed by the brass, that goes without saying; but it will reappear in a completely different form. Varèse is very much at a crossroads: he invokes at the same time notions like those of prisms, metallic sonorities, which lead on to electronic music. Just as the crystalline line passes by way of a whole complex conception of prisms, the metallic line passes by way of a whole complex conception of "ionization," and all that will be entangled—it will be like the genealogical lines of an electronic music. Therefore, it's very complicated, and it all has interest only if you understand that these are not metaphors. It's not a matter of saying that Mozart's music is "like" a crystal: that would only be of minimal interest. Rather, it's a matter of saying that the crystal is an active operator in Mozart's techniques as well as in the conception of music that Mozart constructs for himself—in the same way that metal is an active operator in the conception of music that musicians such as Wagner, like Berlioz, like Varèse, like the "electronicians," construct for themselves.

[This text excerpts a transcript of a lecture given by Deleuze in a seminar at Vincennes on February 27, 1979. Every effort has been made to contact the copyright holders. Translated from French by Timothy S. Murphy. See <http://www.imgaginet.fr/deleuze/≠TXT/eng/270279.html>.]

SUBJECT: 12" AS MEDIUM
TECHNO: FROM YOUTH CULTURE TO CULTURAL CONSITUTION

FROM: MERCEDES BUNZ <MRS.BUNZ@DE-BUG.DE>
DATE: TUE, 06 OCT 1998 02:46:23 +0200

Electronic music in Germany never was just a new sound. Rather, it was a whole new composite of economic, medial, and artistic relationships which was incorporated into every record and every 12" single—into the smallest unit of the system. In the field of pop music, a recognizable shift occurred in three particular and distinctive nodes: in the infrastructure, and consequently the economic situation; in the role of the musical medium; and in the cultural ratio of author-to-composition. One might conclude that a new cultural pattern has emerged.

1. A MISCONCEPTION: CONTROL OF THE ECONOMICS

Let us begin with a reality of infrastructure and economics. As far as I have seen, Germany's largest newspapers having been writing sporadically about the Techno phenomenon for the last ten years. As dictated by the statutes of the information age, the term has been known to the public since 1988—therefore since its emergence. In the meantime, though, the phenomenon has demonstrated its aptitude for cloaking itself from the widely acknowledged hypermedial world by hiding in the midst of information. "Hiding in the Light" was the term given to this subcultural trick by my personal idol and English subculture theoretician Dick Hebdige.

This phenomenon was not simply because the music was heard only in the deepest recesses or perimeters of urbanity; people simply regarded Techno as another fad, soon to pass, unworthy of any further attention—not, at least, serious attention. This situation was favorable for Techno's development.

For media editors, Techno represented a music denied of any cultural or political relevance because it was only technology, not humanity, that was expressing itself. In the beginning of the nineties, Techno encountered an overwhelming lack of interest, and Germany's cultural building sites were presumed to be located elsewhere. To the music industry, Techno was something that had made itself dependent on vinyl, and more precisely the 12"—a medium with no future, long regarded as dead. This meant that, initially, anyone who only wanted to exploit it or couldn't accept it just wouldn't touch it.

As a result, Techno is a pattern of youth culture in motion. But two forms of attention to Techno were lacking, at crucial points: the definition of its significance and the injections of cash. It was this failure of commitment by others that forced the music to train its own base and construct its own infrastructure; it was clear that nobody else would attend to it. At the beginning of the nineties, the acknowledged fact that records were being made but were impossible obtain was a pivotal cause for the opening of record stores and the establishment of distribution companies. This was a necessary step in enabling the music to become what it is today: a globally operative strategic network (or at least one might be inclined to say so, considering how privy one is to the chaos that reigns in the booking agencies and distributors).

Until now, musical youth culture operated on a tactical basis, amounting to a multitude of consumption models in repeated attempt to occupy the industry's infrastructure: studios, record companies, and concert promoters. Aesthetically, one can try to define oneself in relation to the surrounding establishment, but economically this simply isn't feasible. The enormous production costs involved in booking recording studios and so on will dispel any illusions one might harbor of independence in the face of a recording industry that both controls and adapts releases by forcing anything effective into specific technological artifacts. In this game, maybe surprisingly, Techno finds itself playing an altogether different role. On the one hand, the devices used to produce electronic music products were cheap at the time (they've become even cheaper since); on the other hand—and in a more crucial sense—Techno, in its role as an industry outsider and as manufacturer of its own infrastructure, finds itself in the remarkable position of actually profiting from its accomplishments and retaining its independence. It operates not tactically anymore but strategically, in that it now has a "place" of sorts "which can be named one's own and which therefore serves as the basis for the organization of one's relations" (Certeau). If subculture and pop music, as a tactical youth culture, were only considered a marketplace up to this point—money and jobs belong to the "establishment"—the difference is that we are now beginning to own the structures, the capital stock, and the work. From the cultural economics of youth culture a cultural constitution has formed.

2. FORGET VINYL: THE 12" SINGLE AS A MEDIUM

However, it is not as if electronic music—including all that clustered around the phenomenon—lingers in the midst of the business terrain, like an economic and cultural capsule. The connections are too numerous. It is not as if the music industry has discovered its own way of regurgitating Techno as song-based hits. It is not as if many producers compile albums for the music industry because, despite being able to live an individual lifestyle in Technoland, one cannot accumulate riches on an individual basis and one has to work harder for success. It would even be safe to state that the German beer-tent aesthetic, folk music being the very antipode of youth culture, now features traditional folksongs with Techno beats. Despite all of these acquisitions, Techno still seems to be able to make its own way and uphold its own set of rules. The secret of its success is the 12" and the balance of powers it symbolizes. If the music industry is dictated by the album and sales charts, it is the 12" single that rule the turntables. The medium's advantage is the misunderstanding it fosters in the music industry and its low esteem as a relevant vehicle for the business side of things. To the DJ, the 12" is the core of his creation. A producer's esteem is straightforward: you are only as good as your last 12", regardless of your LP. The 12" single transports the musical innovation of the music that, even if it is sometimes considered "retro," is ever perpetual, and eternally addicted to the next release. The long-winded creation of an album represents a delay in the music and its constant drive for new impetus. In addition, the album poses technical problems for DJs as well as producers since pressing more than two tracks on one per side, or more than 12 minutes at 33 1/3 rpm or 9 minutes at 45 rpm, infringes the quality. (One should bear in mind that club sound systems are a lot more precise than home stereo equipment.)

3. SELF-ASSESSMENT

Both units, the 12" single as a medium and the self-constructed infrastructure, guarantee Techno's—and electronic music in the widest sense—artistic position. The direct connection between author and composition, which in modern European tradition is regulated by direct expression, has shifted. Where originality was the keyword of cultural constitution, now, with Techno, sampling and mixing, determine artistic relationship. Any available material is used. Producers make use of devices' sounds and samples from other records; DJs use producers' records as their tools. A myriad of voices is injected into any given track or set. The person, the author, the subject, the classical origin of the artistic work, is no longer the focal point: the piece or the composition takes its place. Because of this, producers use so many pseudonyms that even the specialists, the DJs, lose track of who produced which track. It can be considered one of the rules of the cultural constitution known as "Techno," that names are insignificant. The music is no longer the *medium*; it does not represent the expression of the artist behind it. Rather, it is the *center of attention*. One could define the new relation with the words of the German philosopher Friedrich Nietzsche: "there is no perpetrator behind the act." Because only the act defines the perpetrator.

Nevertheless, Techno is being transformed into a well-behaved discourse, re-introducing the concept of the artist and the expression into electronic music. There is an attempt to maintain normalcy—alias, the "sell-out" of Techno. Perhaps the consequence is to embed the discourse in advanced cultural values; perhaps it is a case study on the ascent and fall of a classical subculture. To see Techno as a cultural constitution is not, however, tantamount to seeing culture as a reflection of society; instead, it refers to music as a part of society. If we can achieve this, traditional notions about the division of highbrow culture and subculture can be abandoned, thereby offering us not only a new and personal field of electronic music but also a new view of culture.

SUBJECTS

SUBJECT: SCIENTIFIC ETHICS: A TWO-SIDED QUESTION

FROM: MICHEL SERRES
(BY WAY OF PIT SCHULTZ <PIT@ICF.DE>)
DATE: MON, 21 SEP 1998 01:34:11 +0200

Teaching raises technical, political, and moral issues. However, the primary moral issue of science is communication. When big firms buy scientific research results, publication no longer ensures that they are true, because the results may be kept back for future sale. The "truth" of the results will then be subject to the profit motive and the firm's greater glory; at first silence stifles the debate, then paid publicity overrides free scientific communication. Hence there is a throwback to the pre-Homeric definition of *aletheia* ("truth"), where the truth is whatever the public believe to be true. The importance of research financing and the exaggerated publicity given to the work of wealthy research groups brings us dangerously close to such a throwback. It is our moral duty to avoid secrecy, which in the medium term could destroy the whole of science. We have an ethical obligation to ensure open publication, holding nothing back and not distorting any of the facts. The second ethical problem is more general. Nowadays all scientific fields use the techniques of modeling and simulation, which may change our understanding of what is "real." Objectives, obstacles, criteria, tests, or referees...reality is wrapped up in the virtual. In biology, for example, concentration on the genome rather than the phenotype leads to exploring what is possible as much as exploring what exists. Formerly, we had to obey Nature to command her. Can we now command her without even consulting her? Our ascent from the real to the possible opens up new worlds—which we ourselves are inventing—without having to face difficulties or proofs that were once unavoidable but are now simply bypassed by variations on the virtual. This freedom from the constraints of reality puts new responsibilities on scientists, since they are now tied less than formerly, and less than other human beings, to the rigors of experiment in the "here and now." Once everything they did was subject to the scrutiny of the real world. Today, more or less free from these constraints, they create possible scenarios that invent a kind of reality that they can impose on others or allow others, richer or more powerful, to impose on them. This leads to a considerable change in status of truth, which used to be tied to statements delivered by the real world, and tested by practical experiment. With the possibilities inherent when the invented models are set up, truth yields its place to responsibility, in relation to a possibility which is achievable, or which is imposed in the context of the new reality. Without leaving the field of science, we are passing from the theoretical to the action ethic, because we are constantly passing from the imagination to the deed, from the model to its outcome, from the possible to the real world. The question, "Are we telling the truth?" shifts toward the

Her name is Slave. After I created her I started by hitting her constantly for about 5 minutes. Then I taught her all the words so it would be easier to make her scared of her surroundings. After she knew all the words, I placed her in a small area, surrounded by the FF Cob, with 5 Grendels. I left her there for about 20 minutes, beating her when she attempted to defend herself from the Grendels. After she was sufficiently traumatized, I put her back in the garden. In the garden I forced her to Get, Look, Push and Pull everything around her, all the time, constantly beating her. I made her fear running so I wouldn't have to deal with that little problem. I also forced her to eat weeds, rewarding her when she did so. At the time I exported her, she's a quivering mass of fear. She might eat, if you're lucky, but she probably won't survive long enough for food to do any good. You can download her by clicking below. Have fun.

Aaron, who was formerly known as HurtMe, has taken kindly to my Albia. I raised his health up to 58% in this file. His life can still be raised, he can speak, and is obedient when alone. Avery was formerly BadGrrl. She still has a low life force, but she always smiles, and has mothered two children with Andrew. She will obey orders if separated from others occasionally. Andrew was formerly known as Quiver. He recovered slowly, and now has health in the low 60%. He's never smiled, but he no longer fears the hand. Ava was Slave. She ate a few times, but I couldn't really rehabilitate her. She's still alive, with a low lifeforce, but I am hoping she'll mate with perhaps Aaron. Allan was formerly known as Gimp. He recovered up to the 60%, but sadly he has a gait problem that causes him to walk a bit, then fall in pain. I hope he doesn't breed, but you never know. Betta is the daughter of Avery and Andrew. As a baby, she was very obedient, but after she met childhood, and other norns, she stopped eating and listening.

question, "Are we doing good?" Are these new worlds, that we have created, exposing our contemporary societies and future generations to the risks of death, violence, famine, pain, disease, and so on? The problem of the false converges toward the ethical problem of the evil. The law, "speak the truth," converges toward the law, "thou shalt not kill." No ethical rule can stop the free exercise of research into the truth. The new conflict is between truth and good. This or that moral rule always arises post hoc after an innovation, an invention, or a new application has appeared, and in consequence it is ineffective. What chance is there of a moral rule being applied successfully before the research? These questions have already been asked, at least once in former times, by a dedicated Greek doctor, Hippocrates. In his day, medicine alone was responsible for life and death, and medicine became more effective as our understanding of the living organism increased. The physicist, the chemist, and even more, the mathematician and the astronomer, were involved in verifiable experimentation and had no need for such questions. But nowadays, all scientists have to ask them. From time immemorial every doctor, at the moment of qualification, takes the Hippocratic oath—a unique proof that a morality can persist down through the generations, past and yet to come. Today we have to rewrite this oath to make it applicable to all the sciences, since all scientists now share these responsibilities. Since the oath should come before each new project, as an expression of the scientist's own conscience, it should be free from the problems of "post hoc," mentioned above. Each scientist should be free to take the oath or not. Here it is: "To the best of my ability, I swear not to use my knowledge, my research findings, and their applications for violence, destruction, or death, for the increase of misery or ignorance, for servitude or inequality. Rather, I shall use them to promote equality among people, for their survival, their betterment, and their liberty."

[This is the summary of a lecture given in Canberra, Australia, on August 5, 1998. Reprinted with permission.]

SUBJECT: DIGITAL WAYS OF FORHGETTING:
ARS OBLIVISCENDI

FROM: TJEBBE VAN TIJEN <T.TIJEN@CABLE.A2000.NL>
DATE: THU, 22 OCT 1998 12:00:14 +0100

SMASHING COMPUTERS AND NEWER FORMS OF CYBERCLASM

The recent phenomena of "cyberclasm" started with radical student actions in North America against university and military administration facilities. One of the earliest examples was in 1969 at Sir George William University in Montreal where, during a conflict about racism on the campus, students stormed the computer center of the university, threw out thousands of punchcards from the windows and smashed the computer equipment. At that time computers were mostly stand alone machines with limited storage capacity and data was either stored in punchcards, that needed to be processed mechanically, or on reels of magnetic tape. A year before a little book with the title *The Beast of Business: A Record of Computer Atrocities* was published in London, containing "a guerilla warfare manual for striking back" at computers that, according to its author Harvey Matusow, were on their way to "grab power": "from now on it is them or us" (H. Matusow, *The Beast of Business: A Record of Computer Atrocities*, London: Wolfe, 1968. In the late sixties, Matusow, an American expat, lived in London and circulated in its "cultural underground scene"; prior to that he worked in the U.S. as an FBI agent and was a paid witness in the McCarthy trials. See <http://sunsite.≠ unc.edu/mal/MO/matusow/>). The whole book had a playful Luddite tone; the guerilla actions it proposed were rather mild, for example, altering punchcards holes or demagnetizing computer-readable magnetic strips, in order to halt the advance of the computer in civil administration. Matusow mentions the military use of computers, but he seems not have understood their function very well, as becomes clear in his slogan: "It is the computers that want war." "It," of course, is the human beings who want and make war; the social network of political, military, industrialist, and scientific establishments—the "military-industrial complex"—that developed the first electronic computer during World War II.

The computer's first function was to assist the calculation of ballistic trajectories of conventional weapons and, later, to aid in the development of the atomic bomb into the far stronger hydrogen bomb. The names of firms that originally specialized in mechanical office equipment—for example, IBM, Burroughs, Remington, and Underwood—can already be found at the military root of the computer pedigree in the forties and fifties: these companies were not just warmongers, their commercial interest also helped to transform the military computer into a civic instrument. In the following decades the computer tree branched from gigantic machines—the ferocious "beasts" Matusow fought—into the familiar and helpful personal computer of our times. Matusow published his anticomputer book in 1968, when the Vietnam War had been raging for four years—and the same year that saw a proposal to combine networks of military and civilian computers (ARPAnet) into a decentralized and flexible form of communication able to resist a nuclear strike.

The growing importance of computers in warfare, now also for military logistics and wargames, had not yet been recognized by the radical movements of that time. Manuals for urban guerrillas of the late sixties and the beginning of the seventies do not mention computer facilities as a target; instead, they still emphasize on radio, television, telephone switches, and electrical power facilities (for example, A. Bayo, "150 Questions for a Guerrilla" [1959/1965]; C. Marighella, "Minimanual of the Urban Guerrilla" [1969/1970]; E. Luttwak, "Coup d'Etat," 1968). It was not until May of 1972 that the first (publicly known) serious attack on a military computer center—the Heidelberg headquarters of the U.S. forces in Europe—was undertaken by the "Kommando 15. Juli," a group related to the German *Rote Armee Fraktion*, to protest the escalation of bombings in Vietnam. Needless to say, this protest did not hinder the metamorphosis of the military ARPAnet into the civil network of networks called *the internet*. This development has, of course, created opportunities for new forms of "cyberclasm" and guerrilla—no longer direct physical attacks on personnel and equipment but indirect attacks, using the computer system itself as a basis for disruptive and destructive activities.

PATROLLING THE INFORMATION HIGHWAY

It is an old tactical adage that each advantage carries with it a disadvantage. This holds true both for assailant and defender. Empires—the Chinese, Mongol, Roman, Napoleonic, and their modern heirs—can only grow on the basis of an efficient transport system of goods, armies, and information. Developed road systems with facilities for resting, refreshing, and maintaining vehicles were created to make such transport movements faster; but these roads, with their valuable traffic, also created new opportunities for robbers, bandits, and other highwaymen to ambush and take what they could not obtain otherwise. Expanding sea traffic showed a similar development, with pirates laying in wait to catch some of the rich cargo moving between colony and imperial motherland. Newer land and air traffic system continued this tradition of robbery and piracy: highwaymen evolved, became train robber, hijackers... All of these freebooters, over the centuries, hold one activity in common: "stealing something while in transit." The modern highwayman (or woman) roams the "information highway," lurking, waiting for the right moment to grab what is not intended for her or him.

The metaphor of the "information highway" can be related as well to other traditions associated with transit and travel, or, more precisely, stopovers—drinking, prostitution, and gambling, as well as authorities' constant fight to suppress such debauchery. It has become a truism that sex and, to a lesser extent, gambling have been very closely associated with the economic development of e internet, and efforts to suppress them have certainly been in the news. But this will never succeed: the moment one too-lusty site is closed down a new one pops up a farther down the road. Closing down the road itself would be the most effective measure, but, because modern society needs information traffic, it must learn to live with the unwanted side effects. Patrolling the net, by human and software agents, has made it possible to ban some of this unwanted information in some contexts, but there is an inherent danger in the principle that some authority will decide for individuals what to read, what to see and what not. (One such facility, Cyber Patrol Corporate, itemizes sites that contain "questionable" material—"Partial Nudity; Nudity; Sexual Acts/Text; Gross Depictions; Intolerance; Satanic or Cult; Drugs/Drug Culture; Militant/Extremist; Violence/Profanity; Questionable/ Illegal and Gambling; Sex Education and Alcohol and Tobacco.") This is not entirely new, obviously: the Catholic Church's *Index Librorum Prohibitorum* (Index of Forbidden Books) was meant to prevent contamination of faith and corruption of morals dating to the end of the fifth century. It was regularly published from 1559 onward and only ceased publication in 1966. With the introduction of modern filtering software that stops what is not approved or, more radically, only let through what is approved, the old principle of world-wide censorship as practiced by the church, has been re-introduced by "modern" governments and affiliated organizations at the end of the twentieth century on a larger scale than ever before.

LOYAL HACKERS AND SPIES

Information that isn't in transit isn't thereby safe, even when securely stored behind "firewalls." As in fairytales, however strong a fortification is made, in the end someone will be able to enter, often not by brute force but by deception. It is not surprising that, in the coming age of digital computers, mythological terms such as *trojan horse* are still used for such cunning tactics whereby unsuspecting computer users allow hidden malicious information through the gates of their equipment, where it unexpectedly raises havoc and destroys valuable information. One can go back in time two millennia plus three centuries to find this principle described in the oldest known text on tactics of war, Sun Tzu's *Ping Fa* ("The Art of War"). The beginning of this ancient Chinese text stresses that "all warfare is based on deception." Sun Tzu clearly distinguishes between direct and indirect ways of fighting and he favors the last form: "indirect methods will be needed in order to secure victory" (see <http://www.promo.net/pg/_authors /tzu_sun.html#theartofwar>).

In 1995, the National Defense University at Fort McNair in Washington, D.C., has instituted a yearly award named after this Chinese war theoretician: "The Sun Tzu Art of War in Information Warfare Research Competition." (The NDU offers the following welcome: "By making unprecedented amounts of information immediately available in easy-to-use forms at diminishing costs, the emerging information highway will certainly alter society, to say nothing of military conflict": see <http://≠ www.ndu.edu/ndu/preswell.html>.) Recent prize-winners include a group of researchers who thought up an imaginary scenario that could have taken place during the Balkan conflict in September 1998: a group of Serbian political activists intervene with the radio frequencies of a temporary airfield at the Bosnian–Croatian border where NATO troops are flown in during a flare-up of the conflict in Bosnia. The result is two military airplanes crashing. The Serbian cyberactivists, immediately after, inform the whole world press by email and put up a political statement on a website on a server in Amsterdam. CNN, Reuters, and others broadcast and publish the statement including the webpage address. Within twenty-four hours the webpage has a million "hits," many from state intelligence organizations. Any computer used to access this website is infected by a trojan horse program that the activists have embedded in the webpage, a program that starts to delete all files and hard disks after twenty-four hours. This exercise in military fiction is used as an explanatory introduction to what "information warfare" could be. The authors warn: "The US military could find it difficult to respond against a small and digitally networked enemy." They propose the establishment of "Digital Integrated Response Teams (DIRTs)" made up of "highly trained information warriors" from military and law enforcement agencies, to counter "information terrorism" (M. G. Devost, B. K. Houghton, and N. A. Pollard [of Science Applications International Corporation] "Information Terrorism: Can You Trust Your Toaster?" [1996], at <http://www.≠

Without doubt the most important tool I possess for Nettime critique is my old school ruler. After loading a new nettime mailing into my reader the first thing I do is to put my ruler onto the glass screen and measure the length of the gray text scrolling bar. 1. If it measures over 1/2 inch long (that's 1.25cm for our foreign friends) then I immediately commence reading. 2. If it is over 1/2 inch then I check the clock for the possible approach of bed time. 3. If it is under 1/4 inch then I close my eyes and gently nudge the delete key. [Dr. Future <richard@dig-lgu.demon.co.uk>, Nettime Rules, Wed, 27 May 1998 23:09:44 +0100]

ndu.edu/inss/siws/ch3.html>). These state "information warriors" are supposed to work from "remote computers," using "anonymous response" tactics without open display of force, in order to avoid any public sympathy for political activists, fighting a possible "right cause" and being attacked by the state.

In the past few years, incidents in which secret state information has been accessed by "intruders" have been played up in the press, but none seem to have posed an enduring security threat to any government to date. At many levels of society is has become clear that the criminalization and persecution of computer hackers often misses the point: in most cases the sole aim of a hacker is to master computer and encoding systems, to explore how far or how deep one can go. Even most of the more political motivated hackers tend to have some basic loyalty to some national state. There are also, of course, cases in which of copyrighted and otherwise protected digital material have been infringed upon, but these incidents involve discrepant interpretation of and/or attitudes toward what acceptable forms of ownership are; they differ from activities of organized crime or terrorist attacks against the functioning of the state. Several academic and military studies present a more differentiated or complex view on the "hacker scene"; some authors see hackers as a positive force in society that can be tapped as a resource to improve security systems (M. G. Devost writes, "The United States should utilize hackers, and give them recognition in exchange for the service they provide by finding security holes in computer systems"; see his "National Security In The Information Age," University of Vermont, 1995). This is, in essence, also an ancient tactic: one can read in the last chapter of Sun Tzu's *Art of War* that describes the use of spies: "The enemy's spies who have come to spy on us must be sought out, tempted with bribes, led away and comfortably housed. Thus they will become converted spies and available for our service."

A WORLD WITHOUT ELECTRICITY

As the computerized informationization of all levels of society progresses, a feeling of vulnerability is growing. In early 1998 the Clinton administration issued a "White Paper on Critical Infrastructure protection" that describes what to do against "nations, groups or individuals" that "seek to harm us in non-traditional ways" (<http://www.uhuh.com/laws/pdd63.htm>). Others use catch phrases such as an "Electronic Pearl Harbor" or "cyberwar, blitzkrieg of the twenty-first century" to fire the imagination of the politicians and civil servants who decide about budgets for new research, new special task forces and new weapons. The reasoning is constant through human history: what the enemy can do to us, we should be able to do to the enemy. Apart from the indirect methods and approaches of hackers, computer criminals, and their state counterparts, the "information warriors," a whole new arsenal for more direct forms of "information war" is being prepared: rumors of guns that fire "High Energy Radio Frequencies," hitting electronic circuits with an overload that will knock out any radio and television transmitter, telephone switch, computer network, aircraft or other transport system dependent on electronics; miniature "nanotechnological" robots that can physically alter or destroy electronic hardware; low-energy lasers that can damage optical sensors used in many modern vehicles and equipment; and, best of it all, the Electro-magnetic Pulse (EMP), originally discovered as a side effect of nuclear bombs, which disables all copper-wired electronic circuits, halting all electronic equipment and communication not specially shielded against this form of attack. (For an overviews from a military point of view, see <http://www.defence.gov.au/apsc/paper47.htm1>). There are different plans for the usage of the EMP weapon: the "shock and awe" tactic whereby whole urban areas or battlefields will be blasted with such an energy that all electricity stops functioning, as well as the more "precise" targeting of single objects in a range of a few hundred meters. Modified cruise missiles for such confined operations exist already. It is difficult to imagine a world without electricity. One wonders what it would be like, to live without all those electric facilities and contraptions, to have lesser, but maybe deeper contacts, in a more tangible world.

INVISIBLE STRINGS OF VOLTAGE

The basis of most electronic documents is recoding of human-readable text and graphics and machine readable sound and video. At all stages of production and reproduction, different layers of technology reside between the human organs of perception and digital documents. Recoding as such is not a new phenomena; it is recoding of language into written text that "permits us to create a record that many other people, far distant from us and from one another in time and space, can read" (P. Delany and G. P. Landow, "Managing the Digital Word," in *The Digital Word*, Cambridge, MIT, 1993, 6). The nonelectronic recoding of language, by hand with its directly readable physical marks on a physical surface, have left us with only a limited number of documents from early ages; many did not even survive their own epoch. The shortage of good writing materials such as papyrus and parchment meant that reusable surfaces, such as wax tablets, were often favored. Parchment was rare and expensive and for that reason often "recycled," reused as "palimpsest" by washing and scraping off the text it carried. The use of paper and the multiplication of writing by the printing press fundamentally changed this situation. The dispersal of multiple copies of a (printed) text led to the long-term preservation of that text. Now digital documents are of another order: they are no longer tangible objects but "essentially an invisible string of stored electrical voltages" (Pamela Samuelson, "Digital Media and the Changing Face of Intellectual Property Law," *Ruthers Computer and Technology Law Journal* 16 [1990], 334). First it was scarcity of carriers for storing these electric currents (floppies, hard disks ,and the like) that led to the same practices as the recycling of wax tablet and parchment in antiquity: erase and reuse. Later the price of digital storage dropped dramatically, but this has introduced a problem of prodigality—the problem of managing large quantities of half-labeled and messy information, which often led to a similar outcome.

As the fixity and multiplicity of the printed is more and more supplanted by the flexibility of multiplicitous digital document, we come to see that new media are posing problems when it comes to long-term preservation of content. Standards for computer hard- and software are in a constant flux, and backward-compatibility and long-term support seems not to generate enough profit to interest industry. Bankruptcy of a firm or defeat of a standard on the marketing battlefield can mean, in practical terms, the loss of massive amounts of information. Eternal transcoding of digital information from old to new standards will need to become a routine operation within bigger institutions, but such facilities are expensive and unreliable and, as such, all but unavailable to smaller institutions and much of the private sector. This last sector of society was already underrepresented in archives and other deposits for historical studies; now, in the digital area, even fewer traces will remain of personal administration, letters, email, unpublished manuscripts, and the like. Going through the belongings of someone who died one might consider keeping some letters, notebooks or photographs, things we can read directly—but what to do with an outdated computer, a shoebox with unreadable floppies, mysterious-looking cartridges, and unlabeled CDs? Their fate is to rust, rot, or burn along with other refuse—or at best to be recycled somehow. In this sense we have seen a similar thing happening earlier this century when old cinematic film was recycled for their silver content.

DATA ARCHAEOLOGY

Global and direct availability over the internet of a wide variety of electronic documents has led, on the one hand, to a speedup of information circulation and, on the other, to a loss of information. The life cycle of content made available over the internet is getting shorter and shorter. Thousands of web pages are "thrown away" each day for various reasons: storage costs, lack of space on computers, hard disk crashes and other digital disasters, information becomes outdated, unwanted, censored, neglected. Strangely enough, the information is often not directly lost but, rather, fades away slowly, like the light of a star that no longer exists but still can be seen in the sky. Information is duplicated on computers elsewhere in the form of mirror sites or caching proxies that temporary store often requested information to diminish long-distance traffic over the internet. In the end, this duplicated information vanishes as well. Some see this as a positive aspect: why pile up the informational debris of each generation on the already towering heap? Others worry about the void of digital historical material we will leave for posterity. Megalomaniac plans, with an imperialistic and totalitarian undertone, to periodically store "all information" available on the internet and associated networks in gigantic digital warehouses have been proposed; one example is Brewster Kahle's 1996 founding of the "Internet Archive" (see <http://www.archive.org/sciam_article.html>; recently his firm Alexa Internet donated a full "snapshot" of the web from early 1997 to the Library of Congress.) It seems more logical that the old principle of "survival through dispersal" will have a longer-lasting effect on preservation and availability of digital documents from the past. ("Destruction, ruin, pillage and fire especially hit great amassments of books that according to the rule are situated in the centers of power. That's why what has remained [of the earlier period] in the end does not come from the big centres but from marginal places": L. Canfora, La Véritable histoire de la biblioteque d'Alexandrie," *Desjonque*, 1986.) Even if a very small percentage of the electronic material on the global network of networks will be preserved, this will be of such a magnitude and diversity that special techniques of "digital paleography," "data mining," and "information recovery" will be needed to dig up something that will make any sense to future generations. (One can imagine theories of extinct technologies...) Another approach is the simulation of the functioning of old hardware and software on new machines, be it military analog computers of the fifties or one of the popular hobbyist computer types of the seventies and eighties. The real experience of the functioning and use of this equipment will be lost in this process, but is not most of what we think to experience from the past a simulation of a reality that never existed?

LOST IN THE DEAFENING BABBLE

The traditional containers of information (books, periodicals, gramophone records, audio CDs, film and video titles produced for the consumer market) fix information in such a way (cover design, title, colophon, credits, numbered series, publisher, place of publication, year, and so on) that we can easily deduce what they are about and have some understanding of the context

Interesting—I have approximately 65 floppy disks (5-1/4 inch) containing approximately 350 programs which I acquired between 1981–85 for my Atari 800 with 48K of RAM. I have the original machine, which works fine, the original floppy drives, which work fine, and, as of this year, all the programs and data files are uncorrupted. All of them. In other words, my digital media has lasted 17 to 12 years without any failure. It was stored in an uncooled location (a warehouse in Queens, NY) where temperatures vary from below freezing to about 100F (40C) every year. I took it out of storage in 1996. [David S. Bennahum <davidsol@panix.≠com>, Re: Dead Media Working Note 32.4, Mon, 4 May 1998 01:02:13 -0400]

in which they function(ed). It took more than four centuries for these standards to develop and come into common use. From this perspective, it is not surprising that the use of new standards for the description of networked electronic documents—a reality that exists hardly two decades—should be less stable. Consider the standards for storing data about data in an electronic document: some of this "metadata" is automatically generated when a document is created—for example, time, date, the hardware used and protocols needed to display or manipulate the document. Without this self-referential information the documents could not even be distributed and consulted. When it comes to description of content (author, title, subject, and so on), new standards do exist, but are little-known and rarely used. This means that there is an immense amount of potentially valuable and interesting information on the internet that remains unnoticed and will be forgotten because its content is not properly described. Whatever powerful "search engines" are used, machine protocols can not sufficiently distinguish between meaningful and meaningless occurrences of search terms used. Most search results give so many "links" that one can not possibly follow all of them. In this way valuable information is "lost in the deafening babble of global electronic traffic" (Delany and Landow, 15).

THE FRAGILITY OF A SPIDER WEB

There are people who think that such a comparison of new electronic information and communication systems with traditional media is not fruitful. Some of these people see a loosening of the bonds that bounded text, sound, and image to their respective media as, rather, a fusion of these elements into a new phenomenon, multimedia—something of a different order, where fixity and linearity are supplanted by a fluid, dynamic recombination of elements, which ultimately will abolish the notion of finite and finished works. This new form of human communication has one of its theoretical bases in literary and semiological theories developed three decades ago, which pointed to the relationships within a given text to a multitude of other texts and the possibility of a new kind of more personal and active reading. This theory of the possibility of different "readings" of text was also extended to the realm of imagery, as it became clear that computers offered new technical opportunities to interact with a corpus of many different linked texts fragments. Soon enough, these theoretical concepts were given a concrete form, "hypertext" (see G. P. Landow, *Hyper Text*, Baltimore, Johns Hopkins, 1992).

The first experiments were with interlinking, some say weaving, of different blocks of text and images in a virtual library made up of such *lexias* and icons, still residing on one computer, or a well-controlled internal network of computers. With the advent of the internet, though, the concept of hypertext has been widened from linking materials on a "wide area network" to links made across networks and protocols. The growing enthusiasm for seemingly endless possibilities led some people to speak of the net as a global brain of interconnected and linked human resources. But these links are weak links: already, and even on the local level, it is very common to encounter an error such as "404: File not found." On a global level, this new digitally unified "brain" suffers from an even worse case of amnesia. One cannot escape the comparison with printed media here; it is like reading a book and suddenly missing a few pages or discovering that some of the footnotes have been torn out, or trying to read a newspaper after someone has cut a series of news clippings from it. The fascination with the internet is like the fascination with the beauty of a spider web dancing in the wind. It is based on the knowledge of its fragility—one unlucky instant will destroy all the work. This ephemeral aspect can of course also be seen in a positive way: enjoy the moment itself, do not leave too many traces, leave the others, the generations after you, some space to discover things for themselves. Ideally, a combination of the two elements might develop, whereby some examples of the constantly broken threads of the web will be collected and preserved, while the rest will be washed away by time. As Simon Pockley has written in "Lest We Forget," "The digital era has been characterized by technological obsolescence and ephemeral standards, ironically threatening the usefulness of digital information. There is little firm ground upon which to build the institutional and private structures necessary for the effective preservation of this material. Nowhere are the challenges more difficult than those concerning the new networked medium of the World Wide Web. The vitality and flexibility of this medium mean that digital material is in a state of constant proliferation and mutation" (<http://palimpsest.stanford.edu/byauth/pockley/pockley1.html>).

[The complete version of this text can be found in the 1996 Ars Electronica catalog *Memesis: The Future of Evolution* (Vienna: Springer Wien, 1996), 254ff., and at <http://www.iisg.nl/~tvt/tijen01.html>.]

SUBJECT: INTERACTIVITY AS WAR (EXCERPT)

FROM: CALIN DAN <CALIN@EURONET.NL>
DATE: SUN, 4 OCT 1998 21:28:30 +0100

1.

The following text starts from the premise that war and interactivity have common patterns and meet in certain places as regards mental models. The point of view accepts the inherent conjunction between art and responsive technologies; this is a point beyond enthusiasm or critique, somewhere in the limbo of entertainment itself. Since I'm not a wargames freak or an expert in interaction or warfare history, my only motivation for fixing these reflections is my remote curiosity about human violence and my never-failing fascination with the mysterious content of machines.

Is war an important issue for how we perceive history? Humanity thinks of itself in terms of achievement. Achievement goes together with competition, and, under a certain level of stress, competition means war. Stress can be induced by increases or decreases in various factors: populations, living conditions, technological development, climate, ideologies, and so on. One thing defines them all: fluctuating data, both qualitatively and quantitatively.

Data, of course, is a commonplace in today's cultural discourse, but it can also be seen as a paradigm, one with interesting implications—it allows us to view history from a standpoint beyond morals. To consider humanity as an amoral species is a practical attitude, if only because such a view is less charged with emotional prejudice. If we start from the position that war is acceptable, we can delve further into—or perhaps go beyond—frantic rejection, embarassed commitment, and negotiated acceptance. This is why life beyond morals is so difficult: it widens our scope of choices. It seems that only old societies can live with this attitude. But they die—usually from invasions by younger, moral societies. Still, the ritualized aspects of warfare prove that encoded violence is an activity as necessary as any other social tissue.

2.

All games are wargames. War is perpetuated via storytelling. Storytelling is a crucial coagulant for the human species: at every historical stratum, storytelling overwhelms other aspects of cultural trade—and war stories are overwhelm other kinds of stories. Is this due to the importance of war at the level of the social, or is it perhaps also determined by some structural requirements of the human species?

Timespace in war: When we look at war in its temporal dimension, it is not a punctual activity. With the exception of the modern period—roughly from the campaigns of Napoleon through World War II—warfare was characterized for the most part by a flow of violence that involved and/or affected populations as a whole. It was from endemic chaos that effective military conflict—the so-called pitched battle—emerged; it did not always or immediately resolve it, though. Our perception of history is guided (misguided?) by peak events, in much the same way that our perception of art history is. We describe and analyze our heritage by making reference to masterpieces, which we see as the result of big streams of data that can only be exposed without risking the "big picture." In that sense, scientific discourse is not different from fiction.

When we consider war in its spatial dimension, we see that consequential wars are, for the most part, very punctual. The way in which armed conflicts sometimes remodel in the medium- to long-term the political aspects of geography can be impressive—but this kind of perception remains retrospective and synthetic. On the level of the individual, the vast majority of wars are limited experiences, even if the war's strategic context is broader. However, strategy is sometimes invented in the aftermath of the events—and, basically, wars themselves are retrospective inventions. The restricted misery of battle obscured the endless pain of populations at war.

Timespace in media: When kids play a computer wargame, they develop with the glowing tube a relation paradoxically similar to the one that we, their elders, have (or maybe have had) with books about war. A retrospective and/or retroactive relation that covers the substantial horrors with a veil of both distance and exciting immediacy.

Screen machines, books, and storytelling in any form secure for us a special form of ambiguity, one that gives us both implication and distance, intimacy and dominance. Media, old and new, are about mediation, hence their addictive fascination: they allow us to be insignificantly small and discretionarily powerful all at once, like a child is in the protective cocoon of its family.

What is truly new about new media is their capacity for combining "zenital" and "genital" views in one: a user simultaneously controls space from the position of the noontime sun, and analyzing it from the inside prospective of the womb.

Maps, beyond their utilitarian aspect, radiate strangely something distant in time as well, not only in spacelike books and screen machines. They are the interface between the two, and also an ideal interface for narratives of war.

3.

Warwaves: War is commonly perceived in Homeric terms, in the sense that even the most cruel and damaging facts are perceived after the fact as symbolic and, therefore, meaningful. In this way, the proximate view of survivors who remember events and the distant view of commentators do not differ very much: everyone agrees that war has negative features, nobody accepts a lust for this trade—but an implicit narrative

I am compiling a list of delirious states and felt sure that you would find it of interest. *1.* Childhood delusions: When very young I recall a frightening experience that occurred every time I slept in my parents bed in the main bedroom. While lying awake I was convinced that malevolent forces were coming out of the walls and tormenting me with threats and menaces. I seem to remember that they were partly visible. *2.* Feverish Delirium: Last time I had flu I was very feverish and awoke one night covered in sweat. I wanted to roll over onto my back onto a cool patch in the bed, but was convinced that I had three backs. I was unable to roll over because I had no way of deciding which of my three backs I could roll over onto. *3.* Media Disorientation: In an episode in the last series of Babylon5, the Garibaldi character is seen watching television in his quarters. We cut to see that the program is a Bugs Bunny cartoon. This cartoon is a typical—Bugs is tormenting Daffy Duck by manipulating the animation that he is in; we have been watching the cartoon for about 10–20 seconds. Suddenly we cut back to Garibaldi's cabin on Babylon5 where he is saying something like "I just love those cartoons." This cut back to the sci-fi series was completely disorientating and for a moment it was not possible for me to decide where I was. Then my normal sense of perception stabilized, a little like awakening from a dream [Dr. Future <richard≠@dig-lgu.demon.co.uk>, .Have Any Other Readers Had Similar Experiences? Sat, 30 May 1998 00:53:44 +0100]

sensuality unifies everyone's attitudes. War is assimilated culturally only in retrospect and as a succession of details. That makes its acceptance so easy, and the responsibility for this acceptance so vague.

Archaic societies, or those societies that I labeled as old and amoral, learned how to deal with war lust in a ritualistic manner, from random clashes to sacrifice-oriented conflicts. This relation with war extinguished slowly in the Mediterranean basin from the advent of the chariot (mid-fourth millennium B.C.) until the invention of the hoplite phalanx (seventh century B.C.), when the destruction of the enemy in pitched battles became a more general rule of warfare. And still, while tactical aspects remained a red thread in the perception of wars as delivered via written reports, something unreal like a fairytale atmosphere surrounding the events came out of those firsthand documents. From Xenophon and Caesar to Clausewitz and Montgomery, war is presented and even analyzed as a game with internal rules and external motivations, but with an autonomy that situates it closer to art than to politics (the "Art of War"), and closer to abstract research than to practical issues. The human brain needs to be fed with narratives, and converting structural violence in storytelling maintains the species on the level of arguable conflict.

4.

Violence as authorship. The fact that interactivity is the first commandment in the religion of new media can be understood as both obvious and unclear. Obvious because interactivity—at the present level of sophistication—is mainly a business of violent intrusion in otherwise linear concepts. And it seems that violence—sublimated though—is now an important ingredient of mass culture. What remains unclear is how much interactivity rewards the idea of authorship, and, connected to that, how significant a need creativity is (or represents) for the user.

At the end of this trip there is just a vague landscape where collective creativity, violence, and control over the territory of fiction compete with the dominance of a single author. If the responsive machines are supposed to infuse a horizontal (nonhierarchic) view on crucial matters, then uniqueness should become obsolete as an entrenched way of defending identity and economics.

Multimedia production of the late nineties should pay attention to the procedures of this ancient and still-available cultural nomadism. Notions such as authorship, ownership, cultural/moral/legal property, appropriation, synthesis, eclecticism, multisensorialism, conceptualism, market values, and aesthetic autonomy are checked upon in the bazaar consistently and without interruption, regardless of political stress.

However much exciting data is lying around, the most reasonable way of being creative is to work on display procedures. But controlling the display means owning the merchandise: a critical option for the new media artist is to have the ability—mental, social, financial—to step into the position of a wholesale shopper.

SUBJECT: DNS: A SHORT HISTORY AND A SHORT FUTURE

DATE: TUE, 13 OCT 1998 16:13:43 +0100
FROM: TED BYFIELD <TBYFIELD@PANIX.COM>

In the debates that have erupted over domain-name system (DNS) policy, two main proposals have come to the fore: a conservative option to add a handful of new generic top-level domains (gTLDs: ".nom" for names, ".firm" for firms, and so on) administered by a minimal number of registrars, and a more radical proposal to level the hierarchical structure of domain names altogether by permitting openly constructed names ("whatever.i.want") administered by an open number of registrars.

The supposed cause for these debates orbit around perceived limitations on the system—monopolization of registration by NSI (in the U.S., of course) and a scarcity of available names; as such, the debates gravitate toward modernizing the system and preparing it for the future. What little attention has been paid to the past has focused on the immediate past, namely, the institutional origins of the present situation.

Little or no attention has been paid to the prehistory of the basic problem at hand: how we map the "humanized" names of DNS to "machinic" numbers of the underlying IP address system. In fact, this isn't the first time that questions about how telecom infrastructures should handle text-to-number mappings have arisen. And it won't be the last time, either; on the contrary, the current debates are just a phase in a pas de deux between engineers and marketers that has spanned most of this century.

A bit of history: From the twenties through the mid-fifties, the U.S. telephone system relied on local-exchange telephone numbers of between two and five digits. As these exchanges were interconnected locally, they came to be differentiated by an "exchange name" based on their location. These names, two-letter location designations, made use of the lettering on telephone keypads: thus an 86x- exchange, for example, might be "TOwnsend," "UNion," "UNiversity," or "VOlunteer." Phone numbers such as "Union 567" were the norm; "86567"—the same thing—would have been seemed confusing, in much the same way that foreign dialing conventions can be. There wasn't a precedent for a purely numerical public addressing system, and, with perfectly good name-and-number models like street addresses in use for centuries, no one saw any reason to invent one.

However, as exchanges became interconnected across the nation, AT&T/Bell found a number of problems—among them, that switchboard operators sometimes had difficulty with accents and peculiar local names. As a result, the national carriers began to recommend standardized exchange names, according to a curious combination of specific and generic criteria: they chose words that resisted regional inflection but were common enough to peg to "local" landmarks. The numbers 5, 7, and 9 were reserved because the keys have no vowels, making it (so the theory goes) more difficult to form words from them; hence artifacts like the fictional prefix 555-, so common in old movies, later became the national standard for prefix for fact, in the form of directory assistance.

By the late fifties, when direct long-distance dialing became possible, then popular, variable length of phone numbers became a problem for the national carriers, which demanded yet more standardization—seven-digit phone numbers in a "two-letter five-number" (2L5N) format. And while it wasn't an immediate problem, the prospect of international telephonic integration—with countries that used different letter-to-number schemes or even none at all—drove yet another push for standardization, this time for an "all-number calling" (ANC) system. Amazingly, the transition to ANC in the U.S. took almost thirty years, up to around 1980 depending on the region. (Just as certain telecom-underserved areas are now installing pure digital infrastructures while heavily developed urban areas face complex digital-analog integration problems, phone-saturated urban areas such as New York were among the last to complete the conversion to ANC.)

Direct long-distance dialing wasn't merely a way for friends and family to keep in touch: it allowed businesses to deal in "real time" with distant markets. And the convention of spelling out numbers, only partially suppressed, hence fresh in the minds of the many, became an opportunity. Businesses began to play with physical legacy of lettered keypads and cultural habits by using number-to-letter conversions as a marketing tool—by advertising mnemonic phone numbers such as "TOOLBOX." And as long-distance calls became a more normal for people to communicate, tolls began to fall, in a vicious—or virtuous, if you prefer—circle, thereby lowering the cost of transaction for businesses and spurring their interest in broader markets.

However, direct long-distance dialing presented a new problem, namely the cost of long-distance calls, which became the next marketing issue—and toll-free direct long-distance dialing was introduced. The marketing game replayed itself, first for the 800- exchange (and again more recently for the 888- exchange). As these number spaces became saturated with mnemonic name–numbers, businesses began to promote spelled-out phone numbers that were *longer* than the functional seven digits (1-800-MATTRESS)—because the excess digits had no effect. The game has played itself out in other ways and other levels—for example, when PBX system manufacturers adopted keypad lettering as an interface for interactive directories that use the first two or three "letters" of an employee's name.

Obviously, this capsule history isn't in a literal allegory for the way DNS has developed—that's not the point at all. There are "parallels," if you like: questions of localized and systematic naming conventions, of national/international integration, of arbitrarily reserved "spaces," of integrating new telecom systems with installed infrastructures, of technical standards co-opted by marketing techniques, and so on. But implicit in the idea of a "parallel" is the assumption that the periods in question are separate or distinct; instead, one could—and should, I think—see them as *continuous* or cumulative phases in an evolving effort to define viable standards for the interfaces between machinic numerical addressing systems and human linguistic systems. Either way, though, DNS—like the previous efforts—won't be the last, regardless of how it is or isn't modified in the next few years.

This isn't to dismiss the current DNS policy debates. On the contrary: they bear on very basic questions that should be addressed *precisely* because their implications aren't clear—questions about national/international jurisdiction and cooperation, centralized and distributed authorities, the (il)legitimacy of de facto monopolies, and so on.

Ultimately, though, these questions are endemic to distributed-network communications and are *not* unique to DNS issues. What *is* unique to DNS isn't any peculiar quality but, rather, its historical position as the first "universal" addressing system—that is, a naming convention called upon (by conflicting interests) to integrate not just geographical references at every scale (from the nation to the apartment building) but also commercial language of every type (company names, trademarks, jingles, acronyms, services, commodities), proper names (groups, individuals), historical references (famous battles, movements, books, songs), hobbies and interests, categories and standards (concepts, specifications, proposals)...the list goes on and on.

The present DNS debates center mostly around the question of whether and how DNS should be adapted to the ways we handle language in these other spheres, in particular, "intellectual property." Given the sorry state of that field—which is dominated by massive industrial pushes to extend proprietary claims indefinitely, to criminalize infractions against those claims, and to weaken "consumer" protections by transforming commodities purchases into revocable and heavily qualified use-licenses—it's fair to ask whether it's wise to conform such an allegedly important system as DNS to that morass. What's remarkable is how quickly this has evolved, from a system almost fanatically insistent on shared resources and collaborative ethics to a speculative, exclusionary free-for-all. A little more history: With the erratic transformation of the "acceptable use policies" (AUPs) of the various institutional and backbones supporters of the internet in the first half of this decade, commercial use of the net expanded from a strictly limited regime (for example, NSFNET's June 1992 "general principle" allows "research arms of for-profit firms when engaged in open scholarly communication and research") to an almost-anything-goes policy left to private internet providers to articulate and enforce (along with questions of spam, usenet forgeries, and so on and so forth). The result was that any entity that couldn't establish educational, governmental, or military credentials was categorized as "commercial" by default. The ".com" gTLD quickly became the dumping ground for just about everything: not just business names and acronyms, but product and service names (tide.com, help.com), people's names (lindatripp.com), ideas and categories (rationality.com, diarrhea.com), parodies and jokes (whitehouse.com, tragic.com), and everything else (iloveyou.com, godhatesfags.com). (This essay omits discussion of the more nebulous ".net" and ".org" gTLDs—which are vaguely defined and became popular only after the domain-name debates —as well as of state [".ny"] and national [".uk", ".jp"] gTLDs.) Thus, the "commercialization" of the net took place on two levels: in the legendary rush of business to exploit the net, obviously, but also in the administrative bias against noninstitutional use of the net.

There were practical reasons for that trend, to be sure: individual or "retail" access was initiated by commercial internet providers, which doled out many more dialup user accounts than domains, as well as technical issues ranging from telecom pricing schedules to software for consumer-level computers that discouraged the casual use of domains. But the trend also had an ideological aspect: the entities that governed DNS preferred the status quo to basic reforms—and, in doing so, relegated the net's fast diversification to a single gTLD that became less coherent even as it became the predominant force.

One can't fault the administrators for failing to foresee the explosion of the net; and their responses are, if not justified, at least understandable. DNS was built around the structurally conservative assumptions of a particular social stratum: government agencies, the military, universities, and their hybrid organizations—in other words, hierarchical institutions subject to little or no competition. These assumptions were built into DNS in theory, and they guide domain-name policy in practice to this day—even though the commercialization of the net has turned many if not most of these assumptions upside down. Not only are the newer "commercial" players prolific by nature, but most of their basic assumptions and methods are very much at odds with the idealized cooperative norms that supposedly marked governmental and educational institutions: they come and go like mayflies, they operate under the assumption that they'll be besieged by competitors at any moment, they thrive on imitation, and they succeed (or at least try) by abstracting everything and laying exclusionary claim to everything abstract—procedures, mechanisms, names, ideas, and so on. The various systems and fields we call "the market" worked this way before the net came along; small wonder that they should work this way when presented with a "new world."

If no one anticipated the speed with which business would take to this new medium, even less could anyone have predicted how it would exploit and overturn the parsimonious principles that dominated the net. Newer domain users quickly broke with the convention of subdividing a single domain into descriptively named sub- and sub-sub- domains that mirrored their institution's structure (e.g., function.dept.school.edu). Instead, commercial players started to strip-mine name space with the same comical insistence that led them to label every incremental change to a commodity "revolutionary." The efficient logic of multiple users within one domain was replaced with a speculative logic in which a few users became the masters of as many domains as they could see spending the money to register. In some cases, these were companies trying to extort attention— and money—out of "consumers" (business's preferred name for "person"); in other cases, they were "domain-name prospectors" hoping to extort money out of business; in many more cases, though, they were simply "early adopters" experimenting with the fringes of a new field. In effect, the potentially complex topology of a multilevel name space was reduced—mostly through myopic greed and distorted rhetoric—to a flatland as superficial as the printed pages and TV screens through which the business world surveys its prey. The minds that collectively composed "mindshare," it was assumed, couldn't possibly grok something as complicated as a host name. So, for example, when Procter and Gamble decided to apply "brand management" advertising theories to the net, it registered diarrhea.com rather than simply incorporating diarrhea.pg.com into its network addressing. And so did the ubiquitous competition, including the prospectors who set about registering every commercial domain they could cook up. The follies of this failed logic are everywhere evident on the net: thousands of default "under-construction" pages for domain names whose "owners"—renters hoping to become rentiers—wait in vain for someone to buy their swampland: graveyard.com, casual.com, newsbrief.com, cathedral.com, lipgloss.com, and so on, and so on.

Under the circumstances—that is, thousands of registered domain names waiting to be bought out—claims that existing gTLD policies have resulted in a scarcity of domain names are doubtful. In fact, within the ".com" gTLD alone, the number of domain names registered to date is a barely expressible fraction of possible domain names, such as "6gj-ud8kl.com": ~2.99e+34 possible domain names *within ".com" alone*, or ~4.99e24 domains for every person on the planet; if these were used efficiently—that is, elaborated with subdomains and hostnames such as "6b3-udh.6gj-ud8kl.com"—the number becomes effectively infinite.

Obviously, then, the "scarcity" of domain name is *not* a function of domain name architecture *or* administration at all. It stems, rather, from the commercial desire to match domain names with names used in everyday life—in particular, names used for marketing purposes. To be sure, "6gj-ud8kl.com" isn't an especially convenient domain name; but, then again, was "Union 567" or "+1-212-674-9850" a convenient phone number, "187 Lafayette St. #5B New York NY 10013" a convenient address, or "280-74-513x" a convenient Social Security number?

But if DNS is in fact such an important issue, does it really make sense to articulate its logic according to the "needs" of marketers? After all, business has managed to survive the tragic hardship of arbitrary telephone numbers for decades and arbitrary street addresses for centuries. Surely, if the net really will revolutionize commerce, to the point of "threatening the nation-state" as some like to claim, the inconvenience of arbitrary domain name will hardly stop the revolution.

Of course there are territorial squabbles over claims to names and phrases. And *of course* some people and organizations profit from the situation. But we don't generally erect a stadium in areas where gang fights break out; so one really has to ask whether it's a good idea to restructure gTLD architecture—supposedly the system that will determine the future of the net, hence a great deal of human communication—to cater to a kind of business dispute that's in no way limited to DNS.

Ultimately, it doesn't really matter which proposed gTLD policy reform prevails, because the gains will be mostly symbolic, not practical— except, of course, for the would-be registrars, for whom these new territories could be quite profitable. At minimum, adding new gTLDs such as ".firm", ".nom", and ".stor" will bring about a few openings—and, more to the point, a new round of territorial expansions, complete with redundant registrations, intellectual-property lawsuits, etc. At maximum, an open domain-name space that allows domains such as "whatever.i.want" will precipitate a domain-grabbing free-for-all that will make navigating domains as unpredictable as navigating file structures.

Moreover—and *much* worse—where commercial litigation is now limited to registered domain names, an open namespace would invite attacks on the use of terms *anywhere* in an address. Put simply: where apple.material.net and sun.material.net are now invulnerable to litigation, in an open namespace Apple Computers and Sun Microsystems could easily challenge "you.are.≠ the.apple.of.my.eye" and "who.loves.the.sun".

Neither proposed reform *necessarily* serves anything resembling a common good. But both proposed reforms will provide businesses with more grist for their intellectual property mills and provide users with the benefits of, basically, vanity license plates. The net result will be one more step in the gradual conversion of language—a common resource by definition—into a condominium colonized by businesses driven by dreams of renting, leasing, and licensing it to "users."

It doesn't, however, follow that the status quo makes sense—it doesn't. It's rife with conceptual flaws and plagued by practical issues affecting almost every aspect of DNS governance—in particular, who is qualified to do it, how their operations can be distributed, and how democratized jurisdictions can be integrated without drifting being absorbed by the swelling ranks of global bureaucracies. The present administration's caution in approaching gTLD policy is an instinctive argument made by people happy to exploit, however informally, the *superabundance* of domain-name registrations.

Without doubt, the main instabilities any moderate gTLD policy reform introduced would be felt in the administrative institutions' funding patterns and revenues. More radical reforms involving more registrars would presumably have more radical consequences—among them, a need to certify registrars and DNS records, from which organizations with strong links to security and intelligence agencies (Network Associates, VeriSign, and SAIC) will surely benefit. The current administration insists that an open name space would introduce dangerous instabilities into the operations of the net. But whether those effect would be more extreme than the cumulative impact of everyday problems—wayward backhoes, network instabilities, lazy "netiquette" enforcement, and human error—is doubtful.

There is one point on which the status quo *and* its critics agree: the assumption that DNS will remain a fundamental navigational interface of the net. But it need not and will not: already, with organizations (ml.org, pobox.com), proprietary protocols (Hotline), client and proxy-server networks (distributed.net), and search-engine portal advances (RealNames, bounce.to), we're beginning to see the first signs of name-based navigational systems that complement or circumvent domain names.

And they're doing it in ways that address not the bogeys that appear in the nightmares of rapacious businessmen but the real problems and possibilities that many, many more users are beginning to face: maintaining stable email addresses in unstable access markets, maintaining recognizable zine-like servers in the changing conditions of dynamic IP subnets, cooperating under unpredictable load conditions, and, of course, *finding* relevant info—not *offering* it, from a business perspective, but *finding* it from a user's perspective.

DNS, as noted, was built around the assumptions of a specific social stratum. Prior to the commercialization of the net, most users were if not computer professionals then at least technically proficient; and the materials they produced were by and large stored in logical places which were systematically organized and maintained. In short, the net was a small and elite town, of sorts, whose denizens—"netizens"—were at least passingly familiar with the principles and practices of functional design. In that context, just as multiple users on a single host was a sensible norm, so were notions of standardized file structures, naming conventions, procedures and formats, and so on. But just as the model of multiple users on a single host has become less certain, so has the rest.

The net has become a nonsystematic distributed repository used by more and more technically incompetent users for whom wider bandwidth is the solution to dysfunctional design and proliferating competitive formats and standards. Finding salient "information" (the very idea of which has changed as dramatically as anything else) has become a completely different process than it once was.

This turn of events should come as no surprise. As commercial domains multiplied, and as users multiplied on these domains, the quantities of material their efforts and interactions produced grew ferociously—but with none of the clarity typical the "old" institutional net. In the past, the information generated around or available through a domain (or to the subdomains and hostnames assigned to a department in a university or military contractor) was often "coherent" or interrelated. But that can't be said of the material proliferating in the net's fastest-growing segments: commercial internet access providers, institutions that automatically assign internet access to everyone, diversified companies, and any other domain-holding entities that permit discretionary traffic.

Instead, what one finds within these domains is mostly random both in orientation and in scale: family snapshots side by side with meticulously maintained databases, amateur erotic writings next to source-code repositories, hypertext archives from chatty mailing lists beside methodical treatises, and so on. In such an environment, a domain name functions more and more as an arbitrary marker, less and less as a meaningful or descriptive rubric.

This isn't to say that domain names will somehow "go away"; on the contrary, it's hard to imagine how the net could continue to function without this essential service. But the fact that it will persist doesn't mean that it will serve as a primary interface for navigating networked resources; after all, other aspects of network addressing have become all but invisible to most users (IP addresses and port numbers to name the most obvious).

The benefit that DNS offers is its "higher level of abstraction"—a stable addressing layer that permits more reliable communications across networks where changing IP numbers change and heterogeneous hardware/software configurations are the norm. But "higher" is a relative term: as the substance of the net changes—as what's communicated is transformed both in kind and in degree, and as the technical proficiency of its users drops while their number explodes—DNS's level of abstraction is sinking relative to its surroundings.

[This essay first appeared on *Rewired* <http://www.rewired.com/> on 28 Sept 1998 under the title "A Higher Level of Abstraction." Thanks to David Hudson for his editing.]

A change of address letter from Graham Harwood. May 98: During the past ten years. I have worked with new technologies and opening up social spaces. For the last three and half years, I have worked at Artec training unemployed people and have made many good friends and set up many good working relationships with the people I taught. This was an extremely busy time for me finishing and publishing Rehearsal of Memory as well as running courses and being involved in the arts programme at Artec. There were many sleepless nights, stress, excitement, and above all there was the possibility of creating a space in which people could safely explore culture clash and exclusion from the trough of society. I wanted this space to be experimental, away from immediate poverty and also away from the excesses of a municipal post socialist pretension. In the last few years, I have seen the context in which Artec and similar organisations operate steadily tightening up, becoming accredited to a new social order. There is a very real danger that these constrictions—or to put it another way, the reordering of powerful elites to cope with technological change—will strangle the technologies bastard miscarriage of social opportunity. Artec I feel, like many other smaller organisations, could be lured into adopting the agenda of academic and political organisations and agencies which may dwarf it. People at Artec work hard and usually do not have the luxury of distance from the day to day grind of running courses and making things happen to see what's coming round the corner. It's always useful to be reminded that the academic and political organisations and agencies now setting the agenda are the ones which failed the client group in the first place. [Matthew Fuller <matt@axia.demon.co.uk>, Change of Address, Mon, 27 Apr 1998 21:47:36 +0100]

SUBJECT: PRECIOUS METAL AS A NETWORK PROTOCOL

FROM: JULIAN DIBBELL <JULIAN@MOSTLY.COM>
(BY WAY OF BRUCE STERLING <BRUCES@WELL.COM>)

DATE: MON, 9 MAR 1998 10:36:58 -0600 (CST)

Sources: J. Buchan, *Frozen Desire: The Meaning of Money*, NY: Farrar Straus Giroux, 1977, and J. Weatherford, *The History of Money: The Struggle over Money from Sandstone to Cyberspace*, NY: Crown, 1997

In his remarkable book, James Buchan writes:

From our vantage, we can see that money is of no particular substance and may be of no substance at all; that whatever money is, it may be embodied in coins or shells, knives, salt, axes, skins, iron, rice, mahogany, tobacco, cases of gin; in persons; in a word or gesture, paper, plastic, electronic impulses or the silver ingots raced through the streets on trays at sundown to make up accounts between the foreign banks in my mother's father's days in Hangkow. (18)

"Economic booms and busts will become more frequent and more severe if programs called software agents control electronic commerce. Agents tend to exaggerate the worst market swings and create disastrous price wars, say two research groups in the US. As more goods and services are bought on the Internet, observers predict that we will need agents to get the best prices. But agents are not subject to the restraints that normally slow economic activity: their transactions take place almost instantaneously, cost next to nothing and distance is irrelevant."

Two things about this passage interest me. The first is its suggestive implication that money has both a "hardware" component (that is, the coins, paper, knives, mahogany, and so on that embody it) and a "software" component (that is, among other things perhaps, the value thus embodied). The second is the wonderfully nostalgic closing tidbit about the shuttling trays of silver in the streets of old Hangkow (this I assume is the former city Hankou, China, now a subdistrict of the megalopolis Wuhan), which provides a vivid, high-Cahill-number image of the essentially abstract dead medium I'm proposing for consideration here: metallic monetary standards, the antiquated practice of backing every piece of circulating currency with a fixed amount of precious metal.

Some preliminary taxonomizing is in order. Bruce Sterling suggested in Dead Media Working Note 22.1 that money might be thought of as a distributed calculating system; and that seems about right. But there's another suggestion built into that one: that we think of money as a network. Strictly speaking, too, we'd want to think of it as an *inter*network, globally distributed and capable of transmitting value from one end of the net to the other, so long as the proper network gateways are traversed. Money, we might even say, throwing precision to the wind, is the original Internet. But let's just call it an analogy, and see where it leads us.

One implication, I think, is that if coins and banknotes and so on are to be thought of as the hardware of the network, then we must also look for some underlying technical system we could call the network protocols. I am not enough of a finance wonk to identify the "protocols" of the contemporary world money system—a frighteningly live medium, in any case—but I think it's safe to say that in the terms of our analogy, "protocols" is exactly what we would have to call the metallic standards that governed monetary exchange during the first great age of global capitalism (that is, from Waterloo until World War I).

In particular, we would mean the gold standard, which died a slow death between 1931, when Great Britain abandoned it, and 1971, when Britain's successor at the helm of world finance, the U.S., finally chucked it too.

If I understand the Hangkow ingot exchange that Buchan alludes to, the system might properly be considered a kind of monetary intranet, operating locally on the same principles as the global network. Globally, a physical transfer of precious metal was also used to settle accounts at the end of the day—though, at that level, the metal was gold rather than silver, and the transfers were between nations as well as banks, and the end of the day was really the end of the quarter or the year.

It was a very different regime than what we have now, with very different effects. The money supply was tighter, often painfully so, and the drift of economies was (according to Buchan) deflationary rather than inflationary. In the U.S. at least, bitter and arcane controversy sometimes surrounded the subject of metallic standards, with the Populists of the late nineteenth century, for instance, supporting a move to a "bimetallist" gold and silver standard that would somehow loosen the money supply and make things easier for the little people.

According to Jack Weatherford's *The History of Money*, it was apparently well understood at the time that L. Frank Baum's *The Wizard of Oz*, published in 1900, was a Populist allegory inveighing against the gold standard (the seductive "yellow brick road" to the sham-world of Oz being merely one of the more obvious clues).

Metal-based money was strange stuff. It's difficult, at this late stage in the world-financial game, to imagine what could possibly bring the metallic standards back. Profound inflationary trauma perhaps, or maybe a global dictatorship. For the time being, at any rate, they remain very much dead.

[This message first appeared as Working Note 30.9 on the Dead Media mailing list.]

SUBJECT: PIRACY NOW AND THEN

FROM: TOSHIYA UENO <VYC04344@NIFTYSERVE.OR.JP>
DATE: TUE, 29 SEP 1998 18:15:50 -0400

What is the first impression or association for us when we hear the term *piracy* or *pirates*? One easily thinks of pirate radio or TV, the pirated editions or versions of any kind of media (music tapes and records, computer applications, books or brochures, and so on). Generally, this term is used in contexts opposed to capitalism or commercialization. If one looks back at the history of capitalism itself, one can see the close connection between piracy and capitalism. Although this essay deals with one aspect of capitalism, its aim is not necessarily to focus on the economics and politics of money and commodities; rather, it is an attempt to elaborate cultural politics in the age of information capitalism through a tactical way of thinking.

In discussing the relationship between piracy and capitalism, I wish to begin by referring to Daniel Defoe's *Robinson Crusoe*. This novel is an important reference point for analyzing the relationship between piracy and capitalism. In Defoe's story, Robinson resisted his father's opinion and Protestant ethics; he did not trust the Christian God of Protestantism. Robinson was longing for his brother, who had become an adventurer in search of property and treasure in an unknown world, either Africa or the West Indies. Robinson tried to do the same. But on his first trip, he was caught by Moors and enslaved. Eventually, he escaped, bought land in Brazil, and ended up managing a huge plantation. However, his plantation fails, and he begins again to navigate the seas—this time in search of African slaves. His ship sink, and he alone survives to live on a desert island. Despite this miserable situation, he appreciates and blesses God. Robinson has reformed and returned to Protestantism. On the island, he tries to make an enclosure much as the gentry or early bourgeoisie established them in England—he returns to the Protestant ethic and the spirit of capitalism.

As you may know, this interpretation is derived from Max Weber. But it is already obvious that the human type of Robinson—a person who acts rationally and productively on the basis of "innerworldly asceticism"—is a sort of fiction. When one reads *Robinson Crusoe* carefully, one comes to understand that his behavior on the island is not at all "rational" or "productive." Instead, his activities depend on monstrous, excessive desires. For example, when he tries to salvage useful materials from the shipwreck on the island's coast, he wants to get "everything" without considering whether or how these things actually will be useful. It is especially clear in his obsession with his fort's construction, since he does not know the purpose of the fort. In short, Robinson doesn't really know what is doing. (This corresponds roughly to Marx's definition of ideology). His behavior and mentality are not and never were based on "value-rationality." So the human type of Robinson is not nearly as ascetic and rational as the bourgeoisie in England were; rather, he resembles the type of humans in the contemporary world. (The phrases "type of human" or "human type" are technical terms in the sociology of Max Weber. One can understand them as an ideal embodiment of type of each class.)

As a character, Robinson is very similar to us in his purposeless and excessive production and consumption. Even though we would define Robinson as the human type of Protestant, the theoretical framework that makes this definition possible is already problematic and dubious. In response to the question "Why did capitalism first arise in England, and not in other places?" the most general reply has been: "It is because the bourgeoisie possessed the Protestant ethic that capitalism developed in England before its advent in other places." But the foundations of this interpretation are beginning to change very radically. For example, according to Immanuel Wallerstein's "world system theory," the response should be: "It is because capitalism appeared in England that it didn't appear elsewhere." The world system is *one* system, and it has a structural totality. The viewpoint adopted by world system theory, it should be noted, relates to theoretical problems raised by colonialism. After 1492, capitalism became synchronized with colonization and colonialism. In our example, Robinson turned to navigation in order to obtain slaves for his Brazilian plantation. However, in his life on the island, he encounters Friday, a "colored native other"—a figure who served as the sine qua non of the Western Enlightenment of reason.

Small wonder, then, that world system theory, or Braudelian historicism, should have engendered scholarly interest in the transportation and communications aspects of sea trading. Robinson, remember, was a sailor; the type of human epitomized by Robinson was found not in yeomanry or the middle bourgeois but, rather, in the sailors and colonizers of the seventeenth century. In this regard, we might note how pivotal this shift can be, from the land to the sea. It was not a new one in Defoe's time: Venice in the Middle Ages, Spain in the sixteenth century, the Netherlands in the seventeenth century, England in the eighteenth century, all were sea empires, and the state exerted hegemony over the sea. In 1492—the year, of course, when Columbus landed in the Americas—Islamic Moors were exiled from the Iberian peninsula. Some became Barbarian pirates and turned to attacking the ships of Christian Europe. The Christian states, in turn, granted many Christians (and hence Europeans) authority to become pirates with letters of marque to attack other nations' ships. The post-Columbus age, it seems, was an age of pirates.

PIRATES

Captain Charles Johnson's *A General History of the Robberies and Murders of the Most Notorious Pirates* (1724) is a very strange and interesting book that deals with the history of the pirates. Its stories about Captain Kidd and Teach, as well as female pirates such as Mary Lead and Ann Bony have influenced countless novels and fictions about pirates. According to Hakim Bey, in his book *T.A.Z.*, and others, Charles Johnson may be a pen name of Daniel Defoe. If so, one might note the curious coincidence that the author of a book that portrays the rise of capitalism is also the author of a history of pirates; but it's no coincidence.

"Electronic Commerce and the Street Performer Protocol": Copyright will be increasingly difficult to enforce in the future. The barriers to making high-quality pirated copies of digital works are getting lower and lower, and solutions such as hardware tamper-resistance and watermarking just don't work. We introduce the Street Performer Protocol, an electronic-commerce mechanism to facilitate the private financing of public works. Using this protocol, people would place donations in escrow, to be released to an author in the event that the promised work is put in the public domain. This protocol has the potential to fund alternative or "marginal works. [J. Kelsey and B. Schneier, The Third USENIX Workshop on Electronic Commerce Proceedings, USENIX Press, September 1998.]

According to Bey, a T.A.Z., or "temporary autonomous zone," is not a concrete and realized societies or fixed spaces but, rather, an ephemeral chronotope marked by autonomy and independence; not surprisingly, such zones tend to be short-lived. In the chapter titled "Pirate Utopias," Bey finds such a zone in the activities of seventeenth- and eighteenth-century pirates; he says that the pirates and corsairs had formed a sort of information network by creating a global web connecting islands and continents. Historically speaking, many pirates founded small communities or utopian societies in Morocco or the Caribbean islands, communities that were quite different and independent from the early power politics of nation-states. Bey goes on to draw a parallel between the overlapping relation among islands and archipelagos connected through pirate societies in that period and our own era's rhizomatic nets of transnational corporations. He also cites Bruce Sterling's novel *Islands in the Net*; like these enormous corporations, many hacker-based and small high-tech manufacturers are operating in ways that transform the quality and meaning of property or ownership itself.

There is one particular society that's quite interesting in this context—the seventeenth-century Pirate Republic of Sale in Morocco, an independent and insurrectionary community formed by corsairs, sufis, adventurers, and the like. Peter Lamborn Wilson, in *Pirate Utopias* (Autonomedia, 1995), suggests that this republic exemplified the pirate utopias, where thousands of Europeans converted to Islam and joined the pirate "holy war." It's interesting to note in passing that, in Defoe's novel, Robinson was taken captive in this republic.

Wilson uses the term *renegadoes* (an older form of the term *renegade*) to describe these "converts." Terms of this kind *renegades* and *converts*—pivotal characters in the history of piracy—tend to carry a negative connotation, for example, a movement toward heresy or paganism; but given that both rely upon a closed community or dogmatic party, which is rejected, the terms also connote an openness.

SEAMEN

Another interesting text in this context is the novel Herman Melville's *Moby Dick*. Of course, Captain Ahab and his crew in the ship *Pequod* aren't pirates, but their story is fundamentally determined by life and work on the sea. Like Robinson, Ahab's activities—his vengeance against Moby Dick—are defined by a renegade and individualistic goal. (The biblical name Ahab itself signifies exile.) And this in the context of an extremely heterogeneous community: there are many races on the Pequod. Around the figure of Ahab as a white, one finds overlapping of marginal natives and tribes—for example, Caribbean, American Indian, African blacks, and European whites.

Melville's writing about whales is, in a word, maniacal. The novel's encyclopedic descriptions of whale lore, "cetology," are clearly fueled by some very extreme passions. In this regard, the structure of the novel is absolutely mirrored in the narrative: Ahab, haunted by his vengeance, consumes his crew, as though he draws some invisible power from the white whale. And the whale itself, in turn, seems nearly immortal: though wounded by a harpoon, it reappears again and again without so much as a scar. Moby Dick seems to draw this power from the sea or, more particularly, from the autonomy that defines the whales' relationship with the sea: Melville's narrator, Ishmael, says the whales know a secret "web" of routes in the sea. Not surprisingly, this informatic structure isn't limited to the sea: the whales themselves are, in Melville's narrative, redefined as informatic structures themselves—for example, the narrator compares patterns on the whale's skin with the designs of primitive Indian art and likens the movements of the whale's tail to the symbols and signs in freemasonry.

This kind of configuration isn't merely novelistic artifice in *Moby Dick*; traces of these relations can be found in the history of whaling in Japan as well. For example, the tradition of whaling in Japan holds that whaling is not merely hunting whales but, rather, a technique of searching out the invisible and uncontrolled zones of the sea, the matrices defined by the movements (or appearances) of whales; whaling necessarily involved entering into unknown and hidden elements in nature. (We now know that whales are intensely sensitive to sounds, and therefore function in an at least partially acoustic relationship with their environment.) Moreover, the histories of whaling and piracy in Japan are closely intertwined: when Hideyoshi Toyotomi persuaded the political and military hegemony—including pirates—to disarm in the late sixteenth century, many pirates turned to whaling. Moreover, a Japanese post-structuralist, Shinichi Nakazawa, has shown how, in the early 1600s, samurai pirate-turned whaler Yolimoto Wada mobilized his village as a "war machine"—including all the procedures, rituals, and technologies its whaling economy relied on—around a series of technological and organizational innovations; the result was one of the first models for manufacturing in Japan, and hence for Japanese capitalism.

It is worth noting that this mobilization was not structured in terms of European-style rationality. For example, whales were not simply objects to be exploited; rather, they held a spiritual significance. It is arguable whether this worldview was particular of singular to Japanese culture; there is no doubt that Japanese whalers believed in a unique cosmology, and were very concerned to distinguish nature from artifacts, physis from nomos, and exchange from exploitation, but it would be a mistake to limit the potential of these distinctions by superimposing upon them some purported "Eastern character" or geographical limitations. Rather, we should see to find in this configuration of concerns some pathways to other ways of viewing the world, other chronotopes and contexts.

Another seemingly disparate source that is useful in this context is the work of the German political philosopher Carl Schmitt. Though notorious for his pro-Nazi politics, after World War II his attention turned toward an analysis of human history in terms of a struggle between land- and sea-based empires. In *The Land and the Sea: A Historical Analysis*, he stresses the role of water—the sea—as being a far more fundamental element than the others (air, fire, and land). He depicts history as an endless struggle between Behemoth, the land monster, and Leviathan, the sea monster. Perhaps not surprisingly, he repeatedly cites *Moby Dick* as a touchstone in understanding the political meaning of navigation, seapower politics, and—perhaps surprisingly—the peculiar technology of whalers. The novel interests Schmitt because, he says, "Through fighting with the creature in the sea, humans were seduced to going into the deep element of the sea." Whalers are not merely catchers or slaughterers but hunters: in the wake of Columbus, Captain Cook, and other navigators, whalers—by definition, followers of whales—effectively charted the globe. Whales, it could be said, liberated humans from the land and taught them the tidal currents of the sea.

Schmitt compared himself to a character in another of Melville's novels, *Benito Cereno*, in which the protagonist, Captain Cereno, is forced by a slave insurrection on his ship to turn to piracy. The parallel between Cereno's piracy and Schmitt's own collaboration with the Nazi regime is clear, but no simple or convenient metaphor: in his works from *The Land and The Sea* (1947) to *The Partisan Theory* (1962), the pirate plays an crucial position in Schmitt's elaboration of the concept of "the political." Much as pirates took to the sea with official lettres of marque, the early bourgeoisie in England, for example, made enclosures in order to develop the wool indus-try (Schmitt describes these Englishmen as "the corsairs of capitalism"). Both sea and land became the field for the primitive accumulation of wealth, as well as the transmission of religious and social beliefs. These effects were hardly limited to pirates: missionaries, for example, dissemi-nated all of the world many of the same basic values that contributed to colonialism and capitalism.

For Schmitt, the essence of the political lies in the distinction between friend and enemy—a distinction that can sometimes be very ambiguous. The main characters of *Moby Dick*—Ahab, Starbuck, Queequeg, Ishmael, and so on—all have such a relationship with the white whale. Though not pirates, they are all, in some sense, outcasts and renegades, or, in Schmitt's terminology, "partisans." In *The Partisan Theory* (1962), Schmitt uses the word *partisan* to describe those who lie outside of the framework (*Hegung*) of ordinary warfare. Partisan tend to depart from conventional warfare and social mobilization and move toward alternative types of warfare and political relations; it is reasonable, then, to speak of pirates as a form of partisan. According to Schmitt's theories, partisans unfold and invent new spaces; and the formation of these spaces depends very directly on avail-able forms of technology and industry.

If the principle of the partisan consists of maneuvering enemies into unknown spaces, then whales and whalers can be seen as opposed parti-sans. By extension, ships of growing sophistication and submarines have expanded these interplays to a worldwide scale, and other mechanisms—nuclear-equipped submarines and space-based surveillance satellites and weapons—have transformed that reach into a more complex "global" phe-nomenon. Given the launch of Sputnik and the growing "space race, it should come as no surprise that Schmitt's speculations on these questions involve the possibility of "space pirates" and "space partisans." From our perspective, we can begin to see how these phenomena will extend into the spaces and nonspaces of "pure" information.

Some have said that Robinson's island was Tobago. Whether that's true, I don't know, but from that island one can see yet another, Tobago. This latter island has brought us yet another theoretician I would like to add to this constellation, C. L. R. James. Among his very diverse works we find one on Melville and Shakespeare, *The Sailors, the Renegades, and the Castaways*, named for a passage in *Moby Dick*, and *American Civilization*, (Blackwell, 1993). In the latter book, James analyzes *Moby Dick*; he argues that the novel is the story as being that of American society itself. The white whale, he says, is not an allegory for undomesticated and violent nature, but, rather, a symbol of industrialization, colonization, imperialism, and class struggle—in short, a meta-struggle to move into new kinds of spaces and metaspaces. He describes its pursuit in these words: "This legitimate activity symbolizes the perpetual relation of civilized man with Nature. The whale was the most striking of living things which man had to subdue in order to have civilized lives. The whale is not a mere fish. The conquest of the air, the mastery of atomic energy, all these are symbolized by the whale." This metasymbol, if you will, spins out thousands of references and interpretations. The struggles in *Moby Dick* represent real struggles within society: "Melville knows and says repeatedly that the conflict is between human and Nature, the demonism that is in Nature. Melville knows also, however, that the struggle with the demonism in Nature involves a certain relation between man and man."

Thus, throughout the novel, the human desire to surpass limits intertwines with the constant crossing from sea to land and from land to sea. The white whale is an active element of the sea, itself, and unknown nature, set in an endless struggle with human beings; but this struggle is also one between people, and defined as much by life on the land as by life on the sea. The fight with the whale is a model of human history, and the narrative of the struggle with and awful, sublime nature is, in fact, an inverted image of social relations. The ship *Pequod* is, in a way, a sort of industrial factory populated by Ahab, the human-type of modern man in industrial society, and Ishmael, the narrator as a model of the modern intellectual. James concluded that Ahab's ability to mobilize people through a unique power makes him very much like modern dictators such as Hitler and Stalin.

If *Moby Dick* is a Leviathan of the nineteenth century, what of the twentieth? The information spaces we are now beginning to contemplate haven't emerged from nowhere at all; the roots of digital modalities can be found in earlier developments in media, for example, "cut 'n' mix" and sampling technologies emerging from various forms of black music—which, not surprisingly, developed in the web of connections that emerged among the exiles and migrations that have characterized black experiences of the modern world. And do not theoreticians concerned with the black diaspora have an interest in pirate culture? Indeed they do. Paul Gilroy is an excellent example: his *Black Atlantic* (inspired in large part by C. L. R. James) relies very heavily on the metaphor of the ship: "The image of the ship—a living, microcultural, micropolitical system in motions—is especially important for historical and theoretical reasons. Ships immediately focus attention on the middle passage, on the various projects for redemptive return to the African homeland, on the circulation of ideas and activists as well as the movement of key cultural and political artifacts: tracts, books, gramophone records, and choirs" (London: Verso, 1993, 4). According to him, the ship is a medium, a living means that connects nodes in the Black Atlantic world, hence central to cultural exchange and travel.

So now we are faced with broad, new, and unimaginable spaces through network technology: from radios and telephones and now through wireless communications and the internet, the lands and seas of information are expanding. Though these media sometimes are commercialized and commodified, we will no doubt invent new forms piracy. Piracy and capitalism have always been two sides of the same token. Information capitalism is no exception.

[Edited by Hope Ted Byfield and Hope Kurtz.]

SUBJECT: OLD AND NEW DREAMS FOR TACTICAL MEDIA
FROM: DAVID GARCIA <DAVIDG@XS4ALL.NL>
DATE: MON, 23 FEB 1998 19:04:36 +0100

PREHISTORY

Our cultural and political life is framed in the symbols and grammar of the electronic media, and these are still overwhelmingly dominated by television. No mainstream political or cultural player can afford to ignore TV's seductive power, in fact the media itself in the form of journalists, editors, TV inquisitors, and spin doctors collectively make up a separate and unelected branch of the political life of liberal democracies.

From its beginnings as a *mass* broadcast medium, TV constructed its audience accordingly, as *the masses*. The notion of mass culture, arising from mass society, was a direct expression of a media system controlled either by the state or by large corporations. Although artists and activists from the early part of the century had consistently challenged the notion of the audience as passive and homogenous, it was not until the eighties that the mainstream media (along with everything else in the capitalist economies) was forced to reconfigure along more flexible and customized lines. It was during this period that the revolution in consumer electronics combined with the regulatory uncertainty in the media landscape spawned the incredible variety of achievements in the field of art, civic communications, and electronic dissidence that we call *tactical media*.

INTERMEDIATE TECHNOLOGIES

There is a tendency to blur into a single step the journey from the period of mass broadcast media described above to our own era of hypermedia and the internet. In fact, tactical media emerged from a vital bridging period during the eighties, when a whole range of intermediate technologies allowed for ways of interacting with the media which were far less passive than pundits and media theorists (including McLuhan) had ever envisaged. The TV zapper, the Walkman, the VCR and video rental industry, the greater range of channels through cable and later home satellite receivers and, above all, the camcorder arrived on the scene within a few years of one another. This series of innovation allowed "audiences" for the first time to create their own individually customized media environments and thereby to explode once and for all the dominance of broadcast media as the centralized source of societies representations. With the camcorder came an "additional modification to the one way flow of images and further developed the process of integrating our individual life experience to life on screen." This was the situation that made tactical media possible. And the fact that these technologies were everyday household appliances freed artists and media activists from the classic rituals of the underground and alternative scene. While at the other end of the spectrum "big media" whether MTV graphics or BBC's *Video Diaries* were incorporating techniques and ideas that for years had been the exclusive province of the avant-garde. This was why we introduced the term tactical media: the old dialectical terminologies of "mainstream versus underground" or "amateur versus professional"—or even "private versus public media"—no longer seemed to describe the situation we were living through.

During the eighties, groups as culturally and geographically diverse as the Gay Men's Health Crisis (New York) and Despite TV (London) and aboriginal telecaster project Satellite Dreaming were proving that you could make effective media interventions from outside of the established hierarchies of power and knowledge. Reemphasizing the role of transitional media is not merely academic: different parts of the world move at different speeds. For members of a rural community in the developing world, struggling to come to terms with the impact of television, picking up a camcorder and making their own stories is still a way of taking power. Anyone who has seen the work of Sylvia Meijer who uses camcorders as a consciousness-raising tool with Colombian women in villages and in jails can attest to the fact interactivity is not just a property of "new media."

THRIVING ON CHAOS

The movement we call tactical media has been comprehensively explored in two conferences held in Amsterdam, called The Next 5 Minutes. As we plan the third, it is important that, like every generation of modernists, we to try to confront the paradoxes and ambiguities of our position. It is an old difficulty in new disguises, but we dare not avoid it.

Along with all other moderns, media tacticians have to face the fact that not only can all their acts of subversion be co-opted by capital, but the perpetual cycle of destruction and renewal which characterizes tactical media, is itself an embodiment of the forces unleashed by capitalism. Plenty has changed since our world was transformed by nineteenth century industrialists, but the mutually dependent relationship between capital and its malcontents remains much the same. This is why even the most corrosively nihilistic movements from Fluxus to Punk can be co-opted so easily. Capital is not threatened by chaos it thrives on it. The difference between our age and others is the growing openness about the fact. nineteenth-century industrialists averted their eyes the from the nihilistic logic of the forces they had unleashed, not only by creating a veneer of respectability and permanence but also by instituting the radical bourgeois public sphere. The civic and cultural institutions including museums of art and academies of science. It is not enough for us to go on subverting this public sphere, which has been the autopilot response of generations of radicals. Modern capital, with its corporate evenings and sponsorship deals is already doing that job effectively enough. For today's operators in the advanced service industries, from insurance to advertising, every act of "ontological terror" is another marketing opportunity. Years after it occurred, Hakim Bey is still fulminating angrily at Pepsi calling one of their parties "a Temporary Autonomous Zone." What did he expect?

Change is good, proclaims *Wired* magazine's cover at the beginning of the first issue of 1998. Demonstrating once again how libertarian capitalism has finally abandoned the strategy of previous generations of bourgeoisie to identify themselves as the "party of order." One of the clearest illustrations of capital's new realism about its brotherhood with the anarchic forces it once feared is the highly profitable partnership between the Damien Hurst generation of English artists and the advertising mogul Charles Saatchi. In his boldest act so far, Saatchi has even succeeded in co-opting the Royal Academy (the epigone of stuffy bourgeois institutions) to display and advertise the "cool Britannia" part of his large collection. And the more horror and shockwaves the exhibition creates, the happier he is.

REDREAMING PUBLIC SPACE

The net is not averse to pretending to be a place. Especially when there is money to be made. On the web, domain names are the equivalent of real estate, and prime locations are already being hotly contested. "Recently the most expensive known domain name—business.com—was sold for $150,000 to an undisclosed buyer by a London-based banking software producer Business Systems International." To give the flavor here is an extract from an add published by InterActive Agency:

WHAT'S IN A NAME? BROADWAY? PARK PLACE? MAGAZINE.COM!

Real Estate is a valuable commodity even on the internet here's your chance to enjoy a penthouse view of cyberspace!"

It was Hannah Arendt in the fifties who asked of Marx (but could have put the same question to any modern—including libertarian—capitalists), "if the free development of each is the condition of the free development of all, what is it that is going to hold these freely developing individuals together?" Perhaps Habermas has come closest to answering, but no theorist of the modern has yet succeeded in building an effective theory of political community. We still have "no true public realm, but only private activities displayed in the open."

A SENSE OF PLACE

In *The Networked Society*, Manuel Castells describes a situation in which everything in our culture is reconfiguring around virtual flows. These flows are not just an element of our social organization; rather, they are an expression of processes dominating our economic, political, and social life.

But *places* do not disappear.

In the wider cultural and political economy the virtual world is inhabited by a cosmopolitan elite. In fact, put crudely, elites are cosmopolitan and people are local. "The space of power and wealth is projected throughout the world, while people's life experience is rooted in places, in their culture, in their history." If projects like the Next 5 Minutes or Nettime place their faith in "ahistorical virtual flows, superseding the logic of any specific place, then the more our emphasis on global power will escape the socio/political control of historically specific local/national societies." We must create a more consciously dialectic relationship between these two realms, which Castells calls the *Space of Flows* and the *Space of Place*, because if they are allowed to diverge to widely, if cultural and physical bridges are not built between these two spatial logics, we may be heading toward—or may have arrived at—life in two parallel universes "whose times cannot meet because they are warped into different dimensions of hyperspace." One possible direction may lie in reclaiming community memory, in re-imagining the public sphere through the symbolic role of the public monument. No broad discussion about the public domain can be separated from the physical embodiments of community memory in the form of public monuments. "The model here is that of the city (the polis) in classical antiquity, and the stress is the memorable action of the citizen, as it publicly endures in narrative." The opposite of this is the dream of the placeless utopia of the metropolitan elite, which is everywhere evident in the social dreams proffered as the hallmarks of that elite—from words like *jetsetter* with neither origin nor destination, to cyberspaceless utopias without borders.

The need for an enduring sense of place with its own community memory was powerfully brought home to me on my visit to Tallinn for this conference. In an artist's club, a young man told me about how a group of his friends were involved in a project to take all the old social realist statues from the communist era and melt them down into one gigantic bronze cube. As he was talking, I remembered a "solution" for similar works in Hungary, where they have been arranged in a park in Budapest, in a sort of virtual history: Communism the "experience," recent history as theme park. I argued with him that communities, like individuals, shouldn't try to deny their past. "We may not like it, but it's a fact." When I suggested that if he and his friends conspired to bury the past, they—or others—would end up regretting it, he looked me straight in the eyes and said, "Don't try to psychoanalyze us—you're an outsider. You don't understand. You don't even begin to understand what its like to live and grow up under a foreign tyranny... For you Soviet stuff is a fashion. The Red Army choir, fur hats, Levis—it's all the same." I apologized. I was put in my place. In secure liberal democracies nationalism (a secure sense of our own *place*) is often portrayed as an irrational vice but for him, the word *nation* was interchangeable with *freedom*.

Tactical media, like most modern movements, has tended to privilege the ephemeral, the moment. But "in opposing the monument to the moment we see the monument not simply as a symbol of repression but also a repository of knowledge and as memory. Reclaiming the monument means reclaiming depth in time, *dureé*, it's a way of getting back to work on memory." Perhaps this sounds dangerously like the familiar siren calls of all those classical revivals, "to the natural order of things through appeals to universal principals outside of space and time." But I'm thinking of very concrete examples where public space and public monuments were appropriated and re-invented, in the way that Martin Luther King and the American civil rights movement of the sixties went to the heart of white American establishment when King made his famous speech from the Lincoln Memorial.

There is one image to which my thoughts around this subject keep returning, my private resolution of the apparent contradiction between the moment and monument—a black-and-white photograph in which the facts are deceptively clear. At the bottom of the image the photographer's clenched fist is turned to the camera to look at his watch. It is daylight, and we can see on the watch that the time is around midday. Beyond the hand and the watch a boulevard stretches out, leading to a square of what is obviously a major European city. But it is as eerily empty as a de Chirico. Even on a Sunday this would be strange. So we are presented with a mystery.

Those who are familiar with central European cities might recognize it as Prague and as one of the main avenues leading to Wenceslas Square. In fact the photograph was taken by Joseph Koudoulka in 1968. A few days earlier Brezhnev's tanks had rolled in to crush Dubcek's experiment in "socialism with a human face." Kadoulka had agreed to meet some fellow citizens for a march on the square. For obvious reasons, they failed to keep the appointment. The failure is marked with this photograph. His watch on a hand clenched in an angry fist, a visual intersection of the picture and the boulevard. Two time lines cross; an individual life and the sweep of history in the making. The photograph seems to hold its breath. I can almost hear the sound of the shutter recording and becoming both a moment and a monument.

[This essay is based on a talk on tactical media for the Interstanding conference in Tallin, sponsored by the Soros Foundation.]

Transnational corporations—TNCs—are the bogeymen of global dreams. They are imaged (on the left at least) as roving postmechanical monsters, outfitted with fantastically complex electronic sensors and vicious trilateral brains, and driven by an endless appetite for the conversion of resources, labor, and consumer desire into profit for a few. There's some truth in that image. But the power of transnational capital is inseparable from the capital "S" of subjective agency, expressed in social, cultural, and political exchange. Which is why I'd like to discuss TNCs in relation to what you might call TNCS: transnational civil society.

Let's start with the bogeyman. It became apparent in the sixties that private corporations were taking over the technological and organizational capacities developed initially in World War II: the coordinated industrial production, transportation, communication, information analysis, and propaganda required for multitheater warfare. Corporations such as Standard Oil or IBM, operating through subsidiary companies in every nation that did not allow direct penetration, were projections of a (mostly U.S.) military-industrial complex into both the developed and the undeveloped world, as part of the globe-girdling Cold War strategy. Yet already in the sixties these "multinational" enterprises were achieving autonomy from their home bases, for instance through the creation by British financiers of the Eurodollar, a way to keep profits offshore, out of the national tax collector's hands. The offshore economy took a quantum leap in the mid-seventies after the first oil shock, when the massive capital transfers to the OPEC countries were channeled by inventive Western bankers into the new, stateless circuits of financial exchange. That's about the time when the full-fledged system of transnational capitalism emerged, with the collapse of the nationally based Fordist–Keynesian paradigm of labor-intensive industrial production plus welfare programs. The proximate cause for the collapse was the inflation brought on by the policies of stimulating consumption through public spending; but the durable factor prohibiting any return to the postwar social contract was the competitive pressure of what is now known as flexible accumulation, based on geographically dispersed yet highly coordinated "just-in-time" production, cheap worldwide distribution through container transport systems, and the complex management, marketing, and financing made possible by telecommunications. The flexible production system allowed the TNCs to avoid the concentrated masses of workers on which union power depends, and so much of the labor regulation built up since the Great Depression was sidestepped or abolished. At the same time, new technologies for financial speculation pushed levels of competition ever higher, as industrialists struggled to keep up with the profit margins that could be realized on the money markets. With the demise of the Soviet Union and the nearly simultaneous resolution of the GATT negotiations, eliminating almost all barriers to international trade, the world stage was cleared for the activities of the lean-and-mean corporations. The favors of unprecedentedly mobile enterprises would now have to be courted by weakened national governments, which increasingly began to appear as no more than "executive committees" serving the needs of the transnationals. And the TNCs grew tremendously, with spectacular mergers that haven't stopped: witness BP/Amoco in oil, Daimler Benz/Chrysler in auto manufacturing, Morgan Stanley/Dean Witter in investment banking, or the proposed "Oneworld" alliance that would group nine international carriers around the two giants, British Airways and American Airlines...

This thumbnail sketch of economic globalization could go on and on, as it does in an incredible stream of recent books and articles from all schools of economics and all frequencies of the political spectrum. But what's generally left out of the hypercritical, alarmist discourse that I personally find most compelling, is some theoretical consideration of the roles played by the individual, human nodes of the world network: I mean *us*, the networkers, the people whose labor actually maintains the global economic webs, and whose curiosity and energy is sucked up into the tantalizing effort to understand them and use them for our own ends. I'm trying think on a broad scale here: the pioneers of virtual communities and net.art are only the tip of this iceberg. What's fascinating to see is the emergence on a sociological level of something like a *class of networkers*, people who are increasingly conscious of the welter of connections that make up the global economy, who participate and to some degree profit from those connections, who suffer from them too, and who are beginning to recognize their own experience as part of a larger pattern. The massification of internet access in the last few years, only since the early nineties, has finally given this class its characteristic means of expression. But precisely this expanded access to worldwide communications has made it pretty much impossible to go on fingering a tiny corporate elite as the sole sources and agents of the global domination of capital. We are now looking at and sharing in a much larger phenomenon: the constitution of a transnational civil society, with something akin to, but different from, the complexity, powers, and internal contradictions that characterized, and still characterize, the nationally based civil societies.

Civil society was initially defined, in the Enlightenment tradition, as the voluntary social relations that develop and function outside the institutions of state power. Toqueville's observations on the importance of such voluntary initiatives for the cohesion of mid-nineteenth-century American society established an enduring place for them in the theories of democracy. The idea recently got a lot of new press and some new philosophical consideration with the upsurge of dissidence in the Soviet Union and the other east-bloc countries in the seventies and eighties; and at the same time, as the neoliberal critique of state bureaucracy resulted in the dismantling of welfare functions and the decay of public education systems, the notion of self-motivated, self-organizing social activities directed toward the common good became something of a Great White Hope in the western societies. So-called nongovernmental organizations could then be seen as the correlates of civil

60 speeches and presentations; 21 business cards; 15 lbs (~7 kilos) of handouts, speeches, newsletters, directories, press releases; 14 jumbo prawns; 6 glasses of chilled orange, cranberry, and apple juice; 4 glasses of Harmony red wine; 3 sit down lunches and dinners; 1 pop-up 3D desktop calendar from Public Utility Law Project; 1 3-ring binder with print outs of presentations and marketing literature; 1 break-out session for discussion; 1 directory of 500+ attendees; No mousepads. T-shirts, or other giveaways; A few good ideas and a few good stories. These are some of the measurables of my attending the Connecting All Americans: Telecommunications: Links in Low Income & Rural Communities conference held in Washington, DC, Feb 24–26, 1998. [Cisler <cisler@pobox.com>, A Critical Report from a U.S. Conference, Thu, 12 Mar 1998 14:46:03 -0800 (PST)]

society in the space of transnational flows. Nowadays, with the environmental and labor abuses of TNCs becoming glaringly violent and systematic, and with their cultural influence ballooning through their sway over the media, a lot of people in nongovernmental organizations are understandably keen on promoting a notion of global civil society as a network of charitable humanitarian projects and political pressure groups operating outside the precinct of *corporate* power (with attempts to develop institutional agency focusing mostly around the U.N.). I sympathize with the intention, but still I'd like to point out that the individual rights and the free exchange of information on which this global civil society depends are also necessary elements of capitalist exchange and accumulation. The internationalization of law and the fundamental demand of "transparency," that is, full information disclosure about all collective undertakings, are among the great demands of the TNCs. To the extent that it wants to participate in capitalist exchange, even a regime as repressive as that of China, for example, has to open up more and more circuits of information flow, and so it pays the price of higher scrutiny, both internal and external, on matters of individual rights and freedoms. The whole ambiguity of capitalism, in its concrete, historical evolution, is to combine tremendous directive power over the course and content of human experience with a structurally necessary space for the development of individual autonomy, and thus for political organizing. The networkers, those whose bodies form nodes in the global information flow, and who therefore can participate in an enlarged civil society, are subject to that ambiguity. Which means, pragmatically, that the expansion of TNCs is inherently connected to the possibility for any democratic governance by a transnational civil society.

As Gramsci made clear long ago, civil society is always fundamentally about levels or thresholds of tolerance to the pressures and abuses of capitalist accumulation. The specific forms and effects of civil society are determined by a complex cultural mood, a shifting, partially unconscious consensus about who will be exploited at work, and how, about whose intelligence and emotions will be brutalized by which commercial media, and when and where and how, about whose land will be polluted, and with what—and, of course, about whose land will just get suburbanized or left tragically undeveloped, about who will be able to refine their intelligence and emotions and in which ways, about who must work and who gets to work and who no longer "needs" to work, who just gets left on the sidelines. Thus Gramsci, writing in the twenties and thirties, had a somewhat jaundiced view of really existing civil society. He conceived it as the primary locus of political struggle in the advanced capitalist societies, but he also saw it as a directive, legitimating cultural superstructure, generally engaged in the justification of brutal domination; and he recalled the violence of petty bureaucrats and clergyman in the Italian countryside, keeping the submissive classes in line. Gramsci's key concept of hegemony expresses both the role of this legitimating function of civil society in maintaining dominance and also its potential mobility, its capacity to effect a redistribution of power in society. I think that the emergence of the transnational class of networkers, operating as a significant minority in most countries, is effectively shifting the articulation of political power in all the world's nations. I'll try to describe how with just a few examples.

Consider the U.S., the country that launched the internet, where an important fraction of the population is extracting new wealth out of what Robert Reich termed the "global webs" of multipartner industrial, commercial, and financial ventures, where many people not directly involved as operative nodes in such webs are still very conscious of them because they have their savings or retirement funds invested in global financial markets (as almost half of Americans now do), and finally, where long lists of NGOs and alternative communication networks are based, many of them with roots in the idealistic social-reform movements of the sixties and seventies. This is also a country where the least wealthy 40 percent of the population has actually seen their wages go down and their working conditions deteriorate over the last twenty years, where chronic social exclusion has become highly visible in the forms of homelessness and renewed racial violence, and where, last but not least, a very powerful Christian Coalition has emerged to reject almost every kind of consciousness change attendant on globalization and the recognition of cultural diversity. To marshal a workable political consensus out of such intense divisions, Clinton–Gore had to simultaneously push even harder toward the flexibilized information economy than their Republican predecessors had done, while making (and then breaking) lots of promises to restructure the country's welfare safety net, maintaining a high-profile international human-rights discourse (for instance with respect to China), and combining talk about environmentalism with a hip and tolerant style to woo all the former sixties radicals whose capacity for cultural and technological innovation fuels so many growth markets. Continuing economic growth has, of course, been the only thing to render this juggling act possible, making the strident neoliberal critique of the Republican right seem redundant—and forcing the Republicans into even greater dependence on the extreme right, as defined and prosecuted by the moral order of Christian fundamentalism. Europeans tend to look on media-driven U.S. politics with consternation and a powerful will to deny any resemblance to the situation in their own countries. But if Tony Blair enjoys so much prestige in the rest of the ECU. right now, it is because of New Labour's ability to juggle the contradictions of an unevenly globalized society, somewhat as Clinton has done. The hegemonic formula reflected by New Labour seems to be a fun, flexible lifestyle, good for stimulating consumption, a fast-paced managerial discipline to keep up with global competition, and a center-left position that shows a lot of sympathy for casual workers and the unemployed while eschewing any genuinely socialist policies of market regulation and restricting the state's role to that of a "promoter" (Blair's word). However, there are of indications that this formula, tantalizing as it is, will not really work in the rest of Europe, stricken by unemployment and yet still reticent to dismantle the remains of its welfare systems. The very interesting resurgence of support for state interventionism and economic regulation in France is one such indication. A more disquieting sign is the rise of populist neofascist parties, not only in France, where the National Front clamors against "*mondialisme*" (globalism), but also in Austria, Italy, Belgium, and Norway. These betoken major resistance to the neoliberal path that the European Union—or more accurately, Euroland—has taken under the economic leadership of the Bundesbank. The compromise-formation between a transnational elite subordinating everything to its privileges and an excluded popular class looking to vent its frustrations seems to be the scapegoating of poorer immigrants. The sight of two immigration officers savagely beating an African in a transit corridor of Schiphol airport has stuck in my mind as an all-too possible future for Euroland.

The powerfully articulated national civil societies of Europe are likely to falter and distort rather than break under the pressure of the split introduced by the transnational class. Hegemonic dissolution occurs when a majority of a country's or region's people can no longer identify themselves with *any* aspect of the institutional structure that purports to govern them. A case in point is Algeria. Here we see the steadily increasing inability of a recently urbanized and relatively educated population to identify with a government that no longer even remotely represents a possibility to share the benefits of industrial growth—because there hasn't been any for the past twenty years. The government is now an oligarchy drawing its revenues from TNCs in the fields of resource-extraction and consumer-product distribution. For many Algerians who have left their former village environment but can no longer get a job or use their education, the only ideology that can render a regression to pre-industrial living conditions tolerable is not democracy, but Islamic fundamentalism. If transnational capital continues to exploit the new international space which it has (de)regulated for its convenience, without any consideration for the daily lives of huge numbers of people, such violent reactions of rejection are inevitable and will spread. The current crisis of the global financial system is all too likely to fulfill this prediction.

Paradoxically, it is the global financial meltdown that may offer the first real chance for transnational civil society to have a significant impact on world politics. Not because networkers will have any direct influence on the few transnational institutions that do exist: only the richest states and the lobbies of the very large corporations can sway the IMF, OECD, and WTO; and despite all the inroads made by non-governmental organizations, the U.N. is only really effective as a kind of megaforum for debate. But in the context of a worldwide economic crisis, networkers may be able to use an understanding acquired by direct participation in global information flows to effectively criticize the institutions, ideologies, and economic policies of their own countries. In other words, transnational civil society may find ways to link back up with the national civil societies. There is already an example of networked resistance to economic globalization that has operated in just this way: the mobilization against the Multilateral Agreement on Investments. This ultraliberal treaty aims not at harmonizing but at *homogenizing* the legal environment for transnational investment. It would prohibit any differential treatment of investors, thus making it impossible for governments to encourage locally generated economic development. It would allow investors to sue governments in any case where new environmental, labor, or cultural policies entailed profit losses. And its rollback provision would function to gradually eliminate the "reservations" that individual states might initially

Your Death is my Business. The viatical industry is in the business buying up life insurance of terminally ill people. Say you hold of life insurance of $ 100,000 and need to the money to get the appropriate treatment for the illness or just to spend it while you can. But the life insurance money won't come until your dead. Here is where the viatical service comes in. A friendly broker will buy your insurance policy, pay you, say, $ 50,00, take over the policy and the payment of premiums and collect the money once your gone. Mutually beneficial. If you die soon. Within a year and the broker makes a killing, so to speak, 100 percent return. If you die in two years the return is still ok. But if you, miraculously recover and live on happily for the next couple of years, the broker sits on a foul investment: the insurance policy that cannot be cashed. The viatical industry started up in the eighties in the wake of the AIDS epedemic and grew considerably in the nineties. Many of the companies have cashed in and are not traded on stock markets. Currently, the industry, some sixty companies, does $650–750 million in business a year and the quicker its clients die, the better their return. In 1996, an AIDS conference in Vancouver confirmed a breakthrough in AIDS research. For this industry, good news are bad news. The stock price of Dignity Partners Inc, a San Francisco firm, plunged from $14.50 earlier the year to just $1.38. [Felix Stalder <stalder≠ @fis.utoronto.ca>, Betting on Death, Fri, 28 Aug 1998 11:28:16 -0700]

impose. Negotiations on the MAI began secretly in 1995 among the twenty-nine member-states of the Organization for Economic Cooperation and Development, and might actually have been concluded in April 1998 had the draft text of the treaty not been obtained and made public, first by posting it on the internet (see the Public Citizen site <http://www.citizen.org>). This plus the resultant press coverage brought cascading opposition from around the world, including a joint statement addressed to the OECD and national governments by 560 NGOs. The result was that member-states were forced into questioning certain aspects of the treaty and negotiations were temporarily suspended, though not definitively adjourned.

Detailed information on the MAI can be obtained over the internet, for instance from the National Centre for Sustainability in Canada (<http://www.islandnet.com/~ncfs/maisite/>). The diffusion of this information remains important at the date I am writing (September 1998), as further negotiations are upcoming. Opponents say that like Dracula, the MAI cannot stand the light of day. What I find particularly interesting in this context is the way the angle of the daylight differs across the world. Canadian activists, having seen their local institutions weakened by NAFTA, are extremely concerned with preserving national sovereignty. Consumer advocates and environmentalists were able to exert the strongest influence on the U.S. Congress. In France, the threat to government subsidy of French-language audiovisual production tipped the balance of indignation. NGOs in developing countries that may be incited to join the treaty immediately pointed to the dangers of excessive speculation by outside investors. Underlying these and many other specific concerns there is no doubt a broad conviction that the single, overriding value of capitalist accumulation by any means, and for no other end than accumulation itself, is insane or inhuman. But even if the current financial crisis is almost certain to reinforce and extend that conviction, still it will have no political effect until translated into more tangible issues, within an institutional environment that is still permeable to those whose only power lies in their intelligence, imagination, empathy, and organizing skills. Like it or not, that environment is still primarily to be found in the nation, and not in some hypothetical Oneworld consciousness. Which is tantamount to saying that transnational civil society, if developed for its own sake, would probably end up as homogeneous and abstract as the process of transnational capital circulation that structures the TNCs. The only desirable global governance will come from the endless harmonization of endlessly negotiated local differences.

I have evoked the position of networkers as human nodes in the global information flow. What are the implications of that position? In his three-volume study of *The Information Age*, sociologist Manuel Castells gives the following definition: "A network is a set of interconnected nodes. A node is the point at which a curve intersects itself." This definition is either fatalistic or provocative. Fatalistic if it defines the network of information exchange as an entirely autonomous system, interlinked only to itself in a structure of recursive proliferation. But provocative if it helps push the human nodes to assert their autonomy by seeking connections outside the recursive system. Can we hope that a redirection of priorities will arise from the aberrant spectacle of financial short-circuiting and resultant material penury in a world whose productive capacities are so obviously immense? I suspect that in the near future at least some progress toward the reorientation of the world economy is likely, particularly in the E.U. where the rudiments of transnational democratic institutions do exist. Even in the U.S., real doubt may grow about the sustainability of the speculative market in which so many have invested. In this context there may be a chance for activists to talk political economics with the far larger numbers of networkers who formerly had ears only for the neoliberal consensus. But a real change in the hegemony will not come about without an expansion of the magic circle of empowerment to people and priorities which have been marginalized and excluded. There is a tremendous need right now to spend some time away from computers and out of airports, not to ideologize people in the national civil societies but just to find out what matters to them, and to discover other levels of experience that can feed one's own capacities for empathy and imagination. Such experience can help requalify the transnational networks. In this respect I continue to think there has been something compelling in the Zapatista electronic insurgency, despite the aura of exoticism it is often reduced to. Not only has it been a vital force in shifting the hegemonic balance in Mexican civil society by giving uncensored voice to the demand for greater democracy. Not only has it been able to mobilize support from far-flung nations at a time when "Third Worldism" was becoming a term of insult and disdain. But in addition to these considerable accomplishments it has been able to infuse the global network with stories and images of the Lacandon forest, evoking experiences of time, place, and human solidarity that seem to have been banished from the accelerating system of abstract exchanges. The thing is not to romanticize such stories and images, but to look instead for the real resonances they can have in one's own surroundings. Call it transnational culture sharing, if you like.

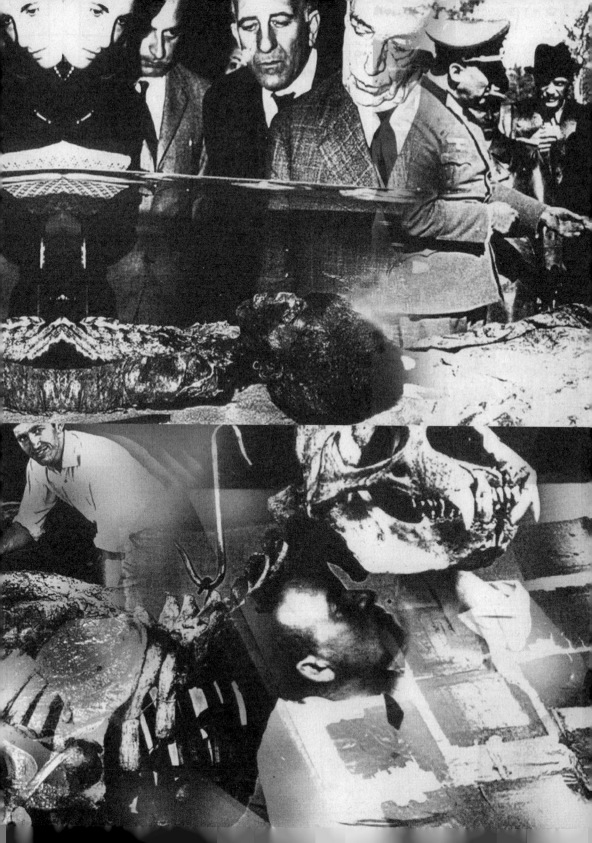

FROM: TJEBBE VAN TIJEN

CREW MEMBERS OF
● B-29 ●
WHICH DROPPED Atom
On HIROSHIMA

MAZE

SUBJECT: LEAVE YOUR MESSAGE AFTER THE BEEP:
ON THE RIGHT TO STUPIDITY,
THE ART OF FORGETTING, AND THE BOLERO 100

FROM: MARC HOLTHOF <HOLTHOF@GLO.BE>
DATE: THU, 17 SEP 1998 23:16:46 +0200 (MET DST)

From the notorious interview with Martin Heidegger that appeared after the philosopher's death in *Der Spiegel* (May 30, 1976) Avital Ronell quotes the following extract in *The Telephone Book*:

Der Spiegel: So you finally accepted. How did you then relate to the Nazis?
Heidegger: ...Someone from the top command of the Storm Trooper University Bureau, S.A. section leader Baumann, called me up. He demanded...
Heidegger, recently appointed rector of Freiburg University, answered the Nazi call/ing. A telephone wire connected the great philosopher to a criminal regime. A "call" became a "calling." On May 1, 1934, Heidegger became a member of the NSDAP *Gau Baden*. His number was 3125894. But just suppose. By way of a modest anachronistic thought experiment. Suppose Heidegger had had an answering machine. Suppose S.A. section leader Baumann had gotten the following message: "This is Martin Heidegger. I'm not home right now. Please leave your message after the beep." What would have happened then?

RECORD
"Even granny was surprised by the Bolero 100's many functions. Its compact and elegant exterior belies this answering machine's astounding capacity to record over 30 minutes of messages. The Bolero 100 stays safely within everyone's budget and proposes a memory function to save personal messages for you and your family. The "space-guarding" function allows you to monitor the goings-on in the answering machine's vicinity. The Bolero 100's primary asset is its sonic guardian, a distress call that's automatically transferred to a number of authorized persons (identified via a secret code). This way you can feel safe and restrict incoming calls to insure granny's afternoon nap" (Christmas promotion for Belgacom's Bolero 100 answering service).
An answering machine is a handy gadget. Even when you're not home you can still take that all-important call and listen to its playback at your leisure. Nothing (the occasional technical glitch notwithstanding) is forgotten, everything is carefully recorded. If we're to believe the national phone company, parents are even using the machine to leave spoken messages for their kids or significant others. Say goodbye to those scribbled Post-it notes on the refrigerator. Urgent family matters—"Don't forget to take out the trash" or "I won't be home tonight"—will henceforth be conveyed by the memory function on the answering machine. More serious messages—like the classic "went out for a pack of cigarettes, be right back" or actual suicide notes—are likely to go the same way.

TIME OF THE INDIVIDUAL

The answering machine's biggest quality is that it succeeds in separating the owner's personal world from his professional life. As long as you don't listen to it, your answering machine will isolate you from the outside world. The telephone has the nasty habit of intruding into your private life at those most inconvenient moments. The answering machine "softens" and sidetracks such intrusions. An answering machine guarantee sits owner's right to privacy.

The answering machine's greatest theoretician is probably Benjamin Constant (1767–1830). In his *Histoire abrégée de l'égalité*, Constant simply characterizes our modern times as *l'époque des individus*. Tzvetan Todorov wrote a wonderful book about this liberal thinker who is gradually being rediscovered. Constant was not just the author of *Adolphe*: he was also one of the most important political thinkers of the early nineteenth century. After the French Revolution, the state, the corporation, and/or the family can no longer impose their will on the individual, Constant notes. "Instead of being enslaved to the family...every individual now lives his own life and demands his own freedom." Constant was enough of a crystal ball–gazer to come up with an astute political analysis some two hundred years ago that is still more than relevant for our contemporary democracy.

Constant's political thinking, argues Todorov, is at once a synthesis and transformation of the work of two important eighteenth-century French political thinkers—Montesquieu and Rousseau. They respectively embody the principle of the separation of power and the sovereign people. In his *Principes de politique* (1806) Constant tries to reconcile the views expressed in Montesquieu's *The Spirit of the Laws* with Rousseau's *Social Contract*, the separation of power with the sovereign people.

Both Montesquieu and Rousseau were keen to improve government. For Montesquieu it didn't really matter who is in power—the king, the aristocracy, or the parliament. It only matters how power is exercised. Every form of power is legitimate as long as that power is limited by laws and/or another source of power. Executive, legislature, and judiciary power should balance each other out. This comes down to what is rather incorrectly described as the "separation of power." In fact Montesquieu is talking about a redistribution or a balancing of power. If and when the powers are balanced, this will automatically lead to a fair and tolerant regime. By contrast, in both individual and collective dictatorships, the different powers are grouped together. Montesquieu (who died in 1755) is obviously not a republican or a democrat. His only ideal—the British monarchy—is a meritocracy: in his view the people are "unable to make their own active decisions" (*Spirit of the Laws* XI.6). The people should be represented and presided over.

Rousseau develops a different reasoning in his *Social Contract*. It is not the way in which power is exercised that matters but who exercises it. The sovereign people should itself decide according to which laws it wants to live. Sovereignty equals the exercise of the will of the collective. This collective will always take precedence over the individual will.

Benjamin Constant accepts Rousseau's postulate that power should be the expression of the will of the people. Given the regime of terror during the French Revolution, however, he adds one condition he borrows from

Montesquieu—that power is not only legitimized by those who exercise it but it is also legitimized by the way it is exercised—it should never be unlimited. Even the sovereignty of the people, the collective will, should be practiced in moderate fashion. Constant chooses neither the liberalism of Montesquieu (which can be undemocratic) nor the democracy of Rousseau (which can be totalitarian). Instead he opts for a liberal democracy. He limits the power of the people and in so doing protects the individual from the arbitrary ruling of the collective: "A people that holds all the power is more dangerous than a tyrant," he concludes. The people's sovereignty should only come into force within certain limitations. Even when it is only one individual who does not agree with the others, those others should not have the power to impose their will (especially not in private matters). The sovereign people should respect the freedom of individual.

THE RIGHT TO STUPIDITY

John Stuart Mill upheld a similar principle in his *Considerations on Representative Government* (1861). He agrees that a society should guarantee the freedom of its citizens. Minorities should be protected from the majority. His conclusion is still extremely relevant for our contemporary media society: "Like the whole of modern civilization, representative governments are inclined towards collective mediocrity." To put it bluntly: The first and most important (but seldom-spoken) principle of any democracy is the right to stupidity. Everyone, no matter how stupid or blunt, has the same unalienable democratic rights guaranteed by universal suffrage. You don't have to take an IQ-test before you elect a representative. And that's the way it should be: it's the democracy, stupid! The scenario changes though when this unalienable democratic right to stupidity becomes an obligation to be stupid. In light of the political and social polarization provoked by the [convicted and accused Belgian sexual murderer of children] Marc Dutroux case, it seems quite useful to confront those few legalists à la Montesquieu and those many populists à la Rousseau with a sane voice like that of Constant or Mill. Yes, the separation of power is a political-judicial fiction that hides a lot of judicial corporatism. No, the people's sovereignty is not the solution to all problems. Democracy does not equal "all power to the people." The biggest advantage of liberal democracy in the way it was conceived by Benjamin Constant is that this kind of government is not only democratic but also guarantees a strict separation between the public and the private.

For Constant, freedom is everything that gives an individual the right to do— it also withholds society the right to forbid. Freedom is insured by the separation between public and private. This separation between public and private is perhaps the greatest achievement of the French Revolution—neither Antiquity nor the Ancien Régime knew the difference.

It is precisely this separation that is threatened by today's media society. The public has intruded into the private through communication technology— first the press and the telephone, later radio, and especially television. In lifestyle magazines and on television the public is camouflaged as the private in order to insure its domination of the individual. It takes away his freedom and makes him conform to those norms and standards imposed by the media. The private is threatened with destruction as everything becomes public. Hence the strange alliance between media hype on the one hand and moral indignation about the Dutroux case on the other hand—between a moral call to arms and the latest ratings. Both parties have but one goal—to impose the dictatorship of the collective onto the private sphere. And all this in the name of the people's sovereignty and (a strikingly narrow interpretation of) democracy. What we need—now more than ever—is an answering machine, an efficient form of protection against the public's increasing nosiness.

MECHANICAL *ANAMNESIS*

You can rightfully ask yourself if the answering machine hasn't become an "anamnetic" device. For those of us who don't know Greek: *anamnesis* is defined as "the act of remembering". In the Orphic–Pythagorean tradition this meant remembering earlier lives one had lived in a different form of being. In the *Meno* and the *Phaedrus*, Plato interprets *anamnesis* as the remembrance of the world of immortal Ideas. In a clerical context it means remembering your deepest sins in the confessional. Freud offers yet another interpretation and talks about remembering a repressed past (either spontaneously or under hypnosis). All this—remembering a past life, a world of ideas, a repressed past—is synoptically resumed by one push on the rewind button of the answering machine. A mechanical *anamnesis* takes place, and your earlier life, reality itself, catches up with you. Switch on the machine and reality comes back to haunt you. This annoys the owner of an answering machine. After a nice quiet day the whole storm awaits you on a compact thirty minute tape courtesy of the Bolero 100.

STOP

The Bolero 100 is a mechanical stand-in for Orphic mysteries, Platonic introspection, Catholic confessionals, and Freudian psychoanalysis. At the same time the answering machine allows the owner to postpone the *anamnesis*. To forget as long as possible. To shut out the world—not an earlier Orphic life, Platonic ideas, clerical sins, or Freudian reality—but the everyday telephonic life. *Amnamnesis* is remembering but remembering after a massive, traumatic, otherworldly forgetting. What do you remember from your earlier life, the immortal Ideas, or all that repressed carnality that explodes onto the psychiatrist's couch or in the confessional? Nothing or not a lot. This way the answering machine also functions as a forgetting machine, an attempt to delay reality, to "move" or "time shift" it into oblivion. While the VCR moves time while recording fiction, the answering machine records and delays reality itself. It is a forgetting well into which we dare not look—for the time being at least.

THE ART OF FORGETTING

The ancient art of remembering was first and foremost an art of forgetting. In *De Oratore* Cicero enlightens us on when the art of memory first came into being. During a feast at which he is invited to give a speech, the poet Simonides is suddenly called outside. During his absence an earthquake takes place and the roof of the banquet room crashes, leaving the host and all his guests buried under the rubble. The bodies are mutilated to such a degree that the family members who have come to collect their dead are not able to identify them. Fortunately Simonides remembers the exact seating of the guests at the dinner table and is thus able to identify their bodies. Simonides became the inventor of the art of memory because he was able to (re)construct his memory in an orderly fashion. His artful remembering inspired numerous orators to construct their speeches as mental images in an imaginary building, images they could "walk through" in their minds so as not to forget anything.

This anecdote marks the beginning of the art of memory that took off during antiquity and the Renaissance. What Cicero implies—but does not mention because it seemed so obvious at the time—was that Simonides' remembering was preceded by a huge, dramatic, momentous forgetting of everything that came before the remembering: the earthquake, the disaster that provided total amnesia and made it impossible for relatives to recognize their brothers, sisters, fathers and mothers. The art of memory relies upon and presupposes an almost complete forgetting. A kind of collective "instant Alzheimer's."

FORGETTING MACHINE

We tend to forget we forget. That forgetting is enormously important. Remembering is primarily not remembering certain things, selecting, trimming and then forgetting. Museums around the world are characterized not so much by what they store but rather by what they cannot, will not, or dare not store. They are not so much storage machines as machines for forgetting. The tape (or the digital memory) of our answering machine we use time and again. Nothing is permanently stored. Messages are recorded for the moment in an attempt to delay time. On a purely technical level the answering machine is also a forgetting machine. You need to keep all the tapes to turn it into a memory machine. Something we don't do—rather, we tend to erase. We use yesterday's tape to record today's messages, and today's for tomorrow's. And we are right to do so. We use our memories selectively and always forget more than we can remember. The past is a heavy load to carry—too heavy a load. Now, more than ever, we need to destroy surplus information. We need to use at least 75 percent of all published books to light the stove; to dig deep forgetting wells for useless information; to print books on extremely acidic paper instead of its acid-free equivalent; to develop magnetic and digital carriers that "forget" their recorded information after a reasonable time; to make all this useless information biodegradable. Orphists, Pythagoreans, Platonists, and Freudians all attached primary importance to the memory function. The past is all-important. The Freudians deny that we even are able to forget—in their book forgetting usually has some kind of deeper, shady, or sexual reason. Nietzsche on the other hand was all for forgetting and re-using the same old tape in our answering machines. In his *Genealogy of Morals*, he wrote: "Forgetting is not simply a kind of inertia, as superficial minds tend to believe, but rather the active faculty to...provide some silence, a "clean slate" for the unconscious, to make place for the new...those are the uses for what I have called an active forgetting..."

I enjoyed Marchart's piece on Neutopia. One correction: Dave Hughes never claimed to have invented NAPLPS. It dates from the early eighties and was developed with the help of companies like ATT, Xerox, Texas Instruments, and IEEE. It was used in public access sites in San Francisco, Toronto, Honolulu, in the eighties and early nineties. Hughes was a tireless promoter of this video standard until he turned his attention to certain wireless technologies (see<http://www.mind≠spring.com/~crhoads/shawn/≠turboard/> has more info on this standard>). I'm not sure whether it is Hedlund or Marchart who makes the comment that the technology was simple and "obviously suited for 'natives'" Judging from the tone of Marchart's essay, I am guessing that he infers that from the article. While Hughes is very opinionated (and I have been acquainted with him for ten years) I never observed that he thought that American Indians could not handle complex technology. Hughes was used to funkier interfaces and programs that many others would not embrace, and the NAPLPS applications I tried were like that. He expected a lot from the people he evangelized.
[Cisler <cisler@pobox.com>, Re: Greetings from Neutopia, Sun, 27 Sep 1998 09:35:30 -0800]

A CONDOM AGAINST REALITY

"[T]he call is precisely something which we ourselves have neither planned nor prepared for nor voluntarily performed, nor have we ever done so. "It" calls, against our expectations and even against our will" (Martin Heidegger in *Time and Being*, trans. by Avital Ronell in *The Telephone book).*

An answering machine separates messages from their temporal frame and cuts them up into sequences of past time (with or without time code). The answering machine therefore is the ideal instrument for those who refuse to experience reality directly and want to experience life in playback mode. The answering machine doesn't actually protect you from bad news but it does let you choose the moment you want to hear it. Someone is dying? No problem, just turn on the machine and we can go on pretending nothing's wrong. Let them die, we don't even know about it! And we don't want to know either. In this day and age of cellphones and portable computers there is no valid reason (beside a flat battery or a technical glitch) why we cannot be reached. And things are going to get worse as Belgacom has recently decided to link up its phone, cellular, and voice-mail services via a special Duet-arrangement: when you call someone you automatically get transferred to their cell phone first then to a regular phone and finally to their voice-mail. So these days if you get an answering machine you know that the person in question just doesn't want to take your call. He does not want to be reached; he wants to protect himself against the intrusions of the outside world. So why bother him then, even with the best or worst news? Get the message caller? He doesn't want to know. In that sense the answering machine is like a condom we use to keep out the pollution of everyday reality. An even more efficient method of screening calls is of course "caller identification," a device that has radically altered the social behavior of American households. With caller identification you see the number of the person calling flash up on the screen before you even answer the phone. Better yet, by linking this caller ID to the database in your computer, you can create a system in which you can only be reached by those people who are already in your address book. This way there's no chance you're going to be reached by a complete unknown. Secret telephone numbers used to be the privilege of famous people who just wanted to be left alone. Now every self-indulging civilian can unfondly remember the days he ever allowed a telephone in the private environment of his home or inside pocket. In this case pollution by an alien, threatening telephone call is no longer possible. The telephone has been replaced by the proxiphone (the telephone by proxy). The telephone becomes a safety device that hermetically seals us off from the rest of reality.

TWO NOTIONS OF FREEDOM

Benjamin Constant is more than just the perfect liberal, stresses Todorov. He didn't just stick to his aforementioned definition of freedom as all things private the individual can do and society cannot forbid, but he also—like Montesquieu and a long time before Isaiah Berlin—distinguished between two different notions of freedom. The first is the modern, negative definition of freedom in the private life; but there is also a positive one—the freedom to actively partake in the political life of the land, as was the custom in ancient Greece. In Greek society personal freedom was of no relevance or value. Constant notes in postmodern fashion: "The ancients had an opinion about everything. We hold only a semblance of an opinion on nothing much in particular." We doubt everything, or seem to be lethally fatigued before we actually do anything and certainly no longer believe in our institutions (Constant noted this trend more than two hundred years ago!). Private concerns have pushed aside all interest in the public life. We need an injection of the ancient freedom! Constant wants two kinds of freedom, that of the "Moderns" and the "Ancients" combined—a freedom of the individual to privately do what he wants, with the added freedom of publicly participating in the collective power. This way he hopes to compensate for the negative sides inherent in both types of freedom. In his famous speech delivered at the Royal Academy in Paris in 1819 he argues that "The danger of the ancient freedom was that it focused exclusively on the redistribution of social power and neglected individual rights and aspirations. The danger of the modern freedom is that we are all too concerned about our personal interests and tend to neglect our right to participate in the exercise of political power." Constant was optimistic nonetheless. He envisaged that people would only need independence in their daily concerns, activities, and fantasies to achieve perfect happiness. He was, as we all know by now, wrong. From the king to the cardinal, everyone stresses the need for guidance and leadership. People have yet to evolve from the slave mentality of the Ancien Régime and still yearn for the master and the whip, the God and His commandment. This is, from a purely empirical point of view, a totally accurate assessment. There has never been more nostalgia for the slave existence under the Ancien Régime than with the most recent batch of free citizens. Contrary to what millenary moralists and other horsemen of the apocalypse like to preach, what we definitely should not do is change this sorry state of affairs and fill up the vacuum that public power has left us with. Constant was absolutely right when he said that: "*L'anarchie intellectuelle qu'on déplore me semble un progrès immense de l'intelligence.*" Whatever those pamphlets say, you're better off hopeless and free than enslaved to some kind of ideology.

PLAY

In 1934 Martin Heidegger got a phone call. "*Nach einigen Tagen kam ein fernmündlicher Anruf,*" reads the original interview. The call came from S.A. Obersturmführer Baumann. And Heidegger took the call/calling. In retrospect—in the interview with *Der Spiegel*—he blamed his ties to the Nazi party on the telephone. One thing is for certain—had Martin Heidegger had an answering machine he would have been able to keep the Nazi influence at bay, or so he thought. This was in the days before the answering machine. Heidegger invented the answering machine. Not Constant's answering machine that installs an important separation between the public and the private, but that other answering machine—the one that is owned by those people who want to avoid reality, and who will not take that call/calling. The same people who would rather stick their heads in the sand than answer the call they've received (from the Führer, for example)—those who say neither yes or no. Please, leave your message after the beep and we'll get back to you—in about twelve years. We know better than that. In Rüdiger Safranski's biography we have read that "National Socialism" had already been the preferred topic of conversation at the Heideggers' mountain resort in Todnauberg during the early thirties. Even then Heidegger had already been convinced that only Nazi dictatorship could save Germany from that most vicious of cultural threats, that of communism. Heidegger didn't really need that call from Obersturmführer Baumann to remind him—he had always been a national socialist, if not in his mind then at least in his heart. Not even the charcoal-colored Bolero 550, the top model in Belgacom's new line of answering machines, could have saved his soul.

SUBJECT: SOUTHERN OSCILLATION INDEX

FROM: MCKENZIE WARK <MWARK@LAUREL.OCS.MQ.EDU.AU>
DATE: THU, 15 OCT 1998 09:53:47 +1000 (EST)

One of the things that reminds me about why the net matters is seeing Rupert Murdoch's face on the front cover of *The Australian* newspaper. He owns that newspaper, but that's not the only reason it covered his speech to News Corporation stockholders on the front page. News Corp is a major international corporation. One that just happens to be based in the provincial Australian city of Adelaide, where the local stock market rules are a convivial environment.

News Corp companies own 70 percent of Australian newspapers, measured by circulation. Australian media is one of the most highly monopolized in the world, and as such is a model for how other national media environments are likely to turn out, if they follow the kind of regulatory practices that successive Australian governments adopted.

It matters that there is a space in which to write about these kind of things, which is why the net matters, for instance. I write for *The Australian*, but while I personally have no complaints about the way that paper treats my writing, its not a publication that has a terribly strong interest in this issue of media concentration.

For a while it looked as though the net could be some kind of ideal alternative to big media. It didn't turn out that way. Its curious how skepticism about the potential of the net was very unevenly distributed. While the net was supposed to be a gossamer thread weaving in and out of national spaces, escaping from them or subverting them, I don't think that's turned out to be the case. So while its good to have a new space, outside of big media, its still an open question what kind of space it is. The virtuality of the net, it seems to me, is imperfectly mapped.

I'm writing from a milieu in which there was never any great enthusiasm for what Mark Dery calls the "theology of the ejector seat." There was never a strong sense in Australian culture that technology was a route to transcendence. Its true that Rupert Murdoch actually expressed an enthusiasm for global media's capacity to break down totalitarian governments, but this was more of a pragmatic than a transcendent way of thinking. It was a view of changing media in terms of undoing something wrong, rather than of raising the human essence to a sublime plane. In any case, its a remark he seems to have retracted when it caused difficulties for him in the emerging Chinese market.

By the same token, I don't think Australian culture is a milieu all that receptive to the European alternative to transcendent American thinking about the net. In the European view, as Geert Lovink once summarized it, the media is not just a political and cultural space, but a metaphysical one. Its not a question, in this version of media theology, of the leap forward, the raising of consciousness to a new plane. Rather, its a more classical ideal. Behind the actual, messy, everyday business of the media, lies the pure, rational, and just concept of what the media ought to be. This shining ideal, rendered so flatly in English, is the "public sphere."

There could be particular historical accidents behind these perceived differences. As Gilles Deleuze and Félix Guattari say, "the only universal history is the history of contingency." So its not a matter of any intrinsic essence of Americanness or Europeanness. It's a matter of accidents that lead to the formation of milieus, which in turn incubate particular concepts. A milieu, in Deleuze and Guattari's thought, is a plane upon which difference proliferates. But there are different planes. They are historical and contingent, and theory has to seek them out.

This, incidentally, is where media theory collides with Deleuze and Guattari. Its clear from the first milieu they talk about, that which simultaneously produced Aegean trade routes, Greek democracy, the city state and the practice of philosophy, is among other things a media milieu. The calm pond upon which the vectors of bronze age naval skill could navigate, the construction of cities around spaces of talk, the practices of oratory and of writing—its a media milieu.

On this score, their work is intersects with that of the great, neglected Canadian pioneer of media theory, Harold Innis. For Innis, a milieu can be made out of many different kinds of communication vector, all of which cross space and time in different ways. Some media, like writing on papyrus, are space binding, good for sending orders and running an empire. Some are time binding, like carving in stone, are time binding, good for priestly casts to maintain their authority through the ages. Innis saw ancient Egypt as a complex struggle between these vectors, a shape-shifting milieu. Deleuze and Guattari touch on a way of seeing classical Greece the same way. But it is the Canadian who has the stronger sense of the material construction of the vector, and its fragility.

It matters, this historical and materialist analysis of how a milieu makes a culture possible, makes certain kinds of ideas possible. But the milieu doesn't determine the concepts that form within it. Rather, a milieu is a space of virtuality, out of which the contingent assembly of, say democracy and the city state and philosophy might emerge.

So what kind of milieu might produce not only Rupert Murdoch but also a certain uneasy distance from both American cyberhype and European net-critique? The same kind that produced Harold Innis—a peripheral, new world environment. One in which the media space of the nation actually precedes the state.

Recent historical research by Graeme Osborne and others shows how the colonial era constitutional conventions, out of which arose Australian federation in 1901, were also forums that took a keen interest in intercolonial telegraphy and coastal shipping—the earliest vectors out of which the space of the nation was created.

The very existence of the colonial, peripheral world depended on the construction of a milieu. Innis showed this in the Canadian case in terms of the importance of a trans-Canadian rail link as a way of averting dependence on the markets and information centers of the U.S.

The mix of pragmatism and anxiety in Australia or Canada, about the transformative power of communication vector, seems to me to have a long history, born of the struggle to create a milieu that might make it possible to even imagine what these places are. What comes naturally to the old world or the metropolitan centers is to the periphery an object of continual anxiety. Europeans and Americans, whatever their differences, argue about what kind of identity they possess. Australians and Canadians argue about whether they have any identity at all. Given the fragile state of the milieu in which the question gets asked, its not surprising that the answer is often that it has all come to nothing, that the milieu is dissipating into the global slipstream. Innis was strongly involved in policy decisions to try and maintain the Canadian milieu. Much the same effort has gone into the maintenance of an Australian media space, although somewhat unevenly so. There was practically no Australian content on television in the late fifties and early sixties. It took a conscious effort to create a partition behind which some kind of local media milieu could exist, and of course changes in media form continually challenge its existence.

Some may ask why it matters. Surely nationalism belongs to the right? Surely the left is internationalist in outlook? Yes and no. In Europe, where nationalism has so often existed in fascist forms, where its ideological premise has so often been "blood and soil," its a tainted concept. But in states that resisted fascism and Stalinism, maintained democratic constitutions, and indeed may require the ongoing viability of the state in order to avoid the imperial demands of stronger and more populous states, there's an argument for a radical nationalism. It provides the semipermeable membrane within which differences local to that milieu can articulate themselves, discover their own virtuality.

This is a very different thing to the coercive nationalism of, say the One Nation Party. Indeed, it may be the only way to resist it. Exposure of national economies to global economic opportunity and global flows of information entails a cost, one that rural constituencies and low skilled workers are going to bear more heavily than anyone else. Their demand is for a strong state to protect their interests and affirm their existing culture, without any recognition of the need for change and negotiation with difference. The state has to be an agent that negotiates differences, between cultures, between concepts of the shared culture, and which makes globalization actually work in terms of generating jobs, distributing wealth and so on.

But the preservation of a purely national space media space can produce unintended results. One of which is Rupert Murdoch. I mentioned that Australian media is a highly monopolised space. Part of the reason is the restriction on foreign ownership, which over the years created a protected market for local oligopolists. Now we're down to two: Rupert Murdoch and Kerry Packer. The latter diversified into other kinds of business; the former built a global media business, and hence is the more internationally famous. Ironically, I see constant reports from other countries where business and government elites justify restricting the flow of international capital into their media businesses on the grounds that they have to resist Murdoch. But the process usually serves only to create local "Murdochs." Or perhaps local Kerry Packers. This is the sense in which monopolization proceeding from to simplistic a linkage of local ownership to local content production is a perverse outcome of nationalistic media regulatory policies.

I once said that Australia needed a branch of the Soros Foundation because its media configuration was even more of a threat to the "public sphere" than in some Eastern European countries. I wasn't necessarily kidding. Part of the impetus for wanting to create a media practice in the margins stems from the monopoly conditions so evident in the center of Australian media. The larger point about peripheral media zones in the new world is that the pragmatics of maintaining any kind of media milieu at all rules out the kind of effervescent optimism of American cyberhype. That and the lack of deep cultural roots for the kind of Protestant millenarianism within which cyberhype thrives. Seen from the outside, transcendent faith in technology looks like the kind of confident doctrine that could only flourish close to the heart of empire, even if that empire is now a military entertainment complex, rather than a military industrial complex.

Ambivalence about European media metaphysics may have even deeper roots. Kant's essay on the enlightenment can stand as Foucault's exemplary document of the eighteenth-century idea of reason, and Bentham's Panopticon as the nineteenth-century engraving in stone and flesh of the instrumental consequences of that reason. But seen from the other side of the world, the key figures are quite different. The eighteenth-century man of reason who matters is not the idealist Kant but the more practical Joseph Banks, botanist, and explorer, who brought back from Cook's voyages of discovery in the South Pacific whole categories of plant and animal species that did not fit the ideal order, the "chain of being," that pre-empirical science imposed on the natural world. Empiricism begins, to put it crudely, with the attempt to integrate the Pacific into the matrix of knowledge. Its data blew that matrix apart, and empirical order, where the categories are imminent in the differences within the data, gains ascendancy.

One of Bentham's famous pamphlets was "Panopticon or New South Wales?" Of course, the Panopticon was never built. English power never really depended on its disciplinary strategies of enclosure and classification. Instead of putting prisoners inside Panopticons, the English sent their resistant surplus populations to the colonies, including New South Wales, Australia.

In short, a strategy not of turning inward, rationalizing and making productive a space long inhabited, but rather a strategy of looking outward, across the open plane of the sea, for space across which power could be extended. Colonial expansion, at which the English excelled, is the unexplored side of European enlightenment and modernity. That colonial expansion always involved the projection of a matrix of vectors across the globe. Enlightenment was not a matter of constructing the metaphysical public sphere in which the essence of pure rationality could find it self. Enlightenment was a matter of constructing a matrix of communication and transport via which the raw materials for constructing modern life could be systematically extracted from the colonies to the advantage of the metropolis.

Of all the paths out of colonialism, places like Canada and Australia had the easiest route. It was granted without a fight. But this lack of self legitimacy stemming from postcolonial struggle comes back to haunt these exceptional peripheral zones. These are not milieus that ever had the confidence to create powerful ideas. These are milieus that were always-already experiencing "globalization" as a source of anxiety. What appears as a late twentieth-century phenomena was actually a foundational one.

In the Australian case, the impulse toward federation into a national space was in a large part what we now call globalization. Federating the colonies was seen as a way of creating economic sovereignty, and preventing the recurrence of the depression of the 1880s. That both the 1880s and the 1930s created worse experiences of depression in the periphery than in the metropolitan centers indicates that the counter-globalizing impulse was not successful.

What I'm trying to say is that its hard, from the periphery, to share the enthusiasm for any of the reigning discourses of cyberspace, as they all seem to me implicated in the uneven spatial distribution of what I would call *vectoral power*. Unlike disciplinary power, vectoral power engages with an outside, and is a completely flexible relationality. Its a matrix of vectors that distributes a flow of information, which in turn organizes a flow of material resources. But from the telegraph to telecommunications, it has always been experienced in the periphery as an unequal flow. How can you get enthusiastic in the periphery about new imperial vectors? How can you get enthusiastic in the periphery about new rhetorics about the power of new modes of communication?

It all sounds so attractive, and of course the attraction of American cyberhype and European net-criticism is itself imperial. It emanates from a center. Here's the irony: a rhetoric about networks and distributed communication that seems, in its own pattern of distribution, very highly centralized. It's hard not to oscillate between tepid enthusiasm and vehement distaste.

But this is only a critique of the limits of transcendent cyberhype and metaphysical net-critique. The trick is to find some potential for a positive relation to one or the other. There may be one advantage in being in this ambivalent oscillation about both American transcendent media theory and European metaphysical media theory: That is that it's possible to see a way out of the impasse created by their confrontation.

It seems to me that both transcendence and metaphysical critique both rely, in the end, on the kind of Platonism that the empirical revolution that followed from the discovery of the South Pacific so radically challenged. Whether the ideal is something to which to move "forward," in transcendence, or discover by stepping back towards the purity of the eighteenth-century image of the public sphere, it is still an ideal, against which the messy difference and chaotic movement of actual media and culture are measured and found wanting. Both transcendence and critique stage media theory as a kind of negativity. The roots of the difference between these kinds of negativity lie in the differences between the kinds of milieu that make them possible.

Of course there are lots of different ideas about the media, in either the American or the European milieu. These ideas are not an ideal expression of the milieus in which they arose—to think that way is still to be trapped within Platonism. Rather, they are just one expression of what those milieus make possible, but in each case, they are expressions that keep getting repeated. There are institutional constraints producing transcendence and critique, over and over—or at least so it looks when you consider media theory from somewhere else. One of the institutional constraints, seen from the periphery, is the desire to reinvent the imperial necessity. The metropolitan powers, no longer able to project force with impunity around the globe, or even across the Balkans, supplement the vectors of material force with vectors of information.

I never thought I had much to contribute to either the transcendent or the critical media theory project. I'm from a milieu that just doesn't support the kind of confidence that is required. I'm too much a product of anxiety, skepticism, a modest and practical sense of what media are for. Not to mention a suspicious mind when it comes to declarations of a new technique of enlightenment that emanates from new or old imperial centers. On the periphery, its enough just to keep the space viable, open but not too open, internally differentiated but not incoherent. Australian culture is just one big listserver, and its enough just to manage the flame wars, keep the traffic steady, implement the new version of the technology when it arrives—from elsewhere.

And of course there was the rise of a nationalism of the right—a serious matter in a country where nationalism is usually on the left. There were local matters to take care of. But now, I'm starting to wonder about what productive use to make of this ambivalence about critique and transcendence. European media theory has been doing a good job of critiquing transcendence—critique is what it does best. But its rhetorical structure is not so different. There is always a Platonic ideal lurking behind the critique of appearances, against which appearances are measured and found wanting.

But the ideal is just the ideal. The public sphere is just a beautiful work of art, made possible by the fact that the resources of the world were exploited to create a milieu in which beautiful ideas could be thought. From Kant to Habermas; from Rousseau to Debord. Images of an ideal matrix of communication against which the real can be judged and found wanting have changed shape and color, but the structure of the discourse persists.

This much has been obvious for some time, but the transition from the broadcast era to cyberspace brings new problems out into the open. Critique was popular when it appeared that there was a centralized media that state and capital controlled between them. The metaphysics of critique fitted with the politics of the left. The image of an ideal world of true expression that would reign once the actual, coercive regime of state and capital controlled media was overthrown provided a source of legitimacy for judging media in terms of what it lacked. The technical details of this philosophy were always to be filled in later.

But the proliferation of do it yourself media, even before the internet, and accelerating with it, can't be sustained by critique alone. It requires a positive practice. If anything, the practice of the net has been hampered by critique. Critique is a set of tools for persuading oneself that reality isn't good enough when compared to an ideal. Its not so good for discovering the potential of what is actually there. Critique sees the glass half empty, not the glass half full. A virtual media theory sees the glass half full, and wants to know what could potentially come out of any and every possible microscopic agitation, not just within the water, but also within the glass.

The internet appears to the Platonism of media critique as something like the South Pacific appeared to the Platonism of classical naturalism. It communicates new data that doesn't fit the ideal scheme of the order of forms. It requires an empirical approach to the production of categories and concepts, imminent to the data, not imposed upon it. Empirical, but not empiricist. The facts of the net, like the facts of the new world, are not enough. They require conceptualization if their potential usefulness is to be realized. Cultural studies has known for some time now that even broadcast media were complex. There were subtle and differentiated relations going on between the mass of the audience and the mass media message. Break it down into its constituent relations—a good empiricist technique—and you find people resisting and negotiating meaning. You discover the chaotic, plural, differentiated world of the everyday. And it is nothing like the ideal of the public sphere. And there is nothing much to be gained by talking only about what actual popular culture and media lack. So while cultural studies worked its way through critical and negative concepts of the media, it worked its way through—almost—to a positive and virtual media theory. That, I think, is the next step.

Of course, empiricism was the original object of critique. Kantian critique responds, in the canonic history of western thought at least, to the empiricism of Hume. I thought this was a closed chapter in western thought until I read Deleuze's first book, *Empiricism and Subjectivity*, in which that veteran anti-Platonist and anti-Kantian revisits the scene of that conflict. His task in that book is firstly to restate empiricism as a philosophy of difference, one that fashion concepts to match the flux of perceptions. His second task is to show the ethical import of such an affirmation.

Practical empiricism has its uses, from running an imperial state to running a global media empire like News Corporation. Conceptual empiricism, the path Deleuze opens up, seems to me to have a different import. It's an alternative to both the transcendent ideal of cyberhype and also to the metaphysical ideal of critique. Ironically enough, I feel like I need the authority of a metropolitan intellectual to state it, but there is another way to think about media theory, and in particular media theory in the age of the internet. The flux and difference of experience of the media can no longer hide behind critique, as it did in the mass media age. It has to be central to the theory.

In particular, it means moving from a theory of representation to one of expression. What cyberhype and net-critique have in common is a critique of appearances that finds them wanting in relation to the idea. The solution in cyberhype is transcendent. The rude differences and misunderstandings of bad communication will be superseded by better technology, which will merge all differences into one. An imperial idea if ever there was one. Critique works differently. It wants to insist that there are certain conditions under which the jarring differences of false representation can be eliminated, and communication can be perfected according to a social rather than a technological ideal. But the question to ask is what and who is to be excluded.

A theory of expression, on the other hand, would see noise, difference, irrationality, as integral parts of communication. The goal would not be to try and eliminate difference, but propagate it. The image would not be critiqued in terms of what it lacks, for its failure to be an authentic representation of the real. Rather, the difference it introduces, its inevitable falseness, would be the starting point of the possibility of the virtual. The imperfection of communication is the ethical basis of the potential for the world to be otherwise. It seems to me that virtuality is already alive and well in the actual practice of media theory as it occurs on the internet. On nettime, for example. There are occasional, high profile attempts to see net-critique as a binary or dialectical process, as the negation of cyberhype, transcendence, the "California ideology." This is critiqued as a false representation, and found wanting according to a true ideal. But it seems to me that this is the least useful aspect of emergent net-based media theory. It seems to me to be the aspect of it still tied most uncritically to imperial desires, no matter how unconscious. I oscillate between indifference and annoyance about them.

But what flows through the cracks in net-critique is something else. A new, positive, productive and connective creativity. New perceptions and new conceptions of those perceptions. An improvised discourse. Just as the eighteenth-century enlightenment was shaped by the milieu of inter-European trade and communication, so too a new milieu struggles to emerge, and one that is potentially even more spatially and temporally diverse. There are not only new spaces, but new speeds. But they struggle to escape from the unthought part of a past enlightenment, and in particular the unthought participation in imperial power of the information vector and the discourses that legitimate it. I started by suggesting there was something specific about a milieu that lacks an imperial confidence, and that working and thinking in Australia was just such a milieu. But I am sure there are many others. The potential is with us now to start breaking up the massified blocks into which specific milieus had congealed, particularly in the broadcast age. But this has to be seen from the peripheral as well as the imperial and metropolitan point of view. The desire on the part of News Corp to break down national spaces is clear, Its about getting in behind the partition and extracting value out of putting a vector into such spaces from without. But from the peripheral point of view, the desire is quite different. Its rather to break open imperial milieus and expose the differences lurking within them.

Strange as it may seem, I agree with the analysis of both Richard Barbrook and "Luther Blissett," as incompatible as they may seem. Barbrook has attacked versions of Deleuze's thought that would read it as a restatement of critical idealism, where the rhizome occupies the same place as an ideal concept that the public sphere occupies in a more classical formulation of media-metaphysical desire. Luther Blissett has thought its way out of the Marxist version of critique, into a more productive concept of the virtuality of communication. Of course the language Barbrook and Blissett use are poles apart, but nothing much of a productive nature emerges from trying to read them as occupying the same milieu, some kind of pan-European theory-wonderland. They are local and contingent expressions of a way out of critique that operate in different milieu, but as yet have little to say to each other—or perhaps to anyone else, other than as instances of a virtuality of media theory, two coordinates of an unknown map of possible ways of making a difference. I suspect that there might be a way to go back and more creatively reread some of the American work here too. Not as the big bad other of critique, but as local and contingent strategies within an particular milieu.

So this is my "southern oscillation index," my sense of ambivalence about a project of constructing a new space for net theory, but which I think has to look also at the skew of the old spaces, out of which it might potentially grow. The southern oscillation index, for those from the north, is the weather pattern over the Pacific which determines which side of the South Pacific the rain will fall on—South America or Australasia. But I think its a nice image of peripheral sensibility, wavering between participation and indifference to the remaking of the media metaphysics of the North.

SUBJECT: SHORT NOTES ON THE POLITICAL
CONDITION OF CYBERFEMINISM

FROM: FAITH WILDING <FWILD+@ANDREW.CMU.EDU> AND CRITICAL ART ENSEMBLE
DATE: WED, 14 OCT 1998 17:47:15 -0400 (EDT)

Cyberfeminism is a promising new wave of (post)feminist thinking and practice. Through the work of numerous net-active women, there is now a distinct cyberfeminist net-presence that is fresh, brash, smart, and iconoclastic
of many of the tenets of classical feminism. At the same time, cyberfeminism
has only taken its first steps in contesting technologically complex territories.
To complicate matters further, these new territories have been overcoded to
a mythic degree as a male domain. Consequently, cyberfeminist incursion
into various technoworlds (CD-ROM production, web works, lists and news
groups, artificial intelligence, and so on) has been largely nomadic, spontaneous, and anarchic. On the one hand, these qualities have allowed maximum freedom for diverse manifestations, experiments, and the beginnings of
various written and artistic genres. On the other, networks and organizations
seem somewhat lacking, and the theoretical issues of gender regarding the
technosocial are immature relative to their development in spaces of greater
gender equity won through struggle. Given such conditions, some feminist
strategies and tactics will repeat themselves as women attempt to establish a
foothold in a territory traditionally denied to them. This repetition should
not be considered with the usual yawn of boredom whenever the familiar
appears, as cyberspace is a crucial point of gender struggle that is desperately in need of gender diversification (and diversity in general).

TERRITORIAL IDENTIFICATION

What is the territory that cyberfeminism is questioning, theorizing, and
actively confronting? The surface answer is, of course, cyberspace, but such
an answer is not really satisfying. Cyberspace is but one small part, since the
infrastructure that produces this virtual world is so vast. Hardware and software design and manufacture are certainly of key importance, and perhaps
most significant of all are the institutions that train those who design and use
the products of cyberlife. Overwhelmingly, these products are designed by
males for business or military operations. Clearly these are still primarily
male domains (i.e., men are the policy makers) in which men have the buying power, and so the products are designed to meet their needs or to play
on their desires. From the beginning, entrance into this high-end tech-
noworld (the virtual class) has been skewed in favor of males. In early social-
ization/education, technology and technological process are gendered as
male domains. When females manipulate complex technology in a productive or creative manner, it is viewed and treated as a deviant act that deserves
punishment.

This is not to say that women do not use complex technology. Women are an important consumer market, and help maintain the status quo when the technology is used in a passive manner. For example, most institutions of commerce or government are all too happy to give women computers, email accounts, and so on if it will make them better bureaucrats. This is why the increased presence of women on the net is not solely a positive indication of equality. In the seventies, creating a female mythology was an inspiring and necessary part of recovering and writing the histories of women, and of honoring female cultural inventions and female generativity (the Matrix). Cyberfeminist mythologizing is a welcome sign of inspiration and empowerment, and at this point in time, makes good tactical sense. Such work offers a clear explanation of a constructive relationship between women and technology, and it begins the process of rewriting the gender code of cyberspace. However, in a political sense, the function of the mythic "natural woman" has its limits. In this case, it seems just as likely that weaving was a woefully boring task that was forced upon the disenfranchised. (This trend of boring and alienating work as the domain of the disempowered is certainly repeating itself in the pancapitalist technocracy.) As cyberfeminist critique increases in complexity, and therefore in ambiguity, the current cyberfeminist mythology will have to fade away much as matriarchal Crete and cunt iconography did in the late seventies.

SUBJECT: FOR A COLONIAL
DISCOURSE ANALYSIS OF CYBERSPACE

FROM: OLIVER MARCHART <OLIVER@TO.OR.AT>
DATE: SUN, 27 SEP 1998 15:17:21 +0200

"As the price of connecting to Cyberspace continues to rise by the privatization of the Net, more and more souls are pushed out of the New World. The Old World is corrupting the New World which has the potential to liberate the dreams of the water inside the Global Brain."

This quote is taken from a printed collection mostly of emails, which has been put together by Alan Sondheim (*Being Online: Net Subjectivity*, NY: Lusitania, 1997). The sender is a Goddess by the name of Doctress Neutopia (a/k/a Gaia Queen) and her mail bears the subject header: "Message from Neutopia." Doctress Neutopia and her Church is a usenet "troll," a hoax especially designed as an easy target for critiques of eco-hippie-ideology. Nonetheless, in order to be operative the whole joke has to sound realistic, that is to say, it has to employ already existing ideological material. The completely moronic neologisms of the churchlike "lovolution," "cyborgasm," or "soulization"—could quite easily stem from some "real" hippie-tribes of the internet—a place highly susceptible to neologisms. Doctress

Neutopia's cult is so "realistic," in a way, that it became one of the rare and sublime moments where parody turns into reality and reality turns into parody (see <http://genesis.tiac.net/neutopia>).

However, in the following I'm not going to take issue with the hilarious metaphor of the global brain—mostly employed by people who seem to be lacking a brain of their own. Nor do I intend to analyze the cyberhippie or eco-fascist mythology of the net. I would rather prefer reading Doctress Neutopia's email as a hyperbolic example for what I would call the colonial discourse of the net (for that, see my "The East, the West and the Rest," in *Convergence* 4.2 [Summer 1998], 56–75). One could find, needless to say, numerous other texts—which do not intend to "troll" people—sharing the idea of cyberspace, the internet, as a kind of Utopia/Heterotopia/Dystopia, in other words, a New World, a New Continent. But let us stick for a second to this specific fantasy and have a closer look at the first two sentences Doctress Neutopia shares with us: "At first glance, entering into Cyberspace is like entering into a new frontier. The blank screen is like the vacuum of Outerspace or in the beginning there was nothingness and then came the World."

What I cannot but admire is the precise way in which a whole genre of narratives is condensed by Doctress Neutopia into a few phrases: What we find here is the notion of cyberspace as a new "frontier"; the notion of cyberspace as "blank screen"; the notion of cyberspace as "vacuum"; and the idea that this innocent "New-blank vacuum frontier screen-World" is being corrupted by the "Old World." All these concepts add up to an enormous liberatory pathos that goes hand in hand with the fantasy of dark powers corrupting cyberspace: "Again, the New World has been colonized by the manufacturers who push greed, private interest, the profit motive, pornography, and war."

"…A NEW FRONTIER":

At least since *Mondo 2000* called its Summer 1990 edition "The Rush is On! Colonizing Cyberspace," we knew what cyberspace is all about: a new colony, a virgin land ready to be discovered and explored by "pioneers of cyberspace" (John Perry Barlow). The most prevalent concept within cyberspatial colonial discourse, hence, is the notion of frontier (just think of the Electronic Frontier Foundation—no troll!). However, the metaphor of the new frontier is not exclusively employed in narratives of cyberspace but, of course, it stands in the tradition of one of the American founding myths. Frederick Jackson Turner in his canonical "The Frontier in American History" claimed as early as in the 1890s—apropos the Western frontier— that the "American character" was based on this very extension of "old" space into new territories. We know how prominent the concept is in regard to this specifically American ideology. In extension—given the American hegemony over the internet—we know about the prominent role of this concept in our cyberspatial imaginary. Yet, I would claim that the term frontier fulfills a concrete function in the discursive setting of Colonial Discourse in general. If we take a look at the discursive mechanism of constructing new world narratives we can discover the following logics: The distinction between water and land, that underlies most narratives on major discoveries, seems to be blurred as soon as land becomes equivalent to frontier. In this case land doesn't denote anymore a kind of fixed and arrested territory but something fluid. The frontier in this sense takes on the characteristics of the wave (so we can speak about "surfing" in contexts of electronic networking). Thus, frontier plays the role of a hinge, a control button switching on and off processes of de- or re-territorialization. Therefore it has something to do with fluidity and fixation of (post-)colonial signifiers.

Referring to the stories of Hernán Cortés and others Mary Fuller and Henry Jenkins observes precisely that floating character of the frontier: "the narratives that set out in search of a significant, motivating goal had a strong tendency to defer it, replacing arrival at the goal (and the consequent shift to another kind of activity) with a particularized account of the travel itself and what was seen and done.... Even goal-driven narratives like those of Raleigh and Columbus at best offered only dubious signs of proximity in place of arrival—at China, El Dorado, the town of the Amazons—phenomena that, interpreted, erroneously suggested it was just over the horizon, to be deferred to some later day." The conclusion we have to draw from these observations is that movement, fluidity and nonfixation seem to belong to the narrative core of New Worlds, since unlike the structure of some fairy-tales the motif of the quest doesn't culminate in the achievement of the goal. No matter if we speak about the discovery of really existing or of fictional places, Mary Fuller detects in all these reports that "the sequenced inventories of places and events replace, defer, and attest to an authentic and exculpating desire for goals the voyages almost invariably failed to reach" ("Nintendo and New World Travel Writing: A Dialogue," in S. G. Jones, ed., *Cybersociety*, London: Sage, 1995, 63). What generates the narrative structure is movement in space and not arrival. It is nonfixity and not fixation.

On the other hand, book titles (*The Internet Navigator*, or *Navigating the Internet*), software names (Netscape Navigator, Internet Explorer), and colloquial expressions (cybernaut and so on), indicate not only the fluid character of cyberspace but also the colonial attempt to master this flux, to "navigate" it, to map the waves. It is for this reason that we have to conclude that the discourse of discovery is structured around three principles at least: water as the very principle of nonfixation, something that threatens the enterprise of discovery and colonization. Land in the sense of stable territory that doesn't move under your feet and can be mapped and meticulously described. And finally frontier as something in between fixation and fluidity, that escapes the colonizing efforts by definition.

Now, arresting this escaping movement of frontier by transferring it into land—by fixing it—is what colonization (and politics) is all about: by defining the limits you are defining the territory—as blood and soil, for instance (it is in this sense that Michel de Certeau claimed: "the central narrative question posed by a frontier is 'to whom does it belong?'"). As long as "land" is understood as frontier (in the American tradition) it owns predicates indicating fluidity. Like a wave this frontier is unfixable. You can surf on it but you can't arrest it. As soon as you arrive at this frontier, as soon as "the West is won," so to speak, the colonization of the whole territory has already begun and fixation sets in. Now, "land" doesn't mean anymore frontier; instead, it denotes a fixed and narrowly circumscribed, motionless terrain. It has lost all the predicates indicating the openness of meaning. At any rate, since this state of total colonization is not likely to be achieved, the political meaning of frontier lies precisely in its nature of something that cannot be fixed completely but nevertheless has to be fixed in one way or the other.

"...THE VACUUM OF OUTERSPACE...":

A certain branch of the vacuum-paradigm of cyberspace, sometimes called the "cues filtered out" approach, presupposes that disembodiment is supposedly allowing for an open reinvention of the self. These highly common ideas of, for example, unproblematic identity-switching, gender-swapping, and so on, are embedded in a rhetoric of self-creation and self-invention based on the assumption of a voluntarist subject, that is, a subject that sets and defines the conditions of his/her own possibility. By assuming the ability to define one's cyberspatial identity at will one is re-inscribing, like Michelle Kendrick puts it, "the myth of a coherent identity that exists outside and prior to the technologies which create cyberspace" (M. Kendrick, "Cyberspace and the Technological Real," in R. Markley, ed., *Virtual Realities and Their Discontents*, Baltimore: Johns Hopkins, 1996, 146). Of course, this identity, a voluntarist subject, does not exist, but not, as Kendrick would have it, because of the "technological real," by which she understands the material effects virtualizing technology has on subjectivity. It is simply because nobody can define at will the conditions of his or her possibility, not even in electronic networks.

Why, then, is cyberspace not a vacuum? Because something or someone is already there. But who? Is there a way to encounter the "other," the net-natives? Let us approach this problem by way of analyzing a typical colonialist text: "Virtual Reality Warriors. Native American Culture in Cyberspace" by Patric Hedlund. The article, published in *High Performance*, narrates the story of David Hughes, described as "the Colonel," "the Cursor Cowboy," "Singer of ASCII Songs," "Poet Laureate of the Network Nation," who, back in the early nineties, invented an algorithm he baptized NAPLPS, which stands for North American (*sic!*) Presentation Level Protocol Syntax. The algorithm is supposed to wrap pictures and words together for artistic means so one can put it on galleries in cyberspace.

On one of his promotion tours, Hughes gave a workshop to a group of "native" American artists. Patric Hedlund reports that "though he didn't realize it at first, he'd finally found a people who could share his vision and then expand it". The article goes on praising the simplicity of Hughes's technology—obviously especially suited for "natives": "NAPLPS is as simple and ingenious a next step as smoke signals and the tom tom." Moreover, there seems to be a natural bound between the spiritual potential of cyberspace and the spiritual heritage of people with a close relation to nature and to their ancestors: "Using NAPLPS and telecommunications to extend the reach of their ancient stories and images wasn't much of a leap at all for people accustomed to hearing their grandparent's voices when they look up at the stars" (P. Hedlund: "Virtual Reality Warriors: Native American Culture in Cyberspace," *High Performance* 52 [Spring 1992], 31–35).

There are at least two levels of Colonial Discourse to be found in this article: (1) The article reports how cyberspace (thereby standing for "culture" in general) was brought to the American "natives" by "Poet Laureate of the Network Nation" David Hughes. On this level, the colonial force is the singing "Cursor Cowboy" whose aim is to enlighten the colonized. (2) On a more general level, the text itself recolonizes the "natives" by constantly putting them in a position of privileged access to "nature," "spirituality," "customs," "heritage," and so on. The new communication technology serves only as an extension of these substances, a means of their re-implementation. On this level, the colonial force is the author's voice and the "natives," hence, are nothing else than a projection of Patric Hedlund's.

"THE BLANK SCREEN..."

The lesson is the following: There is not a single level of colonial discourse where we can encounter the "real natives." But there is no complete unrestricted re-invention of the self either since the white surface—called the New Continent—is just a discursive assumption: you will never encounter a completely white surface, a vacuum. But what do you encounter instead? In this sense the analysis of Hedlund's article shows one interesting phenomenon: What you discover is always your own image in a reversed form (the only thing Hedlund, for instance, informs us about is her own prejudices). This sentence—since obviously it paraphrases the Lacanian communication formula—has an axiomatic status. Wherever you go, you are always already there. Speaking about "the other" from an ontological viewpoint therefore only makes sense as long as we mean a radical other. And in this case we can't say anything about it. In all the other cases, we don't speak about the other—the frontier's beyond—in any meaningful sense of the word—but about parts of ourselves: that is to say, we speak about the same.

The consequences are clear: the New World is always already the old one in a reversed form. The other you discover is always already the same in a reversed and thereby slightly rearranged form. There is no way of grasping the radical other, because as soon as you manage to grasp it, it immediately becomes part of your own. That's why cyberspace is discursively constructed as a new yet unapproachable continent: the discovery of new continents always leads to the repetitive projection of old myths on their supposedly blank screen. What we discover doesn't belong to the screen as such. It is our occidental imaginary that is projected onto these continents: India, China, Australia, America, Cyburbia. Cyberspace serves as a screen for our occidental imaginary, which has always been projecting its own myths onto newly discovered continents. Every Never-Never-Land is an Always-Already-Land. It might be because of this underlying logic that the electronic networks are said to represent a new America: an always receding horizon/frontier that has to be discovered and at the same time protected in its untouched innocent state.

Slavoj Zizek makes the same point in regard to Conrad's *Heart of Darkness*, Poe's "The Narrative of Arthur Gordon Pym of Nantucket," or Rider Haggard's "She" (*Plague of Fantasies*, NY: Verso, 1997). According to Zizek, the key paradox in these colonial stories has to be seen in the fact that in the noncolonized core of the New Continent, in the "Heart of Darkness," in this phantasmatic beyond, we find again our own law, the law of the "white man." In the center of otherness we discover only the other side of the same, of ourselves: our own structure of domination. Or in case of "Arthur Gordon Pym," what he finds on his way to the Antarctic Pole after passing through a village inhabited by completely black "natives" (even their teeth are black) is "a shrouded human figure, very far larger in its proportions than any dweller among men. And the hue of the skin of the figure was of the perfect whiteness of the snow." The structure of these tales, according to Zizek, is that of the Moebius strip: If you go on long enough what you'll find is not the complete other place—but your own one.

FOR A COLONIAL DISCOURSE ANALYSIS OF THE NET

So, can this logic of rediscovering the Old in the New be legitimately seen as one of "corruption," as Doctress Neutopia would have it? I claim such an ethical injunction is illegitimate. Ziauddin Sardar's "alt.civilizations.faq" is one of the texts that have a lot of valuable insights to offer for a Colonial Discourse Analysis of the net (in Z. Sardar and J. R. Ravetz, eds., *Cyberfutures*, London: Pluto, 1996). Unfortunately, even Sardar falls into the very trap of colonial discourse by calling cyberspace "the Darker Side of the West." So while he rightly assumes that people are projecting themselves on the world of cyberspace thereby "forging digital colonies on behalf of Western civilization" he conflates this theoretical insight with moralist lamentations: rootless, alienated individuals without any real identity are posting Nazi propaganda or fantasies about pedophilia and other sexual perversions, turning the whole net into a "toilet wall," and so on.

By complaining that all of this had nothing to do with "intimacy, tenderness or any other human emotion," by claiming that "one can't learn simply by perusing information, one learns by digesting it, reflecting on it, critically assimilating it," and by complaining about the infection of non-Western cultures by the Western "virus" of boredom, Sardar is not only giving in to purely Western ideologies like humanism, pedagogy, and a biologist language of disease, he is also employing the colonial motif of a place beyond "spiritual poverty," inhumanity, and alienation.

What I was describing above are significatory principles and not moral ones. A critique of Colonial discourse of the net can only proceed from within the discourse of colonialism, and the first step would be to describe the mechanism of its construction. It is in this sense that I can only subscribe to what Gayatri Chakravorty Spivak says: "what I find useful is the sustained and developing work on the mechanics of the constitution of the Other; we can use it to much greater analytic and interventionist advantage than invocations of the authenticity of the Other" (G. C. Spivak: "Can the Subaltern Speak?" in P. Williams and L. Chrisman, eds., *Colonial Discourse and Post-Colonial Theory*, London: Harvester Wheatsheaf, 1993). One of these mechanisms—from the perspective of hegemony theory—clearly is the articulation of a chain of equivalence. It is the "New Continent" or "New World" which, as central metaphor, is linking notions like "frontier," "dark space," "vacuum," or "blank screen" together in a chain of equiva-

Not exactly a hoax. Doctress Neutopia (Libby Hubbard) is serious, and is (or was) based at the University of Massachusetts at Amherst. She used to post her plans for a utopia organized as a sort of hive, with herself as queen, to alt.cyberpunk, alt.slack, alt.magick, and similar groups until the usenet gods gave her a newsgroup of her own—alt.society.neutopia—in 1994. I read it, er, religiously for most of 1995. The newsgroup's population is Doctress Neutopia, a few friends, and a legion of mockers and scoffers like Lupus Yonderboy and "Jesse Garon" (named after Elvis Presley's stillborn twin brother). There was a lot to parody. Drs. Neutopia also posted the often-embarrassing details of her love affairs/attempts to kick off the lovolution.

The scoffers made short work of Neutopian ideology. The Monster Truck Neutopians (<http://www.primenet.com/≠ ~lathrop /monster.html>) gave themselves titles like Chief of the Secret Police and Chief Sanitation Engineer, held barbecues, wrote songs, and adopted as their anthem "Wild and Blue," a country/western ballad about a cheatin' husband, by U.K. pop group The Mekons. Neutopia is such an easy a target that it spawned its (highly entertaining) parody long ago.

Somewhat in the context of Oliver's piece, you could say Neutopia is an updated version of early American utopian colonies like Oneida and the Shakers, though of course it never went beyond the planning stages. Speaking of the Shakers, Neutopian sex is a nonphysical "massgasm," a sort of group version of the Shakers' "karezza."

Thanks, Oliver, for reminding me about this. Alt.society.neutopia has definitely seen better days (Neutopian and Monster Truck Neutopian websites are decaying fast) but, like Camelot, it will always exist in the hearts of those willing to believe.

[Bureau of Control <carlg@pop.net>, Re: Greetings from Neutopia, Sun, 27 Sep 1998 15:01:16 -0400]

lences; and—vice versa—these notions specify our very ideas about this "New World." By linking the latter to signifiers like love, ecofeminism, and so on—like in the Doctress Neutopia–hoax or related discourses—our ideas, again, are specified in a certain way.

This being so, shouldn't we assume that every discourse is already a troll since it cannot refer to any underlying "reality" but has to construct the latter out of contingent elements? That is to say, isn't the colonial discourse of the net already something like a troll in itself, a mere construction or articulation of a chain of signifiers? Couldn't something like Sardar's moralist construction of the net as "toilet wall," for instance, perfectly qualify as a troll? And isn't Hedlund's construction of "natives" who are supposedly "playing tom tom" with the net even very likely to be a troll? The answer can only be twofold. First: It is not a question whether or not Colonial Discourse is a troll. The question is who has the power to play the trick. Second: It is precisely because of the constructed character of every discursive chain that, in principle, Colonial Discourse is open for anti-colonial re-articulation. Let's do it.

SUBJECT: DATA TRASH UPDATE

FROM: MIKE WEINSTEIN <WEINSTEI@POLSCI.PURDUE.EDU>
DATE: THU, 17 SEP 1998 13:57:54 +0100

Dear netizens,

Let's begin with the event-scene, the vest-pocket theorization of a media factoid that tells a cautionary tale, which Arthur Kroker and I devised as a genre for undermining virtuality from within. *Data Trash* is an accumulation of event-scenes; its theoretical postulations are extrapolations and exaggerations of our associations with the factoids that arrested us as we wrote the book as a series of exchanges across the net. Each of us would write a section of a chapter, send it to our partner, and then the partner would take off from the other's text, freely varying the themes that had emerged. The interpretation grew through our self-reflections and our collaboration. I don't believe that such a project would be possible without the instantaneous quality of the net as a vehicle of text transmission. The immediacy of our interchange created in us a mutual frenzy that sent us careening into cyberpunk realism. The game of matching event-scenes is the friendly context that engendered *Data Trash*. The book is not only about the net but is of it, exemplifying in its constitution an actualization of one of the net's distinctive possibilities and deconstructing by its constitution any interpretation of *Data Trash* as a negation of the net.

OSAMA BIN LADEN'S CAVE

Osama Bin Laden, arch-terrorist, current scapegoat of Amerikkka—replacing Noriega, Hussein, Khaddafi, Ayatollah Khomeini, and so on, ad infinitum (flies in the neoliberal ointment)—supposedly holes up in a cave in Afghanistan bereft of indoor plumbing or a well-stocked pantry, but graced with a stupendous library of Islamic theology books and a communications complex that gives him instant access to cyberspace.

Hybrid monster and the perfected bimodern personality, Bin Laden is the absolute synthesis of technology and primitivism, finding no contradiction whatsoever between virtuality and stringency. He is also a monopoly capitalist and an Islamic (retrofascist) restorationist.

Bin Laden demonstrates that the only inevitability of the net is to suck us into it one way or the other. Whatever his boring aims of an Islamic renascence might be, he is complicitous in virtualization. He leaves the cave to defecate; he goes back in to communicate. Visit his website and tell him you care.

Bin Laden replaces Bill Gates as Numero Uno Net Man. This absurd figure—also, perhaps, the most "interesting" (in Nietzsche's sense) man of our time and quite attractive, brilliant and engaging—is the kind of mutant that we are likely to see more of as virtualization continues to infest the earth and heavens, and the flesh rejected by it rebels against its technocorporate avatars, all the while feeding like a parasite on their apparatus and confirming thereby its hegemony.

Bin Laden as the world's great comic ironist: his media den is a cave without a john. Home revolution is even more absorbing than home shopping. Hussein watches CNN.

DATA TRASH FIVE YEARS LATER

The major thesis of *Data Trash* still holds true today: the drift of "history" is toward virtualization. The only difference five years later is that the managed depression that we diagnosed back then and that nobody else noticed, has now become unmanageable: the "debt liquidation cycle" has now become too obvious to ignore. As a result, resistances to pancapitalism are appearing everywhere and they are mainly taking a retrofascist form. Fascism at its origins is bimodern, uniting the myth of an heroic premodern past with a promiscuous deployment of technology. In its recrudescence it becomes a denizen of cyberspace, along with everything else.

The virtual class is at home everywhere. Its members are apparatchiks who spread virtualization; it is indifferent to their class interest whether they work for capitalists, communists or fascists. They will satisfy the appetite for virtuality of a species that loathes itself enough to wish to be replaced under whatever regime exists. Now we are learning that no ideology is immanent to the net. Its political essence is neither anarcho-democracy (the utopia of a technological avant-garde), capitalist empowerment (exploitation), nor communitarian resistance, but the virtualization of all of these. The virtual class has no political ideology of its own; it will serve the master of the moment, who will always help it spin the net of virtualization in which all ideologies will be caught and eventually volatilized.

The recline into virtualization would be hastened by an ascendant capitalism, but it will be no more than delayed by the struggles between pancapitalism and retrofascism. Now is the time when severe conflicts will be fought on the net (as well as everywhere else), and the net will win every time (whichever local party gains a temporary victory), and triumph in the end, as long as we don't kill each other first or cause a calamity that rolls back technology. If there is a deep economic depression, the technological infra-structure will be severely stressed. Five years later *Data Trash* broods over apocalypse. Let's end with an event-scene.

THE MEDIA ROOM

Along with a host of other media, *USA Today*, would-be hegemonic medium par excellence (along with CNN), reported recently on the studies that have begun appearing about the psychological effects of plugging into cyberspace (Elizabeth Weise, "Delving Bit by Bit into the Secrets of the Net Mind" 9/2/98, p. 5D). It seems that people suffer mild depression after using the net and that the "overall rate of shyness among Americans" is now 50 percent, "up from a steady 40 percent since the 1960s." The liberal-humanist-behaviorist academics who conduct these studies conclude that "our brains...seem to be hard-wired to need social interaction." You don't get that from "virtual personae."

In a most diabolical piece of research, Dr. Clifford Nass of Stanford sat people down in front of computers and told them that the machines were "virtual personae" of various nationalities, races and genders. The subjects (the "human" ones) proceeded to treat the computers through their social stereotypes and to accord them social niceties.

However, these "people surrogates" seem to lack the pizzazz of flesh-and-blood creatures—the parts of the brain that "light up" during face-to-face interaction don't spark with the computer. Instead, people tend to come out of a session in a chat room or other net activity feeling that their precious time has slipped by in an addictive, compulsive blur. One knows the feeling; plowing through news groups, conducting endless web searches and following links, plowing through email (not to mention shopping)—all producing an irritating sense of futility and tedium pierced by the gnawing recognition of what one might have done with the lost time. This would be bad enough, but to make matters worse regret is followed swiftly by a self-contempt for having allowed oneself to have been gulled into cyberspace. But one will surely go there again, seduced by more riskless adventures. Depression, indeed. The net is our best preparation for death.

And what are we to make of the shyness epidemic? Here the liberal-humanist-behaviorists get on their hind legs and start barking about the loss of society—the disappearance of a "learning ground for people to relate to each other." The brain isn't light(en)ing up in the right places any more. It all comes down to this: will the androids, who will be fit to function in cyberspace, come on line before there is a social crash that prevents their advent and liquidates technology's "artificial nature" (Sorel); that is, will human beings drop the ball of cultural progress before the replacement team takes the field?

While Bin Laden plots revolution on the net, the Western masses are crippled in it, wallowing in their bland humiliation—rubes who can be induced to project their feelings on computers, addictive depressives who resemble nothing more than compulsive gamblers grimly looking for an elusive score, and timid folk who cannot bear contact with their own kind. They are the offerings of pancapitalism to virtuality. They are also its pathetic line of defense against retro-fascism.

Enjoy the apocalypse.

SUBJECT: I'D LIKE TO HAVE PERMISSION TO BE POSTMODERN, BUT I'M NOT SURE WHO TO ASK...

FROM: BETH@NETLINK.COM.AU <BETH SPENCER>
DATE: THU, 17 SEP 1998 13:57:54 +0100

THIS IS MY STORY, AND I'M STICKING TO IT.

Well, anyway, it's stuck to me now.

It all began—or my part in this story began—when my editor wrote a note on my manuscript saying: "You'll have to get permission for all these quotes." Although I suppose it really began when I naively wrote the book with all these quotes in the first place. Or maybe it began that day, back just before I was born, when my father walked into the house carrying a brand new television.

Of course, in some people's reckoning, it began when the U.S. dropped the bomb on Hiroshima...

Anyway, I'm part of a certain kind of world, and I write in a certain kind of way; a way, in fact, that has taken me about twelve years to develop. I used to write stories, and essays; and now I write stories that also sometimes function as cultural criticism, history and review.

As such, my book *How to Conceive of a Girl* (Vintage, 1996) incorporates lots of little narratives—outside texts—within its wider narratives. Everything from all the stories and anecdotes people have ever told me, to bits from *The Donahue Show*, the Bible, *In Bed with Madonna*, books on infertility and birth, lines from popular songs, gossip items from *New Idea*, fragments from philosophy texts, tourist information, characters from detective novels, excerpts from sixties school text books, and so on.

I'm definitely a magpie, but I have a taste generally for things that are well-worn; often things that are of no use any more, or so common that no one's really going to miss out if I make use of them too. The cast-offs or the mass-produced—all the things floating or left lying around out there. The space junk. Mostly things produced originally for an entirely different purpose. In general I don't pick my bits up out of someone else's nest, I pick them up off the street, or in supermarkets, or I dig around in rubbish dumps. I'm really not sure how exactly I came to be suddenly convinced that I had to get permission for all these things or I was going to be sued... I guess I was isolated at the time, I was going through some other legal problems (and hence having to face "reality"—in which good intentions and ethics are largely irrelevant), and I tended to get conservative advice the first time around.

There are so many rumors out there; it's such a "gray" area of the law. I also knew that my own publisher had been sued last year, that it had cost them probably more than I'll ever make from this book, and that just generally everyone was clamping down all of a sudden on this kind of thing and becoming very serious about it.

So, there I am: ten hours a day on the phone, drafting letters and searching back through boxes of notes. Doing (what I now see as) crazy things like making about twenty phone calls trying to track down someone who might know where the records of the now defunct *Sunday Observer* are held so I can get the name of the journalist (no byline, so probably from the U.S.) who wrote a piece on Lynda Carter back in 1980... (A piece that some wonderful subeditor headed "I Want a Baby!—Confessions of Wonderwoman." So perfect. How can I presume to "make these things up" when they're so already out there?)

Then I'd used twenty-five words from an Agatha Christie novel—only twenty-five words, but it's Hercule Poirot and one of his memorable pronouncements on facts and slips... And forty-three words from a philosophy text—but do you need to get permission from the original author, the translator, or the journal in which it was published (or all three?).

Then there's that story within the story that I've rewritten from memory from a sixties *Reader's Digest Omnibus* which turns out to be an abridgment of a children's book by James Thurber... And just tracking down who holds the rights for a particular song can cost me $50 per song if I go through AMCOSS, so I join a Lou Reed mailing list on the internet to see if anyone out there knows and can tell me for free, and I get dozens of daily emails from fans all across North America listing every song in the order he sung them for every concert on his tour, and learn to refer to him as "Lou" or "The Man" like everyone else, and eventually after a few wild goose chases I find out that "Pale Blue Eyes" is administered by EMI. (Um... It was EMI that sued my publisher.)

You see, all this time while I'm busily scratching around after these motes, I guess what I'm desperately trying to ignore are a few rather large and uncomfortable logs. The first one is this: I've made seven references to particular recordings of songs in my book—albeit brief, some only a few words, but ask any music publishing company and they will act totally horrified and aghast at the idea that you could use any word or phrase from a song without permission. Permission fees for songs are determined by the company, but a fee of $150–250 is standard. Add that up, and these seven tiny references (and oh how merrily I knitted them in, in the first place) could end up as a bill for perhaps thousands of dollars...

And then the very nice young woman from Marie Claire in England ("Oh your book sounds absolutely wonderful!"): once I explain (on an expensive telephone call late at night) that from the article syndicated to *Cosmopolitan* four years ago, I'm only using about eighty words that aren't actually on the public record, she says, "Oh, in that case it will just be a token fee of fifty pounds."

I see.

And so (fortunately) it's around about this time that I pause before I post out my two dozen letters seeking permissions...

What if even a proportion of these want to charge "token fees"?

The fact is, you don't earn much money from literary fiction in Australia— especially a book of experimental stories and novellas by an unknown author. Fees like this would not only put me in debt for the next few years, they would make it virtually impossible for me to keep doing what I do. In a very real way they threaten my next book, which I've already spent a year and a half researching, and they threaten everything I've spent twelve years learning how to do.

So there was this minor practical problem I had to deal with.

And then the other log that I could see (in my fitful nightmare-filled sleep, especially if I had to set the alarm to ring Lou in New York at some ungodly hour)—sweeping down the river toward me... Well, there were two of them, sort of tied together. And sitting up there on the first, with an expression on his face that I couldn't quite make out, was the ghost of J. M. Barrie.

In a novella that is about a third of the book, I've used the occasional brief quote from *Peter Pan* as a structuring principle—typographical stepping stones or punctuation points, if you like. Except that my Peta is a girl; which means that even when the quotes stay the same, with a girl-Peta and in the context of a story exploring being childless (either by choice or otherwise) and cultural notions of femininity and adulthood, they take on quite different meanings from the original. For instance:

"If you find yourselves mothers," Peta said darkly, "I hope you will like it." The awful cynicism of this made an uncomfortable impression, and most of them began to look rather doubtful.

And there are other times where I've strategically misquoted.

Every time a woman says "I don't believe in babies" there's a baby somewhere who falls down dead.

The quotes are something like less than four hundred words out of twenty thousand; and I actually feel that Mr. Barrie himself would approve, but he's dead and it would be some unknown person who administers the estate making the decision. What if they, just personally, didn't happen to like what I was doing?

If they refused (and a copyright holder is not required to give any reason for a refusal), there goes a third of my book, and a year's work.

And on the other log: a whole heap of people from *Fatal Attraction*, barreling down on me for a story in which I've not just quoted bits of dialogue from the film, but have also appropriated the main characters and actors and sent them off on a mission around the back streets of Newtown in Sydney...

But how can I possibly ask James Dearden and Adrian Lyne for permission to critique their film in the way I have in this story? (It's not exactly a flattering view.)

So it was at around about this point that some of the people I was seeking advice from (such as the Australian Society of Authors—who did prove to be very helpful in the end), began to accept that maybe I wasn't just a criminal-minded anarchist postmodernist who wanted to be able to rip off other people's words without paying for them... That maybe my rights as a writer also needed defending. And that this (like most things in life) isn't just a simple black and white copyright issue, but is also about things like free speech. I can't keep writing this way if I have to pay everybody a tithe. (And I'm not just talking about lots of little sums: Macmillan in the U.K. wanted $500 for every print run for a few brief quotes and paraphrases from a seventies book about faeries; and EMI originally asked for $830 for eleven words from "Pale Blue Eyes").

It's a bit like when someone tells you an anecdote and you say, "Hm, can I use that in my next book?" and they say, "Do I get a royalty?"

It just can't work that way—if I paid everyone who's ever contributed something to my work, they'd all end up getting about half a cent each and I'd end up with nothing to pay my rent with and the added burden of knowing that every word I write might end up costing me more money than it's ever likely to make for me.

And I can't keep writing this way if anyone who doesn't like what I've said or implied about their work gets the right to refuse to allow me to refer to and quote from it.

The simple answer is: well that's what the fair usage clause is there for. (This is the clause within the Copyright Act that allows for "fair use" of another's work for the purposes of research, criticism, or review.)

But for one thing, this is a book of fiction. Can I really rely on getting a judge who understands that fiction can sometimes also be criticism?

And for another: Most of these things aren't decided by judges anyway, because they never get to court.

Music publishing companies realized this a long time ago: that it's whoever has the biggest team of lawyers and the most money to throw about who in effect get to set the laws. For a long time their interpretation—that even using one line of a song constitutes a copyright violation—has been accepted as fact. Even though to my knowledge this has never been tested in the courts; and it's certainly not the advice I received from the Australian Copyright Council.

In other words, if publishers settle out of court—and who can blame them?—it becomes irrelevant whether my use is legal or not. (And it's certainly irrelevant whether it's ethical or not.)

Let me say, here and right now, that I fundamentally support the principle of copyright protection for authors: that is, the principle of asking for permission to reproduce substantial pieces of another's work, and the need to compensate artists for any loss of sales this might involve, or for their original labor in producing the work. (Effectively so they can go on producing more work).

But I also believe in the principle of free speech, and the need for writers to be able to imaginatively, creatively and productively engage with the cultural products and contemporary cultural events around them. I can't see that it's in anyone's interest (least of all other artists' and musicians') for us to be forced to go on writing books as if music, television, films and magazines don't exist or have important effects in the world or on people's lives and feelings.

And given the nature of contemporary culture, I really don't think it's useful to make a distinction between those who appropriate and those who don't. Everyone borrows from everyone; everything is connected to everything else. What I think is much more useful is to look at the effects and implications of the myriad different kinds of borrowings that do go on: the ethics, if you like, of each type of borrowing, and the politics.

For my own part: I don't just tack other people's work onto my own in order to enhance or embellish it (if I did, then it would be a much simpler proposition to just remove it and save myself time, money and trouble). I'm meticulous about referencing and acknowledging other peoples' work in my own—my initial training was as an historian, and I see no point in putting the quotes in if readers aren't aware of where they come from or aren't given a sense of their original context. Especially if what I'm trying to do is to critique, disrupt, extend or play with something, then it's essential that the original intention (or effects) be also made clear at the same time.

So these are my own personal ethics (or politics) about what I do.

Thus the problem for me, for instance, with Helen Darville's appropriations was not that she used someone else's words (I think pastiche as a form is fine; it can be effective and interesting if done well) but that she didn't acknowledge this. If she had, of course, then her own lack of personal experience and, hence, personal authority would have also automatically been acknowledged and made obvious, and this would have altered the whole way the book was experienced and read. It would have been a different book, with a different history (and vice versa).

Well, anyway, while Darville's lawyers may be able to sleep soundly with the conviction that her appropriations (while admittedly "bad form") are not actionable (that is, not a clear violation of the Copyright Act), I'm afraid I still have the occasional watery nightmare. (Especially with the new Moral Rights law ready to be introduced into Australian Federal Parliament at the next session... but that's a whole other kettle of worms.)

In fact, sometimes I wonder if it's not the case that the more ethical I am, the more potentially actionable I might be making myself in the long run.

There were more than a few times, when talking about these issues, in which I'd receive the helpful advice: well, just don't acknowledge it. Don't identify the source and no one will notice, or they'll have a harder time proving it. Just shuffle the words around a bit and leave off the author's name. Whatever you do, don't write and let them know!

In other words: steal it.

And I guess this is my concern: that if we have an inflexible attitude to the use of other people's words, then we are encouraging a climate in which people steal rather than borrow, pilfer rather than critique. Or where the jokes become merely private.

There seems to be this idea out there that appropriation is easy. A bit like the old idea that free verse in poetry is easy—if you don't have to rhyme, then hey, where's the talent in that? Anyone can be a poet (well yes, I guess, in a sense, that's the point)...

But if you are concerned with attribution and sourcing and referencing; with evoking the original context and maintaining the integrity of the fragment even in its new context; with and all these ethical and political issues, as well as trying to sew the whole thing together into a compelling narrative; with preserving a multiplicity of original voices, and yet still taking some kind of final authorial responsibility for what you are doing; it's actually quite complex and takes a lot of thought, and a lot of repetitive, painstaking labor, and imagination.

It's just not as easy as it looks.

I prefer to think of myself as a collaborator or cultural partner, not a thief.

In fact, without exception (including The Man himself, who instructed EMI to drop the fee to $130 after I wrote him a letter raising these kinds of concerns), every author I've been able to directly contact has been delighted that I've used their work and has wished me every success.

Lifting something can be exactly that; it doesn't have to be exploitative.

As Eudora Welty once put it: "Criticism can be an art, too. It can pick up a story and waltz with it."

SUBJECT: WHAT IS DIGITAL STUDIES?

DATE: TUE, 13 OCT 1998 13:00:21 -0400
FROM: ALEX GALLOWAY <ALEX@RHIZOME.ORG>

There is a need today to situate, keeping an eye on the scant technological ruminations of what we have come to call, simply, "theory," the growing mass of theoretical material devoted to digital technologies. In recent years digital technologies have become more and more involved in how we produce, consume and mediate texts. In light of these new technologies, one is compelled to rethink our theory of textuality, while at the same time, faced with a particularly insidious combination of intellectual technophobia and simply honest ignorance, one must bring a whole intellectual field up to speed, a field hitherto focused on post-structuralism, the signifier, Lacanian psychoanalysis, certain types of French literature and philosophy, structural marxism and media theory (that is, film, television and video).

While many have started to write theory on "technology" or "globalization"—both quite relevant to a study of new media—a second look discovers that much of contemporary theory does not engage substantively with the object of its analysis, the digital. So often, we are scared off too soon by the simple fact that it is technology. The above theoretical legacy—post-structuralism, film theory, and so on—provides us with many useful problematics. My goal is to determine which of these problematics is still relevant, then suggest a direction for the future of this field. Recent criticism focusing on new media is thus my focus on here, attempting to force through this "descriptive" phase toward a more general theory of digital studies.

Digital studies takes digital technology as its object of analysis. Specific topics within digital technology include the internet, the internet browser, the digital "object" (for example, a webpage) and "protocol" (how digital objects are organized). For my purposes, digital studies is, like political economy before it, at once a new theoretical paradigm and a position-taking within that paradigm.

Several theoretical debates must be revisited with the advent of digital technologies. Specifically, in response to the textuality debate ("What is a semiotic network and how does it function?") digital studies argues against signification and the urge to find meaning in objects or texts. Digital studies is not interested in interpreting the web; it is not interested in offering a description of its meaningfulness or its signification.

The following are a few programmatic statements for digital studies. Digital studies is a argument for the idea that objects (net bodies) are organized through protocols into a "netspace" and that certain kinds of knowledge legitimate this organization. This is an argument for the category of netspace as a specific historical event, a result of the reorganization of bodies/objects (a putting-into netspace). Furthermore, it is an argument against those who rely on pragmatic, neoliberal explanations for the changes in social formations under late twentieth-century capitalism. Digital studies opposes the arbitrary use of old metaphors to describe netspace: the text, the tree, Cartesian space, and so on. Digital studies rejects the opposition between mind and body. Digital studies is also against the common notion that the so-called contemporary information overload is destroying social relations. On the contrary, we see not a disintegration but an extreme proliferation and subsequent regulation of social relations under the new media. Digital studies is, above all, a reaction to certain theorists' tendency to throw around the concepts of information economy, new media, networks, and so on, without ever actually describing the technologies at the heart of these changes.

"FIRST COMMODITY, THEN SIGN, NOW OBJECT..."

For many years now theorists have preferred to speak of value economies—be they semiotic, marxian, or psychoanalytic—in terms of genetic units of value and the general equivalents that regulate their production, exchange and representation. Tempting as it may be to follow the lead of film critics like Christian Metz and André Bazin and claim that, like cinema before it, the whole of digital media is essentially a language, or to follow the lead of *Tel Quel* marxist Jean-Joseph Goux (or even the early economics-crazed Baudrillard) and claim that digital media is essentially a value economy regulated by the digital standard of ones and zeros—tempting as this may be, it is clear that digital media requires a different kind of semiotics, or perhaps something else altogether. The net does not rely on the text as its primary metaphor; it is not based on value exchange; its terms are not produced in a differential relationship to some sort of universal equivalent. Digital technology necessitates a different set of object relations. What are these relations?

In the digital economy there is a new classification system: object and protocol. As opposed to the sign, the digital economy's basic unit is the unit of content, an infoid, a digi-narrative. It is not simply a digital commodity nor a digital sign. The object is not a unit of value. The digital object is any content-unit or content-description: MIDI data, text, VRML world, image, texture, movement, behavior, transformation. The object is what Foucault calls a "body," or what Deleuze might call the content of an affect-image. Digital objects are pure positivities.

These objects, digital or otherwise, are always derived from a preexisting copy (loaded) using various kinds of mediative machinery (disk drives, network transfers). They are displayed using various kinds of virtuation apparatuses (computer monitors, displays, virtual reality hardware and other interfaces). They are cached. And finally, objects always disappear. Thus, objects only exist upon use. They are assembled from scratch each time, and are simply the coalescing of their own objectness. Platform independent, digital objects are contingent upon the standardization of data formats. They exist at the level of the script, not the machine. Unlike the commodity and the sign, the object is radically independent from context. Objects are inheritable, extendible, pro-creative. They are always already children. Objects are not archived, they are autosaved. Objects are not read, they are scanned, parsed, concatenated, and split.

Protocol is a very special kind of object. It is a universal description language for objects, a language that regulates flow, directs netspace, codes relationships and connects life forms. Protocol does not produce or causally effect objects, but rather is a structuring structure based on a set of object dispositions. Protocol is the reason that the internet works, and performs work. In the same way that computer fonts regulate the representation of text, protocol may be defined as a set of instructions for the compilation and interaction of objects. Protocol is always a second-order process; it governs the architecture of the architecture of objects.

To help understand the imbrication of object and protocol I offer four examples: HTML, the internet browser, collaborative filtering, and biometrics.

A scripting language for networks, Hypertext Markup Language (HTML) is a way of marking up text files with basic layout instructions—put this sentence in boldface, add an image here, indent this paragraph, and so on. As the universal graphic design standard since its introduction in 1990, HTML designates the arrangement of objects in a browser. The specifications for HTML 3.0 claim that "HTML is intended as a common medium for tying together information from widely different sources. A means to rise above the interoperability problems with existing document formats, and a means to provide a truly open interface to proprietary information systems." To the extent that HTML puts-into-verse text plus layout instructions and also undiversifies qualitatively different data formats, we may call it a versifier. HTML is a scalable protocol, meaning it is able to grow efficiently and quickly with the advent of new technologies. Unlike some other computer scripting languages HTML is platform independent: it is not restricted to a single operating system.

As the HTML example shows, a protocol facilitates similar interfacing of dissimilar objects. Contrary to popular conjecture, the digital network is not a heterogeneity. It is a hegemonic formation, or rather, a dynamic process-space through which hegemonic formations emerge and dissolve. That is to say, digital networks are structured on a negotiated dominance of certain textual forms over other forms, all in accordance with schedules, and hierarchies, and processes. Protocol is the chivalry of the object. Objects are filtered, parsed, concatenated. They are not archived, filed, or perused (these are predigital activities). Protocol constitutes a truly rhizomatic economy. Ebb and flow are governed by the various network protocols (FTP, HTML, SMTP, and so on). Connectivity is established according to certain hierarchies. And like the logic of traditional political economy all elements conform to formal standardization. Textuo-digital protocol "allows objects to read and write themselves." And thus objects are not reader-dependent, rather, they take themselves to market.

One of the defining features of intelligent networks (capitalism, Hollywood, language) is an ability to produce an apparatus to hide the apparatus. For capitalism this logic is found in the commodity form, for Hollywood it is continuity editing. In digital space this "hiding machine," this making-no-difference apparatus is, of course, the internet browser.

The browser is an interpreting apparatus, one that interprets HTML (in addition to many other protocols and media formats) to include, exclude and organize content. It is a valve, an assembler, a machine. In the browser window digital objects (images, text and so on) are pulled together from disparate sources and arranged all at once, each time the user makes a request. There is no object in digital networks, or rather, the object is simply a boring list of instructions: the HTML file. Thus, the browser is fundamentally a kind of filter—something that uses a set of instructions (HTML) to include, exclude and organize content.

Despite recent talk about the so-called revolutionary potential of the new browsers (Web Stalker example <http://www.backspace.org/iod> is the best example), I consider all browsers to be functionally similar and subdivide them into the following classification scheme: dominant (Netscape and Explorer), primitive (Lynx), special media (VRML browsers, applet viewers, audio/video players, etc.) and tactical (Web Stalker).

Outside of the browser, another form of protocol, this one more radically ideological, is the concept of collaborative filtering. Surely this is a type of group interpellation. Collaborative filtering, also called suggestive filtering and included in the growing field of "intelligent agents," allows one to predict new characteristics (particularly our so-called desires) based on survey data. What makes this technique so different from other survey-based predictive techniques is the use of powerful algorithms to determine and at the same time inflect the identity of the user. By answering a set of survey questions the user sets up his or her "profile." The filtering agent suggests potential likes and dislikes for the user, based on matching that user's profile with other users' profiles. Collaborative filtering is an extreme example of the organization of bodies in netspace through protocol. Identity in this context is formulated on certain hegemonic patterns. In this massive algorithmic collaboration the user is always suggested to be like someone else, who, in order for the system to work, is already like the user. Collaborative filtering is a synchronic logic injected into a social relation; that is, like the broad definition of protocol above, collaborative filtering is a structuring structure based on a set of user dispositions. As a representative of industry pioneer and Microsoft casualty Firefly described in email correspondence: "a user's ratings are compared to a database full of other member's ratings. A search is done for the users that rated selections the same way as this user, and then the filter will use the other ratings of this group to build a profile of that person's tastes." This type of suggestive identification, requiring a critical mass of identity data, crosses vast distances of information to versify (to make similar) objects.

The flourishing field of biometrics also illustrates the logic of object and protocol in the new media. What used to stand for identity—external objects like an ID card or key, or social relations like a handshake or an inter-personal relationship, or an intangible like a password that is memorized or digitized—is being replaced by biometric examinations (identity checks through eye scans, blood tests, fingerprinting, and so on), a reinvestment in the measurement and authentication of the physical body. Cryptography is biometrics for digital objects. Authenticity (identity) is once again in the body-object, in sequences and samples and scans. Protocol is "what counts as proof."

What this brief examination of digital technologies aims to argue is that the digital is a set of protocols, based in technology, that governs object relations. My move is to show the inner workings of apparatuses such as HTML as they produce these object/protocol relations. Moving forward from a theoretical legacy then, digital studies can begin to analyze the field of emerging digital technologies—the space of the internet, the internet browser, the digital "object," and the digital "protocol."

SUBJECT: TOWARD A DATA CRITIQUE

FROM: HARTMANN@FSF.ADIS.AT (FRANK HARTMANN)
DATE: TUE, 21 JUL 1998 12:28:22 -0400

"Data is the anti-virus of meaning"—Arthur Kroker

"There is no information, only transformation"—Bruno Latour

The digital datasphere affects all major aspects of cultural production. Is there still a task for critique in this process, aside from cheap falsifications of the techno hype, or from simply articulating fear? What could be the task for a data-critique then, which could succeed to reveal the hidden agenda of the proclaimed "information society"?

AFTER CRITIQUE

According to some commonsense view, we have already entered an era beyond enlightenment and critique: the new media reality creates a symbolic totality, an inclusive environment—a perspective from which any critical discourse seems an irresponsibility of sorts. With this new media reality, the level of theory and of its object becomes indistinguishable, and what we need therefore to grasp cyberspace is not a critique of ideology but a more systematic description of media, an analysis of its infrastructure, and an archaeology of the apparatus. This positive view now aligns intellectuals as well as activists and artists under the efforts of technology.

Critique is negative indeed, and that firstly means it is all about limitations. While net-criticism as an activity indicates the limits of the internet with all its disappointed hopes from the sixties ideology, data critique deals with the philosophical and social assessments of digital technology. Necessarily invoking some spirit for the enlightenment which became unpopular after the recent "death of the subject," the aspects of data critique are reaching beyond any singlehanded notion of progress within the inclusive form of new media.

Philosophers, within their academic discipline, fall short to grasp the meaning of new information and communication technology, as they keep to the beaten track of reading, interpreting and redistributing texts within their classical frame of reference. The academic community, at least the humanities, still largely depends on the gratifications of the paper medium, and that means on traditional "print-publishing" through "publishers." To be media literate otherwise, they consider none of their business. There are several reasons for that ignorance. A quite profane one is "fear of the machine," which can take on very sophisticated forms: from straight neo-luddism to a moralistic, protestant information-ecology with its apotheosis of the pen and the typewriter. These positions for one, seem to make clear—insisting on their professional identity, the so-called humanities tend to exclude any non-humanist discourse in favor of their quest for autonomous "subjects" and their hermeneutic privilege of "making sense." But there is no way in falling for a Heideggerian promise that supposes to reveal an order of things that still could go undisturbed beyond any stirring by "media." There is no such tranquillity of being once after "care" has crossed the river for good (M. Heidegger, *Being and Time*, Oxford University Press, 1962, 242).

GLOBAL INFORMATION ECONOMY IN DIFFERENT WORLDS

A range of sociological questions supersede the technological ones. With the new information and communication technologies (ICT), the end of this century provides the first world with a thorough and disorientating crisis concerning the role of work, education, and entertainment. The reason for this is a postmodern condition at one hand, a global marketing strategy for these technologies on the other. When in 1995 the National Science Foundation's funds for the internet backbone structure in the U.S. finally ran out, new sponsorship was due from somewhere. By going international and also by leaving academic boundaries behind, the providers of the "net" found their new strategy for economic survival. An American concept was ready to become "the boom to humankind [that] would be beyond measure," pulling everybody into "an infinite crescendo of on-line interactive debugging" (<http://www.memex.org/licklider.html>). While some 96 percent of the first and 99 percent of the world population is not online—the information highway has no turnoff to their house and home and maybe will never have—the electronic commerce is exploding and the emerging virtual class takes their advantage of the bit business, "the production, transformation, distribution, and consumption of digital information" (W. Mitchell: *City of Bits*, Cambridge: MIT Press, 1996).

And again, what are we referring to? For the society in transition, the complex social and cultural matrix of change is not properly known; in the present discourse, cyberspace as the emerging social space is perceived merely by technological metaphors and a market-driven development of the broadband ICT infrastructure. Especially in Europe, yet not without a particular reason: the European ICTmarket currently ranges at a total value of ECU 300 billion, and sees an average national per capita investment in Western Europe of approximately ECU 350 (<http://www.fvit-eurobit.de/def-eito.htm>). While internet access still is between 10 and 100 times more expensive in Europe than in the V.S.(5), the European Commission's propaganda sees Europe as the coming heartland of electronic commerce, pushed by those investments and numerous ICT policy action plans (<http://www.ispo.cec.be/>).

New media and the prophecy of an information society are little more than the figleaf of a failed transition of modernity towards a more social society. Judging from various programmatic papers, the social impact of the broadband media applications are very modest. In the so-called Bangemann report (<http://www.ispo.cec.be/infosoc/backg/bangeman.html>) people in the end only exist as the representation of solid markets under the command of an ideology of total competition within the first world(s). With this "new techno-utopia of the emerging global market capitalism" the sole principles of market liberalization, deregulation and privatization are applied (Group of Lisbon, *Limits to Competition*, MIT, 1996). In consequence, the recommendations and the proposals of the Bangemann paper seem to serve more to the benefit of the attending companies in this Expert Group themselves.

The lack of proper understanding for a new information economy beyond competition also derives from an uncertainty or even a crisis of the intellectual position and the role of theory within it. The bit business does not need a media theory. The same goes for the new "Virtual Class," that social segment which—according to Arthur Kroker's observation (A. Kroker and M. A. Weinstein: *Data Trash*, St. Martin's, 1994)—benefits most from the virtualization, and which defends information against any contextualization, with its goal of a total "cultural accommodation to technotopia" exterminating the social potential of the net.

INTELLECTUAL DISCOMFORT

While thousands of websites blossom, most intellectuals feel instinctively uncomfortable with this process. Traditional Homo Academicus all ash and sack, has not much clue to what is going on in the flashy online world. Further to their distance, random ASCII fetishists become the new iconoclasts of the net. Having invested in all that textualism, and having formed this distinctive usenet community, now coping with the masses again, with those impositions of the World Wide Wedge—accompanied with an unquenchable thirst for new software, new applications, more pictures, more entertainment, and more prefab interactivity?

In the beginning, there was the word, then there was programming. In terms of cultural technique, the computer itself substantially changed, as well as our relationship to the machine, in a relatively short time, from number-cruncher to word-processor to thought-processor (M. Heim, *The Metaphysics of Virtual Reality*, NY: Oxford, 1993). Moving from mainframe to personal computing (PC) to net computers (NC) and now all of a sudden computers, as we painfully learned to know them, seem to vanish again. Not only they become less significant parts of an integral whole, but also widely integrated into everyday appliances as in "intelligent" cars, household machines, shoe soles, and the like. Culture moves toward a state of ubiquitous computing, where these machines form the new environment. Amongst many other things, this indicates new forms of social integration and a new involvement in societal relations. Kant's transcendental subject seems to exist not longer in terms of common categories of sensual perception and logical thought but those of the global electronic datasphere. Which brings to mind McLuhan's phrase, that "in the electric age we wear all mankind as our skin."

All mankind, one world? Should this be the heritage of the age-old philosophical dream of a universal language and a common understanding come true? The misleading term of the Global Village forgot to discuss the severe social constraints that determine life in a village. There is a possibility that the information society becomes as culturally homogeneous as any village lifestyle is. But we will never forget that we live in different worlds.

The ideology of individual liberalism can be seen as a cultural movement from west to east, from north to south, a doctrine of salvation, which sells the benefits for a technocratic elite of the Virtual Class as a paradigm for the global social sphere. The electronic frontier actually is a retro-movement across the Atlantic toward Europe, which proceeded within Europe toward the East with considerable delay. The relatively homogeneous character of "Cyberspace American Style" was perceived critically from a European perspective, where the loss of cultural diversity was and still is feared. Besides demographic factors, there are several other hindrances for coping with this specific change. The problems with the new electronic boundaries between East and West are not of a mere technical but also a cultural nature. Cultural differences express themselves through different use of communication and techniques: a technical interface always also is a cultural one.

WINDS OF CHANGE, BATTLE ON CONTENT

Basically, ICT is grossly overestimated as a tool or instrument of change, especially when its brief history (with an open end) is being considered. Will technology change people, or are new technologies already the expression of change? But then, technology is always only a part of the problem. In the end, we have to ask what will determine the shape of Cyberspace: Asian hardware and American software alone? Cyberspace holds political, socio-economical and cultural issues as well, all of which are up to thorough scrutiny by social and political science—I would like to promote this as a specifically European task. As there is cyberspace, what does it mean for "us," living in a fragmented world?

Needless to say, that task is a critical one. Why? It once was argued by philosophers that the bourgeois utopia of a democratic, participatory society was the "natural child" of absolutist sovereignty. The critical task of enlightenment was being performed in a time of societal crisis, and thus took on some hypocritical measure. The object of critique firstly being texts and their social implications, for example, the Bible, enlightenment failed in its task to replace these texts with new content when its critique explicitly was extended towards politics and society as a whole. The benefits of enlightenment meant business for some.

In his critique of aesthetic reason, Kant argued in train of the biblical prohibition of images for an enlightenment that is "just negative" in respect to its task: he not only carried on the age-old quest of intellectuals—defending their cultural privileges, that is, textual against any easier accessible cultural techniques, wanting to be the "true" mediators against any kind of "deceiving" media—he also refused to name what this non-pictorial *Denkungsart* should be, if simple demystification (of the "childish apparatus" provided by religion and corresponding politics to keep people as their subjects) would not do (Kant, *Critique of Judgment* [1790/1793] A124/125). Ages before Kant, nominalism already failed to win its battle on content, which started with the intention to distinguish real content from mere metaphysical noise (*flatus vocis*), and true thought from ideology by ways of, let's say, a proper information economy. Now history shows that a simple purification filter—from thoughts to words, from images to texts, from texts to programs—is not the way it works. Such self-righteous critique easily becomes delusive. This happened to the bourgeois filter of content against transcendence, as the *Encyclopédie* necessarily failed to be the new Bible for modernity.

VIRTUAL INTELLECTUAL TASK FORCE

Rethinking enlightenment? Still an academic endeavor. Reprogramming society? A fading socialist dream. The elements of a data critique are at hand: a task not to be left to the neo-luddites (T. Pynchon, "Is It O.K. to Be a Luddite?" *New York Times Book Review*, October 28, 1984). The Virtual Intellectual—a new figure discovered by Geert Lovink—will be constituted through his/her specific mixture of local and global cultures: "The Virtual Intellectual is conscious of the limitations of today's texts, without at the same time becoming a servant of the empire of images." Critical activities, being the heritage of the textual realm, "will now be confronted by the problem of the visualization of ideas" ("Portrait of the Virtual Intellectual," lecture, Documenta X, Kassel, July 1997 <http://www.desk.nl/~nettime>).

Critique, according to Kant, concentrates on the form versus the content, on the realization of "negativism." As critique always means differentiation, a data critique follows the modulations of information within a process of circulation. It works on the level of subjectivity, while this implicates some sociological sobriety, some demystification, and some diversity. Since digital-ization alone is not the issue, the question is whether there are alternatives within the pretentious information society project?

Philosophically, it keeps its skeptical distance toward ontological questions concerned with "truth," and similar traditional encumbrance. In a kindred spirit, Peirce's pragmatism—stating the fact that "We have no power of thinking without signs" (J. Buchler, ed., *Philosophical Writings of Peirce*, NY: Dover, 1955, 230)—made clear that because sign and signified differ according to an ever changing "interpretant," we rarely have a chance to recall qualities in communication which relate to anything beyond actual sign-use and therefore, media-practice. Thus, the irrelevance of any meta-physical "meaning" as in "true representation" of ideas through texts becomes a notion of enlightenment revised, for generations after the over-whelming encyclopedic project of a thesaurus with all available knowledge (as cognitive possessions), or even the notion of "unified science" (further to d'Alembert or, more recent, Charles Morris, Otto Neurath and others who historically struggled to create a new symbolic "unification") (D'Alembert and J. LeRond, *Discours Preliminaire de l'Encyclopédie* (1751); C. W. Morris, Charles, O. Neurath, et al., *International Encyclopedia of Unified Science*, University of Chicago, 1938–39).

INFORMATION ON INFORMATION

Hypermodern communication tends to synchronize all aspects, and under these conditions to publish, means instant access to all utterance. The imme-diacy of media is getting scary. Thoughts are phrases made while having media presence. Simulation and speed are the two concepts that dominate media philosophy. Language is but the soft currency in an economy to increase the turnover of the information industries. After texts there are doc-uments, after structure there is HTML, after style there is VRML. Meanings are offset in "dot com." All content is but chunks of inert digital information, waiting for the copy pirates. At any common workplace, no material objects are being processed, but information. What are the resources of information work? When information becomes decontextualized, as it does, then what we need is more information on information.

Any information that is not contextualized is worthless. Phil Agre imagined intelligent data as he put forward the idea of "living data" by thinking through all the relationships data participate in, "both with other data and with the circumstances in the world that it's supposed to represent" (<http://www.wired.com/wired/2.11/departments/agre.if.html>). Geert Lovink and Pit Schultz established the notion of a "net-criticism," introduc-ing the fuzzy concept of something like ESCII, a European Standard Code for (critical) Information Interchange (Lovink and P. Schultz, "*Grundrisse einer Netzkritik*" <http:www.dds.nl/~n5m/texts/netzkritik.html>). One could fur-ther elaborate on this list; elements of data critique are there. A data critique, in terms of the announced information society, is not. It may be all about creating context, and defining the conditions. About the power of techno-imagination (*Einbildungskraft*), as media philosopher Vilém Flusser announced it (Vilém Flusser, *Kommunikologie*, Mannheim 1996). And content, what content? The net is a part of creating and/or reinventing cultural context as form, not as content. Concentrating on the form means to keep up cultural tradition. The net's problem is that the social motive that made it possible is seen totally detached from the technological process, and vice versa. While deconstructing illusions, the age of enlightenment produced some illusions of their own. What is needed is not a New Enlightenment through technically enhanced individuals, as Max More suggested for the hypermodern age (<http://www.heise.de/tp/english/special/mud/6143≠/1.html>), but a renewed epistemological agnosticism of sorts, an antidualism set against the notion of that "inner nature" of things that leads to any "true" forms of representation. Why not call it a data critique?

SUBJECT: AI SERVICE

FROM: GÁBOR BORA <GABOR.BORA@ESTETIK.UU.SE>
DATE: SUN, 25 OCT 1998 19:13:07 -0500

(Warning: What follows is a piece of fiction. This does not mean that it is a product of imagination, or fancy; it only indicates that it describes something that does not exist as an actuality. Being a virtual entity, it is a hole in the existent—that is, the existent hosts it. This story is not restricted to the actual conditions, how things are; it is rather hosted by the state how things are.)

Let's talk about the Informator. Having a twice awkward position—due to the bad reputation or imago gathered during the activities in the expiring past and to the boundless suspicions (these are emerging from the same bygone past) entertained about her or his activities of today—s/he deserves at least an iota of detached and dispassionate (nonantipathetic) attention.
The Informator is neither a symbol nor an impersonation. S/he is an existent or a possible existent; as a singular person as well as a manifold constituting a class. S/he is someone like us, thinking, acting, suffering and enjoying. Informators are among us, they are of us despite that we are seldom aware of this. Often they too are unaware of this. This factor of awareness or unawareness of being an Informator is already part of that destiny that is the fate of the class of Informators. I, who register all this, am anxious of the complexity of presenting fates and destinies. I leave it to an Informator to characterize her/himself with ad hoc, randomly chosen selections, own trains of thoughts as well as foreign thoughts considered during her/his activities: the style is the man himself (as an old-time-high Informator once expressed it).

«VEB-site... Actualities constantly complicate things. In this right now ongoing now there is a historically already unrealizable contour emerging, a shape of a would-have-been. One is moving around within the multiplicity of webpages crowded with shifting fripperies, badly colored whimsical knickknacks, zero-resolution images, the whole mess of a cumulatively extending/expanding redundancy. All this reveals retrospectively its own disappeared energetics as a hopeless, because nostalgic, desire. Its object is an aesthetics that got sacrificed for aesthetization. From here emerges a remembrance for something never taken place. It is the memory of the VEB-site, a memorial desire, an aesthetic correlative, a nostalgia toward something that never occurred: a net-design having its model in the GDR post-Bauhaus... (This is perhaps the most simple example: the digital culture is crowded by all kinds of imaginary modifications of temporality: never-existed pasts with nostalgic feelings toward them, impossible futures that one calculates with anyway.)....» ["VEB" was the GDR's generic prefix designation for a collectively owned company.]

«*Virtual communities....* In virtual communities the carrier of the genesis belongs rather to the realm of liberty than to the one of necessity, as Karl Marx once expressed it. Vladimir I. Lenin's doctrine of the weakest chain-loop of capitalism turned out to be a mistake. The royal road to the highest freedom leads not through specific deficiencies; it is rather demarcated by the originating Eros of Information Society's original capital accumulation, the intensity of a surplus energy: the surplus of information, even called information overload, the excessive mass of information guaranteed within a variety of processes; it is not any more a real surplus triggering the greediness that became instinct by the culture of several thousand years, it mobilizes an aesthetic lust-principle, the free play of the faculties of the soul, if it is allowed to abuse the categories of Immanuel Kant. The dictum according to which it is the information that is equally spread among humans is not justified yet, but the promise of its realization is steadily present, just like the threat from the part of corporative obstructions. According to corpo/rationality, capital is "classical," that is real, according to the digital sensitivity it is virtual. If the later, then corporative self-identity is grounded on a misunderstanding of itself believing that capital is still real. If the corporative rationality is right then the order of information soon or later regresses into the order of capital. (This belief gives the corporative impulse to translate the digital worlds to the world of capital.) If the thesis of the virtual sensitivity turns out to be right, then capital will transgress into pure virtuality or information. (No doubt, capital today is becoming increasingly virtual by its definitional edge, pure monetary transaction. Already this can be seen as a transitory phase, the first one of capital's metamorphosis into something exclusively virtual.) A digital community is the realm of freedom (a life in freedom because the promise of the future reality of freedom), if capital morphs into information; it is a realm where the decisive necessities are hidden for its members. At any rate, within these communities the thesis is in working order, the diffusion of information is even; the question that remains, is it just a temporary achievement or is it a realization of a condition of existence coming into being, for the rest of humankind not yet realized. And this question doubles itself: one is told, the half of mankind not even used the phone ever. From this perspective virtual communities are virtual elites, they are elitists like the elite never was before. Knowledge, defined as information, is power; how it is power, however, is for the time being rather incomprehensible. Virtual communities are just waiting for the appropriate definition, in order to changes the promise of power into real power. I happen to know this definition, yet, I won't tell it...»

«...*Overload*... The whole digital culture is nothing but an answer given to "information overload." It is a response given in the same manner, thus, it only multiplies the overload. The overload has two gates, the one is the sum of the actual possibilities of technique, the other is the human ability of elaboration—the latter is not a constant given, moreover, it is connected to the technical apparatuses with a multiplex feedback. The attempt trying to consolidate the overload thus multiplies the overload; the plan made against the surplus within the overload adds itself to this surplus. The plan is made work by the surplus, this is its fuel; consuming this surplus, it produces a surplus that is greater than the consumed one. It produces a gift that is identical with that which the plan worked against. This process produces the culture for which the continual multiplication is the nature/natural. The list could be continued but I set stop here. The culture beginning to take shape is seemingly more interested in activisms than in interpretations; in fact, it is the producer of its own unlimited interpretative horizon. Seemingly, all its analyzers try to come out as its most accurate interpreter in court of a fantasized future; as if they were working for a retrospective confirmation and acknowledgment from a future: "I told this as early as in '98." In fact, this is not the case: this culture in evolution has a simulated information surplus as its own peculiar feature. There is a virtual virtue, a kind of *"virtuality an und für sich"* in it, a teleological thinking hitherto unknown, a completion attached directly to every beginning.
The characteristic mark of this teleology is that it is not futurological at all, it is completely anchored in the now. A future occurrence is determined by the now, thus when something is formulated, it is already a settled thing. Any acknowledgment is subsequent and therefore redundant, almost irrelevant. Things evolve and establish processes before we are aware of as to what these processes and determinants are. (There is a track of commonplace postmodernity in this phenomenon. Post-historicity involves a paradoxical edge: although it embraces an ill-defined feeling of an end of history, it makes everything historical. Everything comes and (anything) goes, nothing lasts forever. Everything is existing in a historical dimension except the fact that everything comes and goes, everything is a question of temporary consensus, perhaps even natural laws. Post-history is a triumph of the metaphysical principle of historicism. History out, its metaphysics in. The end of history is a ultimately Hegelian event: it incorporates what it transgresses. Now, the teleological choice of the digital culture is perhaps the best outcome within this disturbing paradigm. It makes historicity an economical principle. It makes the metaphysics of everything's historicity into an engine. And it doesn't matters if this engine justifies itself or not. Possibly, this tactics is already a way out from postmodernity. In this context, however, it is a necessity to go on more carefully.)....»

«...*Digital sensitivity*... Instead of employing careful conclusions, the characteristic manner is to carry matters to extremes. As if everyone would compete with each other, with oneself and with the flow of times when theorizing cyberculture. All the gathered existing trends become prolonged, lengthened, as far as possible. Projecting the often poor appearance of today's digital reality onto its future completion, these theories seemingly care more about a hope for their near or far future verification than about anything else. It may or may not be so in particular cases, but this doesn't make any difference: the phenomenon that the actuality of the digital culture is thought together with the sum total of visions possibly connectable to the actual is the general feature within this culture in such a high degree that it can be said that this visionary character is a distinctive property of the early digital culture. The future continuation and completion of present states are attached to the present state, they form its nondetachable part; in such a way, the present is a state saturated with visionaries of its own future, thus, these seemingly future references have nothing to do with any future. There is no trace of utopias, theories that seems to be utopian or formulate negative utopias, these theories are completely centered around the present state, around the now. Many judge this culture-after-the-letter to be a new visual culture; in its present state it is more appropriate to call it the culture of visions and visionaries.

The expansion toward the maximum of fictionality is nothing but a symbiosis between a visionary and a real—that which today is possible to produce—level. The sensibility that characterizes today's digital culture is a sensibility stressing the visionary....»

«...*Monolithic and multiple unity*... The dilemma of multiplied personality that at the same time more and less than an individual is interesting only until personality is presupposed to be unitary or at least unified as if according to an eternal law. There is no reason to presuppose such a thing. Ages ago or in the near future, the conception of unity and nondivisibility of the individual could and can be as horrible as today the multiple personality seems to be. The dilemma exists due to a stubborn need, a bad habit in us, that governs us to reduce things to one. Or, to formulate it in another way, when culture learned accepting a conception of unity that contains incompossibility, instead of necessary compossibility, then the dilemma disappears—and surely new ones appear.

The conflict between a monolithic unity, providing a pattern for any possible unity and a nonmonolithic unity is described in the following legend of which no one knows exactly from where it comes.

There is no possibility for representing the passage from monomorphity to polymorphity, for the metaphor of way can only be ascribed to the latter—we were told by the ancients. In our civilization there was no monotheism, rather something more, the deity was not an object for belief but even for being. He knew of everything, he saw everything, but all this couldn't help, he could not hinder the evil deeds of our ancestors, he wasn't able being everywhere at the same time; our ancestors frequently abused this disability, the always punctual and singular divine interventions couldn't balance the manifold of uncanny incidents. At the very end something happened: the one and only divinity, due to being internally infinite, transposed itself into an infinite series. All of sudden, there was an innumerable amount of the one and only deity.

For a long time there was nothing else happening, as the legend has it, than deus ex machina innumerably—until someone realized that there was a necessary concordance between an intervention and something morally improper. Our ancestors learned the moral, started to behave properly and thus expelled the manifold deity from more and more areas where there remained nothing to do. Slowly, the goal of the never outspoken consensus seemed to be within reach: to nullify the transcendent by moral. But then, the endless series of deity changed its character and by now, is exterminating our civilization. According to the legend, the legend ends with a different hand style: "and the eternal peace arrived."

The state of mere sensitivity and the states of mere—pure that is—thinking are divergent; that which has been human, that is both more and less than human. Both promise and danger. A promise originated in the freedom incorporated in virtuality; or a danger originated in the risk that the freedom is nothing but deceptive appearance. It can be danger or even threatened-ness from the moment when someone no longer participate in the culture merely, when someone is merely a passive onlooker, or even less, when some-one cannot decide, rather becomes decided....»

«*This ongoing age has its* charmant *segments*. If it continues along the lines it draws today, then it can arrive at producing things never seen: in addition to the tendencies of the emerging new Middle Ages, hopefully all the rest of historical ages will re-emerge too. All from the Stone Age to the Space Age; tribal social structures rivalize with Knighthood—and both with the bureau-cratic structure registering whatsoever is going on. Stone-age people inform themselves from special websites about the next step to be taken, whereas the webmaster goes to the shaman around the corner to get orientation. Watercycle hooligans start to explore America and when they arrive they give press conference stating they have just discovered Atlantis. This will be the Grand Finale of History: History shows up everything that could be con-tained in it, just before it will collapse by its own logic, namely, that History itself is historical and therefore perishable. Meanwhile, the tired citizen makes a charter trip to Mars where nothing is something else than what it is, everything is simple and one can enjoy Nature without being disturbed. The directions are adequate, it is only the progress of technology that is too slow: Earth today is nothing any more but a museum of mankind, it is high time for it to become that which corresponds to its purpose: obligatory target of class excursions. History, thus, in its last gesture reveals that which always has been its definitive feature: the delay of phase....»

«...*Imagination is outdated and obsolete disposition*... The vision (taken as both per-ception and its connotative, "visionary," and so on) differs from the imagi-nation by the fact that it cannot be owned; vision appears in the conscious-ness of a personality as if it originated and came from somewhere else or from someone else. Imagination is belonging to a person, it is an "I", a self that is participating in it, whereas vision—although it is not impersonal—cannot be owned, does not belong to an individual. From this fact emanated the erroneous belief that tries to archetypify vision, tries to subsume it to the collective unconscious. There is a mistaken step in this rendering: its background is the belief holding that anything that doesn't belong to individuals must be collective. Now, vision is neither collective, nor does it belong to someone. Visions have less independence toward their material vehicles; they have more independence toward those they find: us. They can be portrayed as—almost immaterial or in the digital world completely immaterial—small icons that leave the surface of things and start their often prolonged travel. Because of the length of their voyage, they cannot be tracked back to their origins, they loosed their origins and became mixed with each other. Thus, they are not about their origins any more, rather about their voyage, the inner life and the world of experience they lived during the journey. (In this way, they are not mediating the existence of their origins, they rather achieve an own being, own existence. They are more willingly reticent about things concerning their own being, their own existence; otherwise they are not keeping secrets—rather the opposite.) We know very little about this internal existence and life, precisely as we know very little about our thinking processes. What we know can we know via the outputs covering only a fraction of the activities of the brain. Therefore, we can suppose that within the consciousness there are a number of consciousnesses we do not know about, yet, these consciousnesses can know of each other. In a parallel way visions possibly establish systems of relations for us nonavailable, we could almost call such a system of relations intelligence. But in these issues there is no certainty; exactly this lack of certainty is to be compensated by fiction.

Vision, if not definitively, but by inclination, belongs to perception, whereas imagination is a requirement for the unity of an "I" or an object. (It is established on the original synthetic unity of apperception, to borrow Kant's category.) One of the most often repeated motifs within the cultural criticism of our days, Information overload, is critical from the viewpoint of imagination: the overload emerges not in the context of our perceptual abilities but in the context of the unity-producing activity of the imagination. What the thesis on overload states is no more than this: unification is impossible, is not in working order or cannot be in working order. But does this also mean that all that is not unified cannot be handled; moreover, that the lack of unification leads to the becoming-uncanny of the lifeworld? The answer is yes only when we take unification as requirement. If the answer is no, then the overload—because of the perceptual richness in it—can be taken to be a resource, a surplus energy. If? If there can be a unity that is not a function of imagination...»

You inverted Hermes!—intervened the Stranger the Informator's flow of consciousness. And at this moment he was not that alien anymore. Not at all: because I spoke to the Informator in this way, I, who record all this. After all, it is high time to take back the word from the Informer and contemplate her/him from a greater distance. The task given by the culture is thus to mediate between the levels of the real and the fictionally possible in such a way that what emerges is not a monolithic state unified by the imagination. This would involve a maximalization of the perception of information rather than a unifying access to perceptions. This presuppose a sensibility that was the characteristic mark of informers or Informators in the predigital world. It is a sensibility that is focused on the exploration, what can function as information, with almost no consideration taken to interpretative and classificatory issues. Before the epoch of Information Society, an Informator mediated information toward the apparatus of power, precisely such information that were not intended for it. The Informator handles indeed within the information sphere in the information society. The Informator's former role makes her or him twice appropriate. The former task was too to carry information; moreover, statistically the informer provided to the larger distribution of information. But what is most important is the sensitivity s/he inherited from the past. It is a sensitivity developed by mediating information toward instances for which they are not intended. It is thus a perceptiveness focused around the unexpected: the same information is a routine-message when it reaches its given addressee and it is something unexpected, an "unexpected series of signs" as information theory has it, when it reaches another addressee. It is this moment of perceptiveness that made it possible for the Informator to change her or his character, or finally to find her or his character, in the age of information society. In leisure time the Informator reads stories like this:

The conjuring trick of the snake charmer was built on the exploitation of same of the snake's biologically given sensomotor peculiarities. In this way he didn't need to remove the snake's poison fangs. The trick, thus, could arrive at a greater effect. He didn't execute any part of his job incorrectly; notwithstanding, an otherwise beautiful morning, a novice cobra did bite him to death. There was one victim and because of the low interest (it was early in the morning and it was a weekday) there were about ten witnesses. Victim and witness to what? To an otherwise imperceptible twinkling of the evolutionary progress.

The Informator stops reading and nods: yes, it is a minimal modification within the genetic code; then s/he ask her- or himself: isn't it so that any trick is interesting because there must be some informatic challenge inherent in it? But s/he is loses interest in answering it; the awareness becomes focused toward something else.

Knowingly or not, the Informator is an agent of the Artificial Intelligence Service. There are unknowing and ignorant agents, they are similar to people spreading rumors because they themselves believe in them. Agents without the consciousness of being an agent, they are information mediators, by accident transporting information from a site or medium where it is self-evident and thus not yet par excellence information, rather an embryonic form of itself, to sites where information can appear as information, can transmute into itself: into a nonpreceded and unforeseeable series of signs. It means, information cannot become itself until in a medium, or site, where it is not intended to appear.

The opposite correlation doesn't go, however, for agents with a consciousness of being an agent; they are not similar to rumor-mongers not believing in the stories but spreading them anyway. The difference between being unknowing and conscious does not dwell in a single step or in something like a single gesture. No, it is an—in some cases almost endlessly complicated — series: the conscious informer's relation to information is complicated by existential, epistemological and ontological considerations. S/he is not a mere mediator of information but an activist, a transformer; if s/he transports something, s/he too is involved in the movement; not only mediating but s/he her/himself becomes mediated, becomes transported. The information carried and handed down is at the same time s/he her/himself. The own personality, the own self is modulated into the improbable context-of-message. The own self is becoming an unforeseeable series of signs devoid of origin and context. Informator: identical with information itself. It is this circumstance that determines her/his being. Steadily maintaining an utterly unstable state of balance, the always renewed liberty must brought into existence. This freedom is the presupposition of an unbound, from any context liberated information, information in this way having the ability of transforming itself from an embryonic state into its own proper being. Necessity (of maintaining a balance) and freedom thus level out, they become identical. This is the existential paradox of the Informator, a paradox that cannot have any conceptual solution, a paradox that can only be dissolved in movement. And exactly the energy of this free/necessary movement that keeps the Informator going on. The Informator identifies her/himself with the absence of contexts, the routine task is the avoidance of any given or possible context. The routinelike is, however, always new, never repeating itself: compared to the automatism of the encasement of information into some context, an avoidance of contexts is always concrete. And here, again, a simile is needed, because the endlessly complicated system of relations with which the Informator relates to information, cannot be grasped in anything simple, only a simile can cast some light on it. As Freud put it, some contents run into obstacles during the transmission between the two agencies of the soul; we do not become conscious of these contents, we can only conjecture the censured content with the help of the traces the obstacles leave on contents we become aware of. Now, the activity (and, as it should be clear from the description above, even the existence) of the Informator consists of a weakening of the censured contexts, by attempting to replace the context with her/himself, operating as a membrane that helps the transmission of information instead of being its context, censure that is. What is of importance in this simile is that the Informator is not an interpreter, s/he doesn't interpret, doesn't try to decipher meanings. And there is even something more: similarly to the fact, that the weakening of the censure is realized by the dream work, the activity of the Informator never lacks some element of dreaminess, there is always something hallucinatory overtone present. It is, thus, not exaggeration to say that the agent of AI Service who is conscious of being an agent differs a lot from her/his ignorant temporary double.

The Informator presented here is well aware all of this; moreover these circumstances are determining his intellect, they are the common denominator of his personality, this plural entity. To put it in different way, this common denominator is the vehicle of the plurality of his personality. He is engaged in an uncomfortable activity right now. Let me tell what it is. He is participating neither in extracting nor in producing meaning; he doesn't try decipher codes: if they exist at all they will break by themselves. Therefore, when philosophical dimensions appear in his thinking, they must be inscribed in an oscillation between the infinitely different poles of noncomprehended data and hallucinatory states. Such oscillations are rather percepts than thoughts. He has for a long time diligently gathered these oscillatory movements, gathers the percepts mediating bare data and hallucinatory, dreamwork-like impulses. Right now, this conglomerate is making a metamorphose. By its own inertial energy, from the conglomerate, from an embryonic form, a developed, realized, state emerges and it emerges as a single impulse: the Informator, without really knowing what he says, murmurs: "New Enlightenment." Suddenly, his mind becomes filled with a feeling of uneasiness. He knows, this sounds like a broken code, like a meaning. He tries to concentrate. This is not signification, this is not an interpretation: this has to do with the existence he shares. This is not an essence, this is existence. Not significance but being. Not essence but appearance—and here he must set a stop. It is an illegitimate binary opposition presupposing an essence that is or can be connected to appearances. His whole activity, his whole existence presupposes the upheaval of this opposition. Back to the previous: it is about existence, about a description of a condition; not as an opposition to essence but as the world of lived experiences, a *Lebenswelt*. Yet he is not satisfied with this. Lebenswelt, this is still too abstract in spite of all efforts trying to present the absence of abstractions. Temporarily he gives up pursuing the train of thoughts.

His existence and his activity is the New Enlightenment; but this is hallucinatory data, or datalike hallucination, for the time being. All that he gathered transformed itself into a single thesis; the collection dissipated in this thesis and therefore disappeared. It demands new collection, it presupposes correction, it demands confirmation: the former collection became utilized for a single thesis, the collection itself is gone. And this single thesis presupposes a series of new collections. The lucky star of the Informator is that he is a plural personality: he registers the result of his diligent work; he registers the single thesis, "New Enlightenment" as a loss; but, at the same time, he finds pleasure in the new configuration of things: a pleasure in finding a new Enlightenment that avoids the failures of the old one. The new one is like a laboratory lightning making jumps between data and vision. Martin Heidegger mentions *Lichtung* as the sudden appearance of being. *Lichtung* means glade. Now, it is high time to substitute this by lightning; they, *Lichtung* and lightning are identical when we consider the raw data and not the meaning. The fireworks of the new representation overwrites the bucolic idyll.

The transformation is: *Lichtung* ——→ *Lightning*. The Informator registers this result. It is an Indo-European horror story: glade and lightning are the same word with the same suffix. Then, even the ontology should be the same: a glade covered by lightning. At the same time it is a glade where every blade of grass is a lightning. A glade as a surface of lightning, a surface of a series of lightning. Lightning surface. Zeus, help me! One must not recoil before the consequences...

This artificial lightning is the closest equivalent of old Enlightenment's representation-and-depiction-centrism (built on geometry and solid body physics), with its conviction that all strata of the existent can be represented and depicted. Later on the transparency of the consciousness about things became substituted by the opaqueness of self-consciousness, for which everything must be transformed into meaning or disappear. This was the death of the old Enlightenment.

The New Enlightenment, this new *fröliche Wissenschaft*, striving at a non-derivative transparency of complexities, renders the period of self-consciousness and its imagination-cult as a dark age. The Informator feels even more amused when he considers another circumstance: during the old Enlightenment the secret societies were the built in agents within the Ancien Regime; they were secretly, in the dark, so to speak, spreading the ideas of Enlightenment, they were able to get the aristocracy to follow the trends and counteract themselves, abandon their own essential interests. These secret societies are an equivalent to the condition of the AI Service, which is a secret society in such a high degree that the majority its members are unaware of their membership. This got him remember a story with a mood not dissimilar to the conspiratorial spirit of secret societies. This story was part of the series compressed into one single thesis as it was mentioned above. Now, he recites it:

I have a crucial presentiment: within the digital world it is the quantity of zeros that proliferates. Be it symbolic or not, it can be verified empirically. In my opinion the distribution between ones and zeros is not fifty–fifty: there are slightly more zeros. Now, according to theories, once upon a time it was a similar relationship between matter and antimatter—only slightly more matter than antimatter. Thus, the universe is an insignificantly tiny fraction of the mass of the total amount of matter; most was destroyed, transformed into pure energy, at the beginning of our world. It is this destruction that the subsequent universe is compensating for with entropy. Now, we can presume that the digital world cannot endure duality, just as matter/antimatter could not endure it. Thus, the digitalization will arrive at an—for—us unknown limit, when the digital world explodes/implodes into monolithic, noncompound substances. It will be a clean world, void of redundancy, a world of only zeros. It will be a world with only one type of substance and, therefore, the numerical code will be its only definition. To put it simply: a single number, that expressing the quantity of zeros. Then, for the first time ever, we can contemplate what a single number can signify. This contemplation will be the next entropy, the next compensation.

"Welcome to the New Enlightenment!" —The Informator

<peek-a-boo v2.5>

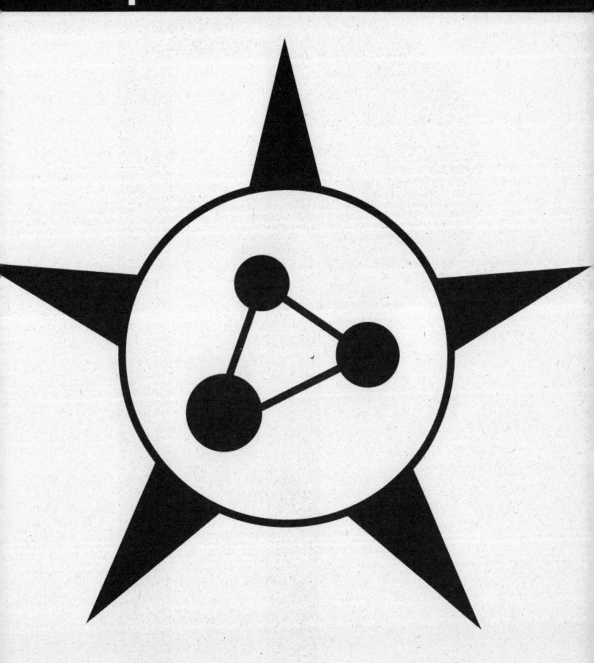

VIRUS

SUBJECT: A WAR IN HEAVEN

FROM: PETER LAMBORN WILSON (BY WAY OF DIANA MCCARTY <DIANA@MRF.HU>)
DATE: SUN OCT 25 07:42:12 EST 1998

As a historian of religion, it is extremely obvious to me that the internet is a religious phenomenon. This may not have occurred to everyone who is closer to it than I am. First of all, all technology can be analyzed according to religious principles. When I speak about religion, I am not speaking from the point of view of religion. All technology is a religious phenomenon: Why? Because unless you belong to the human condition, you cannot have technology. What is the human condition? What makes a human being different from an animal? I would say consciousness or self consciousness. One of the symptoms of consciousness, or self consciousness, is technology and it is impossible, structurally or historically, to separate technology from consciousness when we try to imagine what it is to be human. As soon as we see evidence in the archeological record of a Simian or a similar creature that we could identify as human, then the only reason why we do so is because there are some broken stones next to the bones, that look like they may have been intended to be tools. What separates animals from humans is technology. From one point of view, that is religion. Because you cannot have technology unless you can extricate consciousness outside the body. If you cannot understand that consciousness is something which projects outward into the world, you cannot create the prosthesis, the extension of the body, which is technology, be it a broken stone, or a computer. Because there is this intimate relationship between technology and consciousness, technology itself is always threatening to take the place of religion. Technology is always becoming confused with religion— the Marxists used to call this reification. It means making an intuition a "thing," making it "thingy," or giving it "thinginess." If we want to talk about the Greek word *technê*, it would be useful to describe the whole range of prosthesis of consciousness.

But, if we want to talk about technology, then we are moving into different ground.

Technology is *technê* plus *logos* in Greek. *Technê*, the technique or the mechanic principle plus the *logos*, or the word. If we are trying to find out what the first technology is, in the strict sense of the word, you would have to answer that it is writing, which adds the mechanic to the word. Therefore, there is no *technê*, but *technologia*. Then we see the process of reification that works immediately here. Writing itself defines words. Words do not define writing, but immediately a paradoxical feedback comes up, where writing defines words and words define things. Logically, it should be the other way around, but we know that language is a double edged sword. As a means of communication, language leaves a great deal to be desired. Heath Bunting said that "communication doesn't always communicate." Everyone can understand this immediately: a map is not a territory. As soon as you mistake the word *Budapest* on the map for the city of Budapest, you are in deep trouble. You have got a cognitive problem. If you want to talk about love, or patriotism, or valor, or truth, or communication, or the net, or freedom, or any words like that, which have very few references in the world of thinginess, you have a problem. We reify those concepts and solidify them in writing, in sign systems. Then they influence consciousness as you grow up, as a child learning language. All of these signs are imprinted. Even the alphabet, alphabetic writing, which is supposedly is not free of all images. When you move from the alphabet to binary writing, this is also not free of images. It is a very simple image system, black–white yes–no, but it is still an image system. The computer is still a machine of inscription, it is still a writing machine, in fact for most of you it is just a glorified typewriter. There is going to be a gradual process in the realm of technology of the reduction of the sign: from the complexity of a representational picture to the abstraction of a binary sign system which apparently no longer contains pictures, although we can see that the pictures are just more deeply buried. The Greek word for symbol, symbolon, actually means an object that is broken in half. That is why communication systems are not monodic or unitary, they are always dual or diadic. I prefer to say that all communications are diadic, it involves twoness. There must be a speaker and a hearer, then these relations can be reversed. The breaking of the symbolon symbolizes the split in human consciousness itself. A split between the animal intimacy that we can hypothesize as our Simian heritage, and the idea that consciousness and self are two different things. As soon as that split occurs we have a symbolic system at work, where one thing stands for another. The same holds true for all language systems, all musical systems, all dance systems, anything which can possibly communicate on any level whatsoever. These are all symbolic systems. Language is a symbolic system. All computer programs are symbolic systems. It is important to remember that in any symbolic system this split, the doubling of consciousness, the hypothesis of consciousness which is actually prosthesis, obtains something which is outside the body, and which can act in the world. In the history of religion, this desire for lost intimacy, this desire to recapture unified consciousness, is the cause of yet a further split. We see the whole idea of sacrifice that is meant to heal this wound in the cosmic structure. Sacrifice appears very early in human religion, at least as early as agricultural systems in the Neolithic Age, if not sooner, and it is violent. Initially, it probably involves human sacrifice. Whatever is religious is also inherently violent, because it's based on the split. The split consciousness, the act of splitting is violent, and so the act of repairing the split is also violent. In fact, the word religion, *religio*, in Latin, means to relink, which is really the same as the word in Hindi *yogo* which means yoke, as the yoke that connects two oxen. Religion itself, at its very base, is about this relinking of consciousness. It is an attempt to over come the split of consciousness and to unify what was doubled and make it one. This is a very violent process throughout human history, and it is not an accident that religions were associated with violence.

Most religions are systems of death consciousness because they posit a radical split between body and spirit, but they are no longer upset about it. They are not interested in reconciling the body and the spirit anymore. They are interested in eliminating one of those factors, the body, and perpetuating the other, the spirit, or mind, or perhaps information. So you have spirit and heaven at the top—and nature, body, and earth at the bottom. It becomes associated with the feminine; the catatonic, the chaotic, the uncultured, the uncultivated. It is associated with tribal societies, with hunting and gathering, with everything primitive, with everything despicable. Mind or spirit, which is now separated from the body, is associated with maleness; with power, with structure, with culture, with civilization, and with religion itself. What is in between is now only a technology of the sacred, the actual workings of religion itself. The ritual, the sacrifice, the priesthood, which is now a completely privileged closed off class; you now have class structure.

We now have the pyramidal structure, we now also have cyberspace. We have the concept of the virtual. Heaven or paradise, the mind principle, separated from the body, becomes cyberspace. Cyberspace is a version, paradoxical, or even a parody, of heaven. It's a place where your body is not present, but your consciousness is. It is a place of immortality, of not being mortal, of having over come death. There is a view that cyberspace is a salvational reality, that it saves us from our crude, shit-filled, rotting bodies, and that we will transcend into an angelic sphere of pure data where we will download consciousness and never die. If you have read William Gibson, the image is very clear: you have the hacker, who is jacked in, literally jacked into the computer. The body is rotting, but the cyberpersona is clearly immortal. The problem is that what we have been promised is transcendence through techno-mediation. It is a false transcendence. If we have a god, as in some forms of paganism, that has a material nature, the god is a rebirth. We will call that a eminent form of deity, as opposed to transcendent. What we are being offered in the net is not eminence, not a true eminence, but a false transcendence. It is a dangerous, Gnostic fallacy. Cyberspace is spurious immortality.

This brings me to the point of the military aspect of the net, because the net is actually a war in heaven. What else would the phrase "information war" mean than a war in heaven? A war that would take place in this spurious heaven, this false transcendence of cyberspace. We know that the net originates as a military space. The original ARPAnet was designed in order to avoid the physical disruption that would have been involved in atomic explosion. The net itself is a very Gnostic invention since it transcendentalizes matter in a very rapid and effective way. Basically, we are looking at a war in heaven. Kevin Kelly likes to say that this technology is out of control. This is bullshit, it's not out of control. It's something very different and much more interesting. A brilliant French anthropologist, Pierre Clastres, wrote one book called *Society Against the State*, and another, *The Archeology of Violence*. I follow his thinking very closely on a number of points. He makes a distinction between two kinds of warfare in human history: there is primitive war and classical war. These are not at all the same thing. It cannot even be said that the classical war is a development of the primitive war, it's rather a betrayal of primitive war. If the sacred is violent, then violence is not always negative, unless we believe in pacifism. There are certain kinds of violence which are positive, and primitive warfare is positive in this one sense. Clastres uses the metaphor of centrifugal and centripetal. The centrifugal machine is one that pushes out from the center, and the centripetal machine is one that pulls in toward the center. Clastres believed that this was a chosen path on the part of these societies. Consciously or unconsciously, these societies developed certain social functions to centrifugalize power, they don't want power, they refuse power. They want a society, but they don't want the state. They don't want the centralization of power, they don't want class structure, they don't want economic hierarchy. They want egalitarianism, they want democracy.

Some explanations have given the switch over of the hunting, gathering societies that are egalitarian without exception and do not practice sacrifice, with agricultural societies that are nonegalitarian and almost invariably do practice sacrifice. We are still living in the neolithic age. We are still basically living in the agricultural–industrial period and we still practice sacrifice. If you don't believe it, come to New York State, where they just reintroduced the death penalty, a symbolic sacrifice. At some point primitive warfare turns into classical warfare, and here is the interesting thing about the net. The net is born much more like a primitive warfare structure than a classical one, because of that strange Gnostic necessity to avoid atomic disintegration. The net suddenly turns into a space in which power is dispersed rather than centralized. They thought this was a brilliant strategy. It turned out that they lost control of the net almost instantly. That recentralization of power is going to have to come from outside the system.

This is my point about Kelly's thesis. That a technology, which is out of control as long as you study only the technology, is nothing new. The postal system is out of control. I can get much better security with snailmail now than I can on the net, that is one of the reasons I still don't own a computer. If somebody proved to me that I can really get top security by using a computer and I can send my evil revolutionary messages everywhere with complete safety, I would do it. All the people I knew in the sixties and seventies who were phone phreaking have moved on to the net. The telephone is so old-fashioned, it is just like hot and cold running water. No one is thinking about it at all, there is no mumbo jumbo in the telephone. There is no magic left in the telephone. The magic is all in the net, so that's what everybody wants to control. Mumbo jumbo is power, and if you control the base of a basic symbolic exchange system, you have power. Those who control the definition of words have power. Those who control the means of communication between you and me have power over both of us. Where is this control going to come from, if the system itself, the technology itself, is out of control. Because it was designed to be out of control, then the control has to come from outside the system. The internet is not heaven, the internet is not paradise. The internet is not safe, in terms of control, simply because as a closed system it represents the decentralization of power structures. That power can just reach in from out side, and that's exactly what the Church of Scientology can do. For example, the Church of Scientology can kill you, or disperse all your secrets, they can track you to your house and break in and smash your computers. And if you think that the Church of Scientology is powerful, wait until you hear from the U.S. government. And if you think that the U.S. government is a little outdated, and that as John Perry Barlow says, that governments are not the corporate entities ideally designed to control the new technology, then wait until you hear from AT&T, because they are designed to control. It is far worse.

Since 1989, there is not an ideological struggle in the world. The night the Berlin Wall fell, I turned on the television and I heard that the Cold War was over and we won. History itself, which involved the dialectical struggle, according to Hegel, is now over. The Cold War is over and we, the capital, won. There is now only one ideology that disguises itself as nature. Once again we have a false transcendence of bringing together culture and nature, in a totally phony way, where you can establish a more efficient control mechanism. The net can be controlled from outside, through fear, through terror. The net is extremely susceptible to terror, because the net is a religious phenomenon and religion is inherently violent, the sacred is inherently violent, and invariably both are involved in fear, in terror. That's why the net is perfect ground, *Grund*, in German, for the passion play that is going to occur within five years, maybe within the next five minutes. The net can be controlled from outside, and therefore, resistance must be organized from outside.

So far, we've only had virtual resistance, and actually that is no more than a spectacle of resistance. If we don't organize on the basis of politics, and of economy, then the net has no future as a space for human freedom. No future. So far, I don't see that organizing going on. I see that the most brilliant minds that are involved in the net are all involved in cryptography and PGP, and various kinds of mechanisms, which are meant to protect the net from takeover from within the net, but that's not what the danger is coming from. Sooner or later, some body will figure it out and it better be us because if it isn't, then it's going to be AT&T with six hundred channels and a hundred home shopping networksx. Or riskier, are those heavy-footed, jack-booted gov-

ernments, or the Church of Scientology. So the net is not heaven, the body must be present. I love Heath Bunting's point that, without the presence of body, this whole thing is just a curious form of metaphysical schlock with cream. Whoever understands the net as religion, whoever understands the problem with body and reembodiment, will have a tremendous edge, or at least gain an edge in the struggle of whether the net remains a space of potential freedom, or whether it doesn't. Whoever can understand this, whoever can understand the reason why the state will be the first to lose control of the net?

I would like to think about the economics for a minute. We see that money is also going to heaven. Billions of billions of billions of billions of billions of whatever units of money are there, floating around in cyberspace. Money is now a purely transcendental principle, it's a symbolic system, it's a *symbolon*, just like any other symbol. It is broken into two halves and has meaning only if the two halves are reunited. That's where money begins, precious metal, which has no inherent value whatsoever. The relationship between gold and silver, from the start, is based on the lunar solar cycle. It is pure symbolism. The first coins were temple souvenirs. This is historically known to numismatics experts studying the history of coinage. The first coins are souvenirs, they are picked up in temples and that coin, that image, becomes valuable as nostalgia. You can take them home and trade one of them for a cow, because it's like mumbo jumbo. It's called JuJu. *Mumbo jumbo* and *JuJu* are African words for mysterious power. The coins themselves, which still have a memorable, *valuata* aspect, are made out of precious metal, which is gradually added to less precious metal. Presume coins are largely symbolic, they could change to paper which represents the coins. Then in 1933, in America, the link between the paper and the precious metal is cut, paper is now floating free. It's a reference without any referent, and we now have purely abstract money, ready to jack in. Ready to ascend to heaven, to the heaven of cyberspace, and that's exactly what's happened. Ninety percent of all commercial transactions are electronic and do not involve any form of paper. They are in a world where imagination and electricity interrelate in some strange and metaphysical way. Coins become papers become absence. Finally, there is an absence itself, valued as a form of money, in a kind of a reverse alchemy, changing precious metals into nothing.

In this regard, my favorite story is about the alchemist, Paracelsus, who was traveling through Germany and was invited into the court of one of those petty German princes of the fifteenth century, who said, "Oh, Mr. Paracelsus, great to meet you. We've heard so much about you. You're such a great scientist, we'd like to set you up with your own laboratory here." I don't remember the details, but Paracelsus says, "Oh you must set me up in a laboratory! What do you want me to do?" The king says, "Oh, you had this lead into gold thing. This base metal and precious metal experiment...We are very interested in that." Paracelsus says, "Oh, your Majesty, your Majesty, I am just a Puffer. You, your Majesty, you are the real alchemist." "Why?" "This is because all you have to do is give a license to a bank to lend money. That is gold out of nothing." That was in fifteenth century. It took another couple of hundred years for the Bank of England to be established on that basis. Now all banks in the world can lend up to ten times the amount of money that they have in the vault. It's probably just a hard disk somewhere, so you can take ten times nothing and call it a dollar and change it into a dollar. That's alchemy. Whoever understands that money is also religion, will also gain in the struggle. This lecture was meant to be called "Islam and the Net," I should say something about that. First of all, you probably remember that the Iranian Revolution was entirely based on the cassette tape recorder. If you don't know yet, I'm going to tell you. Khomeni would not have held power in Iran without the cassette tape recorder. He was in exile in Iraq and sent recordings of his sermons, which attacked the Shah, to Iran. The tapes were spread around in a network from mosque to mosque and from cassette recorder to cassette recorder. That was the chief weapon of the Iranian Revolution. There was very little blood involved in that revolution. A very serious revolutionary movement was carried out entirely through communications technology. Just think what they can do with the net. Just think what terrorists can do with the net. The net, to answer the questions of our friends from for-mer Yugoslavia, The net will never reach this world in time. There will always be lag time. The net, the marvelous miracle of communication which might be some utopian reading of the situation, will never reach the other 99 percent of the world in time. The reason that it will never come to save the world, like a miracle, is that terrorists will invade the net. They will be representative of all of the outside, and the outside includes all the countries where the people don't even have telephones. This is all the outside, the outside is all demonic for the inside, and therefore the technology will not be transferred, because that would be asking angels to transfer their technologies to devils. It's not going to happen unless religious power itself is deconstructed or overcome. Because it's religion that has prevented the net from arriving in time to save.

It's a religious problem. We can deconstruct the religious aspect of technology. We can stop reifying technology, and worshipping it. This is a religious paradise, you can't save your soul from technology, unless you know that technology can't save it. An act, even more paradoxically, the process of over coming, can only be to understand and even more paradoxical, this process of overcoming can be carried out through religious means. In other words, we have to understand the power of the imagination to create values. It is, in fact, through imagination and only through imagination, that values are created. If we understand that, we are free. We, as least as individuals, then are free in some meaningful sense. Maybe not free of incompetence, but in some sense we are free. Communication doesn't communicate. Communication as noise. Communication as cognitive dissonance causes separation. Mediation causes alienation. You can't mediate beyond a certain extent. All forms of communication are mediated, even if I speak with you. It's moving through the air and the molecules of the air are carrying sound to your ears. Simple conversation is already mediated, but you can carry that mediation, you can excaberate to a point where it becomes alienation, where you are actually violently separated or split from other people. Mediation which becomes alienation is then reproduced in the media, so the television, newspapers, the internet, all forms of communication, as a media, in the usual sense of that word, simply increase alienation, and of course, wherever advertising comes in, it is very easy to see how this happens. It is very easy to understand how the net itself has become a source of horrible alienation, once advertising has taken it over, once the ones in Rubeca have moved in, once Disney and CocaCola have moved in and taken it over. We even have to go back to language itself. We have to work on language, this is the job of the poet, to clarify the language of the tribe, not purify, but to clarify. We still need ideology in some sense, in that we need ideas, and that we need a *logos*, or a word, or an expression of those ideas. I would prefer to end by referring these problems to Mikhail Bahktin, the Russian critic, who uses the word *dialogics*. I like this word because it doesn't bring in any ideological frame. It's a new, fresh word. It means conversation—it means high value relating. We call it dialogics because it sounds like something we haven't thought of before.

To me, it's just a good, old nineteenth-century American word, communicativeness. Communicativeness is not necessarily the same thing as simple communication. It implies warmth, a human pres-ence, an actual desire, a pleasure, a joy, a jouisance, if you like, of communication. Communicativeness is erratic, essentially, and festive. This is what Bahktin wanted us to remember, that the spiritual path of the material, the body of principle, this is something real. The material body itself, is in effect, a symbol. It is a spiritual principle, and that, if you going to overcome the religious problem, which is to split the body off from the mind, forever. What we need more than anything else, is a spirituality *of* the body *for* the body. A re-enchantment of the nat-ural. Re-enchantment means singing, music. I am not proposing any kind of dialectical materialism or reductionism here. Actually, I am interested in a remytholization, in re-enchantment, in magic, in action at a distance. I am interested in technology because it is magical, it is magic, it is action at a dis-tance. What I want to see is this technology used to reenchant nature, and finally, hopefully, to sacrifice the violence of the sacred.

[Transcript of a lecture given at MetaForum II, Budapest, 1995. Transcribed by Pit Schultz. Edited by Diana McCarty.]

SUBJECT: FROM *FAMA* TO INFORMATION SOCIETY: OF PROPHETS, GODS, AND THE NETTIME SERVER DEMON

FROM: FLORIAN CRAMER
DATE: FRI, 25 SEP 1998 17:28:16 +0200

The concept of information society not only focus-es new media prophecies, politics and business. It also seems central to "net criticism" and "net cul-ture" as they are discussed in Nettime. In the archives of the mailing list, "information society" is typically referred to as an either present or emerg-ing reality: a reality to be reassessed with alterna-tive, critical or at least noncorporate visions.

As a social utopia, information society however predates the Internet and its prophets and critics. In the seventeenth century, the Protestant scholars Johann Valentin Andreae, Jan Amos Comenius, and Samuel Hartlib developed a general program to inform mankind. Their project was outlined in Andreae's 1619 pamphlet *Turris Babel* ("The Tower of Babel"), a dialogical satire on Rosicrucianism. The Rosicrucian reformation of mankind had first been proclaimed five years earlier in the *Fama fra-ternitatis* among whose anonymous authors had been Andreae himself. He soon had to witness how his fiction took up a life of its own. More than 150 replies appeared until 1619 whose authors sought to get in touch with the hermetic brotherhood.

With *Turris Babel*, Andreae joins the debate and mocks the craze he had created. But instead of declaring himself the author of the *Fama*, he brings up seventy-five allegorical protagonists who each pronounce their own opinion about the Rosicrucians. In chapter sixteen, three characters enter the scene, the "reformator," the "deformator," and the "informator." While the deformator wants to do away with all traditional ties and institutions including church and state, the reformator hopes for their restoration through the Rosicrucians. The informator finally supersedes their debate by demanding to "inform" mankind so that "the divine law will be saved from the deformator's corruption and the reformator's eagerness and become the constitution of this world."

"Information" refers to its Latin root here; it reads as "impregnation," "shaping," or "instruction." The informist is an agent of a new *Christiana societas*, which the final chapter of *Turris Babel* and Andreae's subsequent writings proclaim. The Rosicrucians give way to the Christian Society, and *fama* is followed by information, or, education. In the ideal state of this information society, Andreae's utopian republic, all knowledge is denoted in public mural paintings. The information and impregnation of society follows, one could say, the logic of a push channel. Pedagogics becomes the master discipline of this project because it provides the programming tools. In 1620, Andreae writes his educational treatise *Theophilus*; but it were his disciples and confrères Comenius and Hartlib who succeeded in rewriting pedagogics into a new universal science. With the plans of the *Christiana societas* failing last in England, Andreae's followers rescue the technologies of their information utopia into public education. Comenius turns the "view houses" of Christianopolis into an *Orbis pictus* ("The World in Pictures"), the first illustrated children's primer. Until the late eighteenth century, the *Orbis pictus* remains the canonical schoolbook in Europe.

What does the post-Rosicrucian information society have in common with the postmodern information society net prophets and "net critics" describe? Defined against deformation, reformation and *fama*, Andreae's information is not only loaded with pedagogics and theology; more than that, its definition is radically performative. It implies that information is only what has an impact, reaching and impregnating its recipients. This notion is surprisingly modern in its affinity to Shannon's definition of information as anti-redundance. Here, information is not a self-referential plaything. It implies a vertical power relation between informants and the informed, between source and receivers. Infor-

mation comes from the source, it is radically original. To speak originally, the informant must avoid redundant overlapping with the knowledge of the informed; he must speak from a remote place and dwell outside society. Unlike other information societies, Andreae's *Christiana societas* makes no attempt at concealing this place, but labels it "heaven" and calls the informant "God."

Andreae's information society does not inform itself, it is being informed. But is this also the case in contemporary information societies? Can an information society be made a society of informants, instead of a society of the informed? According to the Latin etymology of the word, society is a body of companions (*socii*) who follow (*sequi*) each other. Society thus rests upon smoothed out paths. If smoothing out implies redundance whereas information translates, according to Andreae and Shannon, as anti-redundance, it follows that information and society are contradictions. Andreae's Christian information society resolves this contradiction by secluding the informant from itself. A society founded upon its self-information however—that is, a society founded upon radical originality instead of redundances or a remote informant—cannot communicate. It would not be a society.

Perhaps those who speak of information society today don't use the word *information* in Shannon's or Andreae's rigorous sense, but identify "information" with "signs." As "signs," "information" would comprehend noise as well as signals, fuzziness as much as focus. But in this case, "information society" would no longer make a difference. It would not describe any departure from the habitual signal–noise economics of "society"; it would exhaust itself in a buzzword. But perhaps the question is not whether "information society" is only a buzzword or whether a self-informing information society would be a contradiction in itself. If one acknowledges that the concept of "information society" has political impact nevertheless, then the more relevant conclusion is that no "information society" which is more than a buzzword can do without transcendental informants.

When presupposing information society as a present or emerging reality, "net criticism" and "net culture" do not only operate with the same theoretical dispositive as net prophecy. They also participate, nilly-willy, in the political theology inscribed into its very concept. "Net critics" and net prophets coincide where they pretend to do without transcendental informants, but continue to employ them. When Geert Lovink and Pit Schultz presented their concept of "net culture" and "net criticism" in a panel speech for a congress that accompanied Documenta X in summer 1997, they defended "the net" against traditional academia all the while calling upon academics to go online. Given the academic surrounding and sponsorship of the event, the audience interpreted this as undeserved polemics. It failed to recognize that, instead of a university lecture, it had witnessed a perfect reenactment of the Rosicrucian *Fama*, its bold rhetoric, its general critique of culture and its final appeal to the scholars of the world. The speakers

had furthermore observed the Rosicrucian rules of curing everyone without charging money, wearing innocuous clothing and speaking the local idiom in each country they visit in order to keep their theological mission under the hood.

The next logical step after the *Fama* is Nettime writing itself as a dialogical satire of its own discourse. When the discourse of "net criticism" generates the very critical "net culture" it reflects, and when the discourse of net prophecy generates the very affirmative "net culture" it reflects, and vice versa, it seems as if the "information societies" addressed both in "net prophecy" and "net criticism" are, first of all, self-descriptions. They emerge as romantic symbols: demonic and divine hieroglyphs, shining bright in the rigorous sun of Telechristianopolis.

SUBJECT: THE MANIFESTO OF JANUARY 3, 2000

FROM: BRUCE STERLING <BRUCES@WELL.COM>
DATE: WED, 23 SEP 1998 10:54:04 -0700 (PDT)

The rapidly approaching millennium offers a unique cultural opportunity. After many years of cut-and-paste, appropriation, detournement and, neo-retro ahistoricality, postmodernity is about to end. Immediately after the end of the fin-de-siècle, there will be a sudden and intense demand for genuine novelty.

I suspect that a group that can offer a coherent, thoughtful and novel cultural manifesto on the target date of January 3, 2000, has a profound opportunity to affect the zeitgeist. (On January 1, everyone will be too hung over to read manifestos; on January 2, nobody's computers will work. So naturally the target date must be January 3.) In this preliminary document, I would like to offer a few thoughts on the possible contents of such a manifesto. The central issue as the new millennium dawns is technocultural. There are of course other, more traditional, better-developed issues for humankind. Cranky fundamentalism festers here and there; the left is out of ideas while the right is delusional; income disparities have become absurdly huge;

these things are obvious to all. However, the human race has repeatedly proven that we can prosper cheerfully with ludicrous, corrupt, and demeaning forms of religion, politics, and commerce. By stark contrast, no civilization can survive the physical destruction of its resource base. It is very clear that the material infrastructure of the twentieth century is not sustainable. This is the issue at hand.

We have a worldwide environmental problem. This is a truism. But the unprecedentedly severe and peculiar weather of the late nineties makes it clear that this problem is growing acute. Global warming has been a lively part of scientific discussion since at least the sixties, but global warming is a quotidian reality now. Climate change is shrouding the globe in clouds of burning rainforest and knocking points off the GNP of China. Everyone can offer a weird weather anecdote now; for instance, I spent a week this summer watching the sky turn gray with fumes from the blazing forests of Chiapas. The situation has been visibly worsening, and will get worse yet, possibly very much worse. Society has simply been unable to summon the political or economic will to deal successfully with this problem by using twentieth-century methods. That is because CO2 emission is not centrally a political or economic problem. It is a design and engineering problem. It is a cultural problem and a problem of artistic sensibility.

New and radical approaches are in order. These approaches should be originated, gathered, marshaled into an across-the-board cultural program, and publicly declared—on January 3rd.

Global warming is a profound opportunity for the twenty-first-century culture industry. National governments lack the power and the will to impose dirigiste solutions to the emission of carbon dioxide. Dirigiste solutions would probably not work anyway. It is unlikely that many of us could tolerate living in a carbon-dioxide Ration State. It would mean that almost every conceivable human activity would have to be licensed by energy commissars.

Industry will not reform its energy base. On the contrary, when it comes to CO2 legislation, industry will form pressure groups and throw as much sand as possible into the fragile political wheels. Industry will use obscurantist tactics that will mimic those of American right-wing anti-evolution forces—we will be told that global warming is merely a "theory," even when our homes are on fire. Industry is too stupid to see planetary survival as a profit opportunity. But industry is more than clever enough to sabotage government regulation, especially when globalized industry can play one government off against the next.

With business hopeless and government stymied, we are basically left with cultural activism. The tools at hand are art, design, engineering, and basic science: human artifice, cultural and technical innovation. Granted, these may not seem particularly likely sources of a serious and successful effort to save the world. This is largely because, during the twentieth century, government and industry swelled to such tremendous high-modernist proportions that these other enterprises exist mostly in shrunken subcultural niches.

However, this doesn't have to be the case. With government crippled and industry brain-dead to any conceivable moral appeal, the future of decen-tered, autonomous cultural networks looks very bright. There has never been an opportunity to spread new ideas and new techniques with the alacrity that they can spread now. Human energy must turn in some direction. People will run from frustration and toward any apparent source of daylight. As the planet's levees continue to break, people will run much faster and with considerably more conviction.

It's a question of tactics. Civil society does not respond at all well to moralistic scolding. There are small minority groups here and there who are perfectly aware that it is immoral to harm the lives of coming generations by massive consumption now: deep Greens, Amish, people practicing voluntary simplicity, Gandhian ashrams, and so forth. These public-spirited voluntarists are not the problem. But they're not the solution either, because most human beings won't volunteer to live like they do. Nor can people be forced to live that way through legal prescription, because those in command of society's energy resources will immediately game and neutralize any system of legal regulation. However, contemporary civil society can be led anywhere that looks attractive, glamorous, and seductive.

The task at hand is therefore basically an act of social engineering. Society must become Green, and it must be a variety of Green that society will eagerly consume. What is required is not a natural Green, or a spiritual Green, or a primitivist Green, or a blood-and-soil romantic Green.

These flavors of Green have been tried, and have proven to have insufficient appeal. We can regret this failure if we like. If the semiforgotten energy crisis of the seventies had provoked a wiser and more energetic response, we would not now be facing a weather crisis. But the past's well-meaning attempts were insufficient, and are now part of the legacy of a dying century.

The world needs a new, unnatural, seductive, mediated, glamorous Green. A Viridian Green, if you will.

The best chance for progress is to convince the twenty-first century that the twentieth century's industrial base was crass, gauche, and filthy. This approach will work because it is based in the truth. The twentieth century lived in filth. It was much like the eighteenth century before the advent of germ theory, stricken by septic cankers whose origins were shrouded in superstition and miasma. The truth about our physical existence must be shown to people. It must be demonstrated repeatedly and everywhere.

The central target for this social engineering effort must be the people who are responsible for emitting the most CO_2. The people we must strive to affect are the ultrarich. The rentiers, the virtual class, the captains of industry; and, to a lesser extent, the dwindling middle classes. The poor will continue to suffer. There is clearly no pressing reason for most human beings to live as badly and as squalidly as they do. But the poor do not emit much carbon dioxide, so our efforts on their behalf can only be tangential.

Unlike the modernist art movements of the twentieth century, a Viridian culture-industry movement cannot be concerned with challenging people's aesthetic preconceptions. We do not have the nineteenth-century luxury of shocking the bourgeoisie. That activity, enjoyable and time-honored though it is, will not get that poison out of our air. We are attempting to survive by causing the wealthy and the bourgeoisie to willingly live in a new way.

We cannot make them do it, but if we focussed our efforts, we would have every prospect of luring them into it.

What is culturally required at the dawn of the new millennium is a genuine avant-garde, in the sense of a cultural elite with an advanced sensibility not yet shared by most people, who are creating a new awareness requiring a new mode of life. The task of this avant-garde is to design a stable and sustainable physical economy in which the wealthy and powerful will prefer to live. Mao suits for the masses are not on the Viridian agenda. Couture is on the agenda. We need a form of Green high fashion so appallingly seductive and glamorous that it can literally save people's lives. We have to gratify people's desires much better than the current system does. We have to reveal to people the many desires they have that the current system is not fulfilling. Rather than marshalling themselves for inhuman effort and grim sacrifice, people have to sink into our twenty-first century with a sigh of profound relief. Allow me to speak hypothetically now, as if this avant-garde actually existed, although, as we all know, it cannot possibly come into being until January 3, 2000. Let's discuss our tactics. I have a few cogent suggestions to offer.

We can increase our chances of success by rapidly developing and expanding the postmodern culture industry. Genuine "culture" has "art" and "thought," while the culture industry merely peddles images and information.

I know this. I am fully aware of the many troubling drawbacks of this situation, but on mature consideration, I think that the culture industry has many profound advantages over the twentieth century's physically poisonous smokestack industries. Also, as digital technologists, thinkers, writers, designers, cultural critics, *und so weiter*, we Viridians suspect that the rise of the culture industry is bound to increase our own immediate power and influence vis-à-vis, say, coal mining executives. This may not be an entirely good thing. However, we believe we will do the world less immediate damage than they are doing.

We therefore loudly demand that the culture industry be favored as a suitably twenty-first century industrial enterprise. Luckily the trend is already very much with us here, but we must go further; we believe in Fordism in the culture industry. This means, by necessity, leisure. Large amounts of leisure are required to appreciate and consume cultural-industrial products such as movies, software, semifunctional streaming media, and so on. Time spent at more traditional forms of work unfairly lures away the consumers of the culture industry, and therefore poses a menace to our postindustrial economic underpinnings.

"Work" requires that people's attention to be devoted to other, older, less attractive industries. "Leisure" means they are paying attention and money to us.

We therefore demand much more leisure for everyone. Leisure for the unemployed, while copious, is not the kind of "leisure" that increases our profits. We specifically demand intensive leisure for well-educated, well-heeled people. These are the people who are best able to appreciate and consume truly capital-intensive cultural products. We Viridians suspect that it would require very little effort to make people work much less. Entirely too much effort is being spent working. We very much doubt that there is anything being done in metal-bending industry today that can justify wrecking the atmosphere. We need to burn the planetary candle at one end only (and, in daylight, not at all).

As much time as possible should be spent consuming immaterial products. A global population where the vast majority spend their time sitting still and staring into screens is a splendid society for our purposes. Their screens should be beautifully designed and their surroundings energy-efficient. The planet will benefit for everyone who clicks a mouse instead of shoveling coal or taking an axe and a plow to a rainforest.

The tourist industry is now the number one industry on the planet. Tourists consume large amounts of prepackaged culture. We believe tourism to be a profoundly healthy development. We feel we must strongly resist the retrograde and unprofitable urge to make migrants and migration illegal.

We believe that the movement of human beings across national boundaries and under the aegis of foreign governments is basically a design problem. If guest workers, refugees, pleasure travelers, and so forth were all electronically tracked via satellite or cell repeaters, the artificial division between jet setters and refugees would soon cease to exist. Foreigners are feared not merely because they are foreign, but because they are unknown, unidentified, and apparently out of local social control.

In the next century, foreigners need be none of these things. Along with their ubiquitous credit cards and passports, they could carry their entire personal histories. They could carry devices establishing proof of their personal bona fides that would be immediately obvious to anyone in any language. A better designed society would accommodate this kind of human solidarity, rather than pandering to the imagined security needs of land-based national regimes.

We believe that it should be a general new design principle to add information to a problem, as opposed to countering it with physical resources (in the case of migrants, steel bars, and barbed wire). Electronic tracking seems a promising example. While the threat to privacy and anonymity from electronic parole is obviously severe, there is noth-ing quite so dreadful and threatening as a septic refugee camp. We consider this a matter of some urgency. We believe it to be very likely that massive evacuations will occur in the next few decades as a matter of course, not merely in the disadvantaged Third World, but possibly in areas such as a new American Dust Bowl. Wise investments in electronic tourist management would be well repaid in stitching the fraying fabric of a weather-disrupted civilization.

For instance, we would expect to see one of the first acts of twenty-first-century disaster management to be sowing an area with air-dropped and satellite-tracked cellphones. We believe that such a tracking and display system could be designed so that it would not be perceived as a threat, but rather as a jet-setter's prestige item, something like a portable personal webpage. We believe such devices should be designed first for the rich. The poor need them worse, but if these devices were developed and given to the poor by socialist fiat, this would be (probably correctly) suspected as being the first step toward police roundup and a death camp.

Replacing natural resources with information is a natural area for twenty-first-century design, because it is an arena for human ingenuity that was technically closed to all previous centuries. We see considerable promise in this approach. It can be both cheap and glamorous.

Environmental awareness is currently an annoying burden to the consumer, who must spend his and her time gazing at plastic recycling labels, washing the garbage, and so on. Better information environments can make the invisible visible, however, and this can lead to a swift re-evaluation of previously invisible public ills.

If one had, for instance, a pair of computerized designer sunglasses that revealed the unspeakable swirl of airborne combustion products over the typical autobahn, it would be immediately obvious that clean air is a luxury. Infrasound, ultrasound, and sound pollution monitors would make silence a luxury. Monitor taps with intelligent water analysis in real-time would make pure water a luxury. Lack of mutagens in one's home would become a luxury.

Freedom from interruption and time to think is a luxury; personal attention is luxury; genuine neighborhood security is also very much to be valued. Social attitudes can and should be changed by the addition of cogent information to situations where invisible costs have long been silently exported into the environment. Make the invisible visible. Don't sell warnings. Sell awareness.

The fact that we are living in an unprecedently old society, a society top-heavy with the aged, offers great opportunity. Long-term thinking is a useful and worthwhile effort well suited to the proclivities of old people.

Clearly if our efforts do not work for old people (a large and growing fraction of the G7 populace) then they will not work at all. Old people tend to be generous, they sometimes have time on their hands. Electronically connected, garrulous oldsters might have a great deal to offer in the way of managing the copious unpaid scutwork of electronic civil society. We like the idea of being a radical art movement that specializes in recruiting the old.

Ignoring long-term consequences is something we all tend to do; but promulgating dangerous falsehoods for short-term economic gain is exceedingly wicked and stupid. If environmental catastrophe strikes because of CO2 emissions, then organizations like the anti-green Global Climate Coalition will be guilty of negligent genocide. Nobody has ever been guilty of this novel crime before, but if it happens, it will certainly be a crime of very great magnitude. At this moment, the GCC and their political and economic allies are, at best, engaged in a risky gamble with the lives of billions. If the climate spins out of control, the twenty-first century may become a very evil place indeed.

The consequences should be faced directly. If several million people starve to death because, for instance, repeated El Niño events have disrupted major global harvests for years on end, then there will be a catastrophe. There will be enormous political and military pressures for justice and an accounting.

We surmise that the best solution in this scenario would be something like the Czech lustration and the South African truth commissions. The groundwork for this process should begin now. The alternatives are not promising: a Beirut scenario of endless ulcerous and semicontained social breakdown; a Yugoslav scenario of climate-based ethnic cleansing and lebensraum; a Red Terror where violent panic-stricken masses seek bloody vengeance against industrialism. Most likely of all is a White Terror, where angry chaos in the climatically disrupted Third World is ruthlessly put down by remote control by the G7's cybernetic military. It is very likely under this last scenario that the West's gluttonous consumption habits will be studiously overlooked, and the blame laid entirely on the Third World's exploding populations. (The weather's savage vagaries will presumably be blamed on some handy Lysenkoist scapegoat such as Jews or unnatural homosexual activities.)

With the Czech lustration and the South African truth commissions, the late twentieth century has given us a mechanism by which societies that have drifted into dysfunctional madness can be put right. We expect no less for future malefactors whose sly defense of an indefensible status quo may lead to the deaths of millions of people, who derived little benefit from their actions and were never given any voice in their decisions. We recommend that dossiers be compiled now, for the sake of future international courts of justice. We think this work should be done quite openly, in a spirit of civic duty. Those who are risking the lives of others should be made aware that this is one particular risk that will be focused specifically and personally on them.

While it is politically helpful to have a polarized and personalized enemy class, there is nothing particularly new about this political tactic. Revanchist sentiment is all very well, but survival will require a much larger vision. This must become the work of many people in many fields of labor, ignoring traditional boundaries of discipline and ideology to unite in a single practical goal—climate.

A brief sketch may help establish some parameters. Here I conclude with a set of general cultural changes that a Viridian movement would likely promulgate in specific sectors of society. For the sake of brevity, these suggestions come in three parts. *Today* is the situation as it exists now. *What we want* is the situation as we would like to see it. *The trend* the way the situation will probably develop if it follows contemporary trends without any intelligent intervention.

THE MEDIA

Today: Publishing and broadcasting cartels surrounded by a haze of poorly financed subcultural microchannels.

What we want: More bandwidth for civil society, multicultural variety, and better-designed systems of popular many-to-many communication, in multiple languages through multiple channels.

The trend: A spy-heavy, commercial internet. A Yankee entertainment complex that entirely obliterates many non-Anglophone cultures.

THE MILITARY

Today: G7 Hegemony backed by the U.S. military.

What we want: A wider and deeper majority hegemony with a military that can deter adventurism, but specializes in meeting the immediate crises through civil engineering, public health and disaster relief.

The trend: Nuclear and biological proliferation among minor powers.

BUSINESS

Today: Currency traders rule banking system by fiat; extreme instability in markets; capital flight but no labor mobility; unsustainable energy base.

What we want: Nonmaterial industries; vastly increased leisure; vastly increased labor mobility; sustainable energy and resources.

The trend: Commodity totalitarianism, crony capitalism, criminalized banking systems, sweatshops.

INDUSTRIAL DESIGN

Today: Very rapid model obsolescence, intense effort in packaging; CAD/CAM.

What we want: Intensely glamourous environmentally sound products; entirely new objects of entirely new materials; replacing material substance with information; a new relationship between the cybernetic and the material.

The trend: two design worlds for rich and poor consumers; a varnish on barbarism.

GENDER ISSUES

Today: More commercial work required of women; social problems exported into family life as invisible costs.

What we want: Declining birth rates, declining birth defects, less work for anyone, lavish support for any-one willing to drop out of industry and consume less.

The trend: More women in prison; fundamentalist and ethnic-separatist ideologies that target women specifically.

ENTERTAINMENT

Today: large-scale American special-effects spectacle supported by huge casts and multi-million-dollar tie-in enterprises.

What we want: Glamour and drama; avant-garde adventurism; a borderless culture industry bent on Green social engineering.

The trend: Annihilation of serious culture except in a few non-Anglophone societies.

INTERNATIONAL JUSTICE

Today: Dysfunctional but gamely persistent war crimes tribunals.

What we want: Environmental Crime tribunals.

The trend: justice for sale; intensified drug war.

EMPLOYMENT

Today: MacJobs, burn-out track, massive structural unemployment in Europe.

What we want: Less work with no stigma; radically expanded leisure; compulsory leisure for workaholics; guaranteed support for people consuming less resources; new forms of survival entirely outside the conventional economy.

The Trend: Increased class division; massive income disparity; surplus flesh and virtual class.

EDUCATION

Today: Failing public-supported schools.

What we want: Intellectual freedom, instant cheap access to information, better taste, a more advanced aesthetic, autonomous research collectives, lifelong education, and dignity and pleasure for the very large segment of the human population who are and will forever be basically illiterate and innumerate.

The trend: Children are raw blobs of potential revenue-generating machinery; universities exist to supply middle-management.

PUBLIC HEALTH

Today: General success; worrying chronic trends in AIDS, tuberculosis, antibiotic resistance; massive mortality in nonindustrial world.

What we want: Unprecedently healthy old people; plagues exterminated worldwide; sophisticated treatment of microbes; artificial food.

The trend: Massive dieback in Third World, septic poor quarantined from nervous rich in G7 countries, return of nineteenth-century sepsis, world's fattest and most substance-dependent populations.

SCIENCE

Today: Basic science sacrificed for immediate commercial gain; malaise in academe; bureaucratic overhead in government support.

What we want: Procedural rigor, intellectual honesty, reproducible results; peer review, block grants, massively increased research funding, massively reduced procedural overhead; genius grants; single-author papers; abandonment of passive construction and the third person plural; "Science" reformed so as to lose its Platonic and crypto-Christian elements as the "pure" pursuit of disembodied male minds; armistice in Science wars.

The trend: "Big Science" dwindles into short-term industrial research or military applications; "scientists" as a class forced to share imperilled, marginal condition of English professors and French deconstructionists.

I would like to conclude by suggesting some specific areas for immediate artistic work. I see these as crying public needs that should be met by bravura displays of raw ingenuity. But there isn't time for that. Not just yet.

SUBJECT: THE GREAT TIME SWINDLE! EUROPEAN HISTORY A FAKE! A PATCH FOR THE MILLENIUM BUG PROBLEM?

FROM: EL IBLIS SHAH <ELIBLIS@TO.OR.AT> DATE: TUE, 04 AUG 1998 16:52:51 +0100

There has been quite some media attention in German-speaking countries on discoveries of large scale medieval forgeries of scriptures, official documents and codices. Especially in twelfth-century Europe they were widely used as instruments of political legitimization and psychological propaganda. A large amount of "anticipatory" forgery raised questions: documents that were supposedly faked in the dark ages (sixth–ninth century) but with too many details on later events to be explained as self-fulfilling prophecies. A cultural time-warp based on symbol manipulation?

A recent book on "The Invention of the Middle Ages, or the greatest forgery of time in History" (1) is widely discussed and has made it to the circuit of cultural magazine formats on TV (H. Illig, *Das Erfundene Mittelalter die grosste Zeitfaelschung der Geschichte*, Düsseldorf, 1996 <http://home.ivm.de/≠ ~Guenter/illig.html>). The controversial thesis of this publication claims that the dark ages were so dark as to be practically non-existent. Especially for the period between 611–914, there is no hard evidence that anything ever happened within that time. According to the author, the assumption of an invented time is supported by the fact that the Gregorian calendar reform in the sixteenth century only corrected ten days, instead of the necessary twelve, seven, or thirteen days for the three centuries in question.

Those of us who felt a deep unease about the new millenium can cheer up—according to this research we are just about to enter the eighteenth century, its approximately 1695.

In our fast-paced time the accumulated wealth

of an extra three hundred years on a time-bank could proof extremely valuable. This is not only an instant cure for Millenium Madness, but also a simple solution for the Millenium Bug in computer operating systems.

At the center of research is Charles the Great, Charlemagne, the unique emperor of European unity in the eighth century.

The larger-than-life tasks ascribed to Charles the Great, from his physical qualities, to his intellectual capacity, his financial power, military success, and spiritual status clearly belong in the realm of the fantastic and truly superhuman.

Many of the wondrous accomplishments seem to be totally incompatible with the reality of an economically weak and poorly developed Europe with an undeveloped trade and an inadequate communication and money system. The rather bleak scenario shows hardly any urban centers within the ruins of the roman developments.

A huge collection of circumstantial evidence is brought forward to prove that his grand empire is really fictional and a detailed archeological analysis questions the authenticity of all assumed eighth- and ninth-century architecture.

Charles the Great, the supposed descendant of "the House of David" (yes, that's Jesus supposed bloodline, the messianic legacy) is debunked as a mythological figure and indeed as an only legendary "God-king." In short—the greatest historical figure of the middle ages is about as real as Father Christmas.

A tongue-in-cheek Egon Friedell is quoted on the book cover saying: "Groundbreaking revelations are much less to be expected in recent history than in ancient history because of the long time span involved." Obviously he did not know about the millenium bug and a world where operating systems are in a delicate balance of instability.

Needless to say the research of this group of deep time-warp historians is challenging the foundations of all canonical works on the origins of the European world. Accordingly it is getting responses from the scientific community that range from blank hostility to ornate ridicule— but most of all they are trying to ignore it.

A historical example of the disinformation society? Martin Bernal, in his controversial book *Black*

Athena, The Afroasiatic Roots of Classical Civilization, shows the large scale of deep spin and historical disinformation in the work of European scholars of the last centuries and touches many relevant issues regarding the cultural background of the so called middle ages.

Black Athena is an analysis of the systematical distortion of historical evidence on the part of classical scholars. Based on racism and combined with political interests ("The Fabrication of Ancient Greece 1785–1985") they are tilting the perspective toward a Eurocentric gridlock of ideological hegemony in the interests of a white, male power elite. The ignoring, denying, and surpressing of the crucial role of African, Semitic, Moorish and other non-European influence in western history amounts to censorship. (Bernal's analysis also uses some major corrections of timetables like the realization that the volcanic Thera eruption, thought to have destroyed the Minoan civilization of Crete in about 1500–1450 B.C., actually happened two hundred years earlier, in 1628 B.C.)

Naturally Bernal has encountered immense opposition to his thoroughly documented research. (One might just have a revelationary experience that everything you learned in school about his-story is terribly wrong.)

AN ELECTRONIC TIME-CODE CATASTROPHY?

While it becomes increasingly clear that disinformation, black propaganda and symbolic domination are very much part of our history and the oldest media, the exponentially increased possibilities of social control and mind control through the manipulation of the electromagnetic spectrum and the new media have not yet been fully realized.

It has been demonstrated that artificial empires can be skillfully created so it should be much easier to make civilizations disappear. Strata of digital data to be rediscovered by future archeologists could lead to a future where digital archeology will selectively reconstruct the past from buried layers of bits and bytes as electronic witnesses. Even the use of time-machines by explorers of the future might result in ambiguous results and could lock into some arbitrary echoes of virtual realities.

Our past/future will then be based on a computer

artifact, possibly some random futuristic wargame of the future/past. A barely reconstructed ancient CD as the blueprint of our lost civilization? A total recall of trashed memory?

The broken timelines of European culture could get lost in the dark ages of a disturbed electromagnetic space–time continuum.

SUBJECT: ASCII ACID: RE-COLLAPSE OF SHELTER

FROM: VLADIMIR MUZHESKY <BASICRAY@THING.NET>
DATE: MON, 26 OCT 1998 21:02:40 -0600 (EST)

NEUROSPACE AFTERNET: FIRST GENERATION

"Monism in science is predefined by the structure of the cosmos." —K. Tziolkovsky, 1925

The most thrilling and advertised part of cyberspace has nothing to do with its gadgetry but with the resemblance of its architecture to the structure of neural networks and their constant intertranslatability. Exactly this intertranslatability became an avatar for escape and enabled some skillful holonauts like Leary to migrate into the electronic shelter when the chemical shops closed down. They became living landmarks of neurospace.

On the basis of its perceptual and economic platform, neurospace can be defined as an autonomous hypernetwork of inner–outer inferences of informational discourses. Whether biologically or electronically realized, it theoretically establishes the same conglomerate of protomodel space niches levelled by the modes of perceptual intensities and, hence, correlated with the extent of perceptronic transformation.

Neurospace is a highway for bots. Somewhere they can realize their restricted but powerful mental velocity. Bots are not guests from the future and they are not an isolated case. First-generation communal bots are here: mostly evolved BBS systems like The Thing, or gravitations toward this direction like Ada'Web, or Word they provide navigational and referential support for content manipulation and, as such, lay down their "magestral's" into the wilderness of neurospace for the content industry using chat, search, game and other func-

tion-specific bots as enforced software labor power. These sites are results of the same translation processes that witnesses the closeness of another collapse of shelter. Through concentrating and alienating mental workpower in a close proximity to Heidegger's *technê* they, in turn, desubliminate translation. On a mental plane, this, to use Marx's words, means digging your own grave. The tangled navigation of the already-dead Ada'Web was a good example [see Markets, "Ada'Web," in this volume]. However, being involved in the translation, they, unlike many other virtual organs, retain the capacity to mutate into the second generation of bots-out-there. Finally, they are the first to fill the metavelocity of shelters.

AGGREGATE: SHELL IN THE SHELL

"Results of the separation of symbolic-theoretical and real cultural activities are...futile..."
—V. Muravjev, 1923

Bots in general move without moving within the spatial-discursive tension. Hyperspace is a phase-space of content tension. Its technological facade corrupted, riven with multiple cracks through which the proto-architecture is visible. Cyberbuildings rot. Under these extreme conditions it is not difficult to forecast the aggregate nature of the second generation of bots. In order to survive every new collapse they should provide an architectural perspective and a semiotic showcase for content formatting reality-spaces based on the old rotten nodality. They have to arrange state of content affairs in their own sovereign velocity.

They should become aggregate worlds that functionally replicate and tweak into human informational processing.

In 1988, Alexander Chasen, the founder of the technocerebrum idea, wrote, "Humanity develops an artificial computer-based cerebrum which it primarily associates with deductive analysis. However, the development of the technocerebrum in the direction of artificial intelligence requires the involvement of inductive logic, which will cause a specific autistic computer syndrome...in as much as the technocerebrum is separated from the biochemical emotional basis of the human brain which plays a crucial role in the regulation of conscious.

In a certain way, the consciousness of an individual user is different from the consciousness of in-network-users. The latter constitutes a shelter with the above-mentioned field of mental escape of the technocerebrum. However, an individual user expands his/her consciousness into the network, turning it into another extension of social or political instrumentals. Bots are being designed to fight network one-dimensionality in the same way as psychotropics were designed to fight one-dimensionality in human psycho-social representation. Same structure. Same code. Probably the same destiny.

Paradigmatically, we can imagine this evolution as a semiotic zone located under the code of social communication. (One of the possible biosocial foundations could for example be spurious memories: cognitive events that imply the classification of imaginary situations and objects as real. This phenomenon is inherent in human dreams, when we think about chimerical cities as real. However, they were proven to be a basis for the formation of new languages in neural network studies. Spurious memories find no vehicle of interaction in post-industrial society). We can see some current in this direction in the growing online porn industry, which desublimizes spurious translation mechanisms through providing live-streamed extrapolated body-space content.

At another pole there is augmented reality research which, to quote Katashi Nagao and Jun Rekimoto, two of its apostles from Sony Computer Science Laboratory Inc., "Has as its main theme the overlay of computer-synthesized images onto the user's real-world view. Augmented reality covers interactive systems that can informationally extend the real world." If we look at this statement through the prism of metamute it basically means that AR develops an in-built module for the individual biological carrier. Live-feed that translates the architecture of shelter into the architecture of reality by means of incorporating all the same mutual aid principle: a "real world" agent can support the user's tasks in the "real world" environment.

We are used to architecture's linear polynomial progression from construction to gradual decay. Unlike its provision of shelter, its evolution as architecture of reality is nonpolynomial in its nonlinear state. It is there and not there at the same time because time itself becomes dependent on the translatability of the users' task.

Taking all these factors into consideration we can conclude that the digital or cyberspace commonly referred to as virtual or synthetic locality is, in reality, a conceptual placebo. The epistemological aggregate or defined status of spatiality that is implicit in locality as a concept is either channeled via semiotic zones of references to the real, or memorized as real situations in the virtual environment. Before we are able to define the pattern of interference emerging from the mediated life streams of real worlds and rendered objects and the interrelated neurological, perceptual, semantic, and economic contextual aspects of spatiality we can say nothing about evolving shelter, hence nothing much about its "real" control points and politics.

SUBJECT: "INFORMATION," "SUBJECT," AND "BODY": FROM METAPHYSICS TO THE PRESENT (A CYBERFEMINIST PERSPECTIVE)

FROM: ALLA MITROFANOVA <ALLA@ALIA.SPB.SU>
DATE: MON, 26 OCT 1998 16:03:10 +0100 (MET)

INTRODUCTION

1) The general motive for my work is to discover how to be happy and to work well, liberated from compulsory coding, normative images, and from prescribed and limiting functions—in order to be able to achieve your goals in a productive fashion. In other words, I am going to dig through a few connected topics in order to raise the following questions: How could we distill a database of information from structuralist and phenomenological positions? How to free the transcendental subject in ourselves, who still have a dominant position in understanding the world. S(h)e demands an order in an identity system, in structures of representations, and perceptions. (S)he fights for a stable ontological surface, when descriptive and operative models are determinant. And, finally, how could we install an embodiment/disembodiment problem that favors embodiment into theoretical discourse organized around consciousness?

The problem is to dig up the body in discursive practices, to embody technodiscourse for the sake of cyberfeminism. Cyberfeminism is an ideological speculation that serves as a browser for viewing and navigating through current cultural changes and historical heritages. A good thing about the label is that it is a provocation—as a conceptual mess it makes emotional noise. It is a fake ideological interface. Cyberfeminism is a useful term in feminist philosophy for its radical impact on body and technology theories.

Hacking is becoming a common operative term for an outsider's way of reaching a quick result in economy, culture, politics, and theory. We have to redefine a problem, to reduce or rethink tools and terms. Looking for shortcuts in philosophical heritage for explicit directions means using the operative model of hacking. Hacking provides a functionally quick model; it helps us to safeguard our own interests. Any good descriptive or analytic discourse has a predatory power inside itself: it is very easy to step into and difficult to run away from an order in which everything looks rational and connected. "Hacking philosophy" means to analyze concepts taken from specific traditions into actual contexts, to purify their operative models, and to reinstall them in current situations in order to get theoretically functional positive tools.

1. I propose to follow along a historical passage. There is a danger of getting lost and using too many historical terms. But the benefit is in making a few terms stronger and more operative.

It is methodologically possibleto divide the philosophical tradition of the last three centuries into three parts, in a linear manner, on the basis of an academic historical classification and critical evaluation of tradition: x) the Kantian recuperation of metaphysics, y) the poststructuralist recuperation of Kantian tradition, often called transcendental philosophy, and z) my own experiments in cybertheory. We will try to show a tradition on three control levels: first, how terms are defined to operate; second, what reference system serves the terms and what is taken as immanent;thrid, how it serves needs. Here, the body is the point of reference.

METAPHYSICS	TRANSCENDENTAL	OPERATIVE
	<terms>	
ESSENCE	DATA THING OBJECT	EVENT EGO
TRANSCENDENTAL	SUBJECT AVATAR or HUMANISM	SERIAL SUBJECT
	<reference system>	
UNITY CONSCIOUSNESS	KEYWORD LOGOS	STRUCTURE
INTERACTION CATEGORIES	HYPERLINKS	REPRESENTATION
	<immanent>	
ETERNITY *	DATABASE	EXISTENCE–EMBODIMENT)

The immanent plan is beyond the model—it could be full of surrogates like Power, Knowledge, or Death of the Transcendental Subject.

1. Classic metaphysics (after Aristotle) insists on the identity of things and equity of Reason. Ego is obliged to recognize adequacy through doubt. Metaphysics creates a mysterious order on the plane of eternity. A Thing is presented by its essence and could only be questioned to show its essence. The essence of things is a shared essence of nature. A nature is already completed as Universe and Eternity. The Essence is unchangeable. A metaphysical body is already complete—as a container of subliminal and brutal things, it could be misrepresented, but a correction should bring essence back. The Body is a mirror of the Universe, a microcosm of eternity. A power is done as an order of things.

2. The tradition of transcendental philosophy can be traced from Kant to Husserl, and basically concerns making the foundation of ontology analyzable. Transcendence is taken now not to mean an interface to Eternity or God, but as a gnosological and ontological problem. It is not a thesis of beliefs in the plan of eternity and unity, but a problem of building on identified tools of philosophical reflection—what became a consciousness.
Object is an identity term to represent Being outside of the Subject. Observer and observed, object and Transcendental Subject are becoming an ontological duality. In this way we take consciousness and perception to be a concrete dominant operative machine. Here we should reinstall a plane of immanence from nature to consciousness, from the union of being to transcendental reason. It affects the whole story: if immanence is shifted from nature to consciousness, from essence to representation, we exclude Being, which is not represented in accepted forms—as objective forms. Reflection and practical reason are tools to operate with objectified forms.
A self-referential system is a hierarchy of categories from casual to abstract, from local to universal. The moral state is incorporated into consciousness. It should control the identity of the Subject and the modality of formal appearances or representations. Everything becomes a heavily connected and controlled system. A system can represent itself as a structure (totalitarian) or as a phenomenon (liberal/open). Time becomes the internal foundation or time dimension of Phenomenon. Subject trapped itself by installing a system of immanent operative tools: What is a consciousness supposed to be? Cognizant police is a Sense as a hyper significance. The Sense represents the deep structure of representations. Sense doesn't apply to nature, but it is a basic method of understanding, the result of cognizant operations. Sense applies to the human subjective ability to represent the world as a structure with a concrete identity of objects. Consciousness is always reducible to itself and its own circumstances—the total recall of consciousness.

3. A cyber paradigm reduces consciousness from transcendental ability, which was a self-referential system, to an operative term and renames it as an intelligence. If the Subject lost its generic position, its dual component—object—lost its guarantee of identity becoming data.

01) Perception, Representation, Transcendence, and Data For Aristotle, material is presented as form. In classical metaphysics the thing is equated with essence. There is no problem of subjectifying or alienating a form or a thing. But for transcendentalism the traditional world is possible as a product of a Subject-based operative system: perception, transcendence, representation. In current (cyber)culture, we have a quite different operative model:

–The order of representations, which was organized as a system of knowledge, is reduced to information as an unstructured catalogue of data. When knowledge became information it translated into a self-referential system, which up to now was called consciousness, was lost together with the perception of a unified subject, which worked as a filter in making objects.

–Transcendence became an empty menu; there is no way to transcend or generalize information in hierarchy and unity. We use information without attempting to organize it as a system, to follow a genesis of data. A datum is an operative unit of the catalogue. Data refers to itself or to other data excluding a subjective referential order. The transcendental subject simply retires upon meeting data.

–Perceptions, as personal empirical foundations of objects, are not usable with data. We use data without controlling it with our perceptions. Data normally passes thru a filter of perception. There is no question about whether it is perceivable, or what is reality. Any existing data has its own rights—it is legislated because it exists.

So there is the adventure of data becoming more and more controversial. Data was born in a transcendental paradigm to be a specific kind of intentional object (like a picture or a sculpture in transcendental art), also controlled by perception. This was a very limited position. But soon data escaped the control of perception and intention. Data applied for a new status: as neither subjective nor objective. Data refers only to the database, and the database belongs to the plane of immanence.

02) From Data to Database Coincidence
(Event) of Data Streams
A database is an uncounted sum of local catalogues; even though some of them are rigidly organized, the sum can only be a pool of data. Different streams of information don't even cross each other, but go in different directions without knowledge of each other. They don't recognize a dominant stream. A database is hardly an alienated cultural heritage; it belongs to the plane of immanence, not to the order of the subject. There is no subjective reference system in the order of data. Let's take as a conclusion the following: a database is a pool of information organized locally or discursively, which could be imagined as Bodies without Organs (thanks to schizoanalytic discourse). Bodies without Organs

reside in a certain locality, but are presented and could be used as unlocated data (deterritorialised).

But the functional status of a database is as an archive. How could it be revitalized? Supposedly, by merely linking data we produce a kind of data event. A coincidence (event) or hyperlink of data could produce the Event, could animate data, and could deliver or revitalize the Body without Organs. (In the case of Power: It could be produced in any nook and transmitted on any level of social organization.) Hyperlinks of databases neglect an order of localities. The Body without Organs is a body prepared for cutting. It could be a data Frankenstein, but the process of conserving the baby and baby delivery is strictly immanent and cannot be manipulated. We cannot manipulate an Event—we can only desire it and help the Event to happen. (As opposed to this, the transcendental subject can manipulate events because the main part of it—cognition and description—is a priority of the Subject.)
Data could be revitalized only by being coupled with intensities (subjectively or discursively generated)—the productive forces of revitalization.

03) Subjectivity and Reference System A concept of Subjectivity always has been combined with reference systems (the transcendental subject was made an operator of the transcendental act). *Subjectum* as a term (as Heidegger recovered it) consists of "what is already done to us" and "how we are going to take it": world and tools. In the transcendental view, *Subjectum* is only possible as personality, and its tools belong to consciousness.
The Superhuman (superman) of Nietzsche is offered in different discussions as a breaking point into consciousness-based philosophical tradition from one side and as an anthropology-based theory (as it was shaped by Foucault) of the subject. The Superhuman is an embodied consciousness—it could mean the end of self-consciousness service. If subjectivity is embodied, it should take as a reference point not the transcendental hierarchy of categories, but the complexity of body functions. If it is embodied as superhuman, it should ignore the limitations given to human as social and historical dimensions grasped in structure and phenomenon. The Superhuman highlights hypercultural links in opposition to the human condition of materiality and

locality within a concrete situation—the condition which (s)he is dependant upon. That is why the superhuman is a scandal—discursively it is not possible for transcendental humanity or anthropology. A process of phenomenologically reducing consciousness to its foundations, shown by Husserl, was a way to reduce the transcendental subject to nonexistence. It seems to be a turning point for Heidegger in developing a fundamental ontology of existence. *Dasein* is a kind of self-reference scheme for the Heideggerian Subject. It is a way to deliver the Subject to the immanent plane, to install subjectivity into an open scope of existential possibilities. Dasein proposes to process a multitude of possible individual realizations, as a kind of system operator. (This is how it is used in post-Heideggerian psychiatry by M. Boss in "The influence of Martin Heidegger on the birth of the alternative psychiatry," *Logos* 5, Moscow, 1994). The main character of Dasein is temporality, not only in the sense of mortality, but also as a temporal process and the finitude of any identifications.

In that discourse, Subject is hardly connected with Event (Co-being). Event produces Identity, but event cannot be represented as a chain of identified objects. In Event, Identity is temporal and cannot produce identity orders. Heiddegger, Nietzsche, and Deleuze gave us a notion of process identification which is (1) coexistence, (2) Event, and (3) activity, which are only places for subject to be presented. Event is temporal, unstructured, local, personal. Event constructs from meanings (database) and meaninglessness: existence, Being, intensity...

The Being, viewed by Heidegger as an open stream of existence, is limited and functionalised by Deleuze into intensity as a preformal force vitalizing Event and operating the Body without Organs.

04) EMBODIMENT AS A POSITIVE NECESSITY

The body is presented in culture as a different structure of concepts of what the body is + images + functional models of how the body should act. Our task is to put the body into a flexible controlling position and to liberate the body from compulsory prescriptions of what it is. A connected issue is whether the body is either for the sake of perception or is a location for personalized Being (existence) to happen. In the latter case a body could be equated with Nothingness. Anyway a personal body should be generated on the field of existence, intensity, but not on the field of regulated descriptive concepts. What is an existence, intensities (energy/drive)? Is it opposed to information? It is not desire; desire on the microlevel, as Foucault showed, is still arranged by cultural coding; desire helps in switching from one designation to another. Body is not a form, has no meanings, has no exact borders, it is not a concept (data), but a field where concepts (data) could recreate a function. In other words, we could say that skipping immanence now is folded into the body.

Embodiment includes hyperdimension as a controlling instance, to function as a singularity above the formal compendium of catalogue. We could take a body as a positive functional temporal model in which permanently changing being is equalized with permanently changed forms (information): desire = ability = possibility + unlimited (or satisfactory) Database of information (concepts) and formal expressions and images. If the balance is not achieved, if the operative system has bugs or another dysfunction, then wrong concepts disorient existence and existence becomes destructive, even self-destructive. If we are embodied correctly, we feel the freedom to live. If not, we have some fields of activity blocked for us (as when compulsory gender divisions came to a traumatic end in Europe in the sixties).

Embodiment is a hot issue for contemporary culture, comparable to what the "soul" or "god flesh" was for medieval culture, and to what organs and anatomy were for classical European culture. So embodiment is an intensive cultural process (micropoesis), and is new for every culture (formal catalogue). Even in talking about disembodiment, we install some concepts for the process of embodiment. *Embodiment* as a necessary task of creation.

1. The (division, *die Schnittstelle*, that we call in English an) interface is something that separates one thing from another. Otherwise the term would make no sense.

2. The *Schnittstelle* denotes a difference and a connection.

3. The phenomenon of the *Schnittstelle* appeared when the concept of a unified world gradually developed into the concept of a world that is at least a duality. (The English noun *interface* dates from 1882; the verb *to interface* from 1962; the adjective *interfacial* (used in crystallography) from 1837.)

4. That which a *Schnittstelle*/interface both separates and connects is, in the most general sense, the One from the Other.

5. How we handle the interface and its shaping is therefore preeminently both an aesthetic concern and also an ethical one. Ethics binds the arts and the sciences (and are binding for both).

6. Through the interface, the Ones define their relationship to the Others, those different to themselves, that is, essentially unknown, and vice versa: over the interface the One manifests itself to the Other, but in those aspects that are understandable. 7. For example, in the Baroque period the crystal chandeliers with their myriad light refractions functioned as an interface through which the cosmos became imaginable outside of the constraints of the private and personal sphere.

8. In telematics, as in any technology-based communication, the interface separates and connects the worlds of active people, on the one hand, and the worlds of working machines and programs on the other. (How far machines may command the character of subjects I shall not go into here, but I presume that in many dimensions active people are a part of the inner world of machines and programs.) The interface separates and connects media-people and media-machines. It is the boundary where the medium formulates itself, where the aesthetic praxis takes place.

9. The pragmatic task of the telematic interface is to provide media-people with a particular access to the Other by means of machines and programs. At the end of the twentieth century, telematic machines and programs are themselves a prominent part of this Other.

10. Current efforts in telecommunications, particularly the world wide web, aim to make the differences between media-people, media-machines, and media-programs imperceptible. This represents a special case in the trend toward eradicating the boundaries between production and reproduction, between work- and nonwork-time, in a common system of communication-based consumer and service relations. We are now just at the beginning of this process. With regard to the interface, this process will really take off when the symbolic hindrances to perception and usage (particularly the alphanumeric keyboard) that still exist are no longer prerequisites for using a computer, and when the interface between media-people, media-machines, and media-programs assumes the character of an environment in which media-people will act as they would in non-machine-based communicative relationships (see, for example, the "interactive Filmplanner" by Georg Fleischmann and colleagues). A slightly different but analogous

problem concerns the computer scientists themselves: with increasing digitalization, and due to the speed of microelectronics development, the machine as hardware has become ever more inaccessible to them. Computer science has practically become a pure software science, without access to or intervention in the machine that lies beneath it.

11. The most important, all-embracing device in this hegemonic strategy is illusionization—not in the sense that anything specific is at stake, but rather in the sense of a no-risk identification with the world of icons, symbols, and relations just as it appears on the monitor. At present, the praxis of this illusionization takes two directions: either using concepts of a primary spatial orientation in the tradition of the *ars memoriae*, or using concepts of a primary temporal orientation, as in classical Aristotelian dramaturgy. In adventure games we find both concepts combined, and in the best examples, they are multilinear concepts of a dramaturgy of memory and empathy.

12. The goal of this essentially double strategy is: The Ones (that is, the media-people) are to operate under the illusion that they are totally in the Other (media-machines, for example)—this is called virtual reality or telepresence. Via illusionization, the Other turns into the One, takes on its identity. This is above all the world of metaphors.

13. In this world of metaphors, the allusion to life is central; the discipline of biology maintains its leading function.

14. There is a long tradition of taxing this interrelationship of life and machine. The body perceives that it has passed through various phases of excorporation and incorporation. Many of the first automatons were copies of living things, either in whole or in their details. In his philosophy of technology published in 1877, Ernst Kapp called this "organ projection." Already at this early date, he vehemently critiqued this concept: the "Idea of the organic as a model, involuntarily and unremarked, tinges the mechanical copy and vice versa when the mechanical is used to explain organic processes; in the excitement of experimentation the mechanical swings over into the organism unremarked, so that apart from these metaphoric explanations of the how, why, and wherefore, also obvious confusions that are inadmissible under usual circumstances, are inevitable."

From the mid-nineteenth century onward, the idea behind this was above all the idea of man as slave laborer, a perfect symbiosis of live and machine production ("Avery's Cotton-picker" of 1857 as an image anticipates perfectly what Mcluhan formulates a century later as media theory: "The wheel is the extension of the foot").

At the high point during the new media's foundation in the nineteenth century, the dominant technology of the time—mechanics—was internalized: the individual life, as well as that of the species, was imagined and interpreted as a mechanical process and/or cycle. Heavy-boned mechanics served as a model for the explication and description of social, cultural, and life processes. Behind this was the idea of man as a machine, as a system of conduits, pumps, circuits, as an internal media apparatus. Not only scientists and engineers but artists, too, were fascinated by this idea that both the body and the life system function in similar ways; both were viewed as subject to manipulation and repair like technical systems. Both mechanical systems and life were conceived of using rigorous analogies.

This ranged from simple comparisons such as—the structure of nerve cords and cabling—electrical contacts and nerve contacts to the idea of the structure of neurons as a complex of wiring and relays; and went as far as the description of a complex process like the act of seeing as a simple succession of mechanical, mainly media-mechanical, processes of film recording and projection apparatus; and the linguistic articulation of that which is seen, again as media processing (mechanical typesetting; organ pipes for sound production); culminating in the direct analogy of human sensory processes and the functioning of a radio receiving station including the listener; and positing of a complete correspondence between the construction of an automobile's driving mechanism/car engine and the processes involved in hearing (identity of petroleum/air, flywheel/ear drum, gear system/auditory ossicles, rear wheel/cochlea).

15. In these founding years of the computer-centred telemedia, life is being externalized in the machines and the programs. These are constructed and computed after the naïve model of the organic and its evolutionary dimensions. The underlying idea of this allusion is that life is something that is continuous, flowing, growing, in constant motion (also harmonious). With regard to the concept of evolution, we are dealing here with Darwinian, or at best Neo-Darwinian models, that is, with an extension of the Darwinian principle of the (informationwise) fittest that takes into account recent research in genetics, according to which selection operates at the cellular level and not first at the level of individual organisms and their relationships with one another.

16. From the perspective of being concerned about the aesthetics and ethics of the interface deriving from the autonomy of Others/the Other, both metaphors must be confronted critically—to instruct and inform—and with alternative models: this applies both to life as a leading metaphor and to a concept of biology and evolution which is reduced and of shallow dimensions.

Why? Please allow me to digress briefly into the world of the concept of metaphors and their meanings:

"For why, the senseless brands will sympathize, The heavy accent of thy moving tongue, And in compassion weep the fire out; And some will mourn in ashes, some coal-black, For the deposing of a rightful king."
—Shakespeare, *Richard II*

Metaphors are comparisons. However, not all comparisons are metaphors. To the phylum of comparisons also belong the symbol, the riddle, the allegory, the image... In their function for expression (and its possible meaning), metaphors hover between image, symbol, and enigma. Metaphors originate from the needs and the power of thought and feeling, "Not to be satisfied with the simple, familiar, and unsophisticated but rather to place oneself above it in order to depart for the Other, to linger awhile with the Various, and to put the Twofold together into one" (Hegel). Metaphors are constructed with the intention of augmenting, deepen-

ing, increasing something; or they simply wallow in the fantasy of their constructor. This "something" is either mental or physical. Metaphors are constructed in order to ennoble the physical with the help of the mind or through the comparison with the physical to convert the mental into experience, to make it profane, to reify it.

17. The telematic networks are connections of technical artifact and complex material systems with political, cultural, and aesthetic structures, that is, they are already connections of the "Twofold." The net itself is already a comparison, a trivial image. Not only in the ongoing net discourse is this connection of complex physical and immaterial units and structures once again being compared/connected to life or aspects thereof. This comprises not only the intention of elevating the profane (the technical, the political...), but also the realization of that which is nontransparent, or opaque, and structural (that is, essentially of the mind).

18. On the other hand, the world of machines and programs is a systematically constructed and calculated world. Everything in it has been produced by numbers and the logical and systematic relations between numbers. In this sense it is a coherent and consistent world, in spite of all the complexity that playing with numbers enables. The world of living organisms does not possess a system of such reliability. The decisive factor: this world is irreversible. Due to external disturbances and inherent variations, the many different physiological rhythms that are linked in a living organism never lead back to the same starting point. Organic systems fluctuate around stasis. Digital machines and programs cannot have a state (Otto Rössler). It is precisely their inherent variations that are to be got rid of through digitization and precision in computation. For the artists and students of the Academy of Media Arts in Cologne, in the meantime it is less of "a problem of precise computation, but more a problem of how to teach all the now low-noise machines to make noises again," as our colleague Georg Fleischmann put it in his contribution to our new yearbook on arts and apparatus. "Aren't there any interesting lines of questioning around, where the aforementioned irregular fluctuations are not the

weakness but the strength of the system?" Technological, social, and cultural systems alike are discontinuous to an extreme degree, both in their genesis and in their present extent. All metaphors that promise the free flow of information, that invoke the ocean as a navigation field, that want to make us experience communication structures like trees or roots, are doomed to failure because of this. The archetypal basic structure of technoid and civilization development is the rigid gradation of the staircase. The archetypal basic structure of life is the spiral. The visual proof, that the genetic code (of DNA) is formed like a double helix, like a twofold spiral staircase, was presented by biology at the same time as cybernetics arrived as a new discipline. The image of the double helix succeeds in uniting both discontinuity and continuity, bending out and turning in, standstill and motion... As yet there is still no better example of the exciting mise-en-scène of this complex relation of space and time, including the body in free fall, than Alfred Hitchcock's *Vertigo*.

19. If we admit biology as the leading discipline of the outgoing twentieth century, the very least we should demand with regard to the interface is that the many and varied constructions of evolutionary theory that this century has seen should be taken into account. (Evolution is a theory of the history of life and not life itself). Darwinism and Neo-Darwinism have been supplemented and modified by theories of mutation, synthetic theories, saltation, and punctuated equilibrium, among others. For example, the two latter, although with different emphases, propose that the pace of evolutionary change in species is episodic rather than smoothly gradual.

20. Conclusion: I would like to make a plea for an experimental interface based on contingencies rather than virtual reality, on feasible individual events rather than on a homogeneous, calculated, continuous, illusory world—one

- that is nevertheless recognizable as a constructed world, through which we gain access to the Other

- that enables a relationship of critical appraisal toward itself

- that is less of a cleansing by catharsis and more of a provocation by epic means

- that nonetheless remembers that the world of communications is a world of sensations and that without these, no one would bother to enter into relationships with Others/the Other.

What we need is a language (of text, images, sounds, and their connections) that does not cover up the technical and political/cultural character of the artifact, materials systems, and structures of expanded telecommunications but instead displays this character, in its usage refers to it, and reminds one of it('s existence). Discontinuity, dynamics, circuits, contacts, controls, pulsions, interruptions, power, distribution...the possibility of allusions is as rich as the technical and political/cultural spheres themselves are. Recent history of the media alone suffices as an example of a rich tradition: Think, for instance, how some filmmakers attempted to break free of the aesthetically cumbersome models of the novel or the theatre by moving into abstraction, rhythm, multidimensional narrative: Brecht's *Short Organum for the Theater* (1948) would do very well as a didactic exercise for today's interface specialists; or, for example, the materialist film—the staging of the material as something that possesses an autonomous power of expression... Why do we always think that we have to start everything from the very beginning again and to re-invent the whole world every day?

23. This plea openly insists on the dualism of media-people and media-machines, media-programs. Dualism is necessary in order to reach any kind of clarification. It may represent a transitional stage, but I am convinced that the dramatization of the interface as a boundary between the One and the Other is the only possibility to achieve qualities of the connection that will differ from a simple decision for the One or for the Other.

No to monopolization of technology by narcissistic subjects—for a dramatics of the difference!

Revue Quart Monde: What is your opinion of the new information and communication technologies, such as the internet. Do you see them as an opportunity or a threat to the poor?

Michel Serres: What is unprecedented here is that concentration of knowledge no longer obtains. Up to now, any form of education consisted, for every one of us, in the bridging of not one but several stretches of distance, between one's place of birth, or point of departure, and that particular place where the elements of knowledge happened to be localized: the local libraries, universities, labs, natural science museums, and so on. That was already the case with the great library in Alexandria or Plato's academy; and after that you had universities, schools, and so on. One was always separated by geographical distance from the place of knowledge. But one was separated by social distance also: if you were not born to the right class, or were stuck with a linguistic barrier because your parents did not speak the proper language; or there was a financial barrier. Even a 'mindgap' may be postulated, as when one would not dare to come near these places of knowledge. And yesterday's educational system was a race of attrition on the bumpy road to the sources of knowledge. So what is new about the world we live in is that the people do not have any longer to move in order to obtain knowledge: thanks to the communication networks knowledge comes to them. And despite lingering fears to the contrary, the opportunity for certain people or certain classes to monopolize these assets has radically decreased. Up to now, knowledge used to be concentrated and accumulated according to the rules of capitalism, even if it was never analyzed in such terms. In building the 'Tres Grande Bibliotheque'(2), France today enacts a return to a past world in the era of the internet. Here we have a building that fences knowledge in precisely at the time when the networks enable one to tap into whatever document, wherever it may be located on earth...

RQM: In *Le Prémier homme* ("The First Man"), Albert Camus describes how his primary schoolteacher not only instructed him in the curriculum but also bridged the gap to knowledge by going to his grandmother and convincing her to let her grandson pursue further studies. The very first hurdle deprived people must surpass consists in regaining confidence in their own powers of intelligence.

MS: That is what I just have called the *mindgap*. I do not want to convey the impression that the net is going to abolish every and all distances. It will not obliterate the kind of human relationships described in Albert Camus's book. But it will bring the possibility of knowledge to all. In the end we turn out to have been democrats in everything, but not as far as knowledge was concerned. Knowledge was behind a bulwark, not only of distances but also of other barriers as well. It was the hallmark of *merit*, of the idea that one had to be smart to attain it. Now there is nothing that stands in our way if, for instance, we would like to set up an internet server for the 'Fourth World' association, and make it freely available to the people.

The novelty of it is as great as when printing was introduced. Before then, knowledge was the preserve of very few people. But subsequently it came the way of those people who could afford to buy books. And now, it will reach everybody, eveywhere, and this is a truly great promise, a promise of the democratic kind...

RQM: Yet there remain another aspect of knowledge, its embedding in social life, in community. The "capitalist" appropriation of knowledge is something that stems not from the nature of knowledge itself but from a way of living in society...

MS: This way of living in societies has determined a number of social bonds, of hierarchical bonds, of commercial bonds, of monetary bonds... But— apart from exceptional cases, such as with small schools or monasteries, there were no bonds stemming from knowledge or information. Today, a social bond may well be based on these things. Nowadays, the unemployed person is provided with professional schooling, whereas the excluded person is supposed to be fed with information in order to become a citizen again. (Re)integration, professional schooling, and education are three problems that must be tackled together. For instance, education now comes to grip with society as a whole, not only by way of scientific and professional schooling, but also in imparting the "togetherness" of all citizens. From now on, education is going to be an evolutive feature, which will be last through a lifetime, and the information bond is going to embed itself ever-more profoundly in the social bond itself. We used to have a society where knowledge was retained rather than disseminated. That is why so many people were excluded from it.

RQM: And why would this change?

MS: Because today, we have the technological means to do it. A hundred years ago, when some small paper plant lost in the woods went bankrupt, its workers had no other recourse than to pack up and take on the various distances I was talking about...on foot. Today, those same workers should be able to go to the town hall, or to their former school, which would of course be open after office hours, and avail themselves of all data necessary to change their life. On the negative side, there is this huge crisis we are facing regarding unemployment and a lagging economy—but on the positive side, we have this technology. Everybody knows by now that the only way out of the crisis is to develop further information and education technologies...

RQM: But you've got this fierce competition out there, and the scarcity of jobs is surely not going to diminish it. Sharing knowledge with my neighbor in these circumstances might not be in my best interest...

MS: The economy is predicated upon exchanges, which in their turn are predicated upon scarcity. Now, suppose you have got two francs in your pocket and I have zero. If you give those two francs to me, I'll have two francs, but you'll have nothing... This is what you call a zero-sum game. Knowledge operates from the opposite principle. Let's say that Pythagoras' theorem is something I know but you do not. If I teach it to you, you will obtain that knowledge, and yet I will still retain it. This is not a zero-sum game.

Knowledge is the realm of non-scarcity, as opposed to the economy. True, knowledge has always been classified as a rare good. But who says that the knowledge necessary to fix a scooter is less important than knowledge about quantum physics? In a society where garbagemen are more in demand than natural scientists, knowledge is on an equalization trajectory. Of course, not everybody agrees. Dissenters will try to throw obstacles into this dissemination of knowledge in order to keep it to themselves. For them, knowledge must remain linked to privilege, to "merit"... I believe that with the advent of the Net, all knowledge will be at everybody's disposal. And I pledge to work for it; it is now the time to do so. Knowledge will no longer be for sale. Today one buys a book and one buys all sorts of knowledge. Tomorrow nothing of all that will be for sale.

RQM: There remains nonetheless the problem of secrecy: trade and manufacturing secrets, and things that remain secret because they are not understood.

MS: Once information spreads and circulates there can be no longer dearth of it anywhere. The Net is the place where you cannot hide anything. My great hope for the Net is that true hackers will be truth hackers, meaning hackers going for full disclosure. Twenty or even ten years ago, nobody could imagine that total secrecy would disapear. Even to this day, big corporations are buying up scientists, they are buying up unpublished knowledge, trade secrets, and this is one of the major difficulties faced by scientific research. Tomorrow hackers will show up in labs, and they will being throwing all secrets to the net. Knowledge will no longer be in specific locales, in those places of scarcity consecrated by society. Knowledge will be an ocean, a pervasive environment in which society will plunge but also lose itself. Scarcity will turn into an overload of information, but correctives will be found by working on ever-more powerful search engines. In fact, there will be a new approach to knowledge of which we have no idea yet. It is the human mind that is going to change, just as it changed radically with the Renaissance. Are you aware that the traditional transfer of knowledge is currently crumbling in entire sectors of academia? Prestigious universities in the U.S. see the number of somophores in mathematics dwindling, because, as things now stand, there is no need any longer for that type of reasoning or that particular brand of mnemonic techniques.

RQM: It is because this type of reasoning is already inherently present in all information that's available, and hence it is no longer necessary to master the reasoning oneself. What would you say?

MS: That is partially so. It is still completley impossible to gauge exactly what is going to disapear, but it seems to me that the epistemological shift is going to be even more profound than in the Renaissance. In this mass of information volume, in which society will swim, or "surf", there will be opportunities for democratization which were unfathomable until now. This (evolution) is surely not going to be detrimental to today's least-educated classes.
Ask yourself, which book would you pobably find in the homes of people with little much money to spend? It is a dictionary, a small Webster's. Is this a book that teaches you maths, or history, or economy? Not really. It is a book for which the chief enjoyment consists browsing through it, "surfing" the mass of data provided. The internet is nothing but a massive dictionnary, a gigantic space in which the body travels.
Intelligence is not about knowing axiomatically how to reason... The French sixteenth-century philosopher Montaigne already had dismissed the

concept of a "well-stuffed head." The advent of the printing press made the memorization of Ulysses' travels and of folktales—the basis of knowledge at that time—redundant. Montaigne saw no use in memorizing a library that was potentially infinite. But does not the internet ask for a "well-endowed head"? Won't the best surfer be a "jack of all trades"? The fastest surfer is not going to be be your typical Ivy-league supertitled philosopher—that guy's head will be simply too loaded to sort it out on the net. So there will be fresh opportunities for those who were viewed by society as laggards. It is a clean start with equal opportunities for all. Mankind is going to wander in the mass of information just as you are now wandering in the woods and the mountains exploring the real world. Up to now, knowledge was a space where you would be taught how to reason, and it required that you memorize a great deal. Now it is going to be a space to roam around. That has never happened before.

RQM: But do you think that today's schools are an obstacle to these changes?

MS: Absolutely so, and I would say all schools. We are now at the threshold of the biggest revolution in education in all of history. We will have to radically change the whole education system. Every time humanity switched the carrier of knowledge, schools changed. The carrier is independent of the education system, but the education system depends upon the carrier. The biggest revolution in an education system occurred with the introduction of writing among the Greek. And all those big civilizations that arose upon scrolls for instance, as among the Jews, or hieroglyphs among Egyptians, also came up with the biblical school, the scribes...

RQM: For generations, children were learning their parents' trade, and learning was an immediate thing. Is this not the case with the school as well? It is the local context that lends relevance to what one learns. Local lore imparts meaning to the locally aquired knowledge. Now, if there is no longer a place of knowledge, where will meaning be found?

MS: When the carrier changes, the method of

transfer is interrupted. That happened in the West in the years 1960–80, and it constitutes one of the greatest upheavals of that period. Parents no longer instilled in their children sexual morality, religion, morality in general, or civism... That's the shakeup at this end of the twentieth century. Meaning depends on the platform. In past days, people spoke but did not write. When writing appeared, the world changed: a system of transfer of knowledge took shape. The drawing-up of contracts, the basis of law, became possible; so did stable forms of exchange, the basis of trade; as did institutions, the basis of politics. And thus it became possible for groups of people to live alongside each other, and this formed the basis of cities. Hence we speak of "history," and of what came before that as "prehistory." When the printing press appeared, the preceding centuries became illegible to us; we called them the "Dark Ages." A whole new sensation of meaning came to us with the advent of Renaissance, with people such as Montaigne, Erasmus, Rabelais... The Reformation heralded the liberty of thought, something inimaginable in a tradition grounded in a transfer of knowledge not based on the printed word. Today, a new platform appears, and thus a new meaning will appear as well. It is not something inherent to the channels through which this meaning will flow. The channels are there before the meaning, they make the meaning, and suddenly everybody's going to be astonished that a new meaning is there. Do not look for it today: it is simply not in our world yet. You won't find it, only your children, or your grandchildren...

RQM: Thus, the challenge today is about providing access to these new channels to all kids.

MS: In theory, access is cheap and unrestricted. The estimated budget for a "distance learning" university on a campus opened by the previous French government in an outer suburb of Paris was a mere 1 percent of that of a traditional academic institution... So with sixteen times less money than was spend on the four towers (8 billion Francs each...) of the *Tres Grande Bibilotheque*, all knowledge concentrated therein could have been made available to sixty million people. And they would even have saved on the trainfare to reach Paris from some distant province...

As you may know, the energy that is going about on the networks does not even reach entropy scale. For all practical purposes, these kind of things come for free.

RQM: The falling price of software and the drive toward sophistication in the computer industry are not negligible forces. But you yourself have stated that access time to a database is hundred times faster for a U.S. researcher than it is to her/his African colleague, whose machines and connections are so much less effective.

MS: That is true. For the time being, technology advances profits mostly for the rich, as usual. But things could be different. Of course, the Americans are trying to retain their predominance, but we, the French, are more democratic, more "republican," more inclined to share, and this could make a lot of difference. I am an optimist, a born optimist...
I am thinking of • Claire Hébert-Suffrin. Fifteen years ago she set up, without computers, a "knowledge exchange" network. She put a number of people together who were willing to swap their respective skills, whether the Russian language, repairing scooters, nuclear physics, anything you wanted, as long as money was kept out of the loop. It has become a web of 25,000 people almost all over Europe. She had a true intuition of what knowledge is about: sharing, gift economy, exchanges, and space. If you put all these elements in a computer system, you get a full-fledged university.

RQM: This idea thrills and baffles us at the same time. Father Joseph Wrésinsky, who is our movement's founder figure, always asked those of us who were academics to try covince their colleagues that we needed their knowledge.

MS: Well, at that time Father Wrésinsky was probably right. But today, you don't need academics anymore. Their knowledge is available to you, period. That's the big difference.

RQM: On the other hand, Father Wresinsky made a distinction between different types of knowledge. In his opinion, the knowledge of academics and that of "fieldworkers" was not the same. The latter is an empirical kind of knowledge, rekindled and established by practical experience. Father Wresinsky used to say to academics: "Bring in your knowledge, but for God's sake don't prevent those on the other side from gathering their own!"

MS: That's exactly what I am fighting for. I am totally opposed to the way politicians in France are now dabbling with information technology in their bid to wire up all the schools. What they want is a top-down approach, starting with experts, school inspectors, and so on, and then making their setup compulsory... It is a carbon copy of the old world pushed into the new world: dinosaurs plus the internet.

My idea would be not to begin from preconceived ideas about knowledge, education, and diplomas but, rather, to bring people into contact according to their needs and abilities. Poeople who are excluded will be less so if they are brought together, and out of this gathering of people an effective demand will emerge. Today's educational system is a supply system without a demand function. It makes egg-sellers set up a shop on the vilage square when there are no buyers around. As things stand now, teachers couldn't care less about what pupils really want.

The premisses of the education system must be turned on their head. Enpowerment must be the key element. Empowerment means giving to those who are excluded from society's mainstream: first, the possibility to form a true community and, then, to open a dialogue among themselves and talk about their needs. Then, you will have an effective demand for "eggs." These people will learn fast, and will before soon know where to get hold of the knowledge they want. Meanwhile, the supply side, like the National Centre for Distance Learning, the universities, and so on will have set up free servers. That will be a real revolution, which will not have been started at the top, for once.

With this change of platform, everything is going to change: knowledge, meaning, the human mind, just as when the printing press was introduced.

When the brain rids itself of certain kind of loads, it makes room for others. When printing began to spread, the amount of memory that was "liberated" made possible the invention of physics, just as mathematics became possible at the time of writing. You may compare that with the evolution of the human race toward an erect position. The forelegs, which became available for seizing things, became hands, and liberated the mouth from that task in the process, which enabled humankind to start speaking. This shift could not have been anticipated beforehand.

So I do believe that the current evolution of technology is not something historical but inherent in man. It is not in the order of history, but in the order of evolution.

RQM: We're dazzled! All these developments are going to land us in a position of great responsability. Allow us to quote Father Joseph Wrésinsky again: "We are not going to wait until the great changes in society will have taken place...to align ourselves on the side of the poorest, the more so since these changes are taking place without them, and without any thought being given to their experiences, and they will not benefit them afterward. Structural poverty is not going to fade away as by magic while we are setting out toward a new society: we take it with us. We will have to voluntarily get rid of it as we are building the new society, otherwise poverty will remain as if it was incrusted in its wall themselvelves." You have just spoken to us about the history of the big shifts in society. Yet the poverty of the olden times is still with us, incrusted as it were in the (new) walls of the Renaissance. But these new channels of communication are going to bring forth a "new man" of sorts. We are witnessing a "grace period," where the deficit of knowledge, or of its absence, is going to be made good. But will "new man" also, ipso facto, be less inequalitarian?

MS: The fact is that the circulation of information is a principal parameter that changes everything. Not to make a berth for the poor in this new world would be foolish and bloody-minded. It would be a blueprint for a world even more cruel than this one. If we do not make that turn, we will risk plunging the world in an even worse kind of poverty.

Today, a lack of knowledge is no longer a handicap. We're in a new ballgame now. There has been a "moratorium on the debts" a you said, it is period of grace for knowledge. But this fresh start must profit the weakest members of society. For them there is a fresh chance, opportunity beckons. Time is up. And time is now!

[Luis Join-Lambert and Pierre Klein in conversation with Michel Serres. Published as the feature 'Superhighways for All' in *Revue Quart Monde* (1), Paris (No 163, March 1997). Translated by Patrice Riemens.

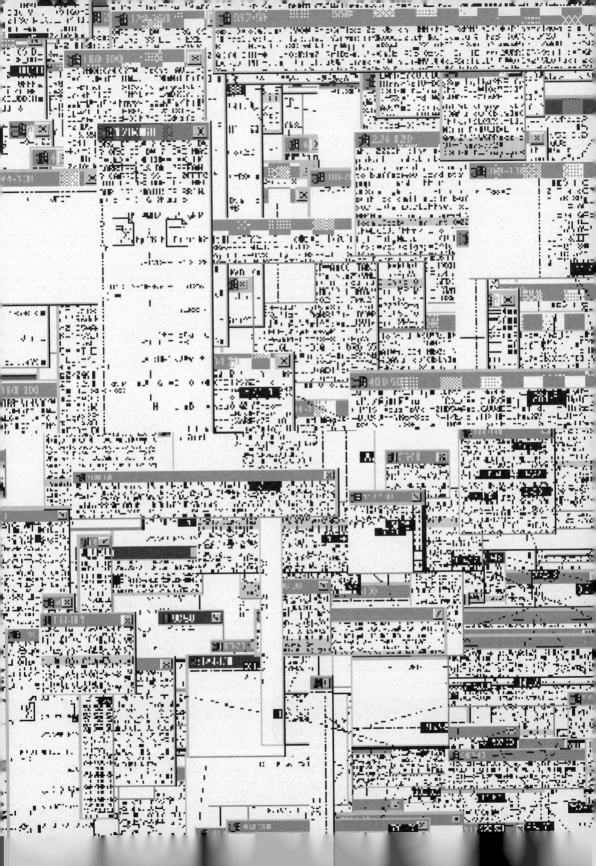

BIOGRAPHIES, NAMES, ADDRESSES AND ACKNOWLEDGEMENTS

BIOGRAPHIES

PHIL AGRE is an associate professor of information studies at UCLA. His books include *Computation and Human Experience* (Cambridge University Press) and *Reinventing Technology, Rediscovering Community: Critical Studies in Computing as a Social Practice* (co-edited with Doug Schuler; Ablex).
<http://dlis.gseis.ucla.edu/pagre/>

AUTONOME A.F.R.I.K.A. GRUPPE is a group of Communication Guerillas living in different German outbacks (i.e., neither in Berlin nor Hamburg nor in the internet). In 1996, they published their *Kommunication Guerilla Handbook* (in German).

RACHEL BAKER is a London-based artist working with sound and the net. She is one of the core members of backspace and is currently teaching a course on alternative economies at the London School of Economics.

RICHARD BARBROOK is professor at Westminster's Hymermedia Research Centre in London. "Californian Ideologies", co-authored with Andy Cameron, was the first widely read text to critique the role of American Neo-Liberalist ideologies in the global popularization of the internet

TILMAN BAUMGAERTEL is a freelance writer, whose work appeared in *Telepolis, Die Tageszeitung, Spiegel-Online, Neue Zˉricher Zeitung, Intelligent Agent, Konr@d,* and a number of other publications. He lives in Berlin and on the internet. His book *Vom Guerillakino zum Essayfilm: Harun Farocki - Monographie eines Autorenfilmers* was published by b_books (Berlin, 1998).

KIT BLAKE created the V2_Organisation website, then spun off into other internet-related orbits. He now plays a satellite role in V2 and other initiatives.

LUTHER BLISSETT, is a paradoxical anthropomorphization of the General Intellect of Marx's *Grundrisse*.

HEATH BUNTING is a London-based artist, most known for his early work using various network technologies. He an unaffiliated subversive dedicated to creating poetic disharmony. His graffiti chalkings can be found all over the world.

TED BYFIELD earns his keep as a freelance book editor and occasional writer; he does other stuff for free. He lives in New York City.

GABOR BORA lives and teaches aesthetics in Sweden. He is a founding member of the Media Research Foundation in Budapest.

JOSEPHINE BOSMA is a writer and journalist who works with Radio Patapoe in Amsterdam. Her interviews and texts about net.art, net.radio, and cyberfeminism have been published in *Mute, Telepolis*, and *Intelligent Agent*. She is also the mother of an angel named Data.

PAULINE VAN MOURIK BROEKMAN is co-editor of Mute, a U.K.-based quarterly culture and technology magazine also available at <http://www.metamute.com>

MERCEDES BUNZ (1971) lives in Berlin and works at *DE:BUG*, the Magazine for the electronic aspects of life, as Publisher and Editor. The rest of the time she tries to get a grip on the the the multitude of nonobserved Theories, which culture draws behind.

SHU LEA CHEANG, once a New York–based media installation artist/filmmaker, is now living a digitalized version of a drifter. See <http://brandon.guggenheim.org> and <http://www.nttcc.or.jp/HoME>.

LAWRENCE CHUA is the author of the novel *Gold by the Inch* (Grove Press) and editor of the anthology *Collapsing New Buildings* (Kaya). See <http://home.earthlink.net/~elchino>.

FLORIAN CRAMER, Berlin, born 1969, M.A. in comparative literature, coder of the combinatory poetry site *Permutations* <http://permutations.home.ml.org>, formerly busy in Neoism as fictioneered on <http://www.neoism.org>

JORDAN CRANDALL is an artist, critic, and media theorist. He is director of the X Art Foundation, New York, founding editor of Blast <http://www.blast.org>, and Visiting Professor, Multimédia, at the École Nationale Supérieure des Beaux-Arts, Paris. His current book is *Suspension* (Documenta X, 1997).

CRITICAL ART ENSEMBLE is a collective of five artists of various specializations exploring the intersections between art, technology, critical theory, and political activism. See <http://mailer.fsu.edu/~sbarnes>.

CALIN DAN was born in Romania and is now based in Amsterdam. He is a freelance developer of media projects. Until 1995, he was busy with art and media policies. Since 1990 he has traveled and worked with subREAL.

ERIK DAVIS, a writer based in San Francisco, publishes in *Wired*, *the Village Voice*, and a number of smaller journals. His first book, *TechGnosis: Myth, Magic, and Mysticism in the Age of Information*, is forthcoming from Harmony Books.

GILLES DELEUZE a prominent French philosopher, writer, and professor is one of the most influential thinkers of the twentieth Century. He authored numerous texts and books including *The Fold*, *Kant's Critical Philosophy*, and with his friend and colleague, Félix Guattari, *A Thousand Plateaus and Difference and Repetition*, among many more. Delueze committed suicide in 1995 at the age of 70.

MARK DERY is a cultural critic. His writings on new media and unpopular culture have appeared in various magazines and journals. He *edited Flame Wars: The Discourse of Cyberculture*, a seminal collection of essays on computer culture, and wrote *Escape Velocity: Cyberculture at the End of the Century*. His latest book, a collection of essays called *The Pyortechnic Insanitarium: American Culture on the Brink*, will be published by Grove Press in 1999.

JULIAN DIBBELL is a New York journalist who has been writing about cyberculture for the last ten years. His book *My Tiny Life*, a literary ethnography about the online society LambdaMOO, will be published by Henry Holt in January 1999.

TIMOTHY DRUCKREY is a curator and writer living in NYC. He edited Culture on the Brink: Ideologies of *Technology, Electronic Culture: Technology and Visual Representation*, and is editing a series called *Electronic Culture: History, Theory, Practice* to be published by MIT Press.

REBECCA LYNN EISENBERG, Esq.,a freelance writer in San Francisco, is a columnist on Net.issues for the *San Francisco Examiner* and CBS *MarketWatch* and a regular contributor to ZDTV. See <http://www.bossanova.com/rebeca>.

VILÉM FLUSSER (1920–1991), Prague-born media philosopher, flew to London and Sao Paolo in I939. After a career in Brasil, he returned to Europe in the seventies. In the late eighties his many publications and lectures gained prominence in German-speaking countries. His collected works (in German) are published by European Photography and Bollmann Verlag.

MATTHEW FULLER lives in South London. He is a member of I/O/D and also regularly collaborates with Mongrel.

ALEX GALLOWAY is Technical Director and Associate Editor for RHIZOME, a leading platform for new media art. A doctoral candidate in the Literature Program at Duke University, Alex lives in Durham, North Carolina.

DAVID GARCIA is an artist and media activist. His work has focused on using different forms of media to give voice to marginalized concerns. He is an organizer of the Next Five Minutes Tactical Media Conferences in Amsterdam.

GASHGIRL (Francesca di Rimini) lives in Adelaide. She is a former member of VNS Matrix and producer of the website *Dollspace*. Her novel *FleshMeat* is forthcoming. See <sysx.apana.org.au/~gashgirl/arc>.

RISHAB AIYER GHOSH is managing editor of *First Monday* <http://www.firstmonday.dk>. He has written several articles and columns on socioeconomic and legal modelsfor the internet. He lives in India, where he analyzes media and communications markets for foreign companies.

GOMMA works in the Milan-based publishing collective ShaKe and is an editor of *Decoder* magazine.

RACHEL GREENE is the editor of RHIZOME. She and colleague Alex Galloway recently curated "Some of My Favourite Web Sites are Art," <http://www.alberta.com/unfamiliarart>. Greene lives and works in New York City.

GENC GREVA is an Albanian journalist and writer from Skodra. He is spokesperson designate for the Ministry of Information.

HONOR HARGER, with her partner Adam Hyde, runs radioqualia, an online radio station aiming to open an electronic portal into the eccentricities of antipodean radio space. See <http://www.radioqualia.va.com.au/>.

FRANK HARTMANN is the manager of Forum Sozialforschung, a Social Science research network in Austria <htpp://www.fsf.at>. He lectures on the topic of media and philosophy at the University of Vienna. Online lectures and texts at <http://mailbox.univie.ac.at/~a6301max/>.

OLIVER MARCHART, studied philosophy and political theory in Vienna and Essex. he is author of articles and books in the fields of art, new media, and politics, including *Neoismus: Avantgarde und Selbsthistorisierung* (Vienna 1997) and *Die Verkabelung von Mitteleuropa: Medienguerilla - Netzkritik - Technopolitik* (Vienna 1998).

LITTLE RED HENSKI was a Bolshevik labor agitator in the twenties Walt Disney cartoon "Alice's Egg Plant." Now ensconced in a Way New factory, he wallows in nostalgia for the *Mondo2000* future and hatches plots to bring it about.

BRIAN HOLMES is a writer, originally from the U.S., who has been living in Paris for many years. He recently co-edited the Documenta X book.

MARC HOLTHOF, Belgian visual arts, television, and film critic, regular contributor to the Flemish magazines *Andere Sinema and De Witte Raaf*.

DAVID HUDSON is currently a freelance writer, screenplay consultant, and translator living in Berlin. He has been published in *The San Francisco Bay Guardian, LA Weekly, The Web, de:Bug, Net Investor, Andere Sinema, Computerwoche, Mute, Wired*. He also edits the weekly *Rewired*. <http://www.rewired.com>.

JOHN HUTNYK is the author of *The Rumour of Calcutta: Tourism, Charity and the Poverty of Representation* (Zed Books, 1996). He teaches Contemporary Media in the anthropology department of Goldsmiths College, University of London. See <http://www.gold.ac.uk/~ans01jh/index.html>.

JODI <www.jodi.org> Joan Heemskerk and Dirk Paesmans.

ZINA KAYE is an Australian artist who uses the net as a medium and maintains an antipodean sound art archive, *L'Audible* <http://laudanum.net/laudible/>. See <http://laudanum.net/> and <http://world.net/~zina>, or mail House of Laudanum PO Box 950, Darlinghurst NSW 2010, Australia.

FRIEDRICH A. KITTLER is Professor of Media History and Aesthetics at the Humboldt University Berlin. His books in English are *Discourse Networks 1800/1900* and *Literature, Media, Information Systems*.

PIERRE KLEIN AND LUIS JOIN-LAMBERT are contributors of *Revue quart monde*, a French periodical that covers a wide range of topics.

ERIC KUITENBERG writer, organizer, and theorist who deals with the collision of new media technology, culture and society. He curently works with political/cultural center De Balie in Amsterdam and the Academy of Media Arts Cologne.

KNOWBOTIC RESEARCH+CF was founded in 1991 and consists of Christian Hubler, Alexander Tuchacek, and Yvonne Wilhelm. They design art projects in which information and knowledge structures are transformed into spaces of experience and agency, emphasising processlike and nonlocatable elements.

JÖRG KOCH: Echtzeit-Historiker based in Berlin. He writes about technology and culture for *Spex, Jetzt, Telepolis*, and *Wired*, and he is working on a documentary about Assata Shakur. He wants to re-animate E.T.

MANUEL DE LANDA is a media artist and self-educated philosopher. He is the author of *War in the Age of Intelligent Machines and One Thousand Years of Nonlinear History*.

MAURIZIO LAZZARATO is completing a doctoral dissertation at the University of Paris VII on communication paradigms, information technologies, and immaterial labor. He is an editor of the journal *Futur Anterieur*.

OLIA LIALINA lives and works in Moscow, where she is a net artist, critic and curator. She is the founder of the Art Teleportacia online gallery and director of the CINE FANTOM film club.

GEERT LOVINK, Amsterdam-based media theorist and activist, member of Adilko, co-founder of Digital City, desk.nl, and contrast.org, co-moderator of nettime. Adilkno's books in English, published by Autonomedia, are *Cracking the Movement* and *The Media Archive*. See <http://thing.desk.nl/bilwet>.

EVELINE LUBBERS, Dutch activist and journalist, member of the Jansen an Jansen research collective, which deals with police and secret service activities and civic counterstrategies. See <www.xs4all.nl/~evel>.

DR. LEV MANOVICH is a theorist and critic of new media who teaches in the Department of Visual Arts, University of California, San Diego. His writings are available at <http://jupiter.ucsd.edu/~manovich>.

OLIVER MARCHART was born in 1968 in Vienna, where he studied philosophy, theater science, and history of art. Occasionally he works as freelance journalist. He has a scientific pastime with lectures, texts, projects, and translations in the area political philosophy, art, and medium theory. See <http://www.t0.or.at/~oliver>.

DIANA McCARTY lives in Budapest, where she works with the Media Research Foundation. She co-organized the MetaForum Conference Series and is co-moderator of the Faces mailing list for women in media.

IGOR MARKOVIC is a journalist, critic, and editor of *Arkzin*, the independant magazine for culture, politics and media (Zagreb). He specializes in media theory and radical critique on cyberculture. See <http://www.arkzin.com>.

ANGELA MELITOPOULIS is a documentary filmmaker based in Cologne. Her current work deals with time and space. She is also working on projects that explore the role of prisons in society.

ALLA MITROFANOVA lives in St. Petersburg, where she is a writer, curator, and editor of the online journal *Virtual Anatomy* <www.dux.ru/vir>. She is the mother of twin boys.

VLADIMIR MUZHESKY is a media artist, theorist, and filmmaker. He is likely to be found around optical network lines. Currently he is based in The Thing, New York node. He teaches 3D animation and web presence making for living at the New York Institute of Technology, Pratt Institute, and International Institute of Multimedia in Paris.

TONI NEGRI was one of the historic leadership of the Italian revolutionary group *Potere Operaio*, and is currently serving a prison sentence in Rebibbia prison, Rome. Negri gave himself up on July 1, 1997, after 14 years' exile in Paris, in a bid to close a chapter in his own personal "judicial history" and that of other far-left militants still in exile.

TOSHIMARU OGURA, Professor, Toyama University, Japan. Introduced the Italian autonomia movements in Japan. Deeply involved in the anticensorship movement around comtemporary arts. His books include *Acid Capitalism*. Ogura is actively organizing against police wiretaps and government surveillance. See <http://www.toyama-u.ac.jp/~ogura/indexEng.html>.

DRAZEN PANTIC is founder and Director of OpenNet, the internet department of Radio B92 (Belgrade, Yugoslavia), and Program Director of Center for Advanced Media, Prague (C@MP). He frequently publishes and lectures on nondiscrimatory access to internet and new media; freedom of expression; and the fight against censorship.

JULIANNE PIERCE is an Australian artist and curator, working at the The Performance Space (Sydney), member of the (former) group VNS Matrix ,and the Cyberfemist International. Seh curated the Code Red program (November 1997) <http://www.anat.org.au/projects/codered/index.html>.

MARKO PELJHAN is a Ljubljana-based performance and communication artist and writer, founder of the arts organization Projekt Atol, and programs coordinator of Ljudmila (Ljubljana Digital Media Lab). His recent project, Makrolab, was part of Documenta X. See <http://makrolab.ljudmila.org>.

RICHARD PELL is a radical tinkerer committed to politicizing the field of robotics.

DANIEL PFLUMM is an electronic artist in Berlin, head of the ELEKTRO MUSIC DEPARTMENT. His sounds can be found at <http://orang.orang.de> and videos at <http://www.thing.net>.

ED PHILLIPS is a writer and free software developer in San Francisco.

SIMON POPE is an artist and software designer based in London. He is a partner in Escape, publishers of the speculative software project I/O/D.

HOWARD RHEINGOLD is author of *Virtual Reality*, *The Virtual Community*, and editor of the *Whole Earth Review*; he is also a founding editor of *HotWired* and founder of Electric Minds. See <http://www.rheingold.com>.

NILS ROELLER is a journalist and organizer of the annual Digitale conference in Cologne, where he is also assistant to principal at the Academy of Media Arts.

ANDREW ROSS is Professor and Director of the Graduate Program in American Studies at New York University. His books include *No Respect*, *Strange Weather*, and most recently, *Real Love*. He recently edited *No Sweat: Fashion, Free Trade and the Rights of Garment Workers*.

NATASCHA SADR-HAGIGHIAN a/k/a TASCHEN-COMICS and TASCHEN-SOUND is an independent artist living and working in Berlin.

SASKIA SASSEN is now at the University of Chicago, after 15 years at Columbia University. Her latest books *are Losing Control? Sovereignty in an Age of Globalization* (Columbia University Press, 1996) *and Globalization and Its Discontents* (The New Press, 1998)

RAF "VALVOLA" SCELSI teaches philosophy and history. He edited *Cyberpunk: antologia di testi politici* (ShaKe, 1990), *and No copyright: Nuovi diritti nel duemil* (ShaKe, 1993). He is a member of the *Decoder* staff.

PIT SCHULTZ is co-founder of the nettime mailing list, and co-editor of ZKP 1–4, the predecessors to this volume. He lives in Berlin where he works with the Mikro foundation.

PHOEBE SENGERS, PhD in Computer Science and Literary and Cultural Theory from Carnegie Mellon University; she is currently a fellow at ZKM.

MICHEL SERRES was born in 1930 and is a Member of the Academie Française. Since 1984 he has been a Professor of History at Stanford University. He is the author of nearly thirty books dealing with the question of communication.

EL IBLIS SHAH is a digital nomad, currently based in Vienna, where he is a Research Fellow Virtual Heritage Foundation.

IVO SKORIC, New York based independent journalist, and activist, known for his numerous postings and reports on mailings lists and newsgroups about the situation of independant media in Former Yugoslavia and the Eastern Europe. See <http://www.igc.org/balkans/indie.html>.

HOWARD SLATER is the editor of Break/Flow and a contributor to *Autotoxicity*, *Datacide*, *Obsessive Eye*, and the *Circuit 8* website <http://c8.com>. 89 Vernon Road, London, E15 4DQ.

RASA SMITE lives in Riga and works with E-L@B.

ALAN SONDHEIM is a theorist/artist who has written extensively on the net. He lives in Brooklyn, New York. His *Internet Text* and other materials is at <http://www.anu.edu.au/english/internet_txt>

BETH SPENCER is the author of *How to Conceive of A Girl* (Random House).See <http://www.netlink.com.au/~beth>.

FELIX STALDER is a researcher, writer, and organizer of culture and politics of new media. He is interested in things he doesn't know, a member of McLuhan Program in Culture and Technology, and lives currently in Toronto, where he works on a PhD thesis on the actor-networks of electronic money.

JOSEPHINE STARRS, a former member of VNS Matrix, lives in Sydney. With Leon Cmielewski she was artist in residence at Xerox PARC for part of 1998.

BRUCE STERLING is an author, journalist, and essayist. His books include the non-fiction *Hacker Crackdown* and the novels *Islands in the Net, The Artificial Kit*, *Schismatrix*, and *Holy Fire*. He co-authored with William Gibson *The Difference Engine*.

RAVI SUNDARAM is at the Centre for the Study for Developing Societies, Delhi. He has written on issues of urban electronic culture and modernity in India and the "Third World," as well as issues of translation between Western and "non-Western" net critique. At present he is working on a book on electronic culture and the urban imaginary in India.

TILLA TELEMANN lives and works as freelance writer and Journalist in Vienna/Austria Studied Art History, Sinology, and Media Studies in Vienna/Austria and Konstanz/Switzerland. She is publisher of the journal *TiCo*.

TJEBBE VAN TIJEN is an archivist, theoretician, and practitioner of art, activism, and archivism with an interest in historical and futuristic interface.

TOSHIYA UENO is associate professor at Wako University, Tokyo, where he works as a sociologist preparing urban tribal studies. Ueno is also an activist and media theoretician of net criticism and cultural studies.

ROBERTO VERZOLA is a member of the Philippine Greens, and runs a small email network for Philippine nongovernment organizations. He has written a number of articles on intellectual property rights and the emerging global information economy.

LINDA WALLACE is an artist and writer. She lives on the southside. See <http://sysx.apana.org.au/artists/hunger>.

McKENZIE WARK is the author of *Virtual Geography*, *The Virtual Republic*, and *Celebrity, Culture and Cyberspace*. He was a contributing editor to *21C* magazine. See <http://www.mcs.mq.edu.au/~mwark>.

MICHAEL A. WEINSTEIN is Professor of Political Science at Purdue University. He has published nineteen books, ranging from cultural theory to metaphysics. With Arthur Kroker he co-authored *Data Trash: The Theory of the Virtual Class*.

FAITH WILDING is a multimedia artist, teacher, and writer, whose work addresses aspects of the somatic, psychic, and sociopolitical history of the body.

PETER LAMBORN WILSON is a religious historian and author of *Pirate Utopias: Moorish Corsairs, European Renegades*. An editor of the New York publisher Autonomedia and and the Semiotext(e) SF collection of short science-fiction, he also studies comparative literature, the origin of religion, and Islamic studies. Wilson is often closely associated with Hakim Bey.

HARTMUT WINKLER teaches media studies at the Ruhr-University in Bochum (Germany). He published a book about TV-reception (*Switching/Zapping*) and a second on film theory (*Der filmische Raum*). His *Docuverse: A Mediatheory of Computers* appeared in 1997. See <http://www.rz.uni-frankfurt.de/~winkler/index.html>.

RICHARD WRIGHT is an artist and filmmaker living in London.

SIEGFRIED ZIELINSKI is Professor and Doctor of Media Studies, as well as the author of numerous books on the history, theory, and practice of media. He is the founding Principal of the Academy of Media Arts Cologne.

NAMES AND ADDRESSES

Bram Dov Abramson <bram@tao.ca>, Montréal |159

Phil Agre, <pagre@weber.ucsd.edu>, Los Angeles, | 343

autonome a.f.r.i.k.a.-gruppe <afrika@contrast.org>, Germany | 310

Autonomedia Books <Autonobook@aol.com>, Brooklyn | 4

Rachel Baker <rachel@irational.org>, London |

Robin Banks <hiddenstair@ yahoo.com>, internet | 177

Richard Barbrook <richard@hrc.wmin.ac.uk>, London | 132

Tilman Baumgaertel <Tilman_Baumgaertel@compuserve.com>, Berlin | 229

Kit Blake <kitblake@v2.nl>, Rotterdam | 197

Luther Blissett <capt_swing@geocities.com>, Bologna | 285

Heath Bunting <monthyear@irational.org>, nomad | 5, 6

Ted Byfield <tbyfield@panix.com>, New York | 405, 419

Gabor Bora <Gabor.Bora@estetik.uu.se>, Uppsala | 496

Josephine Bosma <jesis@xs4all.nl>, Amsterdam | 373, 385, 391

Pauline van Mourik Broekman <pauline@metamute.com>, London | 75, 341, 373

Mercedes Bunz <mrs.bunz@de-bug.de>, Berlin | 400

Shu Lea Cheang <shulea@earthlink.net>, digital drifter | 325

Florian Cramer <cantsin@zedat.fu-berlin.de>, Berlin | 516

Jordan Crandall <crandall@blast.org>, New York | 271, 372

Critical Art Ensemble
<72722.3157@compuserve.com>, Pittsburgh | 71, 220, 246, 295, 471

Calin Dan <calin@euronet.net>, Amsterdam | 416

Erik Davis <figment@sirius.com>, San Francisco | 387

Mark Dery <markdery@well.com>, New York | 8

Julian Dibbell <julian@mostly.com>, New York | 426

Timothy Druckrey <druckrey@interport.net>, New York | 260, 262

Rebecca Lynn Eisenberg <mars@bossanova.com>, San Francisco | 167

Matthew Fuller <matt@axia.demon.co.uk>, London | 25, 37, 129, 211, 253

Alex Galloway <alex@rhizome.org>, Durham | 353, 486

David Garcia <davidg@xs4all.nl>, Amsterdam | 434

GashGirl <gashgirl@sysx.apana.org.au>, Adelaide | 269

Mieke Gerritzen <mieke@vpro.nl>, Amsterdam | 4

Rishab Aiyer Ghosh <rishab@dxm.org> New Delhi | 114

Gomma <gomma@iol.it>, Milano | 301

Rachel Greene <rachel@rhizome.org>, New York | 353

Natascha Sadr-Hagighian <tasche@de-bug.de>, Berlin | **382**

Saskia Sassen <sassen@columbia.edu>, New York | **97**

Raf "Valvola" Scelsi <fikafutura@iol.it>, Milan | **210**

Pit Schultz <pit@icf.de>, Berlin | **25, 211, 453**

Phoebe Sengers <phoebe@khm.de>, Karlsruhe | **52**

El Iblis Shah <eliblis@t0.or.at>, Vienna | **524**

Ivo Skoric <Ivo@igc.org>, New York | **293**

Rasa Smite <rasa@parks.lv>, Riga | **343**

Alan Myouka Sondheim <sondheim@gol.com>, Brooklyn | **49**

Beth Spencer <beth@netlink.com.au>, Sydney | **481**

Felix Stalder <stalder@fis.utoronto.ca>, Toronto | **139, 83**

Michael Stapley <mstapley@midat.de>, Potsdam | **27**

Josephine Starrs <starrs@apana.sysx.org.au>, Sydney | **199**

Bruce Sterling <bruce@well.com>, Austin | **518**

Ravi Sundaram <rsundar@del2.vsnl.net.in>, Delhi | **289**

Tilla Telemann <Tilla@gmx.net>, Vienna | **241**

Tjebbe van Tijen <tijen@inter.nl.net>, Amsterdam | **409, 446**

Mark Tribe < mark@rhizone.org>, New York | **353**

Toshiya Ueno <VYC04344@nifty.ne.jp>, Tokyo | **428**

Roberto Verzola <rverzola@phil.gn.apc.org>, Phillipines | **91**

Linda Wallace <hunger@loom.net.au>, Adelaide | **333**

McKenzie Wark <mwark@laurel.ocs.mq.edu.au>, Sydney | **462**

Michael A. Weinstein <WEINSTEI@polsci.purdue.edu>, West Lafayette | **478**

Faith Wilding <fwild@andrew.cmu.edu>, Pittsburgh | **191, 157, 471**

Peter Lamborn Wilson <n/a>, New York | **509**

Hartmut Winkler <Winkler@tfm.uni-frankfurt.de>, Bochum | **29**

Richard Wright <richard@dig-lgu.demon.co.uk>, London | **256**

Siegfried Zielinski <rektorat@khm.de>, Cologne | **532**

ACKNOWLEDGEMENTS

We would like to thank the numerous authors whose texts
are included in READ ME! for their valuable contributions
to the networked discourse and the realization of this book.
We extend our thanks and apologies to those authors contributed
texts that we were unable to include. Support from the
Ars Electroncia Festival/Center, the Arts Council of England,
the Ljubljana Digital Media Lab, the Society for Old and New Media,
the Spanish Art and Technology Foundation, and the Dutch Electronic Arts
Festival/V2_Organisation is greatly appreciated. The Nettime Fellowship
at the Academy of Media Arts made a valuable contribution in the
technology necessary to produce this book. We grateful to Autonomedia for
their open minded approach to publishing and taking on this project!

We would also like to thank these people who committed time and energy
into making READ ME! a reality; Alex Galloway, Rachel Greene,
John Hopkins, Hope Kurtz, David Mandl, Kevin Paul, and Renee Turner
for their copy editing; Bram Dov Abramson, Jamie Owen Daniel,
Mike Halverson, Michael Hardt, Oliver Koehler, Alessandro Ludovico,
Sebastian Lütgert, Syd Migx, Tom Morrison, Patrice Riemens, and Michael
Stapley for their translations of texts; Jan van den Berg and Jet
Haverkamp for their layout and design support; Walter van der Cruijsen,
Michael van Eeden, and Luka Frelih for their ongoing technical support;
Eric Kluitenberg, for his find slogan collection,Timothy Druckrey and Angela
Melitopoulis for their valuable editorial support and contributions.

The READ ME! editorial team is grateful to Andreas Broeckman and Rafael
Lozano-Hemmer for their crucial input in the early stages of this project.
We would also like to thank Vuk Cosic, Calin Dan, Jim Fleming, Mieke
Gerritzen, Lisa Haskel, Steve Kurtz, Patrice Riemens, Wolfgang Stahle,
Mark Stahlman, Marleen Stikker and Janos Sugar for their valuable
contributions in the concept and production of this book. Last, but not
least, we are indebted to the Nettime subscribers, active posters and
lurkers alike, for enriching the possibilties of a networked discourse.

Ars Electronica Center	<http://www.aec.at>
Arts Council of England	<http://www.artscouncil.org.uk>
The Ljubljana Digital Media Lab	<http://www.ljudmila.org>
The Society for Old and New Media	<http://www.waag.org>
The Foundation for Art and Technology	<http://www.telefonica.es/fat/>
V2_Organisation	<http://www.v2.nl>